Coimbra Mathematical Texts

Volume 5

The Mathematics Department of the University of Coimbra, one of the oldest universities in the world, has a new collection of advanced level mathematical texts, *Coimbra Mathematical Texts*, in collaboration with Springer Nature. This collection aims to publish monographs and conference proceedings of excellent quality, in all branches of mathematics, that will inspire the future generations.

Any researcher interested to contribute to this collection should send an e-mail to cmt@mat.uc.pt with subject "Possible contribution".

The editorial team will perform an initial appraisal of the submitted manuscript based on the quality of the submission, interest and importance of the topic. If the contribution is considered suitable to be sent to peer review, it will be reviewed by specialists of international reputation.

Maria Manuel Clementino · Mai Gehrke ·
Jorge Picado
Editors

Topology, Algebra and Categories in Logic

Lecture Notes of the Coimbra TACL Summer School, 14–18 June 2022

Springer

Editors
Maria Manuel Clementino
Department of Mathematics
University of Coimbra
Coimbra, Portugal

Mai Gehrke
LJAD
Université Côte d'Azur
Nice, France

Jorge Picado
Department of Mathematics
University of Coimbra
Coimbra, Portugal

ISSN 2813-0057 ISSN 2813-0065 (electronic)
Coimbra Mathematical Texts
ISBN 978-3-032-13759-3 ISBN 978-3-032-13760-9 (eBook)
https://doi.org/10.1007/978-3-032-13760-9

This Springer imprint is published by the registered company Springer Nature Switzerland AG
The registered company address is: Gewerbestrasse 11, 6330 Cham, Switzerland

Preface

The biennial TACL—Topology, Algebra and Categories in Logic—conference series was founded in 2003. The purpose of the series is to grow and maintain a more cohesive community by identifying common ground and fostering collaboration across fields, mainly between general and pointfree topology, universal algebra and lattice theory, and category theory by focussing on shared applications in logic and theoretical computer science. Based on a request by the participants at the 2011 meeting in Marseille, an associated summer school was added, starting at the Nashville meeting in 2013. The school is held in the week before the conference and has been immensely successful ever since the first instalment. The purpose of the school is to prepare and initiate young researchers and newcomers to the interdisciplinary research featured at TACL and to foster and strengthen the sense of community, especially among the young members of the community. The program is usually focused around 4 or 5 courses with at least one from each of topology, algebra, and categories as applied in logic. Lecturers are chosen for their leadership in research but also for their excellence in communication. An effort is made to have adequate funding available in order for potential participants without sufficient financial means to be able to participate.

The courses of the 2022 TACL School edition, whose lecture notes comprise this volume, focused on the following topics:

- Pointfree topology and constructive mathematics
- Duality theory
- Metric structures and quantale-enriched categories
- Universal proof theory
- Categorical proof theory.

We would like to express our gratitude to the lecturers for their contribution to the School and to this volume:

- Dirk Hofmann, University of Aveiro, Portugal
- Rosalie Iemhoff and Raheleh Jalali, Utrecht University, Netherlands
- Graham Manuell, Stellenbosch University, South Africa
- Amirhossein Tabatabai, Bernoulli Institute, University of Groningen, Netherlands

– Walter Tholen, York University, Toronto, Canada.

Their invaluable contributions, together with the careful work of the reviewers, made this volume possible.

Finally, we gratefully acknowledge the generous support of:

ERC—European Research Council (DuaLL ERC advanced project 2015–2022)
CMUC—Centre for Mathematics of the University of Coimbra
FCT—Fundação para a Ciência e Tecnologia, Portugal (UIDB/00324/2020)
Department of Mathematics of the University of Coimbra.

Coimbra, Portugal
Nice, France
Coimbra, Portugal
April 2025

Maria Manuel Clementino
Mai Gehrke
Jorge Picado

Contents

About the Editors

Maria Manuel Clementino is Full Professor at the Department of Mathematics and a member of the Centre for Mathematics of the University of Coimbra. Her main research contributions focus on the application of categorical methods in topology and algebra. She serves on the editorial board of several scientific journals and chairs the Steering Committee on Category Theory.

Mai Gehrke is Senior Research Director in Computer Science at the French National Centre for Scientific Research (CNRS) working at the Department of Mathematics of Université Côte d'Azur in Nice, France. Her main contributions are in Stone duality, canonical extensions, and applications in logic and theoretical computer science. She is one of the founders of the TACL series.

Jorge Picado is Full Professor at the Department of Mathematics and a member of the Centre for Mathematics of the University of Coimbra. His research interests include point-free topology and ordered structures. He is the co-author of the books Frames and Locales: Topology Without Points and Separation in Point-Free Topology, and serves on the editorial board of the journals Applied Categorical Structures, Order, and Quaestiones Mathematicae.

Notes on Duality Theory

Dirk Hofmann

Abstract Generally speaking, duality theory is concerned with techniques and results which help to establish an equivalence between a category and the dual of a(nother) category. In these notes we give an overview on some of those results, in particular, we describe the structure of dual adjunction, and present several categorical techniques which we find useful for establishing duality results: a Stone-Weierstraß type theorem, limit sketches and pro-completion, and idempotent split completion.

1 Introduction

One of the most basic principles of category theory is the duality principle. Firstly, to each category C one associates the *dual* (or *opposite*) category C^{op} which has the same objects as C, but the arrows in C^{op} one obtains by reversing all arrows in C, that is, the arrows of C^{op} are precisely the arrows $f^{\mathrm{op}}\colon B \to A$, with $f\colon A \to B$ in C. Secondly, every statement Φ in the language of categories gives rise to the dual statement Φ^{op} which one obtains by reversing the direction of all arrows and of all composites in Φ. By construction, the statement Φ holds in the category C if and only if the statement Φ^{op} holds in C^{op}. This fact implies immediately the *duality principle* (see Mac Lane (1998, Sect. II.1)): if a statement Φ holds in all categories, then so does the dual statement Φ^{op}. Put differently, for every general theorem of category theory we have the dual theorem as well. Clearly, this principle applies also to theorems about classes of categories closed under taking duals, and can be formulated for functors as well.

Whereby the duality principle is very handy as it gives ‘two theorems for the price of one’, the abstract knowledge of the category A^{op} as ‘the category A with all arrows reversed’ does not give us any new insight into the category A itself. However, the situation changes if we are able to identify the category A^{op} with a familiar category B, that is, if we establish an equivalence $\mathsf{A}^{\mathrm{op}} \sim \mathsf{B}$. Such a *dual equivalence* is in particular useful since our knowledge of properties of a category is typically

D. Hofmann (✉)
CIDMA, Department of Mathematics, University of Aveiro, Aveiro, Portugal
e-mail: dirk@ua.pt

M. M. Clementino et al. (eds.), *Topology, Algebra and Categories in Logic*,
Coimbra Mathematical Texts 5, https://doi.org/10.1007/978-3-032-13760-9_1

not symmetric. For instance, many 'everyday categories' admit a representable and hence limit preserving functor to the category Set of sets and functions, therefore in these categories limits are calculated 'as in Set' and therefore their construction is straightforward. However, colimits are frequently much harder to compute; often it is even not obvious whether the category under consideration is cocomplete. In these circumstances, an equivalence $\mathsf{A}^{\mathrm{op}} \sim \mathsf{B}$ together with the knowledge of limits in B help us to understand colimits in A. To illustrate this principle, below we introduce some examples of well-known dual equivalences and of dual concepts. The first two examples will be developed more in detail along these notes.

Example 1 1. We consider first the dual equivalence $\mathsf{Priest}^{\mathrm{op}} \sim \mathsf{DL}$ between the category Priest of Priestley spaces and monotone continuous maps (see Definition 21) and the category DL of distributive lattices and homomorphisms (Priestley 1970, 1972). Being a variety of algebras, we know that the regular epimorphisms in DL are precisely the surjective homomorphisms; but what about the epimorphisms? Is every epimorphism in DL surjective, that is, is every epimorphism in DL regular? It is easy to see that the corresponding dual statement is false in Priest: for every Priestley space $(X, \leq, \tau)$ where $\leq$ is not discrete,

$$(X, =, \tau) \longrightarrow (X, \leq, \tau), \quad x \longmapsto x$$

is a monomorphism in Priest which is not regular.

2. The famous Stone duality theorem (Stone 1936) states that the category BooSp of Boolean spaces and continuous maps (see Definition 16) is dually equivalent to the category BA of Boolean algebras and homomorphisms

$$\mathsf{BooSp}^{\mathrm{op}} \sim \mathsf{BA}.$$

A Boolean space is called *extremally disconnected* whenever the closure of every open set is open. The extremally disconnected Boolean spaces are known to be precisely the projective Boolean spaces. Under Stone duality, extremally disconnected spaces correspond precisely to complete Boolean algebras, therefore we deduce that a Boolean algebra is injective if and only if it is complete. For details we refer to Bell and Machover (1977).

3. Finally, we look at topological groups (Pontrjagin 1934): the category of CHAB of compact Hausdorff Abelian groups and continuous homomorphisms is dually equivalent to the category AB of Abelian groups and homomorphisms

$$\mathsf{CHAB}^{\mathrm{op}} \sim \mathsf{AB}.$$

An Abelian group is torsion-free if and only if its corresponding compact Hausdorff Abelian group is connected (for instance, see Morris (1977)).

Eventually, we will give a proof of Stone duality and of Priestley duality in Sect. 2. We discuss now more in detail an example related to the provability relation $\vdash$ of *intuitionistic propositional logic*. For more information we refer to Sørensen and Urzyczyn (2006).

Theorem 2 *For propositional formulas φ and ψ: if $\nvdash \varphi$ and $\nvdash \psi$, then $\nvdash (\varphi \vee \psi)$.*

By the completeness theorem of intuitionistic propositional logic, $\vdash \theta$ is equivalent to $\models \theta$ for every formula θ; here $\models \theta$ means that $[\![\theta]\!] = \top$ is the top-element of H, for every interpretation $[\![-]\!]$ in a Heyting algebra H. If there are Heyting algebras H_1 and H_2 so that $[\![\varphi]\!]_{H_1} < \top$ and $[\![\psi]\!]_{H_2} < \top$, then we wish to construct a Heyting algebra H and an interpretation in H so that $\varphi \vee \psi$ fails. However, it does not seem to be easy to work with Heyting algebras here; fortunately, there is an alternative way to introduce semantic for intuitionistic propositional logic.

Definition 3 A *Kripke model* is a triple of the form $C = (C, \leq, \Vdash)$ where $(C, \leq)$ is a partially ordered set and $\Vdash$ is a binary relation between elements of C and propositional variables so that

$$\text{if } c \leq c' \text{ and } c \Vdash p, \text{ then } c' \Vdash p.$$

For a Kripke model $C = (C, \leq, \Vdash)$, one extends the relation $\Vdash$ to all propositional formulas in a straightforward manner. For instance, one defines $c \Vdash \varphi \vee \psi$ whenever $c \Vdash \varphi$ or $c \Vdash \psi$, and $c \Vdash \varphi \to \psi$ whenever $c' \Vdash \psi$, for all $c \leq c'$ where $c' \Vdash \varphi$. Furthermore, one writes $C \Vdash \varphi$ whenever $c \Vdash \varphi$ for all $c \in C$, and $\Vdash \varphi$ whenever $C \Vdash \varphi$ for all Kripke models C. The connection with Heyting semantics is provided by the following result.

Theorem 4 *For every propositional formula φ, it holds $\models \varphi$ if and only if $\Vdash \varphi$.*

We discuss the proof of the Theorem 4 in a moment, but first we come back to Theorem 2. Now we can argue as follows. Assume that φ fails in the Kripke frame $C_1 = (C_1, \leq_1, \Vdash_1)$ and ψ fails in the Kripke frame $C_2 = (C_2, \leq_2, \Vdash_2)$. Now we have to deal with orders instead of algebraic structures, and put $C = (C, \leq, \Vdash)$ with $C = C_1 + C_2 + 1$, the order $\leq$ is the sum of $\leq_1$ and $\leq_2$ and, moreover, the additional point becomes a smallest element of $(C, \leq)$. Finally, $\Vdash$ is the sum of $\Vdash_1$ and $\Vdash_2$. Then $\varphi \vee \psi$ fails in $C = (C, \leq, \Vdash)$.

What is the connection between the two ways of defining semantics? Firstly, it is not hard to see that Kripke semantics in $C = (C, \leq, \Vdash)$ is equivalent to Heyting semantics in the Heyting algebra of upper subsets of the partially ordered set $(C, \leq)$ defined by

$$[\![\varphi]\!] = \{c \in C \mid c \Vdash \varphi\}.$$

This proves that $\models \varphi$ implies $\Vdash \varphi$. For the other implication, we note that $\models \varphi$ if and only if $[\![\varphi]\!] = \top$ in every finite Heyting algebra (see Sørensen and Urzyczyn (2006, Theorem 2.4.12), for instance). Let us also note here that every finite distributive lattice is already a Heyting algebra, therefore the restriction of Priestley duality (see Example 1) to finite algebras and finite partially ordered sets tells us that every finite

Heyting algebra is in fact isomorphic to the algebra of upper subsets of a partially ordered set.

Let us describe this correspondence more in detail. To a finite partially ordered set X, one associates the distributive lattice FX of all upper subsets of X, and to a monotone map $f\colon X \to Y$ the monotone map $Ff\colon FY \to FX$ defined by

$$A \longmapsto f^{-1}(A).$$

Similarly, for a finite distributive lattice L, one considers the set GL of all prime ideals with the inclusion order, and for a lattice homomorphism $g\colon L \to M$ between finite distributive lattices one considers the monotone map $Gg\colon GM \to GL$ defined by taking inverse image as well. Equivalently, since M and L are finite, prime ideals correspond precisely to the meet irreducible elements, and, under this identification, Gg corresponds to the restriction of the right adjoint $M \to L$ of g to the meet irreducible elements. This way one obtains a pair of functors

$$F\colon \mathsf{Pos}_{\mathrm{fin}} \longrightarrow \mathsf{DL}_{\mathrm{fin}}^{\mathrm{op}} \quad\text{and}\quad G\colon \mathsf{DL}_{\mathrm{fin}}^{\mathrm{op}} \longrightarrow \mathsf{Pos}_{\mathrm{fin}}$$

between the categories of finite partially ordered sets and monotone maps and of finite distributive lattices and lattice homomorphisms, respectively. Furthermore, for each finite ordered set X one has the monotone map

$$\begin{aligned} \eta_X\colon X &\longrightarrow GFX,\\ x &\longmapsto \{A \mid x \in A\} \end{aligned}$$

and similarly, for each finite distributive lattice L one has the lattice homomorphism

$$\begin{aligned} \varepsilon_L\colon L &\longrightarrow FGL.\\ x &\longmapsto \{A \mid x \in A\}. \end{aligned}$$

The families $\eta = (\eta_X)_X$ and $\varepsilon = (\varepsilon_X)_X$ are natural transformations $\eta\colon 1 \to GF$ and $\varepsilon\colon 1 \to FG$, in fact, they constitute natural isomorphisms (see Sect. 2).

What about the infinite case? One of the key observations in the work of Stone is summarised by the following slogan (Stone 1938a)

> A cardinal principle of modern mathematical research may be stated as a maxim: One must always topologize.

In fact, many duality results involve an algebraic and a geometric side, including the ones mentioned in Example 1. One explanation for this phenomenon might be the following: as in many categories of algebras, every algebra is a filtered colimit of finite algebras, one might expect that in the dual category every object is a cofiltered limit of finite objects, and this property if often guaranteed by adding topology. We give more information about this situation in Theorem 23 and Remark 25.

Duality theory is a very vast topic with many aspects and many deep results. For instance, we find various textbooks on specific duality results such as Pontryagin duality (Morris 1977), the book (Clark and Davey 1998) gives a wealth of techniques in the context of universal algebra, and there are many papers using duality theory in theoretical computer science and logic (see Sambin and Vaccaro (1988), Kupke et al. (2004) and Gehrke (2016), to name only a few). However, in these notes we will focus on general categorical techniques and arguments which, so we believe, are helpful for establishing duality theorems, and do not discuss arguments related to a specific field except for the sake of illustration. In this spirit, let us quote John Isbell on Pontryagin duality (Isbell 1972):

> That fact is a theorem of topological groups. That character groups yield an adjoint connection is a theorem of category theory.

Naturally, throughout this text we will provide references for all main results. However, I would like to acknowledge upfront the huge influence of the paper (Porst and Tholen 1991) and the text books (Johnstone 1986; Clark and Davey 1998) on my understanding and my work on duality theory.

2 Dual Adjunction

2.1 Preliminaries on Category Theory

Let us start with recapitulating some well-known categorical notions, mainly to establish notation. If not indicated otherwise, we refer the reader to Leinster (2014) and, regarding factorisation structures and topological categories, to Adámek et al. (2006).

Definition 5 Let $F\colon \mathsf{A} \to \mathsf{B}$ be a functor. A cone $C = (f_i\colon C \to X_i)_{i\in I}$ in A is said to be *initial with respect to* F whenever, for every cone $\mathcal{D} = (g_i\colon D \to X_i)_{i\in I}$ in A and every morphism $h\colon FD \to FC$ such that $F\mathcal{D} = FC \cdot h$, there exists a unique A-morphism $\bar{h}\colon D \to C$ with $\mathcal{D} = C \cdot \bar{h}$ and $h = F\bar{h}$.

$$\begin{array}{ccccc} C & & FC & \xrightarrow{Ff_i} & FX_i \\ \uparrow{\scriptstyle \bar{h}} & \longmapsto & {\scriptstyle h}\uparrow & \nearrow{\scriptstyle Fg_i} & \\ D & & FD. & & \end{array}$$

Below we present some examples in familiar categories. Throughout this text, Set denotes the category of sets and functions.

Example 6
- In the category Top of topological spaces and continuous maps, a cone $(f_i\colon X \to X_i)_{i\in I}$ is initial with respect to the canonical forgetful functor $|-|\colon \mathsf{Top} \to \mathsf{Set}$ if and only if the domain has the initial topology with respect to this cone, that is, the topology on X is generated by the sets

$$f_i^{-1}(B), \qquad i \in I,\ B \subseteq X_i \text{ open in } X_i.$$

- In the full subcategory $\mathsf{CompHaus}$ of Top defined by all compact Hausdorff spaces, a cone $(f_i : X \to X_i)_{i \in I}$ in $\mathsf{CompHaus}$ is initial with respect to the canonical forgetful functor $|{-}|: \mathsf{CompHaus} \to \mathsf{Set}$ if and only if $(f_i : X \to X_i)_{i \in I}$ is point-separating.
- A cone $(f_i : X \to X_i)_{i \in I}$ in the category Ord of ordered sets and monotone maps is initial with respect to the canonical forgetful functor $|{-}|: \mathsf{Ord} \to \mathsf{Set}$ if and only if, for all $x, y \in X$,

$$x \le y \iff \text{for all } i \in I : f_i(x) \le f_i(y).$$

  Here an *order relation* on a set X is a reflexive and transitive binary relation on X. We call an order relation a *partial order* whenever it is also anti-symmetric, and Pos denotes the full subcategory of Ord defined by all anti-symmetric orders.
- In algebraic categories such as Grp (the category of groups and homomorphisms) and Rng (the category of rings and homomorphisms), every mono-cone is initial with respect to the canonical forgetful functor to Set.

Definition 7 For a faithful functor $|{-}|: \mathsf{A} \to \mathsf{Set}$, a morphism $m: A \to B$ in A is an *embedding* whenever $|m|$ is injective and m is initial.

We note that a morphism f in A is an isomorphism if and only if f is an embedding and $|f|$ is surjective.

Theorem 8 *Let $F: \mathsf{A} \to \mathsf{B}$ be a limit preserving faithful functor and $D: I \to \mathsf{A}$ a diagram. A cone C for D is a limit of D if and only if the cone FC is a limit of FD and C is initial with respect to F.*

Definition 9 A functor $F: \mathsf{A} \to \mathsf{X}$ is *topological* whenever every cone $(f_i : X \to FB_i)_{i \in I}$, with a family $(B_i)_{i \in I}$ of A-objects, admits an initial lifting, that is, an initial cone $(g_i : A \longrightarrow B_i)_{i \in I}$ with $FA = X$ and $Fg_i = f_i$ for all $i \in I$.

$$\begin{array}{ccc} A & \overset{g_i}{\dashrightarrow} & B_i \\ \vdots & & \downarrow \\ X & \underset{f_i}{\longrightarrow} & F(B_i). \end{array}$$

A *concrete category* over a category X is a category C equipped with a faithful functor $|{-}|: \mathsf{C} \to \mathsf{X}$, often written as $(\mathsf{C}, |{-}|)$. We call a concrete category $(\mathsf{C}, |{-}|)$ *topological* whenever $|{-}|: \mathsf{C} \to \mathsf{X}$ is topological. In the sequel we consider typically $\mathsf{X} = \mathsf{Set}$.

We recall now the appropriate notion of 'sameness' for categories.

Definition 10 An *equivalence* between categories A and B consists of functors $F: \mathsf{A} \to \mathsf{B}$ and $G: \mathsf{B} \to \mathsf{A}$ together with natural isomorphisms $1_{\mathsf{A}} \simeq GF$ and $1_{\mathsf{B}} \simeq FG$. We write $\mathsf{A} \sim \mathsf{B}$ if there is an equivalence between A and B.

A useful criterion for establishing an equivalence is the following.

Proposition 11 *A functor $F\colon \mathsf{A} \to \mathsf{B}$ is (part of) an equivalence if and only if F is full, faithful and essentially surjective on objects.*

As we will point out below, equivalences are often constructed starting with an adjunction, a notion we recall next.

Theorem 12 *For functors $F\colon \mathsf{A} \to \mathsf{B}$ and $G\colon \mathsf{B} \to \mathsf{A}$, there is a bijection between*

1. *Pairs of natural transformations $\eta\colon 1_{\mathsf{A}} \to GF$ and $\varepsilon\colon FG \to 1_{\mathsf{B}}$ satisfying*

$$\begin{array}{ccc} F(A) & \xrightarrow{F(\eta_A)} & FGF(A) \\ & {\scriptstyle 1_{F(A)}}\searrow & \downarrow{\scriptstyle \varepsilon_{F(A)}} \\ & & F(A) \end{array} \quad \text{and} \quad \begin{array}{ccc} G(B) & \xrightarrow{\eta_{G(B)}} & GFG(B) \\ & {\scriptstyle 1_{G(B)}}\searrow & \downarrow{\scriptstyle G(\varepsilon_B)} \\ & & G(B) \end{array}$$

   for all A and B, and
2. *Natural Isomorphisms*

$$\mathsf{B}(F-,-) \longrightarrow \mathsf{A}(-,G-).$$

In fact, given natural transformations $\eta\colon 1_{\mathsf{A}} \to GF$ and $\varepsilon\colon FG \to 1_{\mathsf{B}}$ as in 1, then

$$\begin{aligned} \mathsf{B}(FA, B) &\longrightarrow \mathsf{A}(A, GB) \\ h &\longmapsto Gf \cdot \eta_A \end{aligned}$$

defines a natural isomorphism with inverse defined by

$$\begin{aligned} \mathsf{A}(A, GB) &\longrightarrow \mathsf{B}(FA, B). \\ h &\longmapsto \varepsilon_B \cdot Fh \end{aligned}$$

Conversely, for a natural isomorphism as in 2, one obtains η_X as the image of the identity arrow 1_{FX} under

$$\mathsf{B}(FX, FX) \longrightarrow \mathsf{A}(X, GFX),$$

and applying the same reasoning to the inverse

$$\mathsf{A}(-, G-) \longrightarrow \mathsf{B}(F-, -)$$

yields ε.

An *adjunction* between F and G can be defined by either 1 or 2. We write $F \dashv G$ to indicate that there is an adjunction, and say that F is *left adjoint* to G and G is *right adjoint* to F. Furthermore, we say that F *is left adjoint* whenever there is a functor

G with $F \dashv G$; likewise, G *is right adjoint* if there is a functor F with $F \dashv G$. We also recall that adjoints are unique up to isomorphism: if $F \dashv G$ and $F \dashv G'$, then G and G' are isomorphic, written as $G \simeq G'$, and the same holds for the left adjoint.

In the remainder of this subsection, consider an adjunction

$$F\colon \mathsf{A} \longrightarrow \mathsf{B}, \quad G\colon \mathsf{B} \longrightarrow \mathsf{A}, \quad \eta\colon 1_{\mathsf{A}} \longrightarrow GF, \quad \varepsilon\colon FG \longrightarrow 1_{\mathsf{B}}. \tag{1}$$

Let us first recall the following well-known facts.

Remark 13 1. The left adjoint functor F preserves colimits and the right adjoint G preserves limits.
2. The functor F is fully faithful if and only if η is a natural isomorphism. Dually, G is fully faithful if and only if ε is a natural isomorphism (see Mac Lane (1998, Theorem 1 in IV.3)).
3. Assume that G is fully faithful, hence, we may think of B as a full reflective subcategory of A with inclusion functor $G\colon \mathsf{B} \to \mathsf{A}$. Then '$\mathsf{B}$ is as (co)complete as A is'. More in detail, consider a diagram $D\colon I \to \mathsf{B}$ in B. Then a limit cone $(p_i\colon L \to GD(i))_{i\in I}$ in A is up to isomorphism already in B. For a colimit cocone $(c_i\colon GD(i) \to C)$ in A, the cocone $(D(i) \simeq FGD(i) \xrightarrow{Fc_i} FC)_{i\in I}$ is a colimit of D (see Adámek et al. (2006, Propositions 13.27 and 13.30)).

We consider the full subcategories

$$\mathrm{Fix}(\eta) \quad \text{and} \quad \mathrm{Fix}(\varepsilon)$$

of A (resp. B) defined by all objects A in A (resp. B in B) where η_A (resp. ε_B) is an isomorphism.

Theorem 14 *1. The adjunction* (1) *restricts to an equivalence*

$$\mathrm{Fix}(\eta) \sim \mathrm{Fix}(\varepsilon).$$

2. The following assertions are equivalent.

(i) $\mathrm{Fix}(\eta) \hookrightarrow \mathsf{A}$ *is right adjoint with left adjoint* GF *(the monad* $(GF, \eta, G\varepsilon_F)$ *is idempotent).*
(ii) η_G *is an isomorphism.*
(iii) $\mathrm{Fix}(\varepsilon) \hookrightarrow \mathsf{B}$ *is left adjoint with right adjoint* FG.
(iv) ε_F *is an isomorphism.*

Proof See Lambek and Rattray (1979, Theorem 2.0) and (Lambek and Rattray 1975, Theorem 1.1). □

Many (dual) equivalences are constructed using the path indicated above: start with an adjunction (1), and then identify the subcategories $\mathrm{Fix}(\eta)$ and $\mathrm{Fix}(\varepsilon)$.

Remark 15 Assume that $\widetilde{A}$ is a regular cogenerator in a category A with arbitrary powers of $\widetilde{A}$, that is, for each object A in A, the canonical arrow $A \to \widetilde{A}^{\mathsf{A}(A,\widetilde{A})}$ is a regular monomorphism. Therefore, with $I = \mathsf{A}(A, \widetilde{A})$, there is an equaliser diagram

$$A \longrightarrow \widetilde{A}^I \rightrightarrows Y$$

in A. Moreover, since Y is a regular subobject of $\widetilde{A}^J$, with $J = \mathsf{A}(Y, \widetilde{A})$, we obtain an equaliser diagram

$$A \longrightarrow \widetilde{A}^I \rightrightarrows \widetilde{A}^J \tag{2}$$

in A. We conclude that a right adjoint and fully faithful functor $G \colon \mathsf{B} \to \mathsf{A}$ is an equivalence provided that $\widetilde{A}$ is, up to isomorphism, contained in the image of G. To see this, we may think here of $G \colon \mathsf{B} \to \mathsf{A}$ as the inclusion functor of a full reflective subcategory B of A. By Remark 13, the limit of a diagram in B, calculated in A, is up to isomorphism already in B. Therefore, if $\widetilde{A}$ is isomorphic to some object in B, then the whole diagram (2) lies up to isomorphism in B, which tells us that the functor $G \colon \mathsf{B} \to \mathsf{A}$ is an equivalence.

2.2 *Preliminaries on Topology*

Below we recall some notions and results from topology needed throughout these notes.

Definition 16 A compact Hausdorff space is called *Boolean* whenever the simultaneously open and closed subsets (*clopen* for short) form a basis for the topology.

Since the clopen subsets of a topological space X correspond precisely to the continuous maps $X \to \mathsf{2}$ of X into the two-element discrete space, a compact Hausdorff space X is Boolean if and only if the cone $(\varphi \colon X \to \mathsf{2})_\varphi$ is initial with respect to the canonical forgetful functor $|{-}| \colon \mathsf{CompHaus} \to \mathsf{Set}$. We write BooSp for the full subcategory of $\mathsf{CompHaus}$ defined by all Boolean spaces. For every compact Hausdorff space X, consider the (Surjective, Mono-cone)-factorisation

$$(X \overset{e}{\twoheadrightarrow} \overline{X} \longrightarrow \mathsf{2})$$

of the cone of all continuous maps $X \to \mathsf{2}$. By construction, $\overline{X}$ is a Boolean space and $X \mapsto \overline{X}$ defines a left adjoint to the inclusion functor $\mathsf{BooSp} \to \mathsf{CompHaus}$ with unit $e \colon X \to \overline{X}$ (see Adámek et al. (2006, Theorem 16.8)). In particular we note that the category BooSp is complete and cocomplete.

The following important type of spaces was introduced by Nachbin (1950).

Definition 17 A *partially ordered compact space* is a triple $(X, \leq, \tau)$ consisting of a set X, a partial order $\leq$ on X and a compact topology τ on X so that the set

$$\{(x, y) \in X \times X \mid x \leq y\}$$

is closed in $X \times X$ with respect to the product topology.

Remark 18 Every partially ordered compact space $(X, \leq, \tau)$ is necessarily Hausdorff since the anti-symmetry property of the relation $\leq$ implies that the diagonal $\{(x, x) \mid x \in X\}$ is closed in $X \times X$.

Example 19 A simple but important example of a partially ordered compact Hausdorff space is given by the two-element discrete space $\mathsf{2} = \{0, 1\}$ with $0 \leq 1$.

We denote the category of all partially ordered compact spaces and monotone continuous maps by $\mathsf{PosComp}$. The canonical forgetful functor $\mathsf{PosComp} \to \mathsf{CompHaus}$ has a left adjoint which equips a compact Hausdorff space with the discrete order.

Theorem 20 *The category* $\mathsf{PosComp}$ *is complete and cocomplete.*

Proof It is convenient here to consider first the larger category of ordered compact Hausdorff spaces and monotone continuous maps. This category is topological over $\mathsf{CompHaus}$ and therefore complete and cocomplete, and has $\mathsf{PosComp}$ as a reflective full subcategory. For details we refer to Tholen (2009). □

A point-separating cone $(f_i : X \to X_i)_{i \in I}$ in $\mathsf{PosComp}$ is initial with respect to $\mathsf{PosComp} \to \mathsf{CompHaus}$ if and only if, for all $x, y \in X$,

$$x \leq y \iff f_i(x) \leq f_i(y), \text{ for all } i \in I.$$

Since the order of X is anti-symmetric, such a cone is automatically point-separating and therefore $(f_i : X \to X_i)_{i \in I}$ is initial with respect to $\mathsf{PosComp} \to \mathsf{CompHaus}$ if and only if $(f_i : X \to X_i)_{i \in I}$ is initial with respect to $\mathsf{PosComp} \to \mathsf{Set}$. We define now the ordered analogon of Boolean space.

Definition 21 A partially ordered compact Hausdorff space is a *Priestley space* whenever the cone $(\varphi : X \to \mathsf{2})_\varphi$ is point-separating and initial with respect to the canonical forgetful functor $|-| : \mathsf{PosComp} \to \mathsf{Set}$.

We denote the full subcategory of $\mathsf{PosComp}$ defined by all Priestley spaces by Priest. Similarly to the case of Boolean spaces, the inclusion functor $\mathsf{Priest} \to \mathsf{PosComp}$ has a left adjoint, therefore Priest is complete and cocomplete. The underlying compact Hausdorff space of a Priestley space X is a Boolean space since the cone $(\varphi : X \to \mathsf{2})_\varphi$ is point-separating, therefore the forgetful functor $\mathsf{PosComp} \to \mathsf{CompHaus}$ restricts to a functor $\mathsf{Priest} \to \mathsf{BooSp}$. Since every Boolean space equipped with the discrete order is a Priestley space, the functor $\mathsf{Priest} \to \mathsf{BooSp}$ has a left adjoint given by the restriction of the left adjoint $\mathsf{CompHaus} \to \mathsf{PosComp}$ of $\mathsf{PosComp} \to \mathsf{CompHaus}$.

Remark 22 It is not true in general that a partially ordered Boolean space is a Priestley space. A counterexample is given in Stralka (1980).

We end this subsection by recalling a characterisation of cofiltered limits of compact Hausdorff spaces which goes back to Bourbaki (1942).

Theorem 23 *Let* $D\colon I \to \mathsf{CompHaus}$ *be a cofiltered diagram. Then a cone* $(p_i : L \to D(i))_{i\in I}$ *for* D *is a limit cone if and only if*

1. $(p_i : L \to D(i))_{i\in I}$ *is point-separating, and*
2. *for every* $i \in I$,
$$\bigcap_{j\to i} \operatorname{im} D(j \to i) = \operatorname{im} p_i.$$

That is, 'the image of each p_i *is as large as possible'.*

Proof A proof of this result can be found, for instance, in Hofmann (1999). □

Remark 24 If each $p_i : L \to D(i)$ is surjective, then the second condition in Theorem 23 is automatically true.

Remark 25 Recall (for instance, from Adámek and Rosický (1994)) that a cocone $(c_i : D(i) \to C)_{i\in I}$ for a filtered diagram $D\colon I \to \mathsf{Set}$ is a colimit of D if and only if

1. $(c_i : D(i) \to C)_{i\in I}$ is jointly surjective, and
2. for each $i \in I$ and all $x, y \in D(i)$,

$$c_i(x) = c_i(y) \iff \exists (i \xrightarrow{k} j) \in I \quad D(k)(x) = D(k)(y).$$

We note that the second condition above means that the kernel (= coimage) of c_i is equal to the cointersection of the kernels of all $D(k)$ for all I-morphisms $k\colon i \to j$ with domain i. Hence, cofiltered limits of compact Hausdorff spaces and filtered colimits of sets are somehow dual to each other.

Theorem 23 is also very useful for concrete categories $(\mathsf{A}, |-|)$ over $\mathsf{CompHaus}$ where $|-|\colon \mathsf{A} \to \mathsf{CompHaus}$ preserves limits. In this situation, a cone in A is a limit cone if and only if it is initial with respect to $|-|\colon \mathsf{A} \to \mathsf{CompHaus}$ and it is a limit in $\mathsf{CompHaus}$.

In the sequel, we call an object A in A *finite* if the underlying space $|A|$ is finite, and write $\mathsf{A}_{\mathrm{fin}}$ to denote the full subcategory of A defined by all finite objects. For an object A in A, we consider the slice category $A/\mathsf{A}_{\mathrm{fin}}$ and the *canonical diagram*

$$D_A\colon A/\mathsf{A}_{\mathrm{fin}} \longrightarrow \mathsf{A}.$$

$$\left(\begin{array}{ccc} & & A_0 \\ & \nearrow & \Big\downarrow \\ A & & \\ & \searrow & \\ & & A_1 \end{array}\right) \longmapsto \begin{array}{c} A_0 \\ \downarrow \\ A_1. \end{array}$$

Moreover, we have the *canonical cone*

$$(A \longrightarrow A_0)_{A\to A_0} \tag{3}$$

for D_A.

Consider now an object A in A so that this cone is point-separating and initial. Then $A/\mathsf{A}_{\mathrm{fin}}$ is non-empty and indeed cofiltered since finite limits of finite objects are finite in $(\mathsf{A}, |{-}|)$.

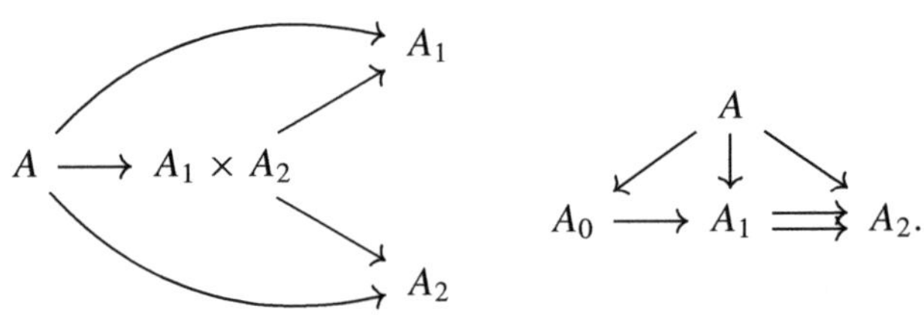

Then (3) is a limit cone for D_A if and only if it satisfies the condition 2 of Theorem 23. This is certainly the case if the concrete category $(\mathsf{A}, |{-}|)$ over $\mathsf{CompHaus}$ has 'image factorisations', that is, for every arrow in A, the (Surjective,Injective) factorisation of the underlying continuous map in $\mathsf{CompHaus}$ lifts to A. Under this condition, the image of every projection $p\colon X \to X_i$ belongs already to the canonical diagram $A/\mathsf{A}_{\mathrm{fin}}$.

$$\text{finite:}\qquad \begin{array}{ccc} X & & \\ {\scriptstyle p}\big\downarrow & \searrow & \\ X_i & \hookleftarrow & \mathrm{im}(p). \end{array}$$

Example 26 1. Every Boolean space is a cofiltered limit of finite spaces.
2. Every Priestley space is a cofiltered limit of finite partially ordered spaces.

If $(\mathsf{A}, |{-}|)$ has 'image factorisations', we can extend the construction above to embeddings. In fact, consider an embedding $m\colon M \to A$ and assume that A is a cofiltered limit $(p_i\colon A \to A_i)_{i\in I}$ of finite objects. For each $i \in I$, consider the factorisation $M \xrightarrow{q_i} M_i \xrightarrow{m_i} A_i$ of $p_i \cdot m$ with q_i surjective and m_i an embedding.

$$\begin{array}{ccc} M & \overset{m}{\hookrightarrow} & A \\ {\scriptstyle q_i}\downarrow & & \downarrow{\scriptstyle p_i} \\ M_i & \underset{m_i}{\hookrightarrow} & A_i \end{array}$$

Then $(q_i : M \to M_i)$ is also a limit cone and $m = \lim_{i \in I} m_i$.

Example 27 1. Every embedding in BooSp is a regular monomorphism since every embedding between finite spaces is a regular monomorphism. Of course, by Urysohn's Lemma we also know that every embedding in $\mathsf{CompHaus}$ is a regular monomorphism.

2. Similarly, every embedding in Priest is a regular monomorphism. The same fact is also true in $\mathsf{PosComp}$ but depends on deeper results of Nachbin (1965) (see also Hofmann et al. (2018)).

2.3 The Structure of Dual Adjunction

In this subsection we investigate the structure of *dual adjunctions* and explain how, under natural conditions, such an adjunction is induced by a *dualising object*. The material of this part is presented more in detail in Porst and Tholen (1991), and see also Dimov and Tholen (1989).

For categories A and B, the term *dual adjunction* between A and B refers to an adjunction which involves the category A and the *dual* of the category B:

$$F\colon \mathsf{A} \longrightarrow \mathsf{B}^{\mathrm{op}}, \quad G\colon \mathsf{B}^{\mathrm{op}} \longrightarrow \mathsf{A}, \quad \eta\colon 1_{\mathsf{A}} \longrightarrow GF, \quad \varepsilon\colon FG \longrightarrow 1_{\mathsf{B}^{\mathrm{op}}}. \qquad (4)$$

It is typically inconvenient to think in terms of B^{op}; therefore, with some abuse of notation, we write $\varepsilon\colon 1_{\mathsf{B}} \to FG$ and refer to both η and ε as the *units* of the adjunction. We also note that the two equations of item 1 of Theorem 12 become now the commutative diagrams

$$\begin{array}{ccc} F(A) & \xrightarrow{\varepsilon_{F(A)}} & FGF(A) \\ & {\scriptstyle 1_{F(A)}}\searrow & \downarrow{\scriptstyle F(\eta_A)} \\ & & F(A) \end{array} \qquad \text{and} \qquad \begin{array}{ccc} G(B) & \xrightarrow{\eta_{G(B)}} & GFG(B) \\ & {\scriptstyle 1_{G(B)}}\searrow & \downarrow{\scriptstyle G(\varepsilon_B)} \\ & & G(B), \end{array}$$

and the natural isomorphism of item 2 of Theorem 12 reads now as

$$\mathsf{B}(-, F-) \longrightarrow \mathsf{A}(-, G-).$$

Example 28 For a category A with an object $\widetilde{A}$ with arbitrary powers, we have the dual adjunction between A and Set defined by

$$\begin{aligned} \mathsf{A}(-,\widetilde{A})\colon \mathsf{A}^{\mathrm{op}} &\longrightarrow \mathsf{Set} & \qquad \widetilde{A}^{(-)}\colon \mathsf{Set}^{\mathrm{op}} &\longrightarrow \mathsf{A} \\ \eta_A\colon A &\longrightarrow \widetilde{A}^{\mathsf{A}(A,\widetilde{A})} & \varepsilon_X\colon X &\longrightarrow \mathsf{A}(\widetilde{A}^X,\widetilde{A}), \end{aligned}$$

where η_A is induced by the cone $(f\colon A\to\widetilde{A})_{f\in\mathsf{A}(A,\widetilde{A})}$ and $\varepsilon_X(x)=\pi_x$, for all $x\in X$.

Theorem 29 *Assume that concrete categories* (A,U) *and* (B,V) *over* Set *with representable forgetful functors* $U\simeq\mathsf{A}(A_0,-)$ *and* $V\simeq\mathsf{B}(B_0,-)$ *and a dual adjunction* (4) *between* A *and* B *are given. Put* $\widetilde{A}=G(B_0)$ *and* $\widetilde{B}=F(A_0)$. *Then the following assertions hold.*

1. $U(\widetilde{A})\simeq V(\widetilde{B})$.
2. $VF\simeq\mathsf{A}(-,\widetilde{A})$ *and* $UG\simeq\mathsf{B}(-,\widetilde{B})$.

Proof To see 1, we calculate

$$U\widetilde{A}\simeq\mathsf{A}(A_0,GB_0)\simeq\mathsf{B}(B_0,FA_0)\simeq V\widetilde{B}.$$

Similarly,

$$VF\simeq\mathsf{B}(B_0,F-)\simeq\mathsf{A}(-,GB_0)\simeq\mathsf{A}(-,\widetilde{A}),$$

which proves the first assertion of 2, the second one can be shown analogously. □

We say that a dual adjunction (4) between concrete categories (A,U) and (B,V) is *represented by the pair* $(\widetilde{A},\widetilde{B})$ of objects $\widetilde{A}$ in A and $\widetilde{B}$ in B, respectively, whenever $VF\simeq\mathsf{A}(-,\widetilde{A})$ and $UG\simeq\mathsf{B}(-,\widetilde{B})$. Thanks to the Yoneda Lemma, such a representation of a dual adjunction is unique up to isomorphism.

To keep notation simple, in the remainder of this section we consider a dual adjunction between concrete categories (A,U) and (B,V) where even

$$VF=\mathsf{A}(-,\widetilde{A})\quad\text{and}\quad UG=\mathsf{B}(-,\widetilde{B})$$

hold, and we 'suppress' the forgetful functors to Set and denote both U and V by $|-|$. Since $|-|$ is faithful, from now on we will not distinguish notationally between f and $|f|$ and simply write f also for the underlying map in Set.

We analyze now the structure of the units η and ε. First we observe that, for every object A in A and every $a\in|A|$, we have the 'evaluation map'

$$\begin{aligned} \mathrm{ev}_{A,a}\colon \mathsf{A}(A,\widetilde{A})=|FA| &\longrightarrow |\widetilde{A}|, \\ \varphi &\longmapsto \varphi(a) \end{aligned}$$

and of course we also have the map

$$\begin{aligned} \mathrm{ev}_{B,b}\colon \mathsf{B}(B,\widetilde{B})=|GB| &\longrightarrow |\widetilde{B}|, \\ \psi &\longmapsto \psi(b) \end{aligned}$$

for every object B in B and every $b\in|B|$.

Theorem 30 *Consider the maps*

$$\overbrace{|\widetilde{A}| \xrightarrow[\eta_{\widetilde{A}}]{} |GF(\widetilde{A})| \xrightarrow[\mathrm{ev}_{F(\widetilde{A}),1_{\widetilde{A}}}]{}}^{\tau} |\widetilde{B}|, \qquad \overbrace{|\widetilde{B}| \xrightarrow[\varepsilon_{\widetilde{B}}]{} |FG(\widetilde{B})| \xrightarrow[\mathrm{ev}_{G(\widetilde{B}),1_{\widetilde{B}}}]{}}^{\sigma} |\widetilde{A}|.$$

Then the following identities hold:

$$\tau \cdot \mathrm{ev}_{A,a} = \eta_A(a), \quad \sigma \cdot \mathrm{ev}_{B,b} = \varepsilon_B(b), \quad \tau = \sigma^{-1}.$$

Proof We verify the first assertion, the proof of the other two is similar. For $\varphi\colon A \to \widetilde{A}$:

$$\begin{aligned}
\tau \cdot \mathrm{ev}_{A,a}(\varphi) &= ev_{F(\widetilde{A},1_{\widetilde{A}})} \cdot \eta_{\widetilde{A}} \cdot \mathrm{ev}_{A,a}(\varphi) \\
&= (\eta_{\widetilde{A}} \cdot \mathrm{ev}_{A,a}(\varphi))(1_{\widetilde{A}}) \\
&= (\eta_{\widetilde{A}} \cdot \varphi(a))(1_{\widetilde{A}}) \\
&= (GF\varphi(\eta_A(a)))(1_{\widetilde{A}}) \\
&= \eta_A(a) \cdot F\varphi(1_{\widetilde{A}}) \\
&= \eta_A(a)(1_{\widetilde{A}} \cdot \varphi) \\
&= \eta_A(a)(\varphi).
\end{aligned}$$

To sum up, for concrete categories $(\mathsf{A}, |-|)$ and $(\mathsf{B}, |-|)$ over Set with representable forgetful functors and a dual adjunction

$$F\colon \mathsf{A} \longrightarrow \mathsf{B}^{\mathrm{op}}, \quad G\colon \mathsf{B}^{\mathrm{op}} \longrightarrow \mathsf{A}, \quad \eta\colon 1_{\mathsf{A}} \longrightarrow GF, \quad \varepsilon\colon 1_{\mathsf{B}} \longrightarrow FG,$$

there are objects $\widetilde{A}$ in A and $\widetilde{B}$ in B with $|\widetilde{A}| \simeq |\widetilde{B}|$ and $|F| \simeq \mathsf{A}(-, \widetilde{A})$ and $|G| \simeq \mathsf{B}(-, \widetilde{B})$. Assuming for simplicity that the last two isomorphisms are even identities, we showed that, up to isomorphism, the units η and ε are of the form

$$\eta_A(a) = \mathrm{ev}_{A,a}, \quad \varepsilon_B(B) = \mathrm{ev}_{B,b}.$$

Remark 31 Let A be an object of A and assume that $\eta_A(a) = \mathrm{ev}_{A,a}$, for every $a \in |A|$. Then $\mathrm{ev}_f \cdot |\eta_A| = |f|$ for every $f\colon A \to \widetilde{A}$,

$$\begin{array}{ccc}
A & \xrightarrow{\eta_A} & \mathsf{B}(FA, \widetilde{B}) \\
 & \underset{|f|}{\searrow} & \downarrow{\scriptstyle \mathrm{ev}_f} \\
 & & \widetilde{A}
\end{array}
\qquad\qquad
\begin{array}{ccc}
a & \longmapsto & \mathrm{ev}_{A,a} \\
 & \searrow & \downarrow \\
 & & f(a)
\end{array}$$

therefore η_A is a monomorphism if and only if the cone $(f\colon A \to \widetilde{A})_f$ is a mono-cone in A.

2.4 How to Construct Dual Adjunctions

We turn now to the question 'How can we construct a dual adjunction between given concrete categories $(\mathsf{A}, |-|)$ and $(\mathsf{B}, |-|)$ over Set?'. Certainly, under mild conditions on the forgetful functors, we have to find objects $\widetilde{A}$ in A and $\widetilde{B}$ in B with $|\widetilde{A}| \simeq |\widetilde{B}|$, for simplicity we shall consider $|\widetilde{A}| = |\widetilde{B}|$. Furthermore, the following should hold.

1. For each object A in A, the cone

$$(\mathrm{ev}_{A,a}\colon \mathsf{A}(A, \widetilde{A}) \longrightarrow |\widetilde{B}|)_{a\in|A|}$$

   admits a lifting

$$(\mathrm{ev}_{A,a}\colon F(A) \longrightarrow \widetilde{B})_{a\in|A|}$$

   such that, for each $f\colon A \to A'$ in A, the map $\mathsf{A}(f, \widetilde{A})$ is a B-morphism $F(f)$.
2. For each object B in B, the cone

$$(\mathrm{ev}_{B,b}\colon \mathsf{B}(B, \widetilde{B}) \longrightarrow |\widetilde{A}|)_{b\in|B|}$$

   admits a lifting

$$(\mathrm{ev}_{B,b}\colon G(B) \longrightarrow \widetilde{A})_{b\in|B|}$$

   such that, for each $g\colon B \to B'$ in B, the map $\mathsf{B}(g, \widetilde{B})$ is an A-morphism $G(g)$.
3. For each object A in A, the map

$$\begin{aligned} \eta_A\colon |A| &\longrightarrow |GF(A)| = \mathsf{B}(F(A), \widetilde{B}) \\ a &\longmapsto \mathrm{ev}_{A,a} \end{aligned}$$

   is actually an A-morphism $\eta_A\colon A \to GF(A)$.
4. For each object B in B, the map

$$\begin{aligned} \varepsilon_B\colon |B| &\longrightarrow |FG(B)| = \mathsf{A}(G(B), \widetilde{A}) \\ b &\longmapsto \mathrm{ev}_{B,b} \end{aligned}$$

   is actually a B-morphism $\varepsilon_B\colon B \to FG(B)$.

If the cones in 1 and 2 admit *initial* liftings, then the other two requirements are automatically satisfied as we point out next.

Theorem 32 *Let $(\mathsf{A}, |-|)$ and $(\mathsf{B}, |-|)$ be concrete categories over Set and let $\widetilde{A}$ be an object in A and $\widetilde{B}$ be an object in B with $|\widetilde{A}| = |\widetilde{B}|$. Furthermore, assume that the following two conditions are satisfied:*

(A) For each object A in A, *the cone*

$$(\mathrm{ev}_{A,a}\colon \mathsf{A}(A, \widetilde{A}) \longrightarrow |\widetilde{B}|)_{a\in|A|}$$

admits an initial lifting

$$(\mathrm{ev}_{A,a}\colon F(A) \longrightarrow \widetilde{B})_{a\in|A|}$$

with respect to $|{-}|\colon \mathsf{B} \to \mathsf{Set}$.

(B) For each object B in B, *the cone*

$$(\mathrm{ev}_{B,b}\colon \mathsf{B}(B, \widetilde{B}) \longrightarrow |\widetilde{A}|)_{b\in|B|}$$

admits an initial lifting

$$(\mathrm{ev}_{B,b}\colon G(B) \longrightarrow \widetilde{A})_{b\in|B|}$$

with respect to $|{-}|\colon \mathsf{A} \to \mathsf{Set}$.

Then $(\widetilde{A}, \widetilde{B})$ induces a dual adjunction.

Proof As an example, consider $f\colon A \to A'$ in A. Then, for every $a \in |A|$, the diagram

$$\begin{array}{ccc} \mathsf{A}(A', \widetilde{A}) & \xrightarrow{\hom(f,\widetilde{A})} & \mathsf{A}(A, \widetilde{A}) \\ & \searrow^{\mathrm{ev}_{A',f(a)}} & \downarrow \mathrm{ev}_{A,a} \\ & & |\widetilde{A}| = |\widetilde{B}| \end{array}$$

commutes, hence, by initiality, $\hom(f, \widetilde{A})$ underlies a morphism $Ff\colon FA' \to FA$ in A. All other properties can be verified similarly. □

Of course, for topological categories these liftings exist, slightly more general one has the following.

Proposition 33 *If* $|{-}|\colon \mathsf{A} \to \mathsf{Set}$ *is mono-topological, then (A) holds.*

But what about the algebraic case? We recall that a *signature* is a pair (Σ, σ) consisting of a class Σ (of operation symbols) and a mapping $\sigma\colon \Sigma \to \mathsf{Card}$ into the class of cardinal numbers (the arity of an operation symbol). Then an *algebra B* for this signature is given by a set $|B|$ together with an operation $\theta^B\colon |B|^{\sigma(\theta)} \to |B|$, for each operation symbol $\theta \in \Sigma$. For algebras B_1 and B_2, a map $f\colon |B_1| \to |B_2|$ is an *algebra homomorphism* $f\colon B_1 \to B_2$ whenever f preserves all operations, that is, for each operation symbol θ and $n = \sigma(\theta)$, the diagram

$$\begin{array}{ccc} |B_1|^n & \xrightarrow{f^n} & |B_2|^n \\ {\scriptstyle\theta^{B_1}}\downarrow & & \downarrow{\scriptstyle\theta^{B_2}} \\ |B_1| & \xrightarrow[f]{} & |B_2| \end{array}$$

commutes. For an algebra B and a set X, we denote by $\mathrm{Cl}_X(B)$ the *clone algebra* of all term functions of arity X, that is, the smallest subalgebra of $B^{|B^X|}$ containing all projection maps $\pi_x \colon B^X \to B$.

Definition 34 The category A is *concretely $\widetilde{A}$-complete* if all powers of $\widetilde{A}$ and all equalisers of pairs of parallel maps between powers of $\widetilde{A}$ exist in A, and these limits are preserved by $|-| \colon \mathsf{A} \to \mathsf{Set}$.

Proposition 35 *Let* $(\mathsf{A}, |-|)$ *be a concrete category over* Set *and* $\widetilde{A}$ *be an object of* A *so that all powers of* $\widetilde{A}$ *exist in* A *and are preserved by* $|-| \colon \mathsf{A} \to \mathsf{Set}$. *Furthermore, let* B *be the category of algebras and homomorphisms, for a signature* (Σ, σ), *equipped with the canonical forgetful functor* $|-| \colon \mathsf{B} \to \mathsf{Set}$, *and let* $\widetilde{B}$ *be an algebra with* $|\widetilde{B}| = |\widetilde{A}|$.

1. *The following assertions are equivalent.*
   - *(i) The condition (A) holds.*
   - *(ii) For every set* X, $\mathrm{Cl}_X(\widetilde{B}) \subseteq \mathsf{A}(\widetilde{A}^X, \widetilde{A})$.
   - *(iii) All operations* $|\widetilde{B}|^n \to |\widetilde{B}|$ *of* $\widetilde{B}$ *are* A*-morphisms* $\widetilde{A}^n \to \widetilde{A}$.
2. *If, moreover,* A *is concretely* $\widetilde{A}$*-complete, then ((A)) implies ((B)).*

Proof Regarding the first assertion, if ((A)) holds, then, for every set X, the set $\mathsf{A}(\widetilde{A}^X, \widetilde{A})$ becomes a subalgebra of $\widetilde{B}^{|\widetilde{B}^X|}$ containing all projection maps, therefore $\mathrm{Cl}_X(\widetilde{B}) \subseteq \mathsf{A}(\widetilde{A}^X, \widetilde{A})$. Clearly, ((ii)) implies ((iii)). Finally, assume ((iii)) and let A be an object of A. For each operation symbol θ with arity n we define

$$\mathsf{A}(A, \widetilde{A})^n \longrightarrow \mathsf{A}(A, \widetilde{A}), \quad (h_i)_i \longmapsto (A \xrightarrow{\langle h_i \rangle} \widetilde{A}^n \xrightarrow{\theta^{\widetilde{B}}} \widetilde{A}).$$

Let $F(A)$ be the algebra $\mathsf{A}(A, \widetilde{A})$ equipped with these operations; hence, $F(A)$ is a subalgebra of $\widetilde{B}^{|A|}$. This proves ((i)).

To prove the second assertion, assume that A is concretely $\widetilde{A}$-complete and that ((iii)) holds. Recall that a map $f \colon |B| \to |\widetilde{B}|$ is an algebra homomorphism if and only if, for every operation symbol θ (with arity n) and every $h = (h_i)_i \in |B|^n$,

$$f \cdot \theta^B(h) = \theta^{\widetilde{B}} \cdot f^n(h). \tag{5}$$

With

$$\pi_{\theta^B(h)} \colon \widetilde{A}^{|B|} \longrightarrow \widetilde{A},$$

the left-hand side of (5) is $\pi_{\theta^B(h)}(f)$. Let $\alpha \colon \widetilde{A}^{|B|} \to \widetilde{A}^n$ be the arrow induced by

$$\begin{array}{ccc} \widetilde{A}^{|B|} & \dashrightarrow & \widetilde{A}^n \\ & \searrow^{\pi_{h_i}} & \downarrow^{\pi_i} \\ & & \widetilde{A}, \end{array}$$

then the right-hand side of (5) can be written as $\theta^{\widetilde{B}} \cdot \alpha(f)$. Therefore $\mathsf{B}(B, \widetilde{B})$ underlies the equaliser of

$$\begin{array}{ccc} |\widetilde{A}|^{|B|} & \overset{\pi_{\theta B(h)}}{\rightrightarrows} & |\widetilde{A}| \\ & \underset{\alpha}{\searrow} \quad |\widetilde{A}|^n \quad \nearrow_{\theta^{\widetilde{B}}} & \end{array}$$

in A. Therefore also (B) holds. □

We say that a dual adjunction is *naturally induced by* $\widetilde{A}$ *and* $\widetilde{B}$ whenever ((A)) and ((B)) hold. We consider now our leading examples.

Example 36 1. We put $\mathsf{A} = \mathsf{CompHaus}$ and $\mathsf{B} = \mathsf{BA}$. By Proposition 35, the two-element compact Hausdorff space $\mathsf{2}$ and the two-element Boolean algebra $\mathsf{2}$ naturally induce a dual adjunction

$$\mathsf{CompHaus} \underset{G}{\overset{F}{\underset{\top}{\rightleftarrows}}} \mathsf{BA}^{\mathrm{op}}.$$

2. For $\mathsf{A} = \mathsf{PosComp}$ and $\mathsf{B} = \mathsf{DL}$, Proposition 35 tells us that the two-element partially ordered compact Hausdorff space $\mathsf{2}$ and the two-element distributive lattice $\mathsf{2}$ naturally induce a dual adjunction

$$\mathsf{PosComp} \underset{G}{\overset{F}{\underset{\top}{\rightleftarrows}}} \mathsf{DL}^{\mathrm{op}}.$$

Remark 37 We consider a dual adjunction

$$F : \mathsf{A} \longrightarrow \mathsf{B}^{\mathrm{op}}, \quad G : \mathsf{B}^{\mathrm{op}} \longrightarrow \mathsf{A}, \quad \eta : 1_{\mathsf{A}} \longrightarrow GF, \quad \varepsilon : 1_{\mathsf{B}} \longrightarrow FG \tag{6}$$

naturally induced by $\widetilde{A}$ and $\widetilde{B}$. Then

$$\eta_A \text{ is an embedding } \iff (f : A \to \widetilde{A})_f \text{ is point-separating and initial.}$$

$$\begin{array}{ccc} A & \xrightarrow{\eta_A} & \mathsf{B}(FA, \widetilde{B}), \\ & \underset{f}{\searrow} & \downarrow \mathrm{ev}_f \\ & & \widetilde{A} \end{array} \qquad \begin{array}{ccc} a & \longmapsto & \mathrm{ev}_{A,a} \\ & \searrow & \downarrow \\ & & f(a). \end{array}$$

Definition 38 Let $(\mathsf{A}, |-|)$ be a concrete category over Set and let $\widetilde{A}$ an object in A.

1. The object $\widetilde{A}$ is called *initial cogenerator* if, for each object A in A, the cone $(f : A \to \widetilde{A})_f$ is point separating and initial.

2. We denote by $\mathrm{InitCog}(\widetilde{A})$ the full subcategory of A defined by all objects A so that the cone $(f : A \to \widetilde{A})_f$ is point separating and initial.

We note that $\widetilde{A}$ is an initial cogenerator of $(\mathrm{InitCog}(\widetilde{A}), |-|)$.

Remark 39 Clearly, the adjunction (6) restricts to the full subcategories $\mathrm{InitCog}(\widetilde{A})$ and $\mathrm{InitCog}(\widetilde{B})$ of A and B, respectively.

Example 40 We continue now the study of our leading examples (see Example 36).

1. For $\mathsf{A} = \mathsf{CompHaus}$ and $\widetilde{A} = \mathsf{2}$, the category $\mathrm{InitCog}(\mathsf{2})$ is precisely the category BooSp of Boolean spaces and continuous maps. By the Prime Ideal Theorem (see Johnstone (1986), for instance), the Boolean algebra $\mathsf{2}$ is a cogenerator in BA, hence $\mathrm{InitCog}(\mathsf{2}) = \mathsf{BA}$. The dual adjunction

$$\mathsf{CompHaus} \underset{G}{\overset{F}{\rightleftarrows}} \mathsf{BA}^{\mathrm{op}}$$

of Example 36 restricts to a dual adjunction

$$\mathsf{BooSp} \underset{G}{\overset{F}{\rightleftarrows}} \mathsf{BA}^{\mathrm{op}}.$$

2. Similarly to the example above, for $\mathsf{A} = \mathsf{PosComp}$ and $\widetilde{A} = \mathsf{2}$, the category $\mathrm{InitCog}(\mathsf{2})$ is precisely the category Priest of Priestley spaces and monotone and continuous maps, and the distributive lattice $\mathsf{2}$ is a cogenerator in DL, hence $\mathrm{InitCog}(\mathsf{2}) = \mathsf{DL}$. The dual adjunction

$$\mathsf{PosComp} \underset{G}{\overset{F}{\rightleftarrows}} \mathsf{DL}^{\mathrm{op}}$$

of Example 36 restricts to a dual adjunction

$$\mathsf{Priest} \underset{G}{\overset{F}{\rightleftarrows}} \mathsf{DL}^{\mathrm{op}}.$$

We finish this subsection by mentioning some further examples.

Example 41 1. We consider now the non-full subcategory Heyt of DL consisting of all Heyting algebras and Heyting homomorphisms, *i.e.* lattice homomorphisms which also preserve the implication. Although the dual adjunction of Example 40 (2) can be restricted to Heyt and to a certain subcategory of Priest, this example does not fit easily into the framework developed so far. Firstly, the implication $\to$ in $\mathsf{2}$ is not a monotone map of type $\mathsf{2} \times \mathsf{2} \to \mathsf{2}$, hence we cannot apply Proposition 35. Secondly, the Heyting algebra $\mathsf{2}$ is not a cogenerator of Heyt. We will discuss duality theory for Heyting algebras in Sect. 4.

2. Let $\mathbb{F}$ be a field and let $\mathsf{A} = \mathsf{B} = \mathsf{Vec}_{\mathbb{F}}$ be the category of vector spaces and linear maps over $\mathbb{F}$. Again by Proposition 35, the vector space $\mathbb{F}$ induces a dual adjuntion

$$\mathsf{Vec}_{\mathbb{F}} \underset{G}{\overset{F}{\rightleftarrows}} \mathsf{Vec}_{\mathbb{F}}^{\mathrm{op}},$$

of course, here $F(V) = G(V) = V^*$ is the dual space of the vector space V. We also note that $\mathbb{F}$ is a cogenerator of $\mathsf{Vec}_{\mathbb{F}}$; however, it is well-known that the canonical map

$$\begin{aligned} V &\longrightarrow V^{**}, \\ v &\longmapsto \mathrm{ev}_v \end{aligned}$$

is an isomorphism if and only if V is finite-dimensional (see Bourbaki (1989)).

3. In continuation of the example above, we assume now that $\mathbb{F}$ is a finite field and consider the category $\mathsf{B} = \mathsf{BooSpVec}_{\mathbb{F}}$ with objects all vector spaces over $\mathbb{F}$ equipped with a Boolean topology making all operations continuous, and arrows all linear and continuous maps. With the discrete topology, we may consider $\mathbb{F}$ as an object of B, and, thanks to Proposition 35, we obtain an adjunction

$$\mathsf{Vec}_{\mathbb{F}} \underset{G}{\overset{F}{\rightleftarrows}} \mathsf{BooSpVec}_{\mathbb{F}}^{\mathrm{op}}.$$

We shall see in Example 99 that this adjunction is indeed an equivalence.

2.5 *Gelfand Duality*

To illustrate the notions and results of the previous subsections, we describe here the well-known Gelfand duality for compact Hausdorff spaces and complex C^*-algebras (see Gelfand (1941) and also Negrepontis (1971), Isbell (1982) and Marra and Reggio (2017)).

Definition 42 A *C^*-algebra* is a commutative unital $\mathbb{C}$-algebra with norm $\|-\|$ and involution $(-)^*$ which is complete with respect to $\|-\|$ and satisfies

$$\|x \cdot x^*\| = \|x\|^2.$$

Example 43 For each topological space X,

$$C^*(X) = \{f : X \to \mathbb{C} \mid f \text{ is continuous and bounded}\}$$

is a C^*-algebra.

We denote by C^*-Alg the category of C^*-algebras and identity and involution preserving $\mathbb{C}$-algebra homomorphisms. The choice of morphisms of C^*-Alg might seem inconsequential as we neglect the norm; however, soon we will see that C^*-algebra homomorphisms are automatically compatible with the norm. In fact, the following proposition contains the key result.

Proposition 44 *For each C^*-algebra B and each element $x \in B$,*

$$\|x\| = \sup\{|\varphi(x)| \mid \varphi \in C^*\text{-}\mathsf{Alg}(B, \mathbb{C})\}.$$

Proof See Gelfand (1941). □

Hence, every homomorphism of C^*-algebras satisfies $\|f(x)\| \leq \|x\|$ and $\mathbb{C}$ is a cogenerator in C^*-Alg. The category C^*-Alg comes naturally with a forgetful functor to Set sending a C^*-algebra to its underlying set, however, this functor is not very convenient as it is not representable. A better choice is the unit ball functor $|-|\colon C^*\text{-}\mathsf{Alg} \to \mathsf{Set}$ which sends a C^*-algebra B to the set $\{x \in B \mid \|x\| \leq 1\}$ and every homomorphism to the restriction to the respective 'unit balls'.

The following facts are now easily shown.

Proposition 45 *The category C^*-Alg is complete and the functor $|-|\colon C^*\text{-}\mathsf{Alg} \to \mathsf{Set}$ preserves limits. Moreover, the C^*-algebra $\mathbb{C}$ is an initial cogenerator of $(C^*\text{-}\mathsf{Alg}, |-|)$.*

Remark 46 In Negrepontis (1971) it is shown that the functor $|-|\colon C^*\text{-}\mathsf{Alg} \to \mathsf{Set}$ is even monadic.

For a topological space X,

$$C^*(X) = \{f : X \to \mathbb{C} \text{ continuous and bounded}\}$$

defines the initial lift of the cone

$$(\mathrm{ev}_{X,x} : \mathsf{Top}(X, \mathbb{D}) \longrightarrow \mathbb{D} = |\mathbb{C}|)_{x \in X}.$$

Therefore we conclude

Corollary 47 *The pair $(\mathbb{D}, \mathbb{C})$ naturally induces a dual adjunction*

$$C^* : \mathsf{Top}^{\mathrm{op}} \longrightarrow C^*\text{-}\mathsf{Alg}, \quad S : C^*\text{-}\mathsf{Alg} \longrightarrow \mathsf{Top}^{\mathrm{op}}$$

with units $\eta\colon 1 \to SC^*$ and $\varepsilon\colon 1 \to C^*S$.

Our aim is now to identify the full subcategories $\mathrm{Fix}(\eta)$ and $\mathrm{Fix}(\varepsilon)$. We start with the following remark.

Remark 48 For each C^*-algebra B, the space $S(B)$ is compact Hausdorff as it can be presented as an equaliser of a pair of continuous maps between powers of $\mathbb{D}$. Therefore $\mathrm{Fix}(\eta)$ is a full subcategory of $\mathsf{CompHaus}$.

Since $\mathbb{C}$ is an initial cogenerator in $(C^*\text{-Alg}, |-|)$, we know already that ε_B is an embedding, for every C^*-algebra B. To identify $\text{Fix}(\varepsilon)$, the classic Stone-Weierstraß Theorem (Stone 1948a, b) gives now the last piece of information.

Theorem 49 (Stone-Weierstraß) *Let A be a compact Hausdorff space and let $M \subseteq C^*(A)$ be a C^*-subalgebra of $C^*(A)$ such that the cone $(f: A \to \mathbb{D})_{f \in |M|}$ separates the points of A. Then $M = C^*(A)$.*

Corollary 50 *For every C^*-algebra B, $\varepsilon_B: B \to C^*(S(B))$ is also surjective. Therefore* $\text{Fix}(\varepsilon) = C^*\text{-Alg}$.

We turn now our attention to the topological side. By Remark 37, for each topological space X, the arrow η_X is an embedding if and only if X is completely regular. In particular, η_X is an embedding for every compact Hausdorff space. The last step of the proof consists of a 'Stone–Weierstraß like' theorem for C^*-algebras.

Theorem 51 *Let B be a C^*-algebra and let $M \subseteq S(B)$ be a closed subspace of $S(B)$ such that the cone $(f: B \to \mathbb{C})_{f \in M}$ separates the points of B. Then $M = S(B)$.*

Proof See Hofmann (2002b, Proposition 2.3). □

Corollary 52 *For every compact Haudorff space A, $\eta_A: A \to S(C^*(A))$ is surjective. Therefore* $\text{Fix}(\eta) = \text{CompHaus}$.

All told:

Theorem 53 $\text{CompHaus}^{\text{op}} \sim C^*\text{-Alg}$.

Remark 54 From Theorem 14 we also deduce that the inclusion functor CompHaus $\hookrightarrow$ Top has a left adjoint $\beta = SC^*: \text{Top} \to \text{CompHaus}$, usually referred to as *Čech–Stone compactification* (Čeh 1937; Stone 1937).

2.6 *Stone-Weierstraß Condition*

In our presentation of the Gelfand duality we stressed the importance of a 'Stone-Weierstraß'-like theorem. In this subsection we formulate a condition on a (contravariant) functor which expresses a 'Stone-Weierstraß' condition and investigate its role in duality theory. We will do this with respect to an abstract choice of 'embeddings'.

Remark 55 Let C be a complete category and let $\mathbb{M}$ be a class of C-morphisms satisfying the following conditions:

1. every regular monomorphism belongs to $\mathbb{M}$ and every arrow in $\mathbb{M}$ is a monomorphism:
$$\text{RegMono}(\mathsf{C}) \subseteq \mathbb{M} \subseteq \text{Mono}(\mathsf{C}),$$

2. $\mathbb{M}$ is closed under composition, stable under pullbacks and, for each family $(m_i : A_i \to A)_{i\in I}$ of $\mathbb{M}$-morphisms, there exists an intersection $d : D \to A$ with $d \in \mathbb{M}$.

We will typically consider $\mathbb{M}$ to be the class of all embeddings of a concrete category $(\mathsf{C}, |-|)$ over Set.

We call an epimorphism $e : A \twoheadrightarrow B$ in C an $\mathbb{M}$-*extremal epimorphism* whenever, for all arrows m in $\mathbb{M}$ and f in C with $e = m \cdot f$, m is an isomorphism. Writing

$$\mathbb{M}\text{-ExtrEpi}$$

for the class of all $\mathbb{M}$-extremal epimorphisms in C, we obtain that $\mathbb{M}$ is part of the factorization structure $(\mathbb{M}\text{-ExtrEpi}, \mathbb{M})$ for morphisms in C. For details regarding factorization structures we refer to Adámek et al. (2006). In particular, we note that an arrow simultaneously belonging to $\mathbb{M}$ and to $\mathbb{M}$-ExtrEpi is an isomorphism, and that $f \cdot g \in \mathbb{M}$ implies $g \in \mathbb{M}$ (see Adámek et al. (2006, Proposition 14.11)).

We also define the following class of small cones of C:

$$\mathcal{M} = \{(f_i : C \to C_i)_{i\in I} \mid I \text{ is a set and } \langle f_i\rangle_{i\in I} : C \to \prod_{i\in I} C_i \in \mathbb{M}\}.$$

Each limit cone belongs to $\mathcal{M}$ since, in this case, $\langle f_i\rangle_{i\in I}$ is a regular monomorphism. Furthermore, a small cone belongs to $\mathcal{M}$ if and only if it contains an $\mathcal{M}$-cone. More in detail, consider families $(g_j : C \to C_j)_{j\in J}$ and $(f_i : C \to C_i)_{i\in I}$ and a map $\varphi : J \to I$ so that, for all $j \in J$, $g_j = f_{\varphi(j)}$. Let $\pi_\varphi : \prod_{i\in I} C_i \to \prod_{j\in J} C_j$ the unique map satisfying $\pi_j \cdot \pi_\varphi = \pi_{\varphi(j)}$, for all $j \in J$. Then the diagram

$$\begin{array}{ccc} C & \xrightarrow{\langle f_i\rangle_{i\in I}} & \prod_{i\in I} C_i \\ & {\scriptstyle\langle g_j\rangle_{j\in J}}\searrow & \downarrow{\scriptstyle\pi_\varphi} \\ & & \prod_{j\in J} C_j \end{array}$$

commutes, hence $\langle f_i\rangle_{i\in I} \in \mathbb{M}$ provided that $\langle g_j\rangle_{j\in J} \in \mathbb{M}$. We also note that $\mathcal{M}$ is closed under composition if and only if $\mathbb{M}$ is stable under products. Finally an object $\widetilde{C}$ of C is called an $\mathbb{M}$-*cogenerator* of C if, for each object C in C, the cone $(f : C \to \widetilde{C})_f$ belongs to $\mathcal{M}$.

Throughout this subsection we consider complete concrete categories $(\mathsf{A}, |-|)$ and $(\mathsf{B}, |-|)$ with limit preserving forgetful functors to Set and a dual adjunction

$$F : \mathsf{A} \longrightarrow \mathsf{B}^{\mathrm{op}}, \quad G : \mathsf{B}^{\mathrm{op}} \longrightarrow \mathsf{A}, \quad \eta : 1_{\mathsf{A}} \longrightarrow GF, \quad \varepsilon : 1_{\mathsf{B}} \longrightarrow FG \tag{7}$$

induced by objects $\widetilde{A}$ in A and $\widetilde{B}$ in B with $|\widetilde{A}| = |\widetilde{B}|$. Furthermore, there are classes $\mathbb{M}_{\mathsf{A}}$ and $\mathbb{M}_{\mathsf{B}}$ of A-morphisms and B-morphisms, respectively, satisfying the conditions of Remark 55 and so that the cones

$$(\mathrm{ev}_{A,a}\colon F(A) \longrightarrow \widetilde{B})_{a\in A} \quad \text{and} \quad (\mathrm{ev}_{B,b}\colon G(B) \longrightarrow \widetilde{A})_{b\in B}$$

belong, respectively, to $\mathcal{M}_{\mathsf{B}}$ and $\mathcal{M}_{\mathsf{A}}$. Finally, we assume that $\widetilde{A}$ is an $\mathbb{M}_{\mathsf{A}}$-cogenerator of A and $\widetilde{B}$ is an $\mathbb{M}_{\mathsf{B}}$-cogenerator of B. Note that the latter condition implies that, for all A in A and $B \in \mathsf{B}$, η_A belongs to $\mathbb{M}_{\mathsf{A}}$ and ε_B belongs to $\mathbb{M}_{\mathsf{B}}$.

We start with the following observation.

Proposition 56 *Assume that our given adjunction* (7) *is already an equivalence.*

1. *The following are equivalent.*
   a. $F(\mathbb{M}_{\mathsf{A}}) \subseteq \mathbb{M}_{\mathsf{B}}$*-ExtrEpi.*
   b. $G(\mathbb{M}_{\mathsf{B}}) \subseteq \mathbb{M}_{\mathsf{A}}$*-ExtrEpi.*
2. *The following are equivalent.*
   a. $F(\mathbb{M}_{\mathsf{A}}$*-ExtrEpi*$) \subseteq \mathbb{M}_{\mathsf{B}}$.
   b. $G(\mathbb{M}_{\mathsf{B}}$*-ExtrEpi*$) \subseteq \mathbb{M}_{\mathsf{A}}$.

Proof As an example, we show that 1a implies 1b. Assume that $F(\mathbb{M}_{\mathsf{A}}) \subseteq \mathbb{M}_{\mathsf{B}}$-ExtrEpi, and let $m\colon B' \to B$ be in $\mathbb{M}_{\mathsf{B}}$. Let $Gm = n \cdot e$ be the ($\mathbb{M}_{\mathsf{A}}$-ExtrEpi, $\mathbb{M}_{\mathsf{A}}$) of Gm. Then $Fe \cdot Fn = FGm \in \mathbb{M}_{\mathsf{B}}$, hence Fn belongs to $\mathbb{M}_{\mathsf{B}}$ and, by hypothesis, Fn is also an $\mathbb{M}_{\mathsf{B}}$-extremal epimorphism. Consequently, Fn is an isomorphism, therefore also n is an isomorphism, and we conclude that Gm is an $\mathbb{M}_{\mathsf{A}}$-extremal epimorphism. □

Remark 57 If

$$\mathbb{M}_{\mathsf{B}}\text{-ExtrEpi} = \{\text{Surjections}\} = \mathbb{M}_{\mathsf{A}}\text{-ExtrEpi}$$

then the conditions 1a and 1b of Proposition 56 are equivalent to $\widetilde{A}$ being $\mathbb{M}_{\mathsf{A}}$-injective and $\widetilde{B}$ being $\mathbb{M}_{\mathsf{B}}$-injective, respectively. Hence, if our given adjunction (7) is already an equivalence, then $\widetilde{A}$ is $\mathbb{M}_{\mathsf{A}}$-injective if and only if $\widetilde{B}$ is $\mathbb{M}_{\mathsf{B}}$-injective.

We come now to the central definition of this subsection.

Definition 58 The functor F of (7) satisfies the *Stone-Weierstraß condition* provided that

(SW) For each object A in A, an $\mathbb{M}_{\mathsf{B}}$-morphism $m\colon M \to F(A)$ is an isomorphism provided that the cone $(m(f)\colon A \to \widetilde{A})_{f\in|M|} \in \mathcal{M}_{\mathsf{A}}$.

By symmetry, there is a corresponding definition for the functor G of (7) which we do not formulate here explicitly.

Remark 59 For every object $A \in \mathsf{A}$, the arrow $\eta_A\colon A \to GF(A)$ belongs to $\mathbb{M}_{\mathsf{A}}$. Furthermore, the cone

$$(\eta_A(a))_{a\in|A|} = (\mathrm{ev}_{A,a})_{a\in|A|}$$

belongs to $\mathcal{M}_{\mathsf{B}}$. Hence, if G satisfies (SW), then $\eta_A\colon A \to GF(A)$ is an isomorphism.

We note that Theorem 49 states that the functor $C^*\colon \mathsf{CompHaus} \to C^*\text{-}\mathsf{Alg}$ satisfies (SW) and Theorem 51 states that the functor $S\colon C^*\text{-}\mathsf{Alg} \to \mathsf{CompHaus}$ satisfies (SW). We shall see in Corollary 66 that it is no coincidence that both functors have this property. Before proceeding, we consider our basic examples.

Example 60 For the natural dual adjunction

$$\mathsf{BooSp} \underset{G}{\overset{F}{\underset{\top}{\rightleftarrows}}} \mathsf{BA}^{\mathrm{op}}$$

of Example 40, we show that the functor $F\colon \mathsf{BooSp} \to \mathsf{BA}$ satisfies (SW) with respect to the class of all embeddings (= regular monomorphisms) in BooSp and the class of embeddings (= monomorphism) in BA. It is here convenient to identify FA with the Boolean algebra of all clopen subsets of the Boolean space A. Accordingly, let $M \subseteq FA$ be a subalgebra such that, for all $x \neq y$ in A, there is some $C \in M$ with $x \in C$ and $y \notin C$. Note that M is closed under complements. Let now D be a clopen subset of A. Let $y \notin D$. For every $x \in D$, we find some $C_{x,y} \in M$ with $x \in C_{x,y}$ and $y \notin C_{x,y}$; by compactness, we find finitely many $x_1, \dots x_n$ so that

$$D \subseteq C_{x_1,y} \cup \dots \cup C_{x_n,y} =: C_y,$$

note that $y \notin C_y$ and $C_y \in M$. Since the open subsets $A \setminus C_y$ cover the compact subset $A \setminus D$, there are finitely many $y_1, \dots, y_m \in A \setminus D$ so that

$$D = C_{y_1} \cap \dots \cap C_{y_m} \in M.$$

We conclude that the functor $G\colon \mathsf{BA}^{\mathrm{op}} \to \mathsf{BooSp}$ is fully faithful.

Example 61 A very similar argument applies to the dual adjunction

$$\mathsf{Priest} \underset{G}{\overset{F}{\underset{\top}{\rightleftarrows}}} \mathsf{DL}^{\mathrm{op}}$$

of Example 40: also here the functor $F\colon \mathsf{Priest} \to \mathsf{DL}^{\mathrm{op}}$ satisfies (SW) with respect to the class of all embeddings (= regular monomorphisms) in Priest and the class of embeddings (= monomorphism) in DL, which implies that the functor $G\colon \mathsf{DL}^{\mathrm{op}} \to \mathsf{Priest}$ is fully faithful.

Proposition 62 *If the functor F in* (7) *satisfies (SW) then $F(\mathbb{M}_{\mathsf{A}}) \subseteq \mathbb{M}_{\mathsf{B}}$-ExtrEpi.*

Proof Let $h\colon A \to A'$ be an arrow in $\mathbb{M}_{\mathsf{A}}$ and consider the factorization

$$\begin{array}{ccc} FA' & \xrightarrow{\;Fh\;} & FA \\ & {\scriptstyle e}\searrow \quad \nearrow{\scriptstyle m} & \\ & M & \end{array}$$

of Fh with $m \in \mathbb{M}_{\mathsf{B}}$ and e being an $\mathbb{M}_{\mathsf{B}}$-extremal epimorphism. Then the cone $(m(f))_{f\in|M|}$ has the subcone

$$(A \xrightarrow{h} A' \xrightarrow{\varphi} \widetilde{A})_{\varphi\in\mathsf{A}(A,\widetilde{A})}$$

and therefore belongs to $\mathcal{M}_{\mathsf{A}}$. Since F satisfies (SW), m is an isomorphism and therefore Fh is an $\mathbb{M}_{\mathsf{B}}$-extremal epimorphism. □

Example 63 In continuation of Examples 60 and 61, Proposition 62 implies that the two-element Boolean space $\mathsf{2}$ is regular injective in BooSp, and that the two-element Priestley space $\mathsf{2}$ is regular injective in Priest. Since both spaces are regular cogenerators in their respective category, using Remark 15 we conclude that both functors

$$G\colon \mathsf{BA}^{\mathrm{op}} \longrightarrow \mathsf{BooSp} \quad \text{and} \quad G\colon \mathsf{DL}^{\mathrm{op}} \longrightarrow \mathsf{Priest}$$

are equivalences.

Remark 64 In Stone (1938b), Stone essentially proves that the category DL is dually equivalent to a category of very special topological spaces and continuous maps, namely the category of spectral spaces and spectral maps. Let us also note that these spaces appear also precisely as the prime spectra of commutative rings (see Hochster (1969)).

Proposition 65 *Assume that the adjunction* (7) *is an equivalence and that* $F(\mathbb{M}_{\mathsf{A}}) \subseteq \mathbb{M}_{\mathsf{B}}$*-ExtrEpi. Then* F *satisfies (SW).*

Proof Let $m\colon M \to FA$ be in $\mathbb{M}_{\mathsf{B}}$ so that the cone $(m(f)\colon A \to \widetilde{A})_{f\in|M|}$ belongs to $\mathcal{M}_{\mathsf{A}}$. Since the adjunction (7) is an equivalence, without loss of generality we may assume that $m = Fe$, for some $e\colon A \to A'$ in A. Hence, for $\varphi \in |B| = \mathsf{A}(A', \widetilde{A})$, $m(\varphi) = \varphi \cdot e$,

$$\begin{array}{ccc} A & \xrightarrow{e} & A' \\ & \underset{\varphi\cdot e}{\searrow} & \downarrow{\scriptstyle\varphi} \\ & & \widetilde{A} \end{array}$$

and therefore $e \in \mathbb{M}_{\mathsf{A}}$. By hypothesis, $m = Fe$ is an $\mathbb{M}_{\mathsf{B}}$-extremal epimorphism and consequently an isomorphism. □

An interesting consequence of the results so far is that either none or both of the functors of (7) satisfy (SW).

Corollary 66 *If the adjunction* (7) *is an equivalence,* G *satisfies (SW) if and only if* F *satisfies (SW).*

We introduce now slightly a weaker condition than (SW).

Definition 67 F satisfies the *clone-condition* provided that the following holds:

(Cl) For each set X, every $\mathbb{M}_{\mathsf{B}}$-morphism $m\colon M \to F(\widetilde{A}^X)$ is an isomorphism provided that the cone $(m(f)\colon \widetilde{A}^X \to \widetilde{A})_{f\in|M|}$ contains all projections.

Remark 68 If B is a category of algebras for a signature (Σ, σ), then the condition above means that

$$|\mathrm{Cl}_X(\widetilde{B})| = |\mathsf{A}(\widetilde{A}^X, \widetilde{A})|.$$

Proposition 69 *If the dual adjunction* (7) *is an equivalence, then F satisfies (Cl).*

Proof Adapting the proof of Proposition 65, we are led to consider an epimorphism $e\colon \widetilde{A}^X \to A$ so that the cone $(\varphi \cdot e\colon \widetilde{A}^X \to \widetilde{A})_{\varphi\in\mathsf{A}(A,\widetilde{A})}$ contains all projection maps. Put differently, for every $x \in X$ there is some $\varphi_x\colon A \to \widetilde{A}$ with $\varphi_x \cdot e = \pi_x$.

$$\begin{array}{ccc} \widetilde{A}^X & \xrightarrow{\;e\;} & A \\ & {}_{\pi_x}\searrow & \downarrow{\scriptstyle \varphi_x} \\ & & \widetilde{A}. \end{array}$$

Let $\varphi\colon A \to \widetilde{A}^X$ be the arrow induced by the cone $(\varphi_x\colon A \to \widetilde{A})_{x\in X}$. Then $\varphi \cdot e = 1$, hence e is an isomorphism and the assertion follows. □

Proposition 70 *If F satisfies (Cl) and $F(\mathbb{M}_{\mathsf{A}}) \subseteq \mathbb{M}_{\mathsf{B}}$-ExtrEpi, then F satisfies (SW).*

Proof Let $m\colon M \to FA$ in $\mathbb{M}_{\mathsf{B}}$ so that $(m(f))_{f\in|M|}$ is in $\mathcal{M}_{\mathsf{A}}$. Let $h\colon A \to \widetilde{A}^{|M|}$ be the arrow induced by the cone $(m(f))_{f\in|M|}$; hence, for every $f \in |M|$ the diagram

$$\begin{array}{ccc} A & \xrightarrow{\;h\;} & \widetilde{A}^{|M|} \\ & {}_{m(f)}\searrow & \downarrow{\scriptstyle \pi_f} \\ & & \widetilde{A} \end{array}$$

commutes. Consider the pullback

$$\begin{array}{ccc} \overline{M} & \longrightarrow & M \\ {\scriptstyle \overline{m}}\downarrow & & \downarrow{\scriptstyle m} \\ F(\widetilde{A}^{|M|}) & \xrightarrow[Fh]{} & FA \end{array}$$

in B. Since pullbacks in B are preserved by $|-|\colon \mathsf{B} \to \mathsf{Set}$, we know that $\overline{M}$ is the inverse image of M and therefore the cone $(\overline{m}(\overline{\varphi})\colon \widetilde{A}^{|M|} \to \widetilde{A})_{\overline{\varphi}\in|\overline{M}|}$ contains all projection maps. We conclude that $\overline{m}$ is an isomorphism, therefore $Fh \cdot \overline{m}$ is an $\mathbb{M}_{\mathsf{B}}$-extremal epimorphism, hence m is so too. Since also $m \in \mathbb{M}_{\mathsf{B}}$, we have shown that m is an isomorphism. □

We finish this part with a direct application of our results to categories of algebras.

Theorem 71 *Let* (Σ, σ) *be a signature and let* $\widetilde{B}$ *be an algebra. Let* B *be the full subcategory of the category of all algebras and homomorphisms with respect to* (Σ, σ) *defined by those algebras* B *so that the cone of all homomorphisms* $B \to \widetilde{B}$ *is point-separating, equipped with the canonical forgetful functor* $|-|: \mathsf{B} \to \mathsf{Set}$. *Furthermore, let* $(\mathsf{A}, |-|)$ *be a concrete category over* Set *and let* $\widetilde{A}$ *be an object in* A *with* $|\widetilde{A}| = |\widetilde{B}|$. *Then the following assertions are equivalent*

(i) $(\widetilde{A}, \widetilde{B})$ *induces an equivalence* $\mathsf{B}^{\mathrm{op}} \sim \mathsf{A}$.
(ii) The following three conditions are fulfilled.
  (a) A *is concretely* $\widetilde{A}$*-complete.*
  (b) $\widetilde{A}$ *is a regular injective regular cogenerator of* A.
  (c) For each set X,
$$|\mathrm{Cl}_X(\widetilde{B})| = |\mathsf{A}(\widetilde{A}^X, \widetilde{A})|.$$

Proof Since A is concretely $\widetilde{A}$-complete, the category A is complete and $|-|: \mathsf{A} \to \mathsf{Set}$ preserves limits; moreover, regular monomorphisms in A are stable under composition and pullbacks. We also note that B is a reflective subcategory of the category of all algebras and homomorphisms with respect to (Σ, σ), hence B is complete and $|-|: \mathsf{B} \to \mathsf{Set}$ preserves limits. Since $|\mathrm{Cl}_X(\widetilde{B})| \subseteq |\mathsf{A}(\widetilde{A}^X, \widetilde{A})|$, $(\widetilde{A}, \widetilde{B})$ induces a dual adjunction (7), and we can chose $\mathbb{M}_{\mathsf{B}}$ as the class of all monomorphisms of B and $\mathbb{M}_{\mathsf{A}}$ as the class of all regular monomorphisms of A.

Since $|\mathrm{Cl}_X(\widetilde{B})| = |\mathsf{A}(\widetilde{A}^X, \widetilde{A})|$, F satisfies (Cl), hence, by Proposition 70, F satisfies also (SW). Therefore $\varepsilon: 1 \to FG$ is an isomorphism, hence $G: \mathsf{B}^{\mathrm{op}} \to \mathsf{A}$ is fully faithful. By Remark 15, G is an equivalence. □

3 Stone-Type Dualities

In this section we investigate conditions which allow to extend a dual equivalence between certain *subcategories* of A and B to the entire categories. Here we consider the subcategories defined by all finite objects of concrete categories over Set. The general strategy employed here is taken from (Johnstone 1986, Chap. VI), but see also Clark and Davey (1998) and Hofmann (2002a).

Specifically, we consider concrete categories $(\mathsf{A}, |-|)$ and $(\mathsf{B}, |-|)$ over Set and a dual adjunction

$$F: \mathsf{A} \longrightarrow \mathsf{B}^{\mathrm{op}}, \quad G: \mathsf{B}^{\mathrm{op}} \longrightarrow \mathsf{A}, \quad \eta: 1_{\mathsf{A}} \longrightarrow GF, \quad \varepsilon: 1_{\mathsf{B}} \longrightarrow FG \tag{8}$$

induced by objects $\widetilde{A}$ in A and $\widetilde{B}$ in B with $|\widetilde{A}| = |\widetilde{B}|$ finite. Furthermore, we assume that the adjunction (8) restricts to an *equivalence* between the full subcategories of A and B defined by all finite objects, that is, we assume that η_A is an isomorphism for every finite object A in A and ε_B is an isomorphism for every finite object B in B.

Under the conditions described above, the adjunction (8) is actually an equivalence provided that

- Each object B in B is a filtered colimit of finite objects.
- Each object A in A is a cofiltered limit of finite objects.
- F sends cofiltered limits of finite objects to colimits.

In fact, under the conditions above, the endofunctor $FG\colon \mathsf{B} \to \mathsf{B}$ preserves filtered colimits of finite objects and, dually, $GF\colon \mathsf{A} \to \mathsf{A}$ preserves cofiltered limits of finite objects. Hence, for an object A in A, there is a limit cone $(p_i\colon A \to A_i)_{i\in I}$ where each A_i is finite, and in the commutative diagram

$$\begin{array}{ccc} A & \xrightarrow{\eta_A} & GFA \\ {\scriptstyle p_i}\big\downarrow & & \big\downarrow{\scriptstyle GFp_i} \\ A_i & \xrightarrow[\eta_{A_i}]{} & GFA_i \end{array}$$

both cones are limit cones. Therefore $\eta_A = \lim \eta_{A_i}$ is an isomorphism. A similar argument justifies that every ε_B is an isomorphism.

To obtain such a situation, we shall assume that our categories A and B are of a certain type, namely full subcategories of model categories of a limit sketch in Set (the category B) and a full subcategory of a model category of a limit sketch in BooSp (the category A). We start by giving more information on the former type of categories.

3.1 Locally Presentable Categories

In this subsection we provide some background material on locally presentable categories and (limit) sketches. For further details we refer the reader to Adámek and Rosický (1994) and Gabriel and Ulmer (1971).

Definition 72 A *finitary limit sketch* is a triple $\mathcal{S} = (\mathsf{C}, \mathcal{L}, \sigma)$ consisting of

- a small category C,
- a set $\mathcal{L}$ of diagrams in C with finite shape, and
- a function σ which assigns a cone to each diagram of $\mathcal{L}$.

A *model* of a finitary limit sketch $\mathcal{S} = (\mathsf{C}, \mathcal{L}, \sigma)$ in a category A is a functor $M\colon \mathsf{C} \to \mathsf{A}$ which sends $\sigma(D)$ to a limit of MD, for each diagram $D\colon I \to \mathsf{C}$ of $\mathcal{L}$. Finally, $\mathrm{Mod}(\mathcal{S}, \mathsf{A})$ denotes the full subcategory of the functor category A^{C} defined by all models of $\mathcal{S}$ in A.

Remark 73 Dually, there is also the notion of a *colimit sketch* where σ assigns to each diagram a cocone, and a model of a colimit sketch sends each selected cocone to a colimit.

Intuitively, limit sketches allow us to specify structures defined by (partial) operations and relations. Below we give some basic examples.

Example 74 Consider the following limit sketch $\mathcal{S} = (\mathsf{C}, \mathcal{L}, \sigma)$:

- C is the category consisting of two objects c_1 and c_2 and, besides the identity morphisms, with three morphisms $o, p_1, p_2 \colon c_2 \to c_1$.
- $\mathcal{L}$ contains only the discrete diagram consisting of two copies of c_1.
- σ assigns the cone $(p_1, p_2 \colon c_2 \to c_1)$ to this diagram.

Then $\mathrm{Mod}(\mathcal{S}, \mathsf{Set})$ is equivalent to the category of magmas and magma homomorphisms, $\mathrm{Mod}(\mathcal{S}, \mathsf{CompHaus})$ is equivalent to the category of 'compact Hausdorff magmas' and continuous homomorphisms and so on.

Example 75 Consider the following limit sketch $\mathcal{S} = (\mathsf{C}, \mathcal{L}, \sigma)$:

- C is the category consisting of three objects c_1, c_2 and r and, besides the identity morphisms, with the morphisms $p_1, p_2 \colon c_2 \to c_1, m \colon r \to c_2$ and $p_1 \cdot m$ and $p_2 \cdot m$.
- $\mathcal{L}$ contains the discrete diagram consisting of two copies of c_1 and the span $r \xrightarrow{m} c_2 \xleftarrow{m} r$.
- σ assigns the cones

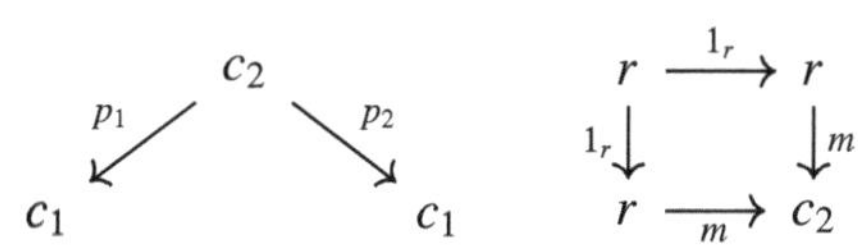

to these diagrams.

Then $\mathrm{Mod}(\mathcal{S}, \mathsf{Set})$ is equivalent to category of sets equipped with a binary relation and relation-preserving maps.

Example 76 For a finitely complete small category C, we may consider the limit sketch $\mathcal{S} = (\mathsf{C}, \mathcal{L}, \sigma)$ where

- $\mathcal{L}$ is the collection of all finite diagrams in C and
- σ assigns a limit to each of these diagrams.

Then $\mathrm{Mod}(\mathcal{S}, \mathsf{Set})$ is the category $\mathrm{Cart}(\mathsf{C}, \mathsf{Set})$ of Cartesian functors and natural transformations.

Theorem 77 *For every finitary limit sketch $\mathcal{S}$, the inclusion functor* $\mathrm{Mod}(\mathcal{S}, \mathsf{A}) \to \mathsf{A}^{\mathsf{C}}$ *has a left adjoint.*

Proof See Kennison (1968) and Freyd and Kelly (1972). □

Corollary 78 *For every finitary limit sketch $\mathcal{S}$, the category* $\mathrm{Mod}(\mathcal{S}, \mathsf{A}) \to \mathsf{A}^C$ *is complete and cocomplete.*

Next we recall an abstract characterisation of model categories of finitary limit sketches.

Definition 79 An object B in a category B is called *finitely presentable* if the covariant hom-functor $\mathsf{B}(B, -)$ preserves filtered colimits. A category B is called *locally finitely presentable* provided that the following conditions hold.

1. B is cocomplete.
2. There exists, up to isomorphism, only a set of finitely presentable objects in B.
3. Each object B in B is a filtered colimit of finitely presentable objects.

Below we list some further important properties of locally finitely presentable categories.

Remark 80 Locally finitely presentable categories are also complete, (co) wellpowered and have a generating set. Moreover, each functor between locally finitely presentable categories which preserves limits and filtered colimits has a left adjoint.

Example 81 • One easily verifies that Set is locally finitely presentable, and that the finitely presentable objects of Set are precisely the finite sets.
- More generally, for a finitary signature, the category of algebras and homomorphisms is finitely presentable. Here the finitely presentable objects are precisely the finitely presented algebras.
- An object of a category is called *finitely copresentable* whenever it is finitely presentable in the dual category. It is shown in Gabriel and Ulmer (1971) that the finitely copresentable compact Hausdorff spaces are precisely the finite ones, and the same holds for Boolean spaces.

Theorem 82 *The model categories of finitary limit sketches in* Set *are precisely (up to equivalence) the locally finitely presentable categories.*

Proof See Gabriel and Ulmer (1971) and Adámek and Rosický (1994). □

Remark 83 More generally, a category of models of a finitary limit sketch in a locally finitely presentable category is locally finitely presentable.

Remark 84 The notion of locally finitely presentable category may be generalised to locally λ-presentable category, for each regular cardinal λ, by substituting 'finite' by 'less than λ' everywhere. We note here that the category of models of a colimit sketch in a locally presentable category is locally presentable (see Adámek and Rosický (1994, Remark 2.63)).

For a limit sketch $\mathcal{S}$, there is in general no canonical forgetful functor $\mathrm{Mod}(\mathcal{S}) \to \mathsf{Set}$, i.e. the category of models of $\mathcal{S}$ might be *many-sorted*. Below we introduce some conditions on a limit sketch $\mathcal{S} = (\mathsf{C}, \mathcal{L}, \sigma)$ making the model category *one-sorted*. To do so, we introduce some notation.

- The class of all $\mathcal{S}$-*monomorphisms* is defined as the composition closure of the class of all C-morphisms $m\colon A \to B$ such that the span $A \xrightarrow{m} B \xleftarrow{m} A$ belongs to $\mathcal{L}$ and σ assigns the cone

$$\begin{array}{ccc} A & \xrightarrow{1_A} & A \\ {\scriptstyle 1_A}\downarrow & & \downarrow{\scriptstyle m} \\ A & \xrightarrow[m]{} & B \end{array}$$

  to this diagram.
- For a full subcategory C_0 of C, we put

  1. $\mathrm{Sub}(\mathsf{C}_0) = \{C \mid C \text{ is an } \mathcal{S}\text{-subobject of an object in } \mathsf{C}_0\}$,
  2. $\mathrm{Lim}(\mathsf{C}_0) = \{C \mid C \text{ is an } \mathcal{S}\text{-limit of an object in } \mathsf{C}_0\}$.

- For an object C in C, we define a chain $\mathcal{G}_n(C)$ $(n \in \mathbb{N})$ of full subcategories of C in the following way:

  1. We put $\mathcal{G}_0(C) = \{C\}$, and
  2. for each $n \geq 0$, $\mathcal{G}_{n+1}(C) = \mathrm{Sub}_{\mathcal{S}}(\mathrm{Lim}_{\mathcal{S}}(\mathcal{G}_n(C)))$.

Definition 85 Let $\mathcal{S} = (\mathsf{C}, \mathcal{L}, \sigma)$ be a finitary limit sketch. An object C_0 in C is called a *sketch-cogenerator* of $\mathcal{S}$ if $C = \bigcup_{n\in\mathbb{N}} \mathcal{G}_n(C_0)$. The sketch $\mathcal{S}$ is called *single-sorted* provided that it has a sketch-cogenerator.

We note immediately the property.

Lemma 86 *Let $\mathcal{S} = (\mathsf{C}, \mathcal{L}, \sigma)$ be a finitary, single-sorted limit sketch with sketch-cogenerator C_0. For each object C in C, there exists a finite subset $M \subseteq \mathsf{C}(C, C_0)$ such that, for each model $F\colon C \to \mathsf{A}$ of $\mathcal{S}$, the cone $(F(f)\colon F(C) \to F(C_0))_{f\in M}$ is a mono-cone in A.*

Corollary 87 *Let $\mathcal{S} = (\mathsf{C}, \mathcal{L}, \sigma)$ be a finitary, single-sorted limit sketch with sketch-cogenerator C_0 and let A be a category. Then the evaluation functor $\mathrm{ev}_{C_0}\colon \mathrm{Mod}(\mathcal{S}, \mathsf{A}) \to \mathsf{A}$ is faithful.*

Furthermore, assume that $|{-}|\colon \mathsf{A} \to \mathsf{Set}$ preserves finite mono-cones and let $F\colon \mathsf{C} \to \mathsf{A}$ be a model of $\mathcal{S}$ in A. Then $|F(C)|$ is finite for each object C in C if and only if $|F(C_0)|$ is finite.

We are now ready to describe the setting we wish to investigate in this section.

3.2 Refining our Situation

We consider single sorted, finitary limit sketches $\mathcal{S}_A = (\mathsf{C}_A, \mathcal{L}_A, \sigma_A)$ and $\mathcal{S}_B = (\mathsf{C}_B, \mathcal{L}_B, \sigma_B)$ with sketch-cogenerators C_A and C_B, respectively. Then we have the following.

- The category $\mathrm{Mod}(\mathcal{S}_B, \mathsf{Set})$ is locally finitely presentable, hence (co)complete and (co)wellpowered and the faithful functor

$$\mathrm{ev}_{C_B} \colon \mathrm{Mod}(\mathcal{S}_B, \mathsf{Set}) \longrightarrow \mathsf{Set}$$

  has a left adjoint and preserves filtered colimits.
- The category $\mathrm{Mod}(\mathcal{S}_A, \mathsf{BooSp})$ is dually equivalent to the category of models of the colimit sketch ${\mathcal{S}_A}^{\mathrm{op}}$ in the locally presentable category BA. By Remark 84, $\mathrm{Mod}(\mathcal{S}_A, \mathsf{BooSp})$ is locally copresentable and therefore (co)complete and (co) wellpowered and has a cogenerating set. In particular, the faithful functor

$$\mathrm{ev}_{C_A} \colon \mathrm{Mod}(\mathcal{S}_A, \mathsf{BooSp}) \longrightarrow \mathsf{BooSp}$$

  has a left adjoint as well.

Furthermore, let $\mathbb{M}_A$ and $\mathbb{M}_B$ be classes of morphisms in $\mathrm{Mod}(\mathcal{S}_A, \mathsf{BooSp})$ and $\mathrm{Mod}(\mathcal{S}_B, \mathsf{Set})$, respectively, closed under composition, pullback and intersection, and containing all regular monomorphisms and contained in the class of all embeddings (see Sect. 2.6).

In addition to the situation described at the beginning of this section, we assume that A is the full subcategory of $\mathrm{Mod}(\mathcal{S}_A, \mathsf{BooSp})$ of all $\mathbb{M}_A$-subobjects of powers of $\widetilde{A}$. Likewise, B denotes the full subcategory of $\mathrm{Mod}(\mathcal{S}_B, \mathsf{Set})$ of all $\mathbb{M}_B$-subobjects of powers of $\widetilde{B}$, and the forgetful functors $|-| \colon \mathsf{A} \to \mathsf{Set}$ and $|-| \colon \mathsf{B} \to \mathsf{Set}$ are defined by 'evaluating at C_A and at C_B, respectively'.

Remark 88 Note that A is an $\mathbb{M}_A$-ExtrEpi-reflective subcategory of $\mathrm{Mod}(\mathcal{S}_A, \mathsf{BooSp})$ with left adjoint $R_{\widetilde{A}} \colon \mathrm{Mod}(\mathcal{S}_A, \mathsf{BooSp}) \to \mathsf{A}$ and B is an $\mathbb{M}_B$-ExtrEpi-reflective subcategory of $\mathrm{Mod}(\mathcal{S}_B, \mathsf{Set})$ with left adjoint $R_{\widetilde{B}} \colon \mathrm{Mod}(\mathcal{S}_B, \mathsf{Set}) \to \mathsf{B}$.

We are immediately able to prove the desired property of the category B.

Theorem 89 *Each object B in B is a filtered colimit of finite objects in B.*

Proof First observe that an object B in B is finite if and only if $\mathsf{B}(B, \widetilde{B})$ is finite. Furthermore, each presheaf F in $\mathsf{Set}^{\mathsf{C}_B}$ is a colimit of representable presheaves. Finally, for a representable presheaf $\mathsf{C}_B(C, -)$,

$$\mathsf{B}(R_{\widetilde{B}}(\mathsf{C}_B(C, -)), \widetilde{B}) = \mathrm{Nat}(\mathsf{C}_B(C, -), \widetilde{B}) = \widetilde{B}(C)$$

is finite. □

In the next subsection we investigate properties of the category A.

3.3 Models in Boolean Spaces

Since A is a reflective subcategory of $\mathrm{Mod}(\mathcal{S}_A, \mathsf{BooSp})$, an object A of A is finitely copresentable in A provided that it is so in $\mathrm{Mod}(\mathcal{S}_A, \mathsf{BooSp})$. The following result gives a useful sufficient condition for this property.

Lemma 90 *Assume that* C_A *is finitely generated. An object* M *in* $\mathrm{Mod}(\mathcal{S}_A, \mathsf{BooSp})$ *is finitely copresentable provided that, for each* C *in* C_A*,* $M(C)$ *is a finite discrete space.*

Proof See Zádori (1995) and Hofmann (2002a), but also Johnstone (1986) for Stone algebras. □

Certainly, the lemma above applies to $\widetilde{A}$.

Corollary 91 $\widetilde{A}$ *is finitely copresentable in* A.

The next result tells us that $F\colon \mathsf{A} \to \mathsf{B}^{\mathrm{op}}$ sends 'sufficiently many' limits to colimits.

Proposition 92 *Let* $D\colon I \to \mathsf{A}$ *be a diagram in* A *with limit cone* $(p_i\colon L \to D(i))_{i\in I}$ *such that each* $\eta_{D(i)}$ *is an isomorphism. Then* $(F(p_i)\colon F(L) \to FD(i))_{i\in I}$ *is a colimit of* $FD\colon I^{\mathrm{op}} \to \mathsf{B}$ *provided that* $\mathsf{A}(-, \widetilde{A})$ *sends* $(p_i\colon L \to D(i))_{i\in I}$ *to a colimit cocone of* $\mathsf{A}(D(-), \widetilde{A})\colon I^{\mathrm{op}} \to \mathsf{Set}$.

Proof See Hofmann (2002a, Proposition 1.2). □

Theorem 93 *Assume that* C_A *is finitely generated. Then, for each object* B *in* B*,* ε_B *is an isomorphism.*

Remark 94 Theorem 93 corresponds to the 'duality compactness theorem' (Clark and Davey 1998, Theorem 2.11).

Our final aim in this section is to ensure that every object of A is a cofiltered limit of finite objects. We will make use here of the results of Sect. 2.2 and in particular the 'Bourbaki criterion' Theorem 23. Under our conditions, for every object A in A, the canonical cone

$$(A \longrightarrow A_0)_{A\to A_0}$$

for the canonical diagram D_A is an initial mono-cone since it contains the cone $(f\colon A \to \widetilde{A})_f$. Putting everything together:

Theorem 95 *Our dual adjunction is a dual equivalence provided that the following hold:*

- C_A *is finitely generated and*
- $(\mathsf{A}, |-|)$ *has 'image factorisations'.*

Example 96 Theorem 95 can be used to obtain a new dual equivalence from a given one by 'structure interchange' since the category $\mathsf{Set}_{\mathrm{fin}}$ of finite sets and functions essentially coincides with the category $\mathsf{BooSp}_{\mathrm{fin}}$ of finite Boolean spaces and continuous maps. To illustrate this principle, consider the classic Stone duality

$$\mathsf{BooSp} \sim \mathsf{BA}^{\mathrm{op}}$$

induced by $(\mathsf{2}, \mathsf{2})$, which of course restricts to a dual equivalence

$$\mathsf{BooSp}_{\mathrm{fin}} \sim \mathsf{BA}_{\mathrm{fin}}^{\mathrm{op}}. \tag{9}$$

Let $\mathsf{BooSpBA}$ be the category of Boolean algebras in Boolean spaces and continuous homomorphisms. Certainly, $(\mathsf{2}, \mathsf{2})$ induces a dual adjunction

$$\mathsf{BooSpBA} \underset{G}{\overset{F}{\rightleftarrows}} \mathsf{Set}^{\mathrm{op}} \tag{10}$$

which restricts to the full subcategories defined by all finite objects; in fact, by (9), this restriction is a dual equivalence

$$\mathsf{BooSpBA}_{\mathrm{fin}} \sim {\mathsf{Set}_{\mathrm{fin}}}^{\mathrm{op}}.$$

Any set with at least two elements is a cogenerator in Set, and Theorem 95 implies that (10) is a dual equivalence provided that $\mathsf{2}$ is a cogenerator in $\mathsf{BooSpBA}$. Fortunately, the following theorem of Johnstone (1986) completes the argument.

Theorem 97 *Let (Σ, σ) be a finitary signature where Σ is finite and $\sigma(\delta) \leq 2$ for each $\delta \in \Sigma$. We put*

$$U = \{\delta \in \Sigma \mid \sigma(\delta) = 1\} \quad \text{and} \quad B = \{\delta \in \Sigma \mid \sigma(\delta) = 2\}.$$

Let $\mathcal{E}$ be a set of Σ-equations and assume that the following conditions are satisfied.

1. *For each $\beta \in B$, the associative law*

$$\beta(x, \beta(y, z)) = \beta(\beta(x, y), z)$$

   is satisfied.
2. *There exists a total order $\beta_1 < \beta_2 < \ldots < \beta_n$ $(n \in \mathbb{N})$ on B such that, for all $1 \leq i < j \leq n$, the distributive laws*

$$\beta_j(x, \beta_i(y, z)) = \beta_i(\beta_j(x, y), \beta_j(x, z))$$

   and

$$\beta_j(\beta_i(x, y), z) = \beta_i(\beta_j(x, z), \beta_j(y, z))$$

are satisfied.

3. *U contains the identity and is closed under composition.*
4. *The 'de Morgan laws' are satisfied, i.e., for each $\alpha \in U$ and each $\beta \in B$, there exist operation symbols $\beta' \in B$ and $\alpha', \alpha'' \in U$ such that*

$$\alpha(\beta(x, y)) = \beta'(\alpha'(x), \alpha''(y))$$

or

$$\alpha(\beta(x, y)) = \beta'(\alpha'(y), \alpha''(x)).$$

Then every algebra in Boolean spaces is profinite, that is, a cofiltered limit of finite algebras.

Proof See Johnstone (1986, Theorem VI.2.9). □

Example 98 From Theorem 97 we deduce that every object in $\mathsf{BooSpBA}$ is a cofiltered limit of finite algebras, hence $\mathsf{2}$ is a cogenerator in $\mathsf{BooSpBA}$.

Example 99 We use now Theorem 95 to deduce that the adjunction

$$\mathsf{Vec}_{\mathbb{F}} \underset{G}{\overset{F}{\rightleftarrows}} \mathsf{BooSpVec}_{\mathbb{F}}^{\mathrm{op}} \quad (F \dashv G)$$

of Example 41 (3) is an equivalence. Firstly, the restriction of this adjunction to finite objects is an equivalence, and $\mathbb{F}$ is a cogenerator of $\mathsf{Vec}_{\mathbb{F}}$ (see Example 41 (2)). Since $\mathbb{F}$ is finite, the algebraic theory of vector spaces over $\mathbb{F}$ is finite: one nullary operation (the zero vector), one binary operation (vector addition) and, for every $\alpha \in \mathbb{F}$, a unary operation (scalar multiplication). One easily verifies that the conditions of Theorem 97 are satisfied, in particular, the discrete space $\mathbb{F}$ is a cogenerator of $\mathsf{BooSpVec}_{\mathbb{F}}$.

4 Kleisli Categories, Splitting Idempotents and All That

In this section we move away from studying only concrete categories over Set and consider in particular categories of *relations*. We shall see that many other dual equivalences can be proven starting with this level of generality.

To start, we consider concrete categories X and A over Set and a dual adjunction

$$\mathsf{X} \underset{G}{\overset{F}{\rightleftarrows}} \mathsf{A}^{\mathrm{op}} \quad (F \dashv G) \tag{11}$$

with units $\eta\colon 1_{\mathsf{X}} \to GF$ and $\varepsilon\colon 1_{\mathsf{A}} \to FG$ induced by $(\widetilde{X}, \widetilde{A})$. If η is not an isomorphism, equivalently, if F is not fully faithful, then we might interpret this fact as 'A has too many morphisms'. One way to improve this situation is to add further structure to A in order to reduce the number of arrows between objects in A. For instance, Johnstone (1986) explains how to lift this adjunction to the category of Eilenberg–Moore algebras for the monad on A induced by the adjunction (11), and gives conditions which guarantee that the 'lifted' functor F is fully faithful. However, in these notes we take the 'dual road' and think that the category X has too few arrows. With $\mathbb{D}$ denoting the monad induced by $F \dashv G$ on X, we 'expand' the category X by passing to the Kleisli category $\mathsf{X}_{\mathbb{D}}$; moreover, the induced functor

$$\mathsf{X}_{\mathbb{D}} \longrightarrow \mathsf{A}^{\mathrm{op}}$$

is fully faithful without further conditions. To make this approach useful, we need to identify the monad $\mathbb{D}$ with a 'familiar' monad $\mathbb{T}$ on X and we need to identify the arrows of X inside the Kleisli category $\mathsf{X}_{\mathbb{T}}$.

Example 100 The powerset monad $\mathbb{P} = (P, m, e)$ on Set consists of the powerset functor $P\colon \mathsf{Set} \longrightarrow \mathsf{Set}$ and

$$e_X\colon X \longrightarrow PX, \quad x \longmapsto \{x\} \quad \text{and} \quad m_X\colon PPX \longrightarrow PX, \quad \mathcal{A} \longmapsto \bigcup \mathcal{A}.$$

One easily sees that the Kleisli category $\mathsf{Set}_{\mathbb{P}}$ is equivalent to the category Rel of sets and relations. Below we describe two fundamentally different ways to recognise functions inside Rel.

1. A relation $r\colon X \relbar\joinrel\mapstochar\joinrel\rightarrow Y$ is a function if and only if r has a right adjoint in the ordered category Rel.
2. A relation $r\colon X \relbar\joinrel\mapstochar\joinrel\rightarrow Y$ is a function if and only if and only if r is a homomorphism of comonoids in the monoidal category Rel, that is, the following diagrams commute.

$$\begin{array}{ccc} X & \xrightarrow{\;r\;} & Y \\ & \searrow^{\top} & \downarrow{\scriptstyle \top} \\ & & 1 \end{array} \qquad \begin{array}{ccc} X \times X & \xrightarrow{\;r \times r\;} & Y \times Y \\ {\scriptstyle \Delta_X}\uparrow & & \uparrow{\scriptstyle \Delta_Y} \\ X & \xrightarrow[\;r\;]{} & Y. \end{array}$$

   Here $X \times X$ denotes the set-theoretical product which can be misleading since it is not the categorical product in Rel.

The first description involves a 2-categorical property; if we want to use it in a duality we must make sure that the involved equivalence functors are locally monotone. In the second case one needs to know the corresponding operation on the other side.

In the considerations above, the Kleisli category $\mathsf{X}_{\mathbb{T}}$ was only introduced as a tool for the study of X; however, in many examples our primary interest lies in $\mathsf{X}_{\mathbb{T}}$.

In this case, a monad $\mathbb{T}$ on X is typically given beforehand, and we wish to find an adjunction so that the monad on X induced by this adjunction is isomorphic to $\mathbb{T}$. For instance, if $\mathbb{T}$ is some variant of the powerset monad, we obtain this way a duality result for categories with certain relations as arrows. We introduce now some monads related to the powerset monad which will play an important role in this section.

Example 101 The upset monad $\mathbb{U} = (U, m, e)$ on Pos consists of the upset functor $U\colon \mathsf{Pos} \to \mathsf{Pos}$ defined by

$$UX = \{A \subseteq X \mid \uparrow A = A\}, \quad Uf\colon UX \longrightarrow UY, \quad A \longmapsto \uparrow f(A)$$

and

$$e_X\colon X \longrightarrow UX, \quad x \longmapsto \uparrow x \quad \text{and} \quad m_X\colon UUX \longrightarrow UX, \quad \mathcal{A} \longmapsto \bigcup \mathcal{A}.$$

A Kleisli arrow between partially ordered sets is a relation $\varphi\colon X \nrightarrow Y$ between the underlying sets which is compatible with the order relation in the following sense: for all $x, x' \in X$ and $y, y' \in Y$,

$$x' \leq x \;\varphi\; y \leq y' \implies x' \;\varphi\; y'.$$

Such a relation is usally called *distributor* or *order ideal*. Furthermore, relational composition of distributors corresponds to composition in $\mathsf{Pos}_{\mathbb{U}}$. Therefore $\mathsf{Pos}_{\mathbb{U}}$ is isomorphic to the category Dist of partially ordered sets and distributors, with relational composition and the identity on X given by the order relation on X.

$$\mathsf{Dist} \sim \mathsf{Pos}_{\mathbb{U}}.$$

We also note that Dist becomes a 2-category with the inclusion order of distributors.

The inclusion functor $\mathsf{Pos} \to \mathsf{Dist}$ corresponds to

$$(-)_*\colon \mathsf{Pos} \longrightarrow \mathsf{Dist}$$

where, for a monotone map $f\colon X \to Y$ between partially ordered sets, we define the relation $f_*\colon X \nrightarrow Y$ by

$$x \; f_* \; y \quad \text{whenever} \quad f(x) \leq y.$$

Finally, $f_*\colon X \nrightarrow Y$ has a right adjoint $f^*\colon Y \nrightarrow X$ in Dist defined by

$$y \; f^* \; x \quad \text{whenever} \quad y \leq f(x).$$

Example 102 The *Vietoris functor* (Vietoris 1922) $V: \mathsf{CompHaus} \to \mathsf{CompHaus}$ is defined by $VX = \{A \subseteq X \mid A \text{ closed}\}$ with the 'hit-and-miss topology' generated by the sets

$$B^\diamond = \{A \mid A \cap B \neq \varnothing\}, \quad K^\Box = \{A \mid A \cap K = \varnothing\} \qquad \text{(for all } B \text{ open, } K \text{ closed);}$$

and $Vf(A) = f(A)$. We obtain indeed a monad $\mathbb{V} = (V, m, e)$ with unit $e_X: X \to VX,\ x \mapsto \{x\}$ and multiplication $m_X: VVX \to VX$ given by union. It is shown in Michael (1951) that this monad restricts to BooSp. Every arrow in the Kleisli category $\mathsf{BooSp}_{\mathbb{V}}$ is in particular a relation between the underlying sets; we refer to the relations corresponding to arrows in $\mathsf{BooSp}_{\mathbb{V}}$ as *continuous relations*. Since relational composition of continuous relations corresponds to composition in $\mathsf{BooSp}_{\mathbb{V}}$, the category $\mathsf{BooSp}_{\mathbb{V}}$ is isomorphic to the category $\mathsf{BooSpRel}$ of Boolean spaces and continuous relations, with relational composition and the identity on X given by the identity relation on X.

Example 103 There is an ordered version of the classic Vietoris functor. We consider now $V: \mathsf{PosComp} \to \mathsf{PosComp}$ defined by $VX = \{A \subseteq X \mid A \text{ upper closed}\}$ with the 'hit-and-miss topology' generated by the sets

$$\begin{aligned} &\{A \mid A \cap B \neq \varnothing\} &&B \text{ lower-open,}\\ &\{A \mid A \cap K = \varnothing\} &&K \text{ upper-closed,}\end{aligned}$$

and $Vf(A) = \uparrow f(A)$ is the up-closure of the image of A. As before, we obtain a monad $\mathbb{V} = (V, m, e)$ with unit e given by $e_X: X \to VX,\ x \mapsto \uparrow\{x\}$ and multiplication as above given by union. Similarly to the classical case, this monad restricts to Priest. For more information we refer to Schalk (1993).

As in Example 102, a Kleisli arrow between Priestley spaces is in particular a distributor between the underlying partially ordered sets. We call a distributor corresponding to an arrow in $\mathsf{Priest}_{\mathbb{V}}$ a *continuous distributor*, and the category $\mathsf{Priest}_{\mathbb{V}}$ is isomorphic to the category $\mathsf{PriestDist}$ of Priestley spaces and continuous distributors, with relational composition and the identity on X given by the underlying order relation of X. The category $\mathsf{PriestDist}$ becomes a 2-category via the inclusion order of relations. We also note that there is a full embedding

$$\mathsf{BooSpRel} \longrightarrow \mathsf{PriestDist}.$$

4.1 Halmos Dualities

In this subsection we present the extension of Stone and Priestley dualities to categories of continuous relations, an idea attributed to Halmos (1956) which in turn is inspired by the work of Jónsson and Tarski (1951). We start with a result that establishes a connection between monad morphisms $j: \mathbb{T} \to \mathbb{D}$ and $\mathbb{T}$-algebra structures on $\widetilde{X}$ *compatible* with the adjunction $F \dashv G$.

Theorem 104 *Let* X *and* A *be categories with representable forgetful functors to* Set, $\mathbb{T} = (T, m, e)$ *a monad on* X *and* $F \dashv G$ *an adjunction*

$$\mathsf{X} \underset{G}{\overset{F}{\rightleftarrows}} \mathsf{A}^{\mathrm{op}}$$

induced by $(\widetilde{X}, \widetilde{A})$. *The following data are in bijection.*

(i) *Functors* $F\colon \mathsf{X}_{\mathbb{T}} \to \mathsf{A}^{\mathrm{op}}$ *commuting with the left adjoints.*
(ii) *Monad morphisms* $j\colon \mathbb{T} \to \mathbb{D}$ *(*$\mathbb{D}$ *induced by* $F \dashv G$*).*
(iii) $\mathbb{T}$*-algebra structures* $\sigma : T\widetilde{X} \to \widetilde{X}$ *such that the map*

$$\widehat{(-)}\colon \mathsf{X}(X, \widetilde{X}) \longrightarrow \mathsf{X}(TX, \widetilde{X}), \quad \psi \longmapsto \sigma \cdot T\psi =: \widehat{\psi}$$

is an A*-morphism* $\kappa_X : FX \to FTX$.

Proof See Hofmann and Nora (2018, Theorem 5.7), for instance. □

For a $\mathbb{T}$-algebra structure $\sigma : T\widetilde{X} \to \widetilde{X}$ such that the map

$$\mathsf{X}(X, \widetilde{X}) \longrightarrow \mathsf{X}(TX, \widetilde{X}), \quad \psi \longmapsto \sigma \cdot T\psi$$

is an A-morphism $\kappa_X : FX \to FTX$, one defines a functor $F\colon \mathsf{X}_{\mathbb{T}} \to \mathsf{A}^{\mathrm{op}}$ commuting with the left adjoints by

$$(\varphi\colon X \to TY) \longmapsto (FY \xrightarrow{\kappa_Y} FTY \xrightarrow{F\varphi} FX).$$

The induced monad morphism $j : \mathbb{T} \to \mathbb{D}$ is given by the family of maps

$$j_X\colon |TX| \longrightarrow \mathsf{A}(FX, \widetilde{A}), \quad \mathfrak{x} \longmapsto (\psi \mapsto \sigma \cdot T\psi(\mathfrak{x})).$$

Remark 105 For every X in X, the diagram

$$\begin{array}{ccc} |TX| & \xrightarrow{j_X} & \mathsf{A}(FX, \widetilde{A}) \\ & \searrow^{\widehat{\psi}} & \downarrow \mathrm{ev}_{FX,\psi} \\ & & |\widetilde{X}| = |\widetilde{A}| \end{array}$$

commutes for every $\psi : X \to \widetilde{X}$. Hence, j_X is an embedding if and only if the cone $(\widehat{\psi} : TX \longrightarrow \widetilde{X})_\psi$ is point-separating and initial.

This construction seems to be more natural if $\widetilde{X} = TX_0$ with $\mathbb{T}$-algebra structure m_{X_0}. Then the functor $F\colon \mathsf{X}_{\mathbb{T}} \to \mathsf{A}^{\mathrm{op}}$ is a lifting of the hom-functor $\mathsf{X}(-, X_0)\colon \mathsf{X}_{\mathbb{T}} \to \mathsf{Set}^{\mathrm{op}}$. Furthermore, interpreting the elements of TX as morphisms $\varphi\colon X_0 \twoheadrightarrow X$ in the Kleisli category $\mathsf{X}_{\mathbb{T}}$ allows to describe the components of the monad morphism j using composition in $\mathsf{X}_{\mathbb{T}}$:

$$j_X\colon |TX| \longrightarrow \hom(FX, \widetilde{A}), \quad \varphi \longmapsto (\psi \mapsto \psi \cdot \varphi).$$

Example 106 We consider $\mathsf{X} = \mathsf{BooSp}$ and $\mathsf{A} = \mathsf{FinSup}$ the category of finitely cocomplete partially ordered sets and maps preserving finite suprema. Furthermore, $\widetilde{X} = \mathsf{2}$ is the two-element Boolean space and $\widetilde{A} = \mathsf{2}$ is the two-element lattice. By Proposition 35, $(\mathsf{2}, \mathsf{2})$ induces a dual adjunction

$$\mathsf{BooSp} \underset{G}{\overset{F}{\rightleftarrows}} \mathsf{FinSup}^{\mathrm{op}}..$$

Furthermore, if we consider the Vietoris monad on BooSp, then $\mathsf{2}$ is a $\mathbb{V}$-algebra $\bigvee\colon V\mathsf{2} \to \mathsf{2}$. We may identify continuous maps $\varphi\colon X \to \mathsf{2}$ with clopen subsets of X; then the map

$$\widehat{(-)}\colon \mathsf{BooSp}(X, 2) \longrightarrow \mathsf{BooSp}(VX, 2)$$

sends and open subset $B \subseteq X$ to $B^{\Diamond}$, therefore $\widehat{(-)}$ preserves finite suprema. The monad morphism j is given by

$$j_X\colon VX \longrightarrow \mathsf{FinSup}(FX, 2).$$
$$A \longmapsto B \mapsto \begin{cases} 1 & \text{if } A \cap B \neq \varnothing, \\ 0 & \text{otherwise.} \end{cases}$$

We observed already that VX has the initial topology with respect to the point-separating source $(\widehat{\varphi}\colon VX \to \mathsf{2})_{\varphi \in \mathsf{BooSp}(X,2)}$, for every Boolean space X. Therefore j is component-wise an embedding. Finally, compactness of X guarantees that j_X is surjective. We denote by $\mathsf{FinSup}_{\mathrm{BA}}$ the full subcategory of FinSup defined by all Boolean algebras, and conclude that

$$\mathsf{BooSpRel} \sim \mathsf{FinSup}_{\mathrm{BA}}. \tag{12}$$

Classic Stone duality can now be obtained by observing that

- a continuous relation is a continuous function between Boolean spaces if and only if item 2 of Example 100 holds, and
- the monoidal structure in $\mathsf{BooSpRel}$ given by the topological product of Boolean spaces corresponds under (12) to the bimorphism representing tensor product in $\mathsf{FinSup}_{\mathrm{BA}}$ (see Hofmann and Nora (2015) for details).

Example 107 Very similar considerations lead us to

$$\mathsf{PriestDist} \sim \mathsf{FinSup}_{\mathrm{DL}},$$

where $\mathsf{FinSup}_{\mathrm{DL}}$ denotes the full subcategory of FinSup defined by all distributive lattices.

Remark 108 Several other simple examples of this procedure are described in Hofmann and Nora (2015). A more involved example relating partially ordered compact spaces and certain *quantale-enriched categories* can be found in Hofmann and Nora (2018); however, this example is beyond the scope of these notes.

Remark 109 The dual equivalences for continuous relations and distributors obtained in this section also lead to dual equivalences for categories of coalgebras (we refer to Adámek (2005) for more information regarding the theory of coalgebras). For instance, it is enough here to observe that a coalgebra $\alpha\colon X \to VX$ for the Vietoris functor on BooSp corresponds to an endomorphism $\alpha : X \nrightarrow X$ in $\mathsf{BooSpRel}$, and a continuous map $f\colon X \to Y$ between Boolean spaces is a homomorphism $f\colon (X, \alpha) \to (Y, \beta)$ of coalgebras if and only if the diagram

$$\begin{array}{ccc} X & \overset{f}{\nrightarrow} & Y \\ {\scriptstyle\alpha}\big\downarrow & & \big\downarrow{\scriptstyle\beta} \\ X & \underset{f}{\nrightarrow} & Y \end{array}$$

in $\mathsf{BooSpRel}$ commutes. Hence, the category of coalgebras and homomorphisms for $V\colon \mathsf{BooSp} \to \mathsf{BooSp}$ is dually equivalent to the category of Boolean algebras equipped with an additional unary operation preserving finite suprema and Boolean homomorphisms preserving this operation (see also Kupke et al. (2004), Petrovich (1996) and Cignoli et al. (1991)).

4.2 *Idempotent Split Completion*

In the last part of these notes we consider the restriction of Priestley duality to Heyting algebras. In fact, the Priestley spaces corresponding under Priestley duality to Heyting algebras are the so-called Esakia spaces: those Priestley spaces X where the down-closure of every open subset of X is again open. This result dates back to the beginning of the 1970's and was independently discovered by M. Adams (unpublished work, see Davey and Galati (2003)) and L. Esakia (see Esakia (1974) and also Bezhanishvili (2014)). Arguably, this dual equivalence appears very naturally in the framework of the previous section since Heyting algebras can be characterised as those distributive lattices which split idempotents in $\mathsf{FinSup}_{\mathrm{DL}}$ between Boolean algebras; likewise, Esakia spaces are precisely those Priestley spaces which split idempotents of Boolean spaces in $\mathsf{PriestDist}$.

We start by recalling the pertinent notions.

Definition 110 An arrow $e\colon C \to C$ in a category C is *idempotent* if $e \cdot e = e$.

Example 111 If $r \cdot s = 1$, then $e = s \cdot r$ is idempotent.

Definition 112 A category C is *idempotent split complete* if every idempotent e is of the form $e = s \cdot r$, for arrows r and s with $r \cdot s = 1$.

Remark 113 Every category C with equalisers is idempotent split complete. In fact, for an idempotent $e \colon C \to C$ in C, consider the equaliser $s \colon E \to C$ of e and 1_C. Since $1_C \cdot e = e = e \cdot e$, there is $r \colon C \to E$ with $s \cdot r = e$.

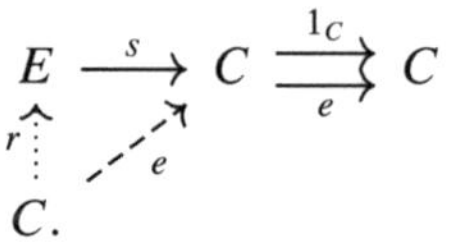

Since

$$s \cdot r \cdot s = e \cdot s = 1_C \cdot s = s \cdot 1_E,$$

we also obtain $r \cdot s = 1_E$.

The dual argument shows that every category with coequalisers is idempotent split complete. In fact, this also follows from the more generic fact below.

Remark 114 The notion of idempotent split complete is self dual, that is, with C also C^{op} is idempotent split complete.

Example 115 The category Rel is not idempotent split complete (see Milius (2000, Example 8.16)): the idempotent relation $<$ defined on the rationals $\mathbb{Q}$ does not split.

Example 116 The category FinSup is complete and therefore also idempotent split complete. It is not difficult to see that a split subobject in FinSup of a distributive lattice is also a distributive lattice, hence the category $\mathsf{FinSup}_{\mathrm{DL}}$ is idempotent split complete. By Example 107, the category $\mathsf{PriestDist}$ is idempotent split compete.

Every category C can be fully embedded into the *idempotent split completion* $\mathrm{kar}(\mathsf{C})$ of C which is given by the following data: the objects of $\mathrm{kar}(\mathsf{C})$ are pairs (C, e) with $e \colon C \to C$ in C idempotent, and an arrow $f \colon (C, e) \to (C', e')$ in $\mathrm{kar}(\mathsf{C})$ is an arrow in C so that $f \cdot e = f = e' \cdot f$. The category C is fully embedded into $\mathrm{kar}(\mathsf{C})$ via

$$C \longmapsto (C, 1_C).$$

Then the category $\mathrm{kar}(\mathsf{C})$ is indeed idempotent split complete and the full embedding $\mathsf{C} \to \mathrm{kar}(\mathsf{C})$ has the expected universal property. It is remarkable that the property of being idempotent split complete can be expressed as representability of distributors between categories (enriched in Set). Moreover, on this level of generality idempotent split complete categories and *Cauchy complete metric spaces* are instances of the same definition (Lawvere 1973). The following observation certainly resembles a well-known fact about Cauchy completion of metric spaces.

Lemma 117 *Let* A *be a full subcategory of* B *and assume that idempotents split in* B*. Let* $\overline{\mathsf{A}}$ *be the full subcategory of* B *defined by the retracts of the objects in* A*. Then idempotents split in* $\overline{\mathsf{A}}$ *and there is a natural equivalence*

$$\mathrm{kar}(\mathsf{A}) \longrightarrow \overline{\mathsf{A}}$$

commuting with the inclusion functors.

4.3 Example: Esakia Duality

In this section we use the equivalence $\mathsf{PriestDist} \sim \mathsf{FinSup}_{\mathrm{DL}}$ to obtain a duality result for Heyting algebras. For technical reasons, we consider here *co-Heyting algebras*, that is, those distributive lattices whose order dual lattice is a Heyting algebra. More details about this approach can be found in Hofmann and Nora (2014).

First some notation. For a map $f: X \to Y$ between partially ordered sets, we define the relations $f_*: X \nrightarrow Y$ and $f^*: Y \nrightarrow X$ as in Example 101; these relations are distributors if and only if f is monotone.

Proposition 118 *Let X and Y be Priestley spaces and $f: X \to Y$ be a map. Then f is monotone and continuous if and only if f_* is a continuous distributor.*

To deal also with the relation f^*, we introduce the following notion.

Definition 119 An arrow $f: X \to Y$ in Priest is called *downwards open* whenever, for every clopen subset $A \subseteq X$, the down-closure $\downarrow f[A]$ of $f[A]$ is open in Y.

Theorem 120 *For an arrow $f: X \to Y$ in* Priest*, the following assertions are equivalent.*

(i) *f is downwards open.*
(ii) *The continuous distributor $f_*: X \nrightarrow Y$ has a right adjoint in* $\mathsf{PriestDist}$.
(iii) *The distributor $f^*: Y \nrightarrow X$ is a continuous distributor.*

Proof See Hofmann and Nora (2014, Theorem 2.7). □

Definition 121 A Priestley space X is called an *Esakia space* whenever, for every clopen subset A of X, its down-closure $\downarrow A$ is open in X.

We denote by $\mathsf{EsaDist}$ the full subcategory of $\mathsf{PriestDist}$ defined by all Esakia spaces.

Theorem 122 *For a Priestley space X, we write X_p for the Priestley space with the same topology but the discrete partial order. Then the following assertions are equivalent.*

(i) *X is an Esakia space.*
(ii) *The monotone and continuous map* $i\colon X_p \to X,\ x \mapsto x$ *is downwards open.*
(iii) *The continuous distributor* $i_*\colon X_p \nrightarrow X$ *has a right adjoint (necessarily given by* i^**).*
(iv) *X is a split subobject of a Boolean space in* PriestDist.

Proof See Hofmann and Nora (2014, Theorem 3.2). □

Corollary 123 *The category* EsaDist *is the idempotent split completion of* BooSpRel.

For a distributive lattice L, we consider its free Boolean extension $j\colon L \hookrightarrow B$ which is given by any epimorphic embedding in DL of L into a Boolean algebra B. We note that j is a completion in the sense of Brümmer et al. (1992). Translated to Priest, the homomorphism j corresponds to the monotone continuous map $i\colon X_p \to X$. Furthermore, every lattice homomorphism $f\colon L_1 \to L_2$ extends uniquely to a homomorphism $\overline{f}\colon B_1 \to B_2$ between the corresponding Boolean algebras.

Theorem 124 *For a distributive lattice L, the following assertions are equivalent.*

1. *L is a co-Heyting algebra.*
2. *The lattice homomorphism* $j\colon L \to B$ *has a left adjoint* $j^+\colon B \to L$ *in* $\mathsf{FinSup}_{\mathrm{DL}}$.
3. *L is a split subobject of a Boolean algebra in* $\mathsf{FinSup}_{\mathrm{DL}}$.

Proof See McKinsey and Tarski (1946) and Došen (1990). □

We write $\mathsf{FinSup}_{\mathrm{coHeyt}}$ to denote the full subcategory of FinSup defined by all co-Heyting algebras. Clearly:

Corollary 125 *The category* $\mathsf{FinSup}_{\mathrm{coHeyt}}$ *is the idempotent split completion of* $\mathsf{FinSup}_{\mathsf{BA}}$.

Putting everything together, we obtain the following.

Theorem 126 *The equivalence* $\mathsf{SpecDist} \sim \mathsf{FinSup}_{\mathrm{DL}}^{\mathrm{op}}$ *restricts to an equivalence*

$$\mathsf{EsaDist} \sim \mathsf{FinSup}_{\mathrm{coHeyt}}^{\mathrm{op}}.$$

Therefore the full subcategory of Priest *defined by all Esakia spaces is dually equivalent to the full subcategory of* DL *defined by all co-Heyting algebras.*

Remark 127 Finally, the result above can be restricted further to obtain a duality result for the category of co-Heyting algebras and co-Heyting homomorphisms, and consequently also for the category Heyt of Heyting algebras and Heyting homomorphisms. In fact, a lattice homomorphism $f\colon L_1 \to L_2$ between co-Heyting algebras preserves the co-Heyting operation if and only if the diagram

$$\begin{array}{ccc} B_1 & \xrightarrow{\overline{f}} & B_2 \\ {\scriptstyle j_1^+}\downarrow & & \downarrow{\scriptstyle j_2^+} \\ L_1 & \xrightarrow[f]{} & L_2 \end{array}$$

commutes; that is, if and only if the corresponding Priestley map $g\colon X_1 \to X_2$ makes the diagram of continuous distributors

$$\begin{array}{ccc} X_1 & \overset{g_*}{\longrightarrow\!\!\!\!\!\!\!+\;\;} & X_2 \\ {\scriptstyle i_1^*}\downarrow & & \downarrow{\scriptstyle i_2^*} \\ (X_1)_p & \underset{g}{\longrightarrow\!\!\!\!\!\!+\;} & (X_2)_p \end{array}$$

commutative. Element-wise: for all $x \in X_1$ and $y \in X_2$ with $g(x) \leq y$, there is some $x' \in X_1$ with $x \leq x'$ and $g(x') = y$.

Acknowledgements This work is supported by CIDMA (https://ror.org/05pm2mw36) under the Portuguese Foundation for Science and Technology (FCT, https://ror.org/00snfqn58), Grants UID/04106/2025 (https://doi.org/10.54499/UID/04106/2025) and UID/PRR/04106/2025 (https://doi.org/10.54499/UID/PRR/04106/2025).

References

Adámek, J., Herrlich, H., Strecker, G.E.: Abstract and Concrete Categories - The Joy of Cats. Repr. Theory Appl. Categ. (17), 1–507 (2006). Reprint of the 1990 original [Pure and Applied Mathematics, John Wiley & Sons Inc., New York]. http://tac.mta.ca/tac/reprints/articles/17/tr17abs.html

Adámek, J., Rosický, J.: Locally presentable and accessible categories, *London Mathematical Society Lecture Note Series*, vol. 189. Cambridge University Press, Cambridge (1994). https://doi.org/10.1017/CBO9780511600579

Adámek, J.: Introduction to coalgebra. Theory Appl. Categ. **14**(8), 57–199 (2005). http://www.tac.mta.ca/tac/volumes/14/8/14-08abs.html

Bell JL, Machover M (1977) A course in mathematical logic. North-Holland Publishing Co., Amsterdam-New York-Oxford

Bezhanishvili, G. (ed.): Leo Esakia on duality in modal and intuitionistic logics, *Outstanding Contributions to Logic*, vol. 4. Springer, Dordrecht (2014). https://doi.org/10.1007/978-94-017-8860-1

Bourbaki, N.: Algebra. I. Chapters 1–3. Elements of Mathematics (Berlin). Springer-Verlag, Berlin (1989). Translated from the French, Reprint of the 1974 edition

Bourbaki, N.: Éléments de mathématique. Part I. Les structures fondamentales de l'analyse. Livre III. Topologie générale. Chapitres III et IV, *Actualités Scientifiques et Industrielles [Current Scientific and Industrial Topics]*, vol. No. 916. Hermann & Cie, Paris (1942)

Brümmer GCL, Giuli E, Herrlich H (1992) Epireflections which are completions. Cahiers Topologie Géom Différentielle Catég 33(1):71–73

Čeh, E.: On bicompact spaces. Ann. of Math. (2) **38**(4), 823–844 (1937). https://doi.org/10.2307/1968839

Cignoli, R., Lafalce, S., Petrovich, A.: Remarks on Priestley duality for distributive lattices. Order **8**(3), 299–315 (1991). https://doi.org/10.1007/BF00383451

Clark DM, Davey BA (1998) Natural dualities for the working algebraist, Cambridge Studies in Advanced Mathematics, vol 57. Cambridge University Press, Cambridge

Davey, B.A., Galati, J.C.: A coalgebraic view of Heyting duality. Studia Logica **75**(3), 259–270 (2003). https://doi.org/10.1023/B:STUD.0000009559.44998.a3

Dimov GD, Tholen W (1989) A characterization of representable dualities. In: Adámek J, MacLane S (eds) Categorical topology and its relation to analysis, algebra and combinatorics: Prague, Czechoslovakia, 22–27 August, 1988. Publ, Teaneck, NJ, World Sci, pp 336–357

Došen, K.: Normal modal logics in which the Heyting propositional calculus can be embedded. In: P.P. Petkov (ed.) Mathematical logic. Proceedings of the summer school and conference dedicated to the ninetieth anniversary of Arend Heyting (1898-1980), held in Chaika, Bulgaria, September 13-23, 1988, pp. 281–291. Plenum, New York (1990). https://doi.org/10.1007/978-1-4613-0609-2_19

Esakia LL (1974) Topological Kripke models. Dokl Akad Nauk SSSR 214:298–301

Freyd, P.J., Kelly, G.M.: Categories of continuous functors. I. J. Pure Appl. Algebra **2**(3), 169–191 (1972). https://doi.org/10.1016/0022-4049(72)90001-1

Gabriel P, Ulmer F (1971) Lokal präsentierbare Kategorien, vol 221. Lecture Notes in Mathematics. Springer-Verlag, Berlin-New York

Gehrke, M.: Stone duality, topological algebra, and recognition. J. Pure Appl. Algebra **220**(7), 2711–2747 (2016). https://doi.org/10.1016/j.jpaa.2015.12.007

Gelfand, I.: Normierte Ringe. Rec. Math. N.S. **9**(1), 3–24 (1941)

Halmos, P.R.: Algebraic logic. I. Monadic Boolean algebras. Compositio Math. **12**, 217–249 (1956). https://eudml.org/doc/88816

Hochster, M.: Prime ideal structure in commutative rings. Trans. Amer. Math. Soc. **142**, 43–60 (1969). https://doi.org/10.2307/1995344

Hofmann, D., Neves, R., Nora, P.: Generating the algebraic theory of $C(X)$: the case of partially ordered compact spaces. Theory Appl. Categ. **33**, Paper No. 12, 276–295 (2018). http://tac.mta.ca/tac/volumes/33/12/33-12abs.html

Hofmann, D., Nora, P.: Dualities for modal algebras from the point of view of triples. Algebra Universalis **73**(3-4), 297–320 (2015). https://doi.org/10.1007/s00012-015-0324-5

Hofmann, D., Nora, P.: Enriched Stone-type dualities. Adv. Math. **330**, 307–360 (2018). https://doi.org/10.1016/j.aim.2018.03.010

Hofmann, D., Nora, P.: Some notes on Esakia spaces. In: M.M. Clementino, G. Janelidze, J. Picado, L. Sousa, W. Tholen (eds.) Categorical methods in algebra and topology: Special volume in honour of Manuela Sobral, *Textos de Matemática/Mathematics Texts*, vol. 46, pp. 201–220. Universidade de Coimbra, Departamento de Matemática, Coimbra (2014). Workshop on Categorical Methods in Algebra and Topology, January 24-26, 2014. https://www.mat.uc.pt/~textos/v46/

Hofmann, D.: A generalization of the duality compactness theorem. J. Pure Appl. Algebra **171**(2-3), 205–217 (2002). https://doi.org/10.1016/S0022-4049(01)00132-3

Hofmann, D.: Natürliche Dualitäten und das verallgemeinert Stone-Weierstraß Theorem. Ph.D. thesis, University of Bremen (1999)

Hofmann, D.: On a generalization of the Stone-Weierstrass theorem. Appl. Categ. Structures **10**(6), 569–592 (2002). https://doi.org/10.1023/A:1020969017551

Isbell, J.R.: General functorial semantics. I. Amer. J. Math. **94**, 535–596 (1972). https://doi.org/10.2307/2374638

Isbell, J.R.: Generating the algebraic theory of $C(X)$. Algebra Universalis **15**(2), 153–155 (1982). https://doi.org/10.1007/BF02483718

Johnstone, P.T.: Stone spaces, *Cambridge Studies in Advanced Mathematics*, vol. 3. Cambridge University Press, Cambridge (1986). Reprint of the 1982 edition

Jónsson, B., Tarski, A.: Boolean algebras with operators. I. Amer. J. Math. **73**(4), 891–939 (1951). https://doi.org/10.2307/2372123

Kennison, J.F.: On limit-preserving functors. Illinois J. Math. **12**(4), 616–619 (1968). http://projecteuclid.org/euclid.ijm/1256053963

Kupke, C., Kurz, A., Venema, Y.: Stone coalgebras. Theoret. Comput. Sci. **327**(1-2), 109–134 (2004). https://doi.org/10.1016/j.tcs.2004.07.023

Lambek, J., Rattray, B.A.: A general Stone–Gelfand duality. Trans. Amer. Math. Soc. **248**(1), 1–35 (1979). https://doi.org/10.2307/1998735

Lambek, J., Rattray, B.A.: Localization and duality in additive categories. Houston J. Math. **1**(1), 87–100 (1975). https://www.math.uh.edu/~hjm/vol01-1.html

Lawvere, F.W.: Metric spaces, generalized logic, and closed categories. Rend. Sem. Mat. Fis. Milano **43**, 135–166 (1973). Republished in Repr. Theory Appl. Categ. **1**, 1–37 (2002). https://doi.org/10.1007/BF02924844

Leinster, T.: Basic category theory, *Cambridge Studies in Advanced Mathematics*, vol. 143. Cambridge University Press, Cambridge (2014). https://doi.org/10.1017/CBO9781107360068

Mac Lane, S.: Categories for the working mathematician, *Graduate Texts in Mathematics*, vol. 5, second edn. Springer-Verlag, New York (1998). https://doi.org/10.1007/978-1-4612-9839-7

Marra, V., Reggio, L.: Stone duality above dimension zero - axiomatising the algebraic theory of $C(X)$. Adv. Math. **307**, 253–287 (2017). https://doi.org/10.1016/j.aim.2016.11.012

McKinsey, J.C.C., Tarski, A.: On closed elements in closure algebras. Ann. of Math. (2) **47**(1), 122–162 (1946). https://doi.org/10.2307/1969038

Michael, E.: Topologies on spaces of subsets. Trans. Amer. Math. Soc. **71**(1), 152–182 (1951). https://doi.org/10.2307/1990864

Milius, S.: Relations in categories. Master's thesis, York University, Toronto (2000)

Morris, S.A.: Pontryagin duality and the structure of locally compact abelian groups, *London Mathematical Society Lecture Note Series*, vol. No. 29. Cambridge University Press (1977). https://doi.org/10.1017/CBO9780511600722

Nachbin, L.: Topologia e Ordem. University of Chicago Press (1950)

Nachbin, L.: Topology and Order. Van Nostrand Mathematical Studies. D. Van Nostrand, Princeton, N.J.-Toronto, Ont.-London (1965). Translated from the Portuguese by Lulu Bechtolsheim

Negrepontis, J.W.: Duality in analysis from the point of view of triples. J. Algebra **19**(2), 228–253 (1971). https://doi.org/10.1016/0021-8693(71)90105-0

Petrovich, A.: Distributive lattices with an operator. Studia Logica **56**(1-2), 205–224 (1996). Special issue on Priestley duality. https://doi.org/10.1007/BF00370147

Pontrjagin, L.: The theory of topological commutative groups. Ann. of Math. (2) **35**(2), 361–388 (1934). https://doi.org/10.2307/1968438

Porst, H.E., Tholen, W.: Concrete dualities. In: H. Herrlich, H.E. Porst (eds.) Category theory at work, *Res. Exp. Math.*, vol. 18, pp. 111–136. Heldermann Verlag Berlin (1991). With Cartoons by Marcel Erné. http://www.heldermann.de/R&E/RAE18/ctw07.pdf

Priestley, H.A.: Ordered topological spaces and the representation of distributive lattices. Proc. London Math. Soc. (3) **24**, 507–530 (1972). https://doi.org/10.1112/plms/s3-24.3.507

Priestley, H.A.: Representation of distributive lattices by means of ordered stone spaces. Bull. London Math. Soc. **2**(2), 186–190 (1970). https://doi.org/10.1112/blms/2.2.186

Sambin, G., Vaccaro, V.: Topology and duality in modal logic. Ann. Pure Appl. Logic **37**(3), 249–296 (1988). https://doi.org/10.1016/0168-0072(88)90021-8

Schalk, A.: Algebras for generalized power constructions. Ph.D. thesis, Technische Hochschule Darmstadt (1993). http://www.cs.man.ac.uk/~schalk/publ/diss.ps.gz

Sørensen, M.H., Urzyczyn, P.: Lectures on the Curry-Howard isomorphism, *Stud. Logic Found. Math.*, vol. 149. Elsevier (2006)

Stone, M.H.: Applications of the theory of Boolean rings to general topology. Trans. Amer. Math. Soc. **41**(3), 375–481 (1937). https://doi.org/10.2307/1989788

Stone, M.H.: The generalized Weierstrass approximation theorem. Math. Mag. **21**(4), 167–184 (1948). https://doi.org/10.2307/3029750

Stone, M.H.: The generalized Weierstrass approximation theorem. Math. Mag. **21**(5), 237–254 (1948). https://doi.org/10.2307/3029337

Stone, M.H.: The representation of Boolean algebras. Bull. Amer. Math. Soc. **44**(12), 807–816 (1938). https://doi.org/10.1090/S0002-9904-1938-06871-1

Stone, M.H.: The theory of representations for Boolean algebras. Trans. Amer. Math. Soc. **40**(1), 37–111 (1936). https://doi.org/10.2307/1989664

Stone, M.H.: Topological representations of distributive lattices and Brouwerian logics. Čas. Mat. Fys. **67**(1), 1–25 (1938). https://eudml.org/doc/27235

Stralka, A.: A partially ordered space which is not a Priestley space. Semigroup Forum **20**(4), 293–297 (1980). https://doi.org/10.1007/BF02572690

Tholen, W.: Ordered topological structures. Topology Appl. **156**(12), 2148–2157 (2009). https://doi.org/10.1016/j.topol.2009.03.038

Vietoris, L.: Bereiche zweiter Ordnung. Monatsh. Math. Phys. **32**(1), 258–280 (1922). https://doi.org/10.1007/BF01696886

Zádori, L.: Natural duality via a finite set of relations. Bull. Austral. Math. Soc. **51**(3), 469–478 (1995). https://doi.org/10.1017/S0004972700014301

Universal Proof Theory

Rosalie Iemhoff and Raheleh Jalali

Abstract These lecture notes survey the emerging area of Universal Proof Theory, which investigates general questions about the existence, equivalence, and characterization of good proof systems for broad classes of logics. In particular, the notes concentrate on the existence problem: for which logics do there exist proof systems satisfying desirable meta-properties (e.g., cut-elimination, analyticity, termination)? After a brief historical and conceptual introduction, we survey different flavors of proof theory (Hilbert systems, natural deduction, sequent calculi) in the context of classical, intuitionistic, modal, and substructural logics. We then develop a general method for obtaining positive and negative existence results, based on interpolation and uniform interpolation techniques, and apply it to a range of logics (intermediate, modal, non-normal, conditional, and substructural). We also discuss variations of the method. As these are lecture notes, proofs are often sketched or omitted, with pointers to papers containing the full proofs. The survey thus aims to chart the scope and challenges of Universal Proof Theory for future work.

1 Origins of Proof Theory

The story of modern proof theory is usually tied to the renowned mathematician David Hilbert (1862–1943), who initiated what is known as *Hilbert's Program* in the foundations of mathematics. Following the discovery of set-theoretic paradoxes, such as Russell's paradox, concerns and uncertainties arose regarding the consistency of various branches of mathematics. In reaction to these emerging doubts at the beginning of the twentieth century, Hilbert put forward a way to provide mathematics with a secure basis by proposing to formalize mathematics in an axiomatic

R. Iemhoff
Utrecht University, Utrecht, The Netherlands
e-mail: r.iemhoff@uu.nl

R. Jalali (✉)
Department of Computer Science, University of Bath, Bath, UK
e-mail: rjk53@bath.ac.uk

M. M. Clementino et al. (eds.), *Topology, Algebra and Categories in Logic*,
Coimbra Mathematical Texts 5, https://doi.org/10.1007/978-3-032-13760-9_2

theory whose consistency can be proved by finitary means. Such an achievement would clearly establish the consistency of mathematics once and for all; this solution would put the nagging doubts about the correctness of our abstract mathematical constructions to rest. To elucidate, Hilbert's proposal was to prove the consistency of more complicated systems in terms of simpler systems. Ultimately, the consistency of all of mathematics could be reduced to basic arithmetic.

As we know, Gödel's Incompleteness Theorems, only a decade after its birth, put an end to Hilbert's program, at least to the program in its strict sense. To understand Gödel's results, suppose a consistent "strong enough" system is given. Gödel's first theorem states that such a system can never be complete. Consequently, it is not possible to formalize all mathematical true statements within a formal system. Gödel's second theorem states that such a system cannot prove its own consistency. Accordingly, Hilbert's assumption that a finitistic system can be used to prove the consistency of itself and more powerful theories, such as set theory, is refuted. To put it another way, if Peano Arithmetic PA is not able to prove its own consistency, how could a finitary theory, much weaker than PA, be able to prove it? This shows that, at least for theories to which Gödel's Incompleteness Theorems apply, Hilbert's aims are unattainable. Still, it is not to be excluded that there are other finitary means available to prove the consistency of PA; means that are not formalizable in PA. But the general opinion is that Hilbert's Program, in its strict sense, did not survive Gödel's famous results.

But there is something that did survive, namely, the study of formal proofs and proof systems, and their use in the foundations of mathematics, as well as in numerous other disciplines, such as computer science, philosophy, and linguistics. Someone who has been very important for this survival is Gerhard Gentzen. In the 1930s, Gentzen showed that although PA cannot prove the consistency of PA, an extension of it by the (suitably coded) statement that the ordinal ε_0 is well-founded[1] does prove that PA is consistent. That Gentzen's proof is not finitary is clear, and in how far one considers Gentzen's theorem as an undoubted proof of the consistency of PA depends on how obvious or undoubted one considers the well-foundedness of ε_0 to be. Gentzen's theorem and, in particular, the methods and proof systems that he developed form the beginning of the systematic study of formal proofs. This field of analyzing proofs, nowadays called *Proof Theory*, is large and has connections to almost all other areas in logic and applications in many others, for instance, in theorem proving and logic programming. One therefore could say that Proof Theory is one of the most successful programs based on a wrong idea.

The subject of these short lecture notes is a recently emerging area within proof theory, called *Universal Proof Theory*. There are at least three fundamental problems in this area worth investigating. The first one is the *existence problem*. Here one is concerned with the existence and nonexistence of good (or useful or applicable or …) proof systems. The second problem is the *equivalence problem*, where one is concerned with investigating natural notions of equivalence of different proof

[1] ε_0 is the ordinal $\omega^{\omega^{\omega^{\cdot^{\cdot^{\cdot}}}}}$, i.e., an ω tower of ω's.

systems. This problem can be regarded as addressing Hilbert's twenty fourth problem of studying the equivalence of different mathematical proofs. The third problem is the *characterization problem*, where one characterizes proof systems via the equivalence relation mentioned before. In these notes, we focus on the existence problem, as the initial step of this evolving area. In the next section, Sect. 2, other areas in Proof Theory are briefly discussed and the idea of Universal Proof Theory is introduced. Sections 3, 4, 5, and 6 introduce the logics and proof systems that these notes are about. In Sect. 7, one of the main methods of Universal Proof Theory is developed and its results for various classes of logics and proof systems are discussed. Section 8 contains nice but unintended consequences of the method, and Sect. 9 is about a related but different method within Universal Proof Theory.

Due to lack of space most topics in these notes are only touched upon briefly and most theorems are given without proof. But ample references are provided for those who wish to know more. The authors wish to mention Amirhossein Akbar Tabatabai as someone whose ideas and actions have greatly helped shape the project summarized in these notes.

2 Flavors of Proof Theory

Looking at the field of Proof Theory today one can distinguish several lines of research. These lines are all concerned with formal proofs and proof systems, but with quite different aims and focusing, not exclusively but in the main, on quite different theories. The areas are not disjoint, but it can be worthwhile to determine them and place proof-theoretic results in one or more of these areas.

What is often called *Classical Proof Theory* is a continuation of Hilbert's Program. It is concerned with the relative consistency and strength of foundational theories and fragments thereof, in line with the early consistency result by Gentzen mentioned above. Set theories and theories of arithmetic are among the main theories that are investigated. These investigations are focused on the ordinal analysis of such theories, their strength with respect to a weak basic theory, as in reverse mathematics, or their relative strength via reductions, what Prawitz (1974) calls *Reductive Proof Theory*.

Prawitz (1974) contrasts this with *General Proof Theory* where proofs, and consequence relations, are the main topic of study, in their own right. This naturally leads to more fine-grained proof systems, since what is provable is less the issue than the nature and structure of proofs.

Structural Proof Theory received its name in Negri and von Plato (2001) and is concerned with the combinatorial analysis of the structure of proofs, and its main methodologies are cut-elimination and normalization. Structural Proof Theory is very close in spirit to General Proof Theory. However, it focuses more on the proof systems that Gentzen developed very early on. As a result of their beauty and the insight that they provide in proofs, Gentzen's systems form the inspiration for many proof systems that are investigated and applied today.

2.1 Universal Proof Theory

Universal proof theory is concerned with classes of logics and the variety of proof systems for them. Rather than on foundational theories, its focus is more on the numerous logics that occur in computer science, philosophy, mathematics, and linguistics. These logics come with their own questions. Here consistency is not an issue but other questions become important. For example, the complexity of the decidability of these logics, (not) having interpolation, and the analyticity of the proofs. Foundational theories also occur as objects of study, but the aspects that one is interested in are mostly of a logical nature and are shared with many theories, such as the tautologies or the admissible rules of a theory. But a large part of the research is about logics: epistemic logics, intermediate logics, modal logics, substructural logics, temporal logics, and so on.

These logics are mostly decidable, and many of the well-known ones have sophisticated proof systems, such as sequent calculi (Sect. 6). Such proof systems can be used to prove properties about the logics and to compare them. Proof systems that stand out in this respect and at the same are close to actual logical reasoning are Natural Deduction and Sequent Calculi, both introduced by Gentzen (1935). These two kinds of proof systems are closely related and both elegant and natural.

In the introduction, we described three of the main aims of Universal Proof Theory. In these notes we focus on the first one, the *existence problem*, which is discussed in detail in Sect. 7. The general idea is simple. Namely, to establish, ideally for large classes of logics, whether these logics have good proof systems. Of course, such a question is fully determined only if we specify what are the proof systems we are considering and when such a system is called "good." As will be clear below, the proof systems that we consider are sequent calculi, introduced in Sect. 6, but the definition of the property "good," which depends on the context, is only provided in Sect. 7. It suffices here to say that many of the cut-free sequent calculi that are in use today have, or almost have that property.

The area originated from the papers Iemhoff (2019a, b) and was developed further in the papers Akbar Tabatabai and Jalali (2025b) and Akbar Tabatabai and Jalali (2018), in which it also received its name. A recent significant extension of the main methods has been obtained in Akbar Tabatabai and Jalali (2025a). Other papers in the area are mentioned in Sect. 7. Before the proper introduction of Universal Proof Theory in Sect. 7, first the logics and proof systems that we focus on in these notes are introduced in Sects. 3, 4, 5, and 6.

We would like to thank the anonymous reviewer, Mai Gehrke, and Iris van der Giessen for their helpful comments.

3 The Logics

We start with some definitions and notations used in the rest of the paper. As we will consider various logics, we have to specify the language in each case. We start with propositional logics. The languages $\mathcal{L}$ of the propositional logics that we consider contain *propositional variables* (or *atoms*), denoted by small Roman letters $p, q, r, \ldots$, the *constants* $\top$ and $\bot$, and a set of *logical operators* which varies depending on the logic in use. For example, in the case of intermediate propositional logics the logical operators are the connectives $\wedge$, $\vee$, $\rightarrow$; for the propositional modal logic K the logical operators consist of the connectives $\wedge$, $\vee$, $\rightarrow$ and the modal operator $\Box$. *Formulas* are defined in the usual way and we denote them by small Greek letters φ, ψ, ... or sometimes capital Roman letters A, B, ... A *subformula* of a formula φ is defined as follows: φ is a subformula of φ; if $\psi \circ \theta$ is a subformula of φ, then so are ψ and θ for $\circ \in \{\wedge, \vee, \rightarrow\}$; if $\Box\psi$ is a subformula of φ, then so is ψ. *Negation* is defined as $\neg\varphi \equiv_{df} \varphi \rightarrow \bot$. The set of formulas in $\mathcal{L}$ is denoted by $\mathcal{F}_{\mathcal{L}}$. A *multiset* of formulas is a collection of formulas where the order of the formulas does not matter but their multiplicity does. We use a *multiset variable* (or *context*) to refer to a generic multiset, i.e., a variable that can be substituted by an arbitrary multiset of formulas. Capital Greek letters Γ, Δ, ... denote multisets of formulas or multiset variables, and it will be always clear which one we are referring to. Later in this chapter predicate logics are briefly mentioned, whose languages do not fall under the above definition. The definition of the language $\mathcal{L}$ for these cases, as well as conditional logics, lax logic, and intuitionistic modal logics, is specified in their corresponding sections.

A *substitution* for $\mathcal{L}$ is a function from $\mathcal{F}_{\mathcal{L}}$ to $\mathcal{F}_{\mathcal{L}}$ that is the identity function on constants, maps every atom to a formula in $\mathcal{F}_{\mathcal{L}}$, and commutes with the logical operators. Given a formula φ, by $\sigma\varphi$ we mean the image of φ under σ. Note that if $p_1, \ldots, p_n$ are all the atoms that occur in the formula $\varphi(p_, \ldots, p_n)$, then we have $\sigma(\varphi(p_, \ldots, p_n)) = \varphi(\sigma p_1, \ldots, \sigma p_n)$. By a *logic* we mean a set of formulas closed under substitution. We say the logic L' *extends* the logic L when $\mathsf{L} \subseteq \mathsf{L}'$. If a formula φ is in the logic L, we denote it by $\vdash_L \varphi$ and if it is clear from the context which logic we are talking about, we simply write $\vdash \varphi$. It is worth noting that the more faithful formalization of logical inference is a consequence relation, namely, a relation on the formulas in the language. However, for this paper the simpler notion, i.e., identifying a logic with a set of formulas (namely, its tautologies), suffices.

The logics that occur in these notes (in Sects. 7 and 8) are either intermediate (i.e., extensions of the intuitionistic propositional logic), normal modal, intuitionistic modal, non-normal modal, conditional, substructural, or are extensions of substructural logics by modal principles. The proof systems mostly used in Universal Proof Theory, and thus in these notes, are sequent calculi, which are introduced in Sect. 6. To put this kind of proof systems in context and to be able to appreciate their special qualities we also introduce in Sects. 4 and 5 two other kinds of proof systems, natural deduction and Hilbert systems. Together with sequent calculi, these are three of the main types of proof systems used in Proof Theory today. We introduce natural deduction proof systems and sequent calculi for Classical Propositional Logic, CPC, and Intuitionistic Propositional Logic, IPC, and their predicate versions Classical Predicate Logic, CQC, and Intuitionistic Predicate Logic, IQC. For reasons of space, we introduce Hilbert systems only for the propositional logics CPC and IPC. After the introduction of these proof systems we will, in Sect. 7, turn to the *existence problem* of Universal Proof Theory which, as mentioned in the introduction, is the main topic of these notes.

4 Hilbert Systems

Considering systems of formal deduction, there are three well-known types: Hilbert system (Hilbert calculus), natural deduction, and sequent calculus (or Gentzen's calculus). We start with introducing Hilbert systems briefly.

A Hilbert system is an axiomatization with axioms and rules of inference. The Hilbert system HJ for IPC has the following axioms:

1. $\varphi \to (\psi \to \varphi)$,
2. $(\varphi \to (\psi \to \theta)) \to ((\varphi \to \psi) \to (\varphi \to \theta))$,
3. $\varphi \to \varphi \vee \psi$,
4. $\psi \to \varphi \vee \psi$,
5. $(\varphi \to \theta) \to ((\psi \to \theta) \to (\varphi \vee \psi \to \theta))$,
6. $\varphi \wedge \psi \to \varphi$,
7. $\varphi \wedge \psi \to \psi$,
8. $\varphi \to (\psi \to (\varphi \wedge \psi))$,
9. $\bot \to \varphi$,

and the modus ponens rule $\dfrac{\varphi \qquad \varphi \to \psi}{\psi}\ (mp)$.

In most Hilbert systems, there is an asymmetry between the number of axioms and rules of inference, namely, there are several axioms and few rules of inference. Hilbert system HK for CPC is obtained by adding *the law of double negation* $\neg\neg\varphi \to \varphi$, or equivalently *the law of excluded middle* $\varphi \vee \neg\varphi$ to HJ.

By a *proof* of φ from a set of assumptions Γ, we mean a sequence of formulas $\varphi_1, \ldots, \varphi_n$ such that $\varphi_n = \varphi$ and each φ_i is either an element of Γ, or is an instance of an axiom, or is derived from φ_j and φ_k for $j, k < i$ by the modus ponens rule.

Example 1 Consider the following axioms:

$$A \to (B \to A)\ (H1) \quad (A \to (B \to C)) \to ((A \to B) \to (A \to C))\ (H2)$$

The following is a proof of $A \to A$ using the axioms $(H1)$ and $(H2)$:

1. $A \to ((A \to A) \to A)$ $(H1)$
2. $A \to ((A \to A) \to A) \to ((A \to (A \to A)) \to (A \to A))$ $(H2)$
3. $(A \to (A \to A)) \to (A \to A)$ $1, 2, mp$
4. $A \to (A \to A)$ $(H1)$
5. $A \to A$ $3, 4, mp$

Hilbert systems are not useful for proof search as explained below: Suppose we want to show that $\bot$ cannot be derived in HJ. In other words, we want to show that HJ is consistent. Let $\bot$ be provable and $\varphi_1, \ldots, \varphi_n$ be its proof. Thus, $\varphi_n = \bot$. As $\bot$ is not an axiom, the last step of the proof is (mp). It means that there is a formula φ such that both φ and $\varphi \to \bot$ are proved. Now, either φ is an axiom and we analyze the proof of $\varphi \to \bot$, or φ is derived from ψ and $\psi \to \varphi$ using (mp). Either ψ is an axiom or it is derived via (mp) and so on. In a backward proof search, each time that we reach (mp), a new formula appears.

It is worth mentioning that the weakness of Hilbert systems is not on the positive side, i.e., in the sense of proving theorems. After proving some, and not many meta theorems, which is the hard part, we can prove theorems rather quickly. Some common meta theorems are

- The deduction theorem: $\Gamma, \varphi \vdash \psi$ if and only if $\Gamma \vdash \varphi \to \psi$.
- Contraposition: If $\Gamma, \varphi \vdash \psi$ then $\Gamma, \neg\psi \vdash \neg\varphi$.

The problem appears when we want to show a formula is not provable. Then you have to consider all the proofs, and as we observed, guessing the structure of the proof is a very combinatorially complicated task. In this sense, Gentzen observed that Hilbert systems are not useful in achieving the goal of proving consistency. Therefore, he introduced two other proof systems, natural deduction, and sequent calculi, to study proofs systematically. Being in a transparent context, they are regarded as elegant systems.

5 Natural Deduction

When it comes to choosing proof systems, two main strategies exist depending on the trade-off between logical axioms and rules of inference. Hilbert-style proof systems include many schemes as logical axioms and as few rules of inference as possible. Natural deduction systems and sequent calculi take the opposite strategy, including many inference rules and few or no logical axioms. Gentzen developed the proof systems *Natural Deduction* (ND) as a faithful representation of the logical reasoning used in mathematics. As opposed to Hilbert-style proof systems, in Natural Deduction, logical reasoning is close to "natural" reasoning: starting from assumptions and reaching conclusions using inference rules. Like in "real" mathematics, derivations in ND contain assumptions and case distinctions.

A *deduction* (or *derivation*) in ND, usually denoted by $\mathcal{D}$ or $\mathcal{D}'$, is a tree whose nodes are labeled with formulas. A deduction is defined inductively as stated in the next paragraph. The formulas at the leaves of the tree (i.e., nodes at the top of a branch) are *assumptions* and the formula at the root (i.e., node at the bottom of the tree) is the *conclusion*. Assumptions can be either *open* or *closed*. An assumption φ that is closed is depicted as $[\varphi]$, usually superscripted by a symbol $a, b, \ldots$ or by a natural number. The closure is indicated by writing the superscript of the formula at the inference rule. When all assumptions in a deduction are closed, we call the deduction a *proof* and the deduction tree a *proof tree*.

Deductions are inductively defined: A single node with label φ is a deduction, where φ is an open assumption and there is no closed assumption. A given deduction can be extended according to the rules given in Fig. 3. In this way, the systems ND for CQC and NDi for IQC are defined. Note the closure of assumptions by the rules $(I \to)$, $(E\vee)$, $(E\exists)$, and $(E_c\bot)$ in Fig. 3. The rule $(E_i\bot)$ is called the *intuitionistic absurdity* rule. The rule $(E_c\bot)$, sometimes also denoted by (RAA) in the literature, is called the *classic absurdity* rule (reductio ad absurdum). This rule embodies proofs by contradiction: if by assuming φ is false we can derive a contradiction, then φ must be true.

Recall that $\neg\psi$ is defined as $\psi \to \bot$. The definition of deduction implies that the following are valid deductions:

$$\dfrac{\begin{matrix}\mathcal{D}\\ \neg\varphi\end{matrix} \qquad \begin{matrix}\mathcal{D}'\\ \varphi\end{matrix}}{\bot}\,E\to \qquad\qquad a\,\dfrac{\begin{matrix}[\varphi]^a\\ \mathcal{D}\\ \bot\end{matrix}}{\neg\varphi}\,I\to$$

When negation is considered as primitive in the language, then the two rules above are added to the proof system.

To appreciate the somewhat complicated handling of assumptions and disjunction elimination, consider the following derivation in ND of the law of excluded middle:

$$a\,\dfrac{\dfrac{[\neg(\varphi\vee\neg\varphi)]^a \qquad \dfrac{b\,\dfrac{\dfrac{\dfrac{[\varphi]^b}{\varphi\vee\neg\varphi}\,I\vee}{\bot}\,E\to}{\neg\varphi}\,I\to}{\varphi\vee\neg\varphi}\,I\vee}{\bot}\,E\to}{\varphi\vee\neg\varphi}\,E_c\bot$$

Let us explain how this proof tree provides a proof of the excluded middle: For a given formula φ, we want to prove $\varphi \vee \neg\varphi$. We start by assuming φ and $\neg(\varphi \vee \neg\varphi)$. Using the assumption φ and following the rules of Fig. 3 we reach $\varphi \vee \neg\varphi$. Now, applying the rule $E \to$ on this formula and the assumption $\neg(\varphi \vee \neg\varphi)$ we reach the contradiction. Finally, applying the rule $E_c\bot$ we get $\varphi \vee \neg\varphi$, as required. Note that the assumptions φ and $\neg(\varphi \vee \neg\varphi)$ are closed when the rules $I \rightarrow$ and $E_c\bot$ are applied, respectively.

Note that the derivation of the law of excluded middle does not go through for NDi, as it should, since it is not valid in intuitionistic logic.

Let us revisit Hilbert systems and compare the two proof systems we have introduced so far. As mentioned in Sect. 4, Hilbert systems are not useful for proof search and it is difficult to use them for reasoning about reasoning. Moreover, they do not convey the meaning of the logical symbols. On the other hand, in natural deduction, the introduction and elimination rules are in harmony and the rules convey the meaning of the logical symbols. However, proof search is still difficult as there may be many possibilities. For instance, there are different ways to prove $\varphi \rightarrow \varphi$ in ND:

$$a\ \dfrac{\dfrac{\dfrac{[\varphi]^a \qquad [\varphi]^a}{\varphi \wedge \varphi}\, I\wedge}{\varphi}\, E\wedge}{\varphi \rightarrow \varphi}\, I\rightarrow \qquad\qquad a\ \dfrac{[\varphi]^a}{\varphi \rightarrow \varphi}\, I\rightarrow$$

The goal of introducing natural deduction was to prove the consistency of propositional logic. Let us analyze the potential scenarios: Suppose $\bot$ is proved in NJ. Considering the last rule applied in the proof, there are several possibilities: the last rule may be $(E \rightarrow)$ and φ and $\varphi \rightarrow \bot$ are proved. Or the last rule can be $(E\wedge)$ and $\varphi \wedge \bot$ is proved, and many more options. This is more complicated than the situation in Hilbert systems. There we only had to deal with (mp), while in natural deduction there are several possibilities. To overcome this problem Gentzen proposed to avoid redundancies or detours in a proof, which led to *normalization*.

5.1 Normalization

Although natural deduction indeed seems to capture part of the logical reasoning in mathematical proofs, as a proof system it has some drawbacks, as observed earlier. One is that it allows for derivations with unnecessary detours, derivations in which formulas that are introduced are eliminated afterward. For example, it allows derivations that contain the following detours[2]:

$$\dfrac{\dfrac{\begin{matrix}\mathcal{D}_1\\ \varphi_1\end{matrix} \qquad \begin{matrix}\mathcal{D}_2\\ \varphi_2\end{matrix}}{\varphi_1 \wedge \varphi_2}\, I\wedge}{\varphi_i}\, E\wedge \qquad a, b\ \dfrac{\dfrac{\begin{matrix}\mathcal{D}\\ \varphi\end{matrix}}{\varphi \vee \psi}\, I\vee \qquad \begin{matrix}[\varphi]^a\\ \mathcal{D}_1\\ \chi\end{matrix} \qquad \begin{matrix}[\psi]^b\\ \mathcal{D}_2\\ \chi\end{matrix}}{\chi}\, E\vee \qquad \dfrac{a\ \dfrac{\begin{matrix}[\varphi]^a\\ \mathcal{D}\\ \psi\end{matrix}}{\varphi \rightarrow \psi}\, I\rightarrow \qquad \begin{matrix}\mathcal{D}'\\ \varphi\end{matrix}}{\psi}$$

conjunction   disjunction   implication

[2] We restrict ourselves to the propositional part of ND, but the observations apply to the quantifiers as well.

Clearly, these derivations can be contracted to derivations without that detour:

$$\begin{array}{ccc} & \mathcal{D} & \mathcal{D}' \\ \mathcal{D}_i & \varphi & \varphi \\ \varphi_i & \mathcal{D}_1 & \mathcal{D} \\ & \chi & \psi \\ \text{conjunction} & \text{disjunction} & \text{implication} \end{array}$$

The detours can be a bit more intricate as well, such as:

$$b,c\,\dfrac{[\varphi\vee\psi]^a \quad \dfrac{[\varphi]^b}{\varphi\vee\psi}\,I\vee \quad \dfrac{[\psi]^c}{\varphi\vee\psi}\,I\vee}{a\,\dfrac{\varphi\vee\psi}{\varphi\vee\psi\to\varphi\vee\psi}\,I\to}\,E\vee \qquad a\,\dfrac{[\varphi\vee\psi]^a}{\varphi\vee\psi\to\varphi\vee\psi}\,I\to$$

Both are proofs of the formula $\varphi \vee \psi \to \varphi \vee \psi$, but the leftmost proof has redundancies.

Normalization is omitting all the redundancies in a proof and the result is a *normal* proof. It is not trivial to see that this task is possible for any given proof. It may be the case that two introduction elimination rules do not occur right after each other, but in the steps of omitting the redundancies, these two rules may occur after one another. Hence, new redundancies may appear. However, one can show, in both ND and NDi, that every derivable formula has a proof in *normal form*.

Theorem 2 *Every formula provable in* ND *(*NDi*) has a proof in* ND *(*NDi*) in normal form.*

Suppose we have proved the normalization theorem, Theorem 2, i.e., it is possible to make any proof normal. So, we are sure that if a formula is provable, there exists a finite and algorithmic process that takes a proof and makes it normal. Is it possible to prove the consistency now? The following theorem helps us find out:

Theorem 3 *Every normal proof, without any open assumptions, ends with an introduction rule.*

Proof (Proof sketch) Suppose on the contrary that there are normal proofs ending with an elimination rule. Take the shortest such proof. We investigate all the possibilities of elimination rules in the last step. For instance, if it is $(E\wedge)$, then as the proof above, leading to this formula is shorter, it must end with an introduction rule. However, this is not possible because it violates the proof of being normal (i.e., an elimination rule occurring right after an introduction rule). □

Proof of consistency, as well as the proof of the disjunction property in intuitionistic logic, is an immediate consequence of the above theorem.

Without defining the notion of normalization in detail, let us quote Prawitz (1971), page 249: *A normal derivation has quite a perspicuous form: it contains two parts, one analytical part in which the assumptions are broken down in their components by use of the elimination rules, and one synthetical part in which the final components obtained in the analytical part are put together by use of the introduction rules.*

For NDi normalization follows from the translations between NDi and Gentzen's LJ and the cut-elimination theorem for the latter. In fact, Gentzen's development of the sequent calculus is a result of his search for a proof of normalization, for both ND and NDi. For ND it was proved only after Gentzen. By now there have appeared various proofs of normalization for ND, the constructive normalization method by Prawitz (1965), who considered the language without disjunction and existential quantifier, is often used in the literature.

What is remarkable is that what initially seemed a by-product, the sequent calculus, has become one of the most studied and applied class of proof systems in logic.

5.2 The Curry–Howard Isomorphism

The Curry–Howard isomorphism (also known as the Curry–Howard correspondence, or the proofs-as-programs and propositions-as-types interpretation) stands as a noteworthy connection between two seemingly disparate realms: intuitionistic natural deduction and type theory. Its essence is the observation that every derivation in the implicational fragment of intuitionistic natural deduction (NDi) corresponds precisely to a term in the simply typed lambda calculus, and vice versa. This profound connection, initially highlighted by the mathematician Haskell Curry and the logician William Alvin Howard, underscores the functional nature of implication. Moreover, it supports the idea that intuitionistic logic captures constructive reasoning. Here, an implication is considered to be a construction or a function that maps proofs of the antecedent to proofs of the conclusion. The isomorphism has been extended to other logical operators as well.

In essence, this correspondence immerses intuitionistic logic with a computational flavor, as it aligns logical derivations with program constructions. As a consequence, the Curry–Howard isomorphism enriches our understanding of logic and enhances our ability to reason about programs. This deep insight has led to advances in a range of disciplines, from type theory and programming languages to formal verification and proof theory. For a gentle introduction to the topic, see Sørensen and Urzyczyn (2006).

6 Sequent Calculi

6.1 Sequents

Consider a language $\mathcal{L}$ as described in Sect. 3, thus for propositional or propositional modal logics. A *sequent* in this language is an expression $\Gamma \Rightarrow \Delta$, where Γ and Δ are finite multisets of formulas in $\mathcal{L}$. The *interpretation* $I(\Gamma \Rightarrow \Delta)$ of a sequent $\Gamma \Rightarrow \Delta$ is defined as the formula $\bigwedge \Gamma \to \bigvee \Delta$, where $\bigwedge \varnothing$ and $\bigvee \varnothing$ are defined as $\top$ and $\bot$, respectively. Given a sequent $S = (\Gamma \Rightarrow \Delta)$, Γ is the *antecedent* of S, denoted by S^a, and Δ is the *succedent* of S, denoted by S^s. The sequent S is *single-conclusion* if S^s is empty or consists of a single formula. Otherwise, S is *multi-conclusion*. When working with sequents with a superscript, such as S^i, then S^{ia} is short for $(S^i)^a$, and similarly for S^{is}. For a multiset Γ and sequent S:

$$\sigma\Gamma \equiv_{df} \{\sigma\varphi \mid \varphi \in \Gamma\} \text{ (considered as a multiset)} \quad \sigma S \equiv_{df} (\sigma S^a \Rightarrow \sigma S^s).$$

We define *Meta-sequents* similarly as sequents, except that in the antecedent and/or succedent we also allow Γ and $\Box\Gamma$, where Γ is a multiset variable. A *substitution* of a meta-sequent is defined similar to the substitution of sequents, mapping formulas to formulas and multiset variables to multisets of formulas. A *partitioned* sequent is an expression of the form $\Gamma; \Pi \Rightarrow \Delta; \Sigma$ in case of multi-conclusion sequents and of the form $\Gamma; \Pi \Rightarrow \Delta$ for single-conclusion sequents. The use of partitioned sequents will become clear in the sections on interpolation.

6.1.1 Complexity Measures

The *degree* of a formula φ is inductively defined as $d(\bot) = d(\top) = 0$, $d(p) = 1$, $d(\bigcirc\varphi) = d(\varphi) + 1$ for $\bigcirc \in \{\Box, \Diamond\}$, and $d(\varphi \circ \psi) = d(\varphi) + d(\psi) + 1$ for $\circ \in \{\wedge, \vee, \to\}$. In the setting of intuitionistic logic we need another order on formulas, the *weight function* $w(\cdot)$ from Dyckhoff (1992) extended to modalities: the weight of an atom and the constants $\bot$ and $\top$ is 1, $w(\bigcirc\varphi) = w(\varphi) + 1$, and $w(\varphi \circ \psi) = w(\varphi) + w(\psi) + i$, where $i = 1$ in case $\circ \in \{\vee, \to\}$ and $i = 2$ otherwise.

We use the following orderings on sequents: $S_0 <_d S_1$ ($S_0 <_w S_1$) if and only if $S_0^a \cup S_0^s < S_1^a \cup S_1^s$, where $<$ is the order on multisets determined by degree (weight) as in Dershowitz and Manna (1979): for multisets Γ, Δ we have $\Delta < \Gamma$ if Δ is the result of replacing one or more formulas in Γ by zero or more formulas of lower degree (weight).

6.2 Sequent Calculi

A *sequent calculus* (or *calculus* for short) is a finite set of (sequent) axioms and (sequent) rules, where an axiom is a single meta-sequent and a rule is an expression of the form:

$$\frac{S_1 \dots S_n}{S_0}$$

where $S_0, \dots, S_n$ are meta-sequents. In each rule, the *principal* or *main* formulas are defined as the formulas in the conclusion of the rule which are not in the context.

A calculus is *single-conclusion* if every meta-sequent in its axioms and rules is single-conclusion. Meta-sequents above the line are called the *premises* and the meta-sequent below the line is called the *conclusion* of the rule.

We define σS to be the σ*-instance* of S and $\sigma S_1 \dots \sigma S_n / \sigma S_0$ to be the σ*-instance* of rule $S_1 \dots S_n / S_0$. Recall the definition of substitution from Sect. 3.

A *derivation* (or *proof*) of S in a calculus G is a tree labeled with sequents in such a way that the root has the label S and there is a substitution σ such that all leaves of the tree are σ-instances of axioms of G and for every non-leaf node in the tree with label S' and immediate successors labeled $S_1, \dots, S_n$, there is a rule in G such that $S_1 \dots S_n / S'$ is a σ-instance of that rule. A sequent S is *derivable* (or *provable*) in G if it has a derivation in G, in which case we write $\vdash_{\mathsf{G}} S$. The *depth* of a derivation is the length of its longest branch. We write $\vdash_m S$ to denote that S has a derivation of depth at most m. A rule R is *admissible* in G, if for any σ-instance $S_1 \dots S_n / S_0$ of R, if for any $1 \leq i \leq n$ we have $\vdash_{\mathsf{G}} S_i$ then $\vdash_{\mathsf{G}} S_0$. The rule R is *depth-preserving admissible* in G, if for all m and $1 \leq i \leq n$ we have $\mathsf{G} \vdash_m S_i$ then $\mathsf{G} \vdash_m S_0$.

We say that a calculus G is a *calculus for logic* L or L *is the logic of* G if the formula interpretation of the sequents provable in G is derivable in L:

$$\vdash_{\mathsf{G}} (\Gamma \Rightarrow \Delta) \text{ if and only if } \vdash_{\mathsf{L}} (\bigwedge \Gamma \to \bigvee \Delta).$$

Specifically, G derives the sequent version of all formulas that are derivable in L:

$$\vdash_{\mathsf{L}} \varphi \text{ if and only if } \vdash_{\mathsf{G}} (\Rightarrow \varphi).$$

Let G and H be two sequent calculi. We say H is an *extension* of G when $G \subseteq H$.

6.3 A Sequent Calculus for Classical Propositional Logic

One of the standard sequent calculi for classical propositional logic CPC is the calculus $\mathsf{G1cp}$ given in Fig. 4. It is not hard to see that all the axioms and rules of this calculus are sound for CPC: clearly, the interpretations of the axioms, which by definition are the formulas $p \to p$ and $\bot \to \bot$, hold in CPC, and if the interpretation of the premises of a rule hold in CPC, then so does the conclusion. For example, if the interpretations of the premises of the rule $(L\to)$ are valid, which means $\bigwedge\Gamma \to \varphi \vee \bigvee\Delta$ and $\bigwedge\Gamma \wedge \psi \to \bigvee\Delta$ are, then so is $\bigwedge\Gamma \wedge (\varphi \to \psi) \to \bigvee\Delta$, which is the interpretation of the rule's conclusion, i.e., $\Gamma, \varphi \to \psi \Rightarrow \Delta$.

The following are two examples of derivations in G1cp, where φ and ψ are atomic, of the law of excluded middle (the left derivation) and of Peirce's law (the right one):

$$\dfrac{\dfrac{\dfrac{\varphi \Rightarrow \varphi}{\varphi \Rightarrow \varphi, \bot}\,RW}{\Rightarrow \varphi, \neg\varphi}\,R\to}{\Rightarrow \varphi \vee \neg\varphi}\,R\vee \qquad \dfrac{\dfrac{\dfrac{\varphi \Rightarrow \varphi}{\varphi \Rightarrow \psi, \varphi}\,RW}{\Rightarrow \varphi \to \psi, \varphi}\,R\to \quad \varphi \Rightarrow \varphi}{(\varphi \to \psi) \to \varphi \Rightarrow \varphi}\,L\to$$

Note how carefully the occurrences of formulas in proofs are documented because of the use of multisets and the structural rules, more on this in the next section.

6.4 A Sequent Calculus for Intuitionistic Propositional Logic

A beautiful fact in Proof Theory is the fact that a sequent calculus G1ip for intuitionistic propositional logic IPC can be obtained by *restricting the sequents in* G1cp *to single-conclusion ones*, see Fig. 5.

Note that the right contraction rule (RC) disappears, as sequents cannot have more than one formula on the right. Moreover, the right disjunction rule $(R\vee)$ disappears for the same reason and note its reformulation in Fig. 5.

Since G1ip is a calculus for IPC it should not prove, for instance, the sequent versions of the law of excluded middle $\Rightarrow \varphi \vee \neg\varphi$ and Peirce's law $\Rightarrow ((\varphi \to \psi) \to \varphi) \to \varphi$. We do not have the tools yet to prove this fact, these will follow in Sect. 6.7. But we can at least observe that the two derivations in G1cp given in the previous section are indeed not valid in G1ip because not all sequents in them are single-conclusion.

6.5 Predicate Logic

In these notes, we will be mostly concerned with propositional logics, but for completeness sake, we briefly discuss the extension of G1cp to its predicate version here. The language of predicate logic is an extension of the propositional language $\mathcal{L}$ with the quantifiers $\exists$ and $\forall$, countably infinite variables $x, y, \ldots$, and n-ary relation symbols and n-ary function symbols for any natural number n. We sometimes call 0-ary relation symbols as *proposition variables* and 0-ary function symbols as *constants*. The terms and formulas in the predicate language are defined as usual. The sequent calculi G1c and G1i consist of G1cp and G1ip, respectively, and the following quantifier rules:

$$\frac{\Gamma, \varphi(y) \Rightarrow \Delta}{\Gamma, \exists x \varphi(x) \Rightarrow \Delta}\, L\exists \quad \& \quad \frac{\Gamma \Rightarrow \varphi(t), \Delta}{\Gamma \Rightarrow \exists x \varphi(x), \Delta}\, R\exists$$

$$\frac{\Gamma, \varphi(t) \Rightarrow \Delta}{\Gamma, \forall x \varphi(x) \Rightarrow \Delta}\, L\forall \qquad \frac{\Gamma \Rightarrow \varphi(y), \Delta}{\Gamma \Rightarrow \forall x \varphi(x), \Delta}\, R\forall$$

In $(L\exists)$ and $(R\forall)$, y is not free in Γ, Δ, φ and is called an *eigenvariable*. In G1i, $\Delta = \varnothing$ in $(R\exists)$ and $(R\forall)$.

The elegant symmetry of the propositional rules is again present in the quantifier rules and the difficulty of the elimination of existential quantifiers in natural deduction is no longer there.

6.6 Structural Rules

The sequent calculi introduced by Gentzen differ from the ones in the previous section in two ways. First, Gentzen's sequents consist of *sequences* rather than multisets of formulas. The difference between a sequence of formulas and a multiset of formulas is that in a sequence the order of formulas matters. Gentzen's calculus[3] LK consists of G1c (but then with Γ and Δ standing for sequences of formulas) plus the following two rules of *Exchange*:

$$\frac{\Gamma, \varphi, \psi, \Pi \Rightarrow \Delta}{\Gamma, \psi, \varphi, \Pi \Rightarrow \Delta}\, LE \qquad \frac{\Gamma \Rightarrow \Delta, \varphi, \psi, \Sigma}{\Gamma \Rightarrow \Delta, \psi, \varphi, \Sigma}\, RE$$

Nowadays, it is more common to use sequents that consist of multisets or even sets, and, in this paper, we consider the multiset version most of the time.

Second, besides the weakening, contraction, and exchange rules, LK contains one other structural rule, the *Cut rule*:

$$\frac{\Gamma \Rightarrow \Delta, \varphi \qquad \varphi, \Sigma \Rightarrow \Pi}{\Gamma, \Sigma \Rightarrow \Delta, \Pi}\, \text{Cut}$$

The formula φ is the *cutformula* of the inference. The Cut rule is structural in that no connective is introduced or appears. But it is special in another sense too: there is a formula, namely, φ, that "disappears" in the conclusion. For all the other rules in the calculus LK, and thus for all the rules of G1c, any formula that occurs in the

[3] *Logistische Kalkül* in German, whence the name LK.

premises of a rule occurs as a subformula in the conclusion, where we consider $\varphi(t)$ to be a subformula of $\exists x\varphi(x)$ and $\forall x\varphi(x)$. This property is the *subformula property*, which G1c thus satisfies and LK does not.

Note that, therefore, in any derivation in G1c of a sequent S, any predicate that occurs in the proof occurs in S. One could see this as a form of *purity of proof*, a notion in (the philosophy of) mathematics, where a proof of a theorem is *pure* if it does not use notions alien to the theorem. An example of a proof that is not pure is the topological proof of the infinitude of primes. For more on the notion of purity, see, e.g., Arana and Detlefsen (2011), Iemhoff (2017), and Kahle and Pulcini (2018).

Thus, the absence of the Cut rule has interesting consequences. But so does having the Cut rule in a sequent calculus. On a very general level, the Cut rule is considered to capture the use of lemmas in mathematics: here, the cut formula φ is the lemma, and if the mathematician shows that assumptions Γ imply the lemma and that the lemma implies ψ, then the mathematician is justified to conclude that ψ follows from assumptions Γ:

$$\frac{\Gamma \Rightarrow \text{lemma} \qquad \Gamma, \text{lemma} \Rightarrow \psi}{\Gamma \Rightarrow \psi}\ \text{Cut}$$

The beautiful and extremely useful fact about the Cut rule is that it can be eliminated from proofs in LK, as shown by Gentzen (1935). If LK^- denotes LK minus the Cut rule, then

Theorem 4 (Gentzen) *If* $\vdash_{\mathsf{LK}} S$, *then* $\vdash_{\mathsf{LK}^-} S$.

Theorem 4, as stated above, shows the admissibility of cut in LK. Gentzen indeed proved something stronger, namely, the *cut-elimination* theorem. He provided an algorithm to transform any proof π of a sequent S in LK into a proof π' of S in LK^-, which means that the cut rule is not used in π'. His proof was based on certain global transformations as well as some local transformation steps, including the permutation of a rule upward or replacing a cut on a formula with some cuts on its immediate subformulas.

In the systems G1c, G1i, and their propositional parts G1cp, G1ip, the same holds, but since the cut rule is not a rule in these systems, one speaks of *cut-admissibility* rather than cut-elimination (G+Cut denotes the sequent calculus G to which the Cut rule is added):

Theorem 5 *For any* $\mathsf{G} \in \{\mathsf{G1c}, \mathsf{G1i}, \mathsf{G1cp}, \mathsf{G1ip}\}$*:* $\vdash_{\mathsf{G+Cut}} S$ *implies* $\vdash_{\mathsf{G}} S$.

A derivation without cuts is *cut-free*.

Because of the admissibility of Cut we can use or omit it whenever the setting requires so. When we easily want to combine derivations to form new ones, Cut comes in handy: when we have shown that $\varphi \Rightarrow \psi$ and $\neg\varphi \Rightarrow \psi$, then we have a derivation of $(\Rightarrow \psi)$ from a derivation of $(\Rightarrow \varphi \vee \neg\varphi)$, an application of $(L\vee)$ and Cut. It may well be that a cut-free derivation of ψ has a very different form and cannot be obtained easily from the given derivations.

On the other hand, suppose we want to show the consistency of CQC by using G1c+Cut. Then, the structural rules and Cut, in particular, are a nuisance because they spoil bottom-up proof search in the calculus. This becomes clear when it is shown that the calculus that is introduced next, G3cp, has good computational properties. For now, let us say that having a calculus with many of the good properties of G1c but without the structural rules is highly desirable. This brings us to the sequent calculus G3c given in Fig. 6. Its intuitionistic variant G3i is given in Fig. 7, and is essentially nothing but G3c with the requirement that all sequents are single-conclusion, and a reformulation of $(R\vee)$ and $(L\rightarrow)$. The first one needs no argument, and the form of the second one will become clear in Lemma 6.

G3cp (G3ip) is the propositional version of G3c (G3i), so for the propositional language and without the rules for the quantifiers. For the curious reader, there are also sequent calculi G2cp and G2ip, which are obtained from G1cp and G1ip, respectively, by considering a generalized version of their axioms and leaving out the weakening rules. There are a few things to note here. Indeed, G3c contains no structural rules. The axioms have changed in that they are generalized to have arbitrary multiset variables on both sides. The rules for the connectives remain unchanged except for $(R\vee)$. For G3i, there are two more differences with G1i, and that is the duplicate of the principal formula in the premise of $(R\exists)$ and $(L\forall)$, which causes a *hidden contraction*. In the proof of the admissibility of the structural rules in G3cp and G3c, Lemma 7, we will see the reason for these adjustments. But first, we need the inversion lemma.

Lemma 6 (Inversion Lemma) *Let* $\vdash$ *denote derivability in* G3c *(for* $\vdash_n$ *see Sect. 6.2). Then,*

$(L\wedge)$ $\vdash_n \Gamma, \varphi \wedge \psi \Rightarrow \Delta$ *implies* $\vdash_n \Gamma, \varphi, \psi \Rightarrow \Delta$;
$(R\wedge)$ $\vdash_n \Gamma \Rightarrow \varphi \wedge \psi, \Delta$ *implies* $\vdash_n \Gamma \Rightarrow \varphi, \Delta$ *and* $\vdash_n \Gamma \Rightarrow \psi, \Delta$;
$(L\vee)$ $\vdash_n \Gamma, \varphi \vee \psi \Rightarrow \Delta$ *implies* $\vdash_n \Gamma, \varphi \Rightarrow \Delta$ *and* $\vdash_n \Gamma, \psi \Rightarrow \Delta$;
$(R\vee)$ $\vdash_n \Gamma \Rightarrow \varphi \vee \psi, \Delta$ *implies* $\vdash_n \Gamma \Rightarrow \varphi, \psi, \Delta$;
$(L\rightarrow)$ $\vdash_n \Gamma, \varphi \rightarrow \psi \Rightarrow \Delta$ *implies* $\vdash_n \Gamma \Rightarrow \varphi, \Delta$ *and* $\vdash_n \Gamma, \psi \Rightarrow \Delta$;
$(R\rightarrow)$ $\vdash_n \Gamma \Rightarrow \varphi \rightarrow \psi, \Delta$ *implies* $\vdash_n \Gamma, \varphi \Rightarrow \psi, \Delta$;
$(L\exists)$ $\vdash_n \Gamma, \exists x\varphi(x) \Rightarrow \Delta$ *implies* $\vdash_n \Gamma, \varphi(y) \Rightarrow \Delta$ *for all* y *not in* Γ, Δ, φ;
$(R\forall)$ $\vdash_n \Gamma \Rightarrow \forall x\varphi(x), \Delta$ *implies* $\vdash_n \Gamma \Rightarrow \varphi(y), \Delta$ *for all* y *not in* Γ, Δ, φ.

The same holds for G3ip, *in which case sequents are single-conclusion, which means that in* G3ip $(R\vee)$ *does not hold, and in the case* $(L\rightarrow)$ *only* $\vdash_n \Gamma, \psi \Rightarrow \Delta$ *can be concluded from* $\vdash_n \Gamma, \varphi \rightarrow \psi \Rightarrow \Delta$.

Proof Left to the reader, or see Proposition 3.5.4 in Troelstra and Schwichtenberg (2000). □

Lemma 7 (Weakening and Contraction Lemma) *Weakening and Contraction are depth-preserving admissible in* G3cp *and* G3c, *and in* G3ip *and* G3i.

Proof We treat G3c, the other three calculi can be treated in the same way. $\vdash$ stands for derivability in G3c, and $\vdash_n S$ means that S has a derivation of depth at most n in G3c.

Weakening We have to show that if $\vdash_n \Gamma \Rightarrow \Delta$, then $\vdash_n \Gamma, \chi \Rightarrow \Delta$ and $\vdash_n \Gamma \Rightarrow \chi, \Delta$, for any formula χ. We prove this by induction on n. If $\Gamma \Rightarrow \Delta$ is an axiom, then $\Gamma, \chi \Rightarrow \Delta$ and $\Gamma \Rightarrow \chi, \Delta$ are instances of the same axioms, which proves the statement. If $\Gamma \Rightarrow \Delta$ is not an axiom, consider the last inference of the derivation. We distinguish by cases according to the rule R of which the last inference is an instance. Suppose R is $L\vee$, then the last inference looks as follows, where $\Gamma = \Gamma', \varphi \vee \psi$:

$$\frac{\Gamma', \varphi \Rightarrow \Delta \qquad \Gamma', \psi \Rightarrow \Delta}{\Gamma', \varphi \vee \psi \Rightarrow \Delta} \; L\vee$$

By the induction hypothesis $\Gamma', \varphi, \chi \Rightarrow \Delta$ and $\Gamma', \psi, \chi \Rightarrow \Delta$ have derivations of at most depth $n-1$. An application of $L\vee$ to these sequents shows that $\Gamma', \varphi \vee \psi, \chi \Rightarrow \Delta$ has a derivation of depth at most n. The case for right weakening is analogous. The other connectives and the quantifiers can be treated in exactly the same way. Note that for the strong quantifiers (i.e., positive occurrences of the universal quantifier and negative occurrences of the existential quantifier), with the condition on the variable, the eigenvariables may have to be changed in order to not interfere with the variables in χ. We leave the details to the reader.

Contraction We have to show that $\vdash_n \Gamma, \chi, \chi \Rightarrow \Delta$ implies $\vdash_n \Gamma, \chi \Rightarrow \Delta$, and that $\vdash_n \Gamma \Rightarrow \chi, \chi, \Delta$ implies $\vdash_n \Gamma \Rightarrow \chi, \Delta$. We treat the second case, contraction on the right, and use induction on n. If $\Gamma \Rightarrow \chi, \chi, \Delta$ is an axiom, then $\Gamma \Rightarrow \chi, \Delta$ is an instance of the same axiom, which proves the statement. If $\Gamma \Rightarrow \chi, \chi, \Delta$ is not an axiom, consider the last inference of the derivation. We distinguish by cases according to the rule R of which the last inference is an instance. Suppose R is $R\vee$, then there are two possibilities for the last inference, either χ is the main formula or not:

$$\frac{\Gamma \Rightarrow \varphi, \psi, \chi, \chi, \Delta'}{\Gamma \Rightarrow \varphi \vee \psi, \chi, \chi, \Delta'} \; R\vee \qquad \frac{\Gamma \Rightarrow \varphi, \psi, \varphi \vee \psi, \Delta'}{\Gamma \Rightarrow \varphi \vee \psi, \varphi \vee \psi, \Delta'} \; R\vee$$

In the first case, the induction hypothesis applies to the premise, and then an application of $R\vee$ gives $\vdash_n \Gamma \Rightarrow \varphi \vee \psi, \chi, \Delta'$, as desired. In the second case, we can apply the Inversion Lemma 6 to the premise $\vdash_{n-1} \Gamma \Rightarrow \varphi, \psi, \varphi \vee \psi, \Delta'$ and obtain $\vdash_{n-1} \Gamma \Rightarrow \varphi, \psi, \varphi, \psi, \Delta'$. By applying the induction hypothesis twice, once to φ and once to ψ, we obtain $\vdash_{n-1} \Gamma \Rightarrow \varphi, \psi, \Delta'$. An application of $R\vee$ proves $\vdash_n \Gamma \Rightarrow \varphi \vee \psi, \Delta'$.

The other connectives can be treated in exactly the same way. In the treatment of $L\!\rightarrow$ for G3ip the use of the main formula in the premise will become clear.

From the quantifiers, we only treat $R\exists$. Thus, the last inference can have the following two forms:

$$\frac{\Gamma \Rightarrow \varphi(t), \exists x\varphi(x), \chi, \chi, \Delta'}{\Gamma \Rightarrow \exists x\varphi(x), \chi, \chi, \Delta'} \; R\exists \qquad \frac{\Gamma \Rightarrow \varphi(t), \exists x\varphi(x), \exists x\varphi(x), \Delta'}{\Gamma \Rightarrow \exists x\varphi(x), \exists x\varphi(x), \Delta'} \; R\exists$$

In both cases, we can apply the induction hypothesis to the premise to remove a χ or $\exists x\varphi(x)$, respectively, and then apply $R\exists$ to obtain the desired conclusion. □

Remark 8 Note that in the proof of Lemma 7 it is essential that in $R\exists$ the main formula is copied in the premise. The same remark holds for $L\forall$.

6.7 Cut-Elimination

We have seen that weakening and contraction are admissible in the four calculi G3cp, G3c, G3ip, and G3i. Also, the other admissible rule, Cut, is admissible in them, but that is far harder to prove. We do not give the proof here, there are many in the literature, for example, in Troelstra and Schwichtenberg (2000).

Theorem 9 (Cut-elimination) *For* X *being any of the calculi* G3cp, G3c, G3ip, G3i, *and for all sequents S:*

$$\vdash_{\mathsf{X+Cut}} S \textit{ implies } \vdash_{\mathsf{X}} S.$$

In other words, Cut is admissible in these calculi. Moreover, there is an algorithm that given a proof of S in X + Cut *provides a proof for S in* X.

Thus, with the G3-calculi we have proof systems without structural rules but with the power of these rules. A striking example is G3cp, in which proof search is terminating without any side conditions on the search (for a definition of termination, see Sect. 7.1). Proof search in G3ip is also terminating, but under an additional side condition that we will not dwell on here.

The cut-elimination has numerous consequences, among which are the following two. Theorem 10 illustrates the usefulness of sequent calculi to prove properties about a logic, a phenomenon we also encounter in the sections on interpolation. Theorem 11 shows that for CPC as well as IPC there exist two sequent calculi, each of which has properties that make it useful in a given context, as explained in Sect. 6.6.

Theorem 10 *Derivability in* G3cp *and* G3ip *is decidable. Thus,* CPC *and* IPC *are decidable.*

Proof For the decidability of IPC via a bottom-up proof search in G3ip see Theorem 4.2.6 in Troelstra and Schwichtenberg (2000). The result for G3cp and CPC is similar and in fact easier. □

Theorem 11 *The sequent calculi* G1cp(G1c) *and* G3cp(G3c) *prove the same sequents. So do* G1ip(G1i) *and* G3ip(G3i).

Proof See Proposition 3.5.9 in Troelstra and Schwichtenberg (2000). □

Theorem 11 implies that G1ip does not derive the law of excluded middle and Peirce's law, as predicted in Sect. 6.4. This follows from the fact that G3ip does not derive these principles, a fact that is not hard to establish, given that G3ip does not contain structural rules: a proof of $(\Rightarrow p \vee \neg p)$ in G3ip necessarily has to end with an application of $R\vee$, but then the premise would be $(\Rightarrow p)$ or $(\Rightarrow \neg p)$, two clearly underivable sequents. The argument for Peirce's law is similar.

6.8 Sequent Calculi for Other Logics

We have seen several elegant sequent calculi for classical and intuitionistic logic. What is a beautiful fact is that there are many logics, be it intermediate, modal, conditional, or substructural, that have sequent calculi that are extensions of or are similar to the calculi we have seen. These calculi are as well-behaved as the ones discussed above, and therefore useful in the study of the mentioned logics.

The discussion below describes sequent calculi for intermediate, modal, and conditional logics. Because of the lack of space, the usual calculi for standard substructural logics are not given. They are as elegant and useful as the calculi that are discussed here, pointers to the literature are given below and see Galatos et al. (2007).

7 Existence of Sequent Calculi

We have introduced proof systems for IPC and CPC, namely, the sequent calculi G1cp, G3cp, G1ip, G3ip, and argued that they are good proof systems, at least in a given context. Thus, the question arises: which other logics have sequent calculi with certain desirable properties, such as the absence but admissibility of structural rules, or termination? A positive answer to that question for a given logic is often provided by developing a proof system for that logic and proving that it has the desirable properties one wishes it to have. But for a negative answer, in case the logic does not have good sequent calculi, it is not as clear how one should obtain such a result. How to prove that no sequent calculus from the class of calculi considered "good" can be a calculus for the logic?

Although the answer to that question naturally depends on the precise definition of "good" that one wishes to use, the general methodology to approach this question as described in this section applies to various interpretations of "good." However, because the method can be better explained when applied to a concrete case, we consider a calculus good when it is terminating, and comment on the general case later on.

7.1 Terminating Calculi

A given calculus G for a (modal) propositional logic is *terminating*[4] if it satisfies the following properties:

finite G consists of finitely many axioms and rules;
instance finite for every sequent S there are at most finitely many instances of rules in G of which S is the conclusion;
well-ordered there is a well-order $<$ on sequents such that
∘ every proper subsequent comes before the sequent in $<$;
∘ for every rule in G the premises come before the conclusion in $<$.
∘ any sequent of the form $(\Gamma, \Pi \Rightarrow \Delta, \Lambda)$, where $\Pi \cup \Lambda$ is nonempty comes before a sequent of the form $(\Gamma, \Box\Pi \Rightarrow \Delta, \Box\Lambda)$.

Thus, we arrive at the question of which logics have a terminating calculus, and in particular at the problem of how to establish, for logics that do not have such calculi, the fact that they do not. The latter results, stating that a certain logic does not have a terminating calculus, we call *negative results*.

In the literature on the proof theory of modal logics, there are several sequent calculi (defined in Sect. 7.5) that are indeed terminating with respect to the order $<_d$ on sequents defined in Sect. 6.1.1: G3cp, G3K, and G3KD. Thus, for the corresponding logics CPC, K, and KD, the question has been answered positively in the literature. For definitions of these logics, see Table 1. On the other hand, G3S4 and G3GL are not terminating, at least not with the order $<_d$, because these calculi are not well-ordered in this order. In fact, we do not know whether there exist terminating calculi for these logics; the method that we are going to develop does not apply to them (yet). For more on proof theory of modal logic, see Negri (2011).

When turning to logics based on intuitionistic logic one naturally turns to the calculus G3ip for IPC and tries to extend it to a calculus for the logic one is considering, it being intermediate or an intuitionistic modal logic. But the fact that G3ip is well-ordered neither in $<_d$ nor $<_w$ blocks that approach. There is, however, a variant of G3ip that is terminating with respect to $<_w$, the calculus G4ip introduced by Dyckhoff (1992) that is given in Fig. 8. This calculus will be the basis for the calculi for the intuitionistic modal and intermediate logics treated in this paper. It is not hard to see that G4ip is indeed terminating. For a proof of its equivalence to G3ip we refer the reader to Dyckhoff's paper. Thus, IPC has a terminating calculus. It is worth mentioning that the terminating calculus for IPC was developed independently by Dyckhoff (1992) and Hudelmaier (1988, 1992, 1993).

[4] In Iemhoff (2019b) the notion is defined slightly differently.

Table 1 Some modal axioms and normal modal logics

Name of axiom	Axiom	Name of logic	Logic
k	$\Box(p \to p) \to (\Box p \to \Box q)$	K	CPC+(k)
d	$\Box p \to \Diamond p$	KD	K+(d)
4	$\Box p \to \Box\Box p$	K4	K+(4)
t	$\Box p \to p$	S4	K4+(t)
löb	$\Box(\Box p \to p) \to \Box p$	GL	K+ (löb)
Grz	$\Box(\Box(\varphi \to \Box\varphi) \to \varphi) \to \varphi$	Grz	K+ (Grz)

7.2 Method Toward Negative Results

How can one prove negative results, results stating that certain logics do not have terminating calculi? Here we introduce a possible method, one that is easy to explain but requires ingenuity to apply. It can be applied not only to the class of terminating sequent calculi, but to any class of sequent calculi $\mathcal{S}$ that one is interested in, and we will formulate it in this general form.

Consider a class of sequent calculi $\mathcal{S}$ (e.g., the terminating sequent calculi) and a class of logics $\mathcal{C}$ (e.g., intermediate, modal, etc.). Suppose that there is a property of logics $\mathcal{P}$ and a subset $\mathcal{C}'$ of $\mathcal{C}$ such that one can show the following:

(I) Any logic in $\mathcal{C}$ that has a sequent calculus in $\mathcal{S}$ has property $\mathcal{P}$.
(II) No logic in $\mathcal{C}'$ has property $\mathcal{P}$.

Then, one can conclude by simple contraposition:

(III) No logic in $\mathcal{C}'$ has a sequent calculus in $\mathcal{S}$ (conclusion).

If the property $\mathcal{P}$ is rare, meaning that not many logics in $\mathcal{C}$ have $\mathcal{P}$, and the class of logics $\mathcal{C}'$ and the class of calculi $\mathcal{S}$ are large, then the conclusion states that there is a large class of logics, namely, $\mathcal{C}'$, such that for no logic in that class can there be a sequent calculus in the large class $\mathcal{S}$. Exactly the kind of negative results we are looking for. Thus, the general aim is to find $\mathcal{S}$ and $\mathcal{C}'$ that are large and a property $\mathcal{P}$ that is relatively rare among the logics in $\mathcal{C}$.

In this paper, the property $\mathcal{P}$ that we are mostly considering is (Craig, uniform, Lyndon) interpolation, defined in Sects. 7.3 and 7.4.1 and Sect. 8. The class of sequent calculi $\mathcal{S}$ is, in most settings, a subclass of the (terminating) calculi, usually the largest class of (terminating) calculi to which the method has been successfully applied, but not ruling out that there are more calculi for which (I) holds. The results of (II) we mainly obtain from the literature. These are the results stating that certain logics do not have (Craig, uniform, Lyndon) interpolation. Since these results are often proved with tools from algebraic logic, the whole method shows a nice interplay between that area and proof theory.

Uniform (Lyndon) interpolation is rarer than interpolation and therefore the better choice for the property $\mathcal{P}$. However, for this exposition, we have chosen to illustrate the method for the case that $\mathcal{P}$ is the property of interpolation. The advantage is that this case still shows the spirit of the method, but without the technical complexities and conceptual difficulties that come with the study of uniform interpolation. References to papers on the full method will be given during the exposition.

7.3 *Intermediate Logics*

We describe the details of the above method, introduced in Sect. 7.2, to obtain negative results for the class of intermediate logics ($\mathcal{C}$). The class of calculi $\mathcal{S}$ that we use is a subclass of the class of terminating sequent calculi, to be defined below. As the property $\mathcal{P}$ we use interpolation.

Let us remind the reader that a propositional (modal) logic L has *Craig Interpolation Property* (CIP), also called *interpolation*, if for every implication $\varphi \to \psi$ that is derivable in L there exists a formula α (the *interpolant*) such that both $\varphi \to \alpha$ and $\alpha \to \psi$ are derivable in L and α is in the *common language* of φ and ψ, meaning that every atom that occurs in α occurs in both φ and ψ.

7.3.1 Sequent Interpolation for G3ip

We start with a proof-theoretic proof of interpolation for IPC. The proof uses the calculus G3ip and the following sequent version from which the standard interpolation property follows. A calculus G has *sequent interpolation* if for every partitioned sequent $\Gamma; \Pi \Rightarrow \Delta$ (partitions are defined in Sect. 6.1) there is a formula α in the common language of $(\Gamma \Rightarrow)$ and $(\Pi \Rightarrow \Delta)$ such that $\Gamma \Rightarrow \alpha$ and $\Pi, \alpha \Rightarrow \Delta$ are derivable in G. Here the common language of $\Gamma; \Pi \Rightarrow \Delta$ is by definition the common language of $\Gamma \Rightarrow$ and $\Pi \Rightarrow \Delta$, which means the atoms that occur in some formula in Γ as well as in some formula of Π or Δ.

The proof of Theorem 14, which states that G3ip has sequent interpolation, uses two lemmas that we prove first.

Lemma 12 *Every partition of every instance of an axiom of* G3ip *has sequent interpolation.*

Proof Let $S = (\Gamma; \Pi \Rightarrow \Delta)$ be a partition of an instance of an axiom of G3ip. In case S is an instance of $L\bot$, then the interpolant can be $\top$ in case $\bot \in \Pi$ and $\bot$ otherwise. In case S is an instance of the other axiom, there are several cases to consider: if $p \in \Pi \cap \Delta$, then $\top$ is an interpolant, and if $p \in \Gamma \cap \Delta$, then p is in the common language and an interpolant. □

Lemma 13 *For every instance of a rule $\mathcal{R}$ in* G3ip, *for every partition $\Gamma; \Pi \Rightarrow \Delta$ of its conclusion there are partitions $\Gamma_i; \Pi_i \Rightarrow \Delta_i$ of its n premises and a formula $\gamma(p_1, \dots, p_n)$ such that $\gamma(\alpha_1, \dots, \alpha_n)$ is an interpolant of $\Gamma; \Pi \Rightarrow \Delta$ whenever α_i is an interpolant of $\Gamma_i; \Pi_i \Rightarrow \Delta_i$ for every $i = 1, \dots, n$.*

Proof We treat the case that $\mathcal{R}$ is $R\vee$, $R\wedge$, and $R\rightarrow$. In the first case, the last inference is

$$\frac{\Gamma, \Pi \Rightarrow \varphi}{\Gamma, \Pi \Rightarrow \varphi \vee \psi}\ R\vee$$

Let $\Gamma; \Pi \Rightarrow \varphi \vee \psi$ be the partition of the conclusion. Consider the partition $\Gamma; \Pi \Rightarrow \varphi$ of the premise and suppose β is its interpolant. Thus, $\Gamma \Rightarrow \beta$ and $\Pi, \beta \Rightarrow \varphi$ are derivable, which implies that so is $\Pi, \beta \Rightarrow \varphi \vee \psi$. Hence, β satisfies the conditions of being an interpolant for the conclusion and let $\gamma(p_1)$ be p_1. In the second case, the last inference is

$$\frac{\Gamma, \Pi \Rightarrow \varphi_1 \qquad \Gamma, \Pi \Rightarrow \varphi_2}{\Gamma, \Pi \Rightarrow \varphi_1 \wedge \varphi_2}\ R\wedge$$

Let $\Gamma; \Pi \Rightarrow \varphi_1 \wedge \varphi_2$ be the partition of the conclusion. Consider the partitions $\Gamma; \Pi \Rightarrow \varphi_i$ of the premises and suppose the α_i are their interpolants. Thus, $\Gamma \Rightarrow \alpha_i$ and $\Pi, \alpha_i \Rightarrow \varphi$ are derivable, which implies that so are $\Gamma \Rightarrow \alpha_1 \wedge \alpha_2$ and $\Pi, \alpha_1 \wedge \alpha_2 \Rightarrow \varphi \wedge \psi$. Hence, $\alpha_1 \wedge \alpha_2$ satisfies the conditions of being an interpolant for the conclusion and let $\gamma(p_1, p_2)$ be $p_1 \wedge p_2$.

In the third case, $R\rightarrow$, the last inference is of the form

$$\frac{\Gamma, \Pi, \varphi \Rightarrow \psi}{\Gamma, \Pi \Rightarrow \varphi \rightarrow \psi}\ R\rightarrow$$

Let $\Gamma; \Pi \Rightarrow \varphi \rightarrow \psi$ be the partition of the conclusion. Consider the partition $\Gamma; \Pi, \varphi \Rightarrow \psi$ of the premise and let β be its interpolant. Hence, we have derivable sequents $\Gamma \Rightarrow \beta$ and $\Pi, \varphi, \beta \Rightarrow \psi$. Thus, β satisfies the conditions of being an interpolant for the conclusion, as β is in the common language of the partition of the premise, and let $\gamma(p_1)$ be p_1. □

Theorem 14 G3ip *has sequent interpolation: For every partitioned sequent* $\Gamma; \Pi \Rightarrow \Delta$ *derivable in* G3ip *there is a formula* α *in the common language of* $(\Gamma \Rightarrow\ )$ *and* $(\Pi \Rightarrow \Delta)$ *such that* $\Gamma \Rightarrow \alpha$ *and* $\Pi, \alpha \Rightarrow \Delta$ *are derivable in* G3ip.

Proof We use induction on the depth of the derivation of $S = (\Gamma; \Pi \Rightarrow \Delta)$. The case that S is an instance of an axiom is covered by Lemma 12. If S is not an axiom and the conclusion of an application of a rule in G3ip, then Lemma 13 applies, which completes the proof. □

The above is a constructive proof of the following well-known theorem for IPC.

Corollary 15 IPC *has Craig interpolation.*

7.3.2 Generalization of Sequent Interpolation for G4ip

A similar result as in Theorem 14 can be proved for the terminating sequent calculus G4ip.

Theorem 16 G4ip *has sequent interpolation: For every partitioned sequent* $\Gamma; \Pi \Rightarrow \Delta$ *derivable in* G4ip *there is a formula* α *in the common language of* $(\Gamma \Rightarrow)$ *and* $(\Pi \Rightarrow \Delta)$ *such that* $\Gamma \Rightarrow \alpha$ *and* $\Pi, \alpha \Rightarrow \Delta$ *are derivable in* G4ip.

Proof We use the analog of Lemma 13 for G4ip. The new cases for the rule $\mathcal{R}$, compared to the proof of Lemma 13, are the rules $Lp\!\to$ and $L\!\to\to$.

In the case $Lp\!\to$, the last inference is of the form:

$$\frac{\Gamma', \Pi', p, \varphi \Rightarrow \Delta}{\Gamma', \Pi', p, p \to \varphi \Rightarrow \Delta}\ Lp\!\to$$

In case $p, p \to \varphi$ belongs to the same partition, the interpolant of the conclusion is the same for the premise. We consider the two remaining cases. First suppose the partition of S is $\Gamma', p; \Pi', p \to \varphi \Rightarrow \Delta$ and let β be the interpolant of the premise which is partitioned as $\Gamma', p; \Pi', \varphi \Rightarrow \Delta$. Since p belongs to the common language of the partitioned S, $\beta \wedge p$ satisfies the conditions of being an interpolant for the conclusion and we take $\gamma(p_1)$ to be $p_1 \wedge p$. In case the partition of S is $\Gamma', p \to \varphi; \Pi', p \Rightarrow \Delta$, partition the premise as $\Gamma', \varphi; \Pi', p \Rightarrow \Delta$ and let β be its interpolant. Since p belongs to the common language of S, $p \to \beta$ satisfies the conditions of being an interpolant for the conclusion and we take $\gamma(p_1)$ to be $p \to p_1$.

In the case $L\!\to\to$, the last inference is of the form:

$$\frac{\Gamma', \Pi', \varphi_2 \to \varphi_3 \Rightarrow \varphi_1 \to \varphi_2 \qquad \Gamma', \Pi', \varphi_3 \Rightarrow \Delta}{\Gamma', \Pi', (\varphi_1 \to \varphi_2) \to \varphi_3 \Rightarrow \Delta}\ L\!\to\to$$

First, suppose the partition of S is $\Gamma'; (\varphi_1 \to \varphi_2) \to \varphi_3, \Pi' \Rightarrow \Delta$ and let α_1, α_2 be the interpolants of the premises, which are partitioned as $\Gamma'; \varphi_2 \to \varphi_3, \Pi' \Rightarrow \varphi_1 \to \varphi_2$ and $\Gamma'; \varphi_3, \Pi' \Rightarrow \Delta$. Thus, $\alpha_1 \wedge \alpha_2$ satisfies the conditions of being an interpolant for the conclusion and we take $\gamma(p_1, p_2)$ to be $p_1 \wedge p_2$, noting that α indeed belongs to the common language of the partitioned conclusion.

Second, suppose the partition of S is $\Gamma', (\varphi_1 \to \varphi_2) \to \varphi_3; \Pi \Rightarrow \Delta$. Then, we consider the partition $\Pi'; \varphi_2 \to \varphi_3, \Gamma' \Rightarrow \varphi_1 \to \varphi_2$ of the left premise with interpolant α_1. For the right premise we consider the partition $\Gamma', \varphi_3; \Pi' \Rightarrow \Delta$ and its interpolant α_2. Thus, the following sequents are all derivable.

$$(1)\ \Pi' \Rightarrow \alpha_1 \qquad (2)\ \alpha_1, \varphi_2 \to \varphi_3, \Gamma' \Rightarrow \varphi_1 \to \varphi_2$$
$$(3)\ \Gamma', \varphi_3 \Rightarrow \alpha_2\ (4)\ \alpha_2, \Pi' \Rightarrow \Delta.$$

From this, follows the derivability of the following sequents:

$$\begin{array}{ll} (5)\ \Gamma', \alpha_1, (\varphi_1 \to \varphi_2) \to \varphi_3 \Rightarrow \alpha_2 & \text{by (2) \& (3)} \\ (6)\ \Gamma', (\varphi_1 \to \varphi_2) \to \varphi_3 \Rightarrow \alpha_1 \to \alpha_2 & \text{by (5)} \\ (7)\ \Pi', \alpha_1 \to \alpha_2 \Rightarrow \Delta. & \text{by (1) \& (4)} \end{array}$$

Thus, $\alpha_1 \to \alpha_2$ satisfies the conditions of being an interpolant for the conclusion and we leave it to the reader to verify that the formula is indeed in the common language of the partitioned S. We let $\gamma(p_1, p_2)$ be $p_1 \to p_2$. □

If one considers the proof of sequent interpolation for G4ip (Theorem 16) then the cases $Lp\to$ and $L\to\to$ seem specific and the others general. This brings us to define a large and general class of rules, the semi-analytic rules.

Definition 17 In the following definitions, expression $\langle\, \langle\, \Gamma_i, \overline{\varphi}_{ik} \Rightarrow \chi_{ik}\, \rangle_{k=1}^{a_i}\, \rangle_{i=1}^{m}$ is short for the sets of premises

$$\langle\, \Gamma_1, \overline{\varphi}_{1k} \Rightarrow \chi_{1k}\, \rangle_{k=1}^{a_1} \quad \dots \quad \langle\, \Gamma_m, \overline{\varphi}_{mk} \Rightarrow \chi_{mk}\, \rangle_{k=1}^{a_m},$$

where $\langle\, \Gamma_i, \overline{\varphi}_{ik} \Rightarrow \chi_{ik}\, \rangle_{k=1}^{a_i}$ is short for the set of premises:

$$\Gamma_i, \overline{\varphi}_{i1} \Rightarrow \chi_{i1} \quad \Gamma_i, \overline{\varphi}_{i2} \Rightarrow \chi_{i2} \quad \dots \quad \Gamma_i, \overline{\varphi}_{ia_i} \Rightarrow \chi_{ia_i}.$$

Likewise for expressions $\langle\, \langle\, \Gamma_i, \overline{\varphi}_{ik} \Rightarrow \Delta_i\, \rangle_{k=1}^{a_i}\, \rangle_{i=1}^{m}$. By $\overline{\psi}$, we mean a multiset of formulas $\{\psi_1, \dots, \psi_n\}$.

A sequent rule $\mathcal{R}$ is *left semi-analytic* if it is of the following form:

$$\frac{\langle\, \langle\, \Gamma_i, \overline{\varphi}_{ik} \Rightarrow \Delta_i\, \rangle_{k=1}^{a_i}\, \rangle_{i=1}^{m} \qquad \langle\, \langle\, \Pi_j, \overline{\psi}_{jl} \Rightarrow \chi_{jl}\, \rangle_{l=1}^{b_j}\, \rangle_{j=1}^{n}}{\Gamma_1, \dots, \Gamma_m, \Pi_1, \dots, \Pi_n, \varphi \Rightarrow \Delta_1, \dots, \Delta_m}\ \mathcal{R} \tag{1}$$

A *right semi-analytic* rule is of the following form:

$$\frac{\langle\, \langle\, \Gamma_i, \overline{\varphi}_{ik} \Rightarrow \chi_{ik}\, \rangle_{k=1}^{a_i}\, \rangle_{i=1}^{m}}{\Gamma_1, \dots, \Gamma_m \Rightarrow \varphi}\ \mathcal{R} \tag{2}$$

Here the $\Gamma_i, \Pi_j, \Delta_i$ are multiset variables, and we have the following respective variable conditions for the left and the right rules:

$$\bigcup_{ik} V(\overline{\varphi}_{ik}) \cup \bigcup_{jl} V(\overline{\psi}_{jl}) \cup \bigcup_{jl} V(\chi_{jl}) \subseteq V(\varphi)\,, \quad \bigcup_{ik} V(\overline{\varphi}_{ik}) \cup \bigcup_{ik} V(\chi_{ik}) \subseteq V(\varphi).$$

Recall that we consider single-conclusion sequents, so that the $\Delta_1, \dots, \Delta_m$ are empty or consist of one formula, meaning that at most one of the Δ_i can be nonempty. A left semi-analytic rule as above is *context-sharing* if $m = n$ and $\Gamma_i = \Pi_i$ for all i. A rule is *semi-analytic* if it is either right or left semi-analytic.

Note that the left and right conjunction and disjunction rules and the right implication rule of G3ip (and thus of G4ip) are semi-analytic rules. The rules $L\to$ of G3ip and $L\to\to$ of G4ip are context-sharing semi-analytic rules. Regarding the other left implication rules of G4ip, the rule $Lp\to$ is not semi-analytic, but the rules $L\wedge\to$ and $L\vee\to$ are.

In the following two lemmas we see how for semi-analytic rules, interpolants for the conclusion can be obtained from interpolants of the premises in a uniform way.

Lemma 18 *For any intermediate logic* L*, any right semi-analytic rule* (2) *and any partition* $\Pi_1, \ldots, \Pi_m; \Pi'_1, \ldots, \Pi'_m \Rightarrow \varphi$ *of its conclusion: if* α_{ik} *is an interpolant for premise* $\Pi_i; \Pi'_i, \overline{\varphi}_{ik} \Rightarrow \chi_i$*, then* $\alpha = \bigwedge_{ik} \alpha_{ik}$ *is an interpolant for the partitioned conclusion.*

Proof By assumption, for all α_{ik} with $1 \leq i \leq m$ and $1 \leq k \leq m_i$:

$$\vdash_{\mathsf{L}} \Pi_i \Rightarrow \alpha_{ik} \qquad \vdash_{\mathsf{L}} \alpha_{ik}, \Pi'_i, \overline{\varphi}_{ik} \Rightarrow \chi_i$$

$$V(\alpha_{ik}) \subseteq V(\Pi_i) \cap \big(V(\Pi'_i) \cup V(\overline{\varphi}_{ik}) \cup V(\chi_i)\big).$$

This implies the derivability of $\Pi_1, \ldots, \Pi_m \Rightarrow \alpha$ and of $\alpha, \Pi'_i, \overline{\varphi}_{ik} \Rightarrow \chi_i$. The form of the semi-analytic rule $\mathcal{R}$ implies that the following is an instance of it:

$$\frac{\langle\langle \alpha, \Pi'_i, \overline{\varphi}_{ik} \Rightarrow \chi_i \rangle_{k=1}^{m_i} \rangle_{i=1}^{m}}{\alpha^m, \Pi'_1, \ldots, \Pi'_m \Rightarrow \varphi}\ \mathcal{R}$$

Hence, $\vdash_{\mathsf{L}} \alpha, \Pi'_1, \ldots, \Pi'_m \Rightarrow \varphi$. As $\mathcal{R}$ is semi-analytic, $V(\overline{\varphi}_{ik}) \cup V(\chi_i) \subseteq V(\varphi)$. Thus, the variable conditions on α_{ik} above imply that

$$V(\alpha) \subseteq \bigcup_I V(\Pi_i) \cap \big(V(\Pi'_i) \cup V(\varphi)\big).$$

Hence, α is the desired interpolant. □

Lemma 19 *For any intermediate logic* L*, any left semi-analytic rule* (1)*, and all partitions* $\Gamma'_i; \Gamma''_i$ *of* Γ_i *and* $\Pi'_i; \Pi''_i$ *of* Π_i*:*

1. *For partition* $\overline{\Gamma'_i}, \overline{\Pi'_i}; \varphi, \overline{\Gamma''_i}, \overline{\Pi''_i} \Rightarrow \Delta_1, \ldots, \Delta_m$*, if* α_{ik} *and* β_{jl} *are interpolants of the partitioned premises* $\Gamma'_i; \overline{\varphi}_{ik}, \Gamma''_i \Rightarrow \Delta_i$ *and* $\Pi'_j; \overline{\psi}_{jl}, \Pi''_j \Rightarrow \chi_j$*, respectively, then* $\bigwedge_{ik} \alpha_{ik} \wedge \bigwedge_{jl} \beta_{jl}$ *is an interpolant for the partitioned conclusion.*
2. *For partition* $\overline{\Gamma'_i}, \overline{\Pi'_i}, \varphi; \overline{\Gamma''_i}, \overline{\Pi''_i} \Rightarrow \Delta_1, \ldots, \Delta_m$*, as it is single-conclusion, w.l.o.g. suppose for* $i \neq 1$*,* $\Delta_i = \emptyset$*. If* α_{ik} *and* β_{jl} *are interpolants of the partitioned premises* $\Gamma'_i, \overline{\varphi}_{ik}; \Gamma''_i \Rightarrow \Delta_i$ *and* $\Pi'_j, \overline{\psi}_{jl}; \Pi''_j \Rightarrow \chi_j$*, respectively, then* $(\bigwedge_{ik, i\geq 2} \alpha_{ik} \wedge \bigwedge_{jl} \beta_{jl}) \to \bigvee_k \alpha_{1k}$ *is an interpolant for the partitioned conclusion.*

Proof 1. Let γ denote $\bigwedge_{ik} \alpha_{ik} \wedge \bigwedge_{jl} \beta_{jl}$. By assumption we have

$$\vdash_{\mathsf{L}} \Gamma'_i \Rightarrow \alpha_{ik} \ , \ \vdash_{\mathsf{L}} \alpha_{ik}, \overline{\varphi}_{ik}, \Gamma''_i \Rightarrow \Delta_i \ , \ \vdash_{\mathsf{L}} \Pi'_j \Rightarrow \beta_{jl} \ , \ \vdash_{\mathsf{L}} \beta_{jl}, \overline{\psi}_{jl}, \Pi''_j \Rightarrow \chi_j .$$

Hence, $\vdash_{\mathsf{L}} \overline{\Gamma}'_i, \overline{\Pi}'_i \Rightarrow \gamma$. The form of the semi-analytic rule $\mathcal{R}$ implies that the following is an instance of it:

$$\frac{\langle\langle \gamma, \Gamma''_i, \overline{\varphi}_{ik} \Rightarrow \Delta_i \rangle_{k=1}^{m_i} \rangle_{i=1}^{m} \qquad \langle\langle \gamma, \Pi''_j, \overline{\psi}_{jl} \Rightarrow \chi_j \rangle_{l=1}^{n_j} \rangle_{j=1}^{n}}{\gamma^x, \Gamma''_1, \ldots, \Gamma''_m, \Pi''_1, \ldots, \Pi''_n, \varphi \Rightarrow \Delta_1, \ldots, \Delta_m} \mathcal{R}$$

We can conclude that γ is the desired interpolant, as the condition on its atoms follows from that of its conjuncts.

The proof of 2. is analogous. □

Similar to the semi-analytic rules we define a class of axioms that includes the axioms of $\mathsf{G3ip}$ or $\mathsf{G4ip}$ and such that every partition of a sequent in the class has sequent interpolation.

Definition 20 A sequent is a *focused* axiom if it has one of the following forms, where $|\Delta| \leq 1$:

$$\varphi \Rightarrow \varphi \qquad \Rightarrow \varphi \qquad \varphi_1, \ldots, \varphi_n \Rightarrow \qquad \Gamma, \varphi_1, \ldots, \varphi_n \Rightarrow \Delta \qquad \Gamma \Rightarrow \varphi$$

where $V(\varphi_i) = V(\varphi_j)$ for every $1 \leq i.j \leq n$. In a classical setting the definition of focused axiom is similar, but there the requirement of Δ is lifted.

Lemma 21 *Every partition of every focused axiom has sequent interpolation.*

Proof Straightforward checking of all cases. □

7.3.3 Positive and Negative Results via Interpolation

Having defined all the relevant notions we are now ready to present the first negative result. Regarding the general method described in Sect. 7.2: here $\mathcal{C} = \mathcal{C}'$ is the class of intermediate logics, $\mathcal{P}$ is Craig interpolation, and $\mathcal{S}$ is the class of (context-sharing) semi-analytic rules and focused axioms. And Theorem 22 and its Corollary 23 form part (I) of the method, Theorem 25 is part (II), and Corollary 26 is part (III), the conclusion, the negative result we are aiming for. The following theorem is obtained in Akbar Tabatabai and Jalali (2025b) and is an extension of a similar result in Iemhoff (2019b). The latter considers extensions of $\mathsf{G4ip}$ rather than $\mathsf{G3ip}$and a set of rules that is a proper subset of the semi-analytic rules.

Theorem 22 (Iemhoff (2019b) and Akbar Tabatabai and Jalali (2025b)) *Any sequent calculus* G *that is an extension of* $\mathsf{G3ip}$ *by (context-sharing) semi-analytic rules and focused axioms has sequent interpolation.*

Proof For G as in the theorem we have to show that for every partitioned sequent $\Gamma; \Pi \Rightarrow \Delta$ derivable in G there is a formula α in the common language of $(\Gamma \Rightarrow)$ and $(\Pi \Rightarrow \Delta)$ such that $\Gamma \Rightarrow \alpha$ and $\Pi, \alpha \Rightarrow \Delta$ are derivable in G. We use induction on the depth of the derivation of $S = (\Gamma; \Pi \Rightarrow \Delta)$.

The case that S is an instance of an axiom is covered by Lemma 21. If S is not an axiom and the conclusion of an application of a rule $\mathcal{R}$, then Lemma 13 applies in case $\mathcal{R}$ belongs to $\mathsf{G3ip}$ and Lemmas 18 and 19 in case $\mathcal{R}$ is a right or left semi-analytic rule not in $\mathsf{G3ip}$. □

Corollary 23 *Any intermediate logic* L *that has a calculus* G *only consisting of focused axioms and (context-sharing) semi-analytic rules has interpolation.*

Proof First, extend the calculus G by adding the axioms and rules of $\mathsf{G3ip}$ and call it G^*. Then, by the definition of a calculus for a logic L (see Sect. 6.2) it is easy to see that since L is an intermediate logic, G^* is a calculus for L. Now, we can use Theorem 22 to get the result. □

Corollary 24 *If an intermediate logic* L *does not have interpolation, then it cannot have a sequent calculus only consisting of focused axioms and (context-sharing) semi-analytic rules.*

Theorem 25 (Maksimova (1977)) *There are exactly seven intermediate logics with Craig interpolation:*

$$\mathsf{IPC}, \mathsf{Sm}, \mathsf{GSc}, \mathsf{LC}, \mathsf{KC}, \mathsf{Bd}_2, \mathsf{CPC}.$$

The frame conditions of the logics with Craig interpolation:

Logic	Class of frames
Sm	Frames of at most two nodes
GSc	The fork
LC	Linear frames
KC	Frames with one maximal node
Bd_2	Frames of depth at most 2

Corollary 26 *If an intermediate logic is not one of the seven in Theorem 25, then it does not have a sequent calculus only consisting of focused axioms and (context-sharing) semi-analytic rules.*

Theorem 27 (Maksimova (1991)) *Among the normal extensions of* $\mathsf{S4}$ *there are at most 37 logics with interpolation. Exactly 6 normal extensions of* Grz *have interpolation.*

Corollary 28 *The following do not have sequent calculi that only consist of focused axioms and semi-analytic rules:*

- *all but the 37 (at most) extensions of* S4 *with interpolation;*
- *all but the 6 extensions of* Grz *that have interpolation.*

As mentioned above, here we have our first negative results: a large (in this case infinite) class of logics, namely, all but the seven intermediate logics with CIP, cannot have calculi in a large class of natural sequent calculi, namely, the class of calculi only consisting of focused axioms and (context-sharing) semi-analytic rules.

7.4 Negative Results via Uniform Interpolation

Having seen the method at work for the class of intermediate logics, the question of whether it can be adapted to other logics naturally arises. What about normal modal logics, substructural logics, non-normal modal logics, and intuitionistic modal logics? To obtain negative results for these logics with the method described in Sect. 7.2, one could still use Craig interpolation as the property $\mathcal{P}$. But as was explained there, the more rare the property $\mathcal{P}$ is for the logics that one considers, the stronger the negative result is. We therefore turn to a strengthening of CIP, uniform interpolation, defined in the next section. For intermediate logics, the two notions coincide, but, as we will see, for modal logics this is no longer the case, and, therefore, using the stronger notion effectively strengthens our results.

7.4.1 Uniform Interpolation

A logic L has the *Uniform Interpolation Property (UIP)* if for any formula φ and any atom p there exist two formulas, usually denoted by $\forall p\varphi$ and $\exists p\varphi$, in the language of the logic, that do not contain p or atoms not in φ and such that for all ψ not containing p:

$$\vdash \psi \to \varphi \Leftrightarrow \vdash \psi \to \forall p\varphi \qquad \vdash \varphi \to \psi \Leftrightarrow \vdash \exists p\varphi \to \psi.$$

Given a formula φ, its *universal uniform interpolant with respect to* $p_1 \dots p_n$ is $\forall p_1 \dots p_n\varphi$, which is short for $\forall p_1(\forall p_2(\dots(\forall p_n\varphi)\dots))$, and its *existential uniform interpolant with respect to* $p_1 \dots p_n$ is $\exists p_1 \dots p_n\varphi$, which is short for $\exists p_1(\exists p_2(\dots(\exists p_n\varphi)\dots))$. The requirements above could be replaced by the following four requirements that explain the names of the uniform interpolants:

$$\vdash \forall p\varphi \to \varphi \qquad \vdash \psi \to \varphi \Rightarrow \vdash \psi \to \forall p\varphi. \tag{$\forall$}$$

$$\vdash \varphi \to \exists p\varphi \qquad \vdash \varphi \to \psi \Rightarrow \vdash \exists p\varphi \to \psi. \tag{$\exists$}$$

In classical logic, one only needs one quantifier, as $\exists p$ can be defined as $\neg\forall p\neg$ and vice versa. Although in the intuitionistic setting $\exists p$ can also be defined in terms of $\forall p$, namely, as $\exists p\varphi = \forall q(\forall p(\varphi \to q) \to q)$ for a q not in φ, having it as a separate quantifier is convenient in the proof-theoretic approach presented here (we follow Pitts (1992), which also uses both quantifiers).

It is easy to see that uniform interpolation implies interpolation. For suppose $\vdash_{\mathsf{L}} \varphi \to \psi$, and let $\bar{p} = p_1, \ldots, p_n$ be all the atoms that occur in ψ but not in φ. Hence, $\forall p_1 \ldots p_n\psi \to \psi$ and $\varphi \to \forall p_1 \ldots p_n\psi$ are derivable in L. Since all atoms in $\forall p_1 \ldots p_n\psi$ have to occur in both φ and ψ, it is indeed an interpolant for $\varphi \to \psi$.

To define uniform interpolants in the setting of sequents, we introduce the notion of a partition, which applies to sequents and rules. A *partition* of a sequent S is an ordered pair (S^r, S^i) (i for *interpolant*, r for *rest*) such that $S = S^r \cdot S^i$, where the *multiplication* of the sequents S_1 and S_2 is defined as $S_1 \cdot S_2 := S_1^a \cup S_2^a \Rightarrow S_1^s \cup S_2^s$. It is a *p-partition* if p does not occur in S^r. A sequent calculus G has *uniform sequent interpolation* if for all sequents S and all atoms p there are formulas $\forall p S$ and $\exists p S$ in $\mathcal{L}$ that do not contain p nor any atoms that do not occur in S, and such that the following holds, where $\vdash$ stands for derivability in G:

($\forall$l) $\vdash S^a, \forall p S \Rightarrow S^s$;
($\exists$r) $\vdash S^a \Rightarrow \exists p S$.
($\forall\exists$) If S is derivable, for all p-partitions (S^r, S^i) of S: $\vdash S^r \cdot (\exists p S^i \Rightarrow \forall p S^i)$.

In the case of a single-conclusion sequent calculus, the last requirement is replaced by

$$\begin{array}{ll} \vdash S^r \cdot (\exists p S^i \Rightarrow \forall p S^i) & \text{if } S^s \neq \varnothing \text{ and } S^{rs} = \varnothing \\ \vdash S^r \cdot (\exists p S^i \Rightarrow\ ) & \text{if } S^s = \varnothing \text{ or } S^{rs} \neq \varnothing. \end{array}$$

Properties ($\forall$l) and ($\exists$r) are the *independent* (from partitions) *interpolant properties*, and ($\forall\exists$) is the *dependent interpolant property*. It is not hard to see that whenever a logic L has a sequent calculus that has uniform sequent interpolation, L has uniform interpolation: for $\forall p\psi$ we take $\forall p(\ \Rightarrow \psi)$ and for $\exists p\varphi$ we take $\exists p(\varphi \Rightarrow\ )$.

Example 29 In classical logic:

sequent S	$\forall p S$	$\exists p S$
if $\vdash S$	$\top$	$\bot$
$p \Rightarrow q$	q	$\neg q$
$q \Rightarrow p$	$\neg q$	q

For intermediate logics, the two notions of interpolation coincide (Ghilardi and Zawadowski (2002)), but for modal logics they do not. For example, $\mathsf{K4}$ and $\mathsf{S4}$ have interpolation but not uniform interpolation (Bílková 2007; Ghilardi and Zawadowski 1995). The difference between the two notions also shows in the difficulty of proving them. For interpolation, these are often relatively straightforward proofs but proofs of uniform interpolation tend to be hard. The proof by Pitts (1992) for IPC is complicated, as are the semantics proofs of uniform interpolation for modal logics that appeared after that (for instance, Shavrukov (1993) and Visser (1996a, b), for K and GL, respectively). Pitts' proof is proof-theoretic and forms the inspiration for the proof-theoretic proofs of the theorems presented in this exposition.

7.5 *Positive and Negative Results for (Intuitionistic) Modal Logics*

Having defined the strengthened notion of interpolation, namely, uniform interpolation, negative results for several classes of logics can be obtained: for classical and intuitionistic modal logics, for substructural logics, and substructural logics extended with modalities. Positive results stating that some logics in these classes actually do have uniform interpolation will be also discussed. For a nice survey on intuitionistic modal logic see Simpson (1994).

First we extend Corollary 23 to uniform interpolation:

Theorem 30 (Iemhoff (2019b) and Akbar Tabatabai and Jalali (2018)) *Any intermediate logic that has a terminating sequent calculus that is an extension of* G4ip *by focused axioms and (context-sharing) semi-analytic rules has uniform interpolation.*

This again leads to the negative result in Corollary 26. Theorem 30 also has a positive corollary, namely, a proof of Pitts' celebrated proof of the uniform interpolation of IPC:

Corollary 31 IPC *has uniform interpolation.*

For the proof of Corollary 31, we apply Theorem 30 on the terminating sequent calculus G4ip. Note that every rule except for $Lp\to$ is semi-analytic and we investigate this case separately.

We turn to modal logics, and in order to obtain the negative results that we aim for we have to first extend the notion of semi-analytic rules to modal rules. As we will see, these are just the standard rules for K and D in a classical or intuitionistic setting. This is a modest extension, and we do think that these rules can be generalized further, but already for these versions the proofs of uniform interpolation are quite complicated, which is why we restrict ourselves to the modal rules just mentioned.

Definition 32 A rule is *modal semi-analytic* if it has one of the following forms:

$$\frac{\Gamma \Rightarrow \varphi}{\Box\Gamma \Rightarrow \Box\varphi, \Delta}\, R_K \qquad \frac{\Gamma \Rightarrow}{\Box\Gamma \Rightarrow \Delta}\, R_D$$

In the case of a single-conclusion calculus, the Δ is empty in R_K and contains at most one formula in R_D. In the context of modal logics, a rule is *semi-analytic* if it is either right or left semi-analytic or a modal semi-analytic rule. Let us stress that the intuitionistic modal logics that we consider here have only one modality, $\Box$, and not a diamond $\Diamond$, which is why they are denoted with a subscript $\Box$, as in $\mathsf{iK}_\Box$, as is common in the literature.

First, we define the intuitionistic normal modal logic without the diamond operator, $\mathsf{iK}_\Box$, by introducing its sequent calculus, $\mathsf{G4iK}_\Box$: take the calculus $\mathsf{G4ip}$ as in Fig. 8 and add the following two rules:

$$\frac{\Gamma \Rightarrow \varphi}{\Pi, \Box\Gamma \Rightarrow \Box\varphi} \qquad \frac{\Gamma \Rightarrow \varphi \qquad \Pi, \Box\gamma, \psi \Rightarrow \Delta}{\Pi, \Box\Gamma, \Box\varphi \to \psi \Rightarrow \Delta}$$

The logic $\mathsf{iKD}_\Box$ is defined by its sequent calculus $\mathsf{G4iKD}_\Box$ as the extension of $\mathsf{G4iK}_\Box$ with the following rule:

$$\frac{\Gamma, \varphi \Rightarrow}{\Pi, \Box\Gamma, \Box\varphi \Rightarrow \Delta}$$

For classical modal logics, the notion of a semi-analytic (modal) rule can be defined for multi-conclusion calculi, analogously to the definition for single-conclusion calculi given in Definition 17. Due to the lack of space, we will not spell out the details. For the definitions of the logics used in the following results, see Table 1 in Sect. 7.1.

First, consider the following results that form part (I) of our method (Sect. 7.2).

Theorem 33 (Iemhoff (2019b) and Akbar Tabatabai and Jalali (2018)) *A classical modal logic that has a terminating calculus that consists of focused axioms and semi-analytic rules has uniform interpolation.*

From a similar result in Iemhoff (2019b), but then for intuitionistic modal logics, we obtain the following positive results, known from the literature (Ghilardi and Zawadowski 1995; Pattinson 2013; Iemhoff 2019b).

Corollary 34 K, KD, $\mathsf{iK}_\Box$*, and* $\mathsf{iKD}_\Box$ *have uniform interpolation.*

But the theorem also implies negative results. For this, we use the following results from the literature about concrete logics not having uniform interpolation, part (II) of our method (Sect. 7.2).

Theorem 35 $\mathsf{S4}$ *does not have uniform interpolation (Ghilardi and Zawadowski 1995).* $\mathsf{K4}$ *does not have uniform interpolation (Bílková 2006).*

Combining the theorems above leads to the following negative result (part (III) of our method, Sect. 7.2) for classical modal logics.

Corollary 36 $\mathsf{K4}$ *and* $\mathsf{S4}$ *do not have terminating sequent calculi that only consist of focused axioms and semi-analytic rules.*

Proof Theorems 33, 35, and 27. □

The intuitionistic modal logics $\mathsf{iK4}_\Box$ and $\mathsf{iS4}_\Box$ are defined as follows:

$$\mathsf{iK4}_\Box := \mathsf{iK}_\Box + (\Box p \to \Box\Box p) \quad \mathsf{iS4}_\Box := \mathsf{iK4}_\Box + (\Box p \to p).$$

In her PhD thesis, van der Giessen (2022) settled that $\mathsf{iS4}_\Box$ and $\mathsf{iK4}_\Box$ do not have uniform interpolation. Thus, we can also conclude:

Corollary 37 $\mathsf{iK4}_\Box$ *and* $\mathsf{iS4}_\Box$ *do not have terminating sequent calculi that only consist of focused axioms and (context-sharing) semi-analytic rules.*

7.6 Extension to Substructural Logics

In Akbar Tabatabai and Jalali (2025b) and Akbar Tabatabai and Jalali (2018), the authors prove a negative result, using the Craig and uniform interpolation properties, for substructural logics, both classical (extensions of $\mathsf{CFL_e}$) and intuitionistic (extensions of $\mathsf{FL_e}$) and allowing modal semi-analytic rules as well. Due to the lack of space, we do not introduce the substructural logics but simply state the results. For the definitions of the logics, see Galatos et al. (2007).

Theorem 38
- *If* $\mathsf{CFL_e} \subseteq \mathsf{L}$ *has a (terminating) sequent calculus consisting of focused axioms and semi-analytic rules, then* L *has (uniform) interpolation.*
- *If* $\mathsf{FL_e} \subseteq \mathsf{L}$ *has a (terminating) sequent calculus consisting of focused axioms and semi-analytic rules, then* L *has (uniform) interpolation.*

Corollary 39 $\mathsf{CFL_e}$, $\mathsf{FL_e}$, *and their modal extensions with* K *and with* KD *have (uniform) interpolation.*

8 Positive Results

The method presented in Sect. 7 can also be used to prove positive instead of negative results, meaning results stating that a logic actually does have uniform interpolation. In this section, we state some of the results of this form.

The flexibility of the method presented here is illustrated by the fact that it can be used to prove positive results (results stating that a logic has UIP) for many logics as well. We state some of the results of this form. In most cases, it appears that we can even prove a strengthening of uniform interpolation, namely, uniform Lyndon interpolation, in which the interpolant respects the polarity of propositional variables (the definition follows below). It first occurred in Kurahashi (2020), where it was shown that several normal modal logics, including K and KD, have that property. We define uniform Lyndon interpolation for sequent calculi, but from that definition, one can easily read off the corresponding notion for logics.

Given a sequent S, let $V^+(\varphi)$ ($V^-(\varphi)$) denote the atoms that occur positively (negatively) in S (defined as usual, reading $\to$ for $\Rightarrow$, but see Definition 1 in Akbar Tabatabai et al. (2021) for the precise definition). A sequent calculus has *uniform Lyndon interpolation* ULIP if for both $\circ = +$ and $\circ = -$, for all sequents S and all atoms $p \in V^\circ(\varphi)$, there are formulas $\forall^\circ pS$ and $\exists^\circ pS$ in $\mathcal{L}$ such that $V^\star(\forall^\circ pS) \subseteq V^\star(S)$ and $V^\star(\exists^\circ pS) \subseteq V^\star(S)$ for all $\star \in \{+, -\}$, and $p \notin V^+(\forall^\circ pS)$ in case $\circ = +$ and $p \notin V^-(\forall^\circ pS)$ in case $\circ = -$, and such that the following holds, where $\vdash$ stands for derivability in G:

(∀l) $\vdash S^a, \forall^\circ pS \Rightarrow S^s$;
(∃r) $\vdash S^a \Rightarrow \exists^\circ pS$;
(∀∃) If S is derivable, for all partitions (S^r, S^i) of S such that $p \notin V^\circ(S^r)$: $\vdash S^r \cdot (\exists^\circ pS^i \Rightarrow \forall^\circ pS^i)$.

The single-conclusion version can be obtained from it in the same way as for UIP in Sect. 7.4.1.

It seems likely that the positive results given below can be generalized in such a way that one can obtain negative results from them as well. However, at this stage, such generalization has not yet been carried out and we therefore only present the positive results that have already been obtained.

8.1 Non-Normal Modal Logics

Non-normal modal logics are modal logics in which the K-axiom, i.e., the axiom $\Box(\varphi \to \psi) \to (\Box\varphi \to \Box\psi)$, does not hold, but a weaker version that is given by the following E-rule does:

$$\frac{\varphi \leftrightarrow \psi}{\Box\varphi \leftrightarrow \Box\psi}$$

Thus, the *minimal non-normal modal logic*, E, consists of classical propositional logic plus the E-rule above. Over the last decades, non-normal modal logics have emerged in various fields, such as game theory and epistemic and deontic logic (Chellas 1980). Two well-known non-normal modal logics that are investigated in this paper are natural weakenings of the principle $\Box(\varphi \wedge \psi) \leftrightarrow (\Box\varphi \wedge \Box\psi)$ that implies the principle K over E. Namely, the two principles:

$$(M) \quad \Box(\varphi \wedge \psi) \to (\Box\varphi \wedge \Box\psi) \qquad (C) \quad (\Box\varphi \wedge \Box\psi) \to \Box(\varphi \wedge \psi).$$

Another principle of the logic K that is often considered is

$$(N) \quad \Box\top.$$

Because the K-axiom holds in the traditional relational semantics for modal logic, non-normal modal logics require different semantics, of which the most well-known is neighborhood semantics. As we do not need semantics in this paper, we refer the interested reader to the textbook (Pacuit 2017).

The method that we used to prove ULIP for intermediate and normal (intuitionistic) modal logics also applies here to non-normal modal logics:

Theorem 40 (Akbar Tabatabai et al. (2021)) *The non-normal modal logics* E, M, EN, MN, MC *have uniform Lyndon interpolation.*

Some of these results have already been obtained in the literature. That the logics E, M, and MC have UIP has already been established in Pattinson (2013) and Santocanale and Venema (2010), but that they have uniform Lyndon interpolation is, as far as we know, a new insight. In Orlandelli (2021), it is shown that E, M, MC, EN, MN have Craig interpolation. The proof that these logics have UIP is not a mere extension of the proof that they have interpolation but requires a different approach.

The proof is constructive in that the uniform interpolants can be constructed explicitly from the proof.

Theorem 41 (Akbar Tabatabai et al. (2021)) *The non-normal modal logics* EC *and* ECN *do not have Craig interpolation, and thus not uniform (Lyndon) interpolation.*

The logics in the theorem above would thus be excellent candidates of non-normal modal logics to obtain negative results for in the style of Sect. 7. But that we leave for future work.

8.2 Conditional Logics

Conditional logics formalize reasoning about conditional statements, which are statements of the form *if* φ *were the case, then* ψ, denoted as $\varphi \rhd \psi$. This includes formal reasoning about counterfactuals, which are instances of conditional statements in which the antecedent is false. Examples of well-known conditional logics are the logics CE and CK. The logic CE consists of the classical propositional logic plus the following *CE*-rule, which could be seen as the conditional counterpart of the E-rule above (by reading the φ_i as $\top$):

$$\frac{\varphi_0 \leftrightarrow \varphi_1 \qquad \psi_0 \leftrightarrow \psi_1}{(\varphi_1 \rhd \psi_1) \leftrightarrow (\varphi_0 \rhd \psi_0)}$$

The conditional logic CK is the result of adding to classical propositional logic the following *CK*-rule, which derives the conditional counterpart of K:

$$\frac{\{\varphi_0 \leftrightarrow \varphi_i\}_{i \in I} \qquad \{\psi_i\}_{i \in I} \to \psi_0}{\{\varphi_i \rhd \psi_i\}_{i \in I} \to (\varphi_0 \rhd \psi_0)}$$

where I is a possibly empty finite set.

$$\mathsf{CEN} = \mathsf{CE} + (CN) \qquad \mathsf{CM} = \mathsf{CE} + (CM)$$

$$\mathsf{CMN} = \mathsf{CM} + (CN) \qquad \mathsf{CMC} = \mathsf{CM} + (CC)$$

$$\mathsf{CK} = \mathsf{CMC} + (CN) \qquad \mathsf{CEC} = \mathsf{CE} + (CC)$$

$$\mathsf{CECN} = \mathsf{CEC} + (CN) \qquad \mathsf{CKID} = \mathsf{CK} + (ID)$$

$$\mathsf{CKCEM} = \mathsf{CK} + (CEM) \qquad \mathsf{CKCEMID} = \mathsf{CKCEM} + (ID)$$

Fig. 1 Some conditional logics

Other conditional logics can be obtained by adding the following conditional axioms to CE, see Fig. 1 for a list of the logics. Note how CX is the conditional counterpart of the non-normal principle X.

$$(\varphi \rhd \psi \wedge \theta) \to (\varphi \rhd \psi) \wedge (\varphi \rhd \theta) \quad (CM) \qquad (\varphi \rhd \psi) \wedge (\varphi \rhd \theta) \to (\varphi \rhd \psi \wedge \theta) \quad (CC)$$

$$\varphi \rhd \top \ (CN) \qquad (\varphi \rhd \psi) \vee (\varphi \rhd \neg\psi) \ (CEM) \qquad \varphi \rhd \varphi \ (ID)$$

Conditional logics have played an important role in philosophy ever since their introduction by Lewis (1973). Another area where they appear is artificial intelligence, in the setting of nonmonotonic reasoning (Friedman and Halpern 1994). The first formal semantics and axiomatization of conditionals appeared in Stalnaker (1968) and in the years after that, the logics were further investigated in Chellas (1980), Lewis (1973), and Cross and Nute (2002).

By using the same technique as for Theorem 40 we can prove the counterpart of that theorem for conditional logics. Also in this case the uniform interpolants can be constructed explicitly from the proof.

Theorem 42 (Akbar Tabatabai et al. (2022)) *The conditional logics* CE, CM, CEN, CMN, CMC, CK, CKID *have uniform Lyndon interpolation.*

Theorem 43 (Akbar Tabatabai et al. (2022)) *The conditional logics* CKCEM *and* CKCEMID *have UIP but not ULIP.*

Theorem 44 (Akbar Tabatabai et al. (2022)) *The conditional logics* CEC *and* CECN *do not have Craig interpolation.*

As in the corresponding case for non-normal modal logics, the last theorem suggests that by generalizing Theorem 42 one can obtain negative results for CEC and CECN. But, again as in the case of non-normal logics, we leave that for future work.

8.3 *Lax Logic*

Finally, we present a positive result for a single modal logic that occurs in several areas of logic. This logic, Lax logic LL, has a single unary modal operator $\bigcirc$ and is axiomatized by the principles of intuitionistic propositional logic IPC plus the following three modal principles:

$$\varphi \to \bigcirc\varphi \ (\bigcirc R) \quad \bigcirc\bigcirc\varphi \to \bigcirc\varphi \ (\bigcirc M) \quad \bigcirc\varphi \wedge \bigcirc\psi \to \bigcirc(\varphi \wedge \psi) \ (\bigcirc S).$$

In type theory, the modal operator describes a certain type constructor, and there exists a Curry–Howard-like correspondence between Lax Logic and the computational lambda calculus (Benton et al. 1998). In algebraic logic, the operator is a so-called nucleus, and in recent work in that area, embeddings of superintuitionistic logics into extensions of LL have been used to provide new characterizations of subframe superintuitionistic logics (Bezhanishvili et al. 2019), while in Melzer (2020) the method of canonical formulas for Lax Logic has been developed. The origins of the modality can be traced back to Curry (1966) but the logic received its name *Lax Logic* in the PhD thesis of Mendler (1993), where the logic was introduced in the setting of hardware verification. In that context, Lax Logic is used to model reasoning about statements of the form φ *holds under some constraint*, denoted as $\bigcirc\varphi$. Further investigation led to Fairtlough and Mendler (1995) and Fairtlough and Mendler (1997), in which, among other things, a sequent calculus for the logic is provided.

To prove that LL has uniform interpolation, one needs another calculus. Let G4LL be the calculus G4ip plus the rules in Fig. 2.

Theorem 45 (Iemhoff (2024)) G4LL *is a terminating sequent calculus for Lax Logic in which Cut is admissible. Lax Logic has uniform interpolation.*

$$\frac{\Gamma \Rightarrow \varphi}{\Gamma \Rightarrow \bigcirc\varphi}\ R\bigcirc \qquad \frac{\Gamma, \psi \Rightarrow \bigcirc\varphi}{\Gamma, \bigcirc\psi \Rightarrow \bigcirc\varphi}\ L\bigcirc$$

$$\frac{\Gamma \Rightarrow \varphi \qquad \Gamma, \psi \Rightarrow \Delta}{\Gamma, \bigcirc\varphi \to \psi \Rightarrow \Delta}\ R\bigcirc^{\to} \qquad \frac{\Gamma, \chi \Rightarrow \bigcirc\varphi \qquad \Gamma, \bigcirc\chi, \psi \Rightarrow \Delta}{\Gamma, \bigcirc\chi, \bigcirc\varphi \to \psi \Rightarrow \Delta}\ L\bigcirc^{\to}$$

Fig. 2 G4LL is G4ip plus all four rules

9 The Method for Other Properties

We have seen the method described in Sect. 7.2 at work for various classes of logics. Recall that in the most general terms, the method aims to prove, for a given class $\mathcal{S}$ of sequent calculi, and given classes $\mathcal{C}' \subseteq \mathcal{C}$ of logics, and a certain property $\mathcal{P}$ of logics:

(I) Any logic in $\mathcal{C}$ that has a sequent calculus in $\mathcal{S}$ has property $\mathcal{P}$.
(II) No logic in $\mathcal{C}'$ has property $\mathcal{P}$.

So one can conclude by contraposition the negative result:

(III) No logic in $\mathcal{C}'$ has a sequent calculus in $\mathcal{S}$.

In Sect. 7, we have applied the method for the case that $\mathcal{P}$ is interpolation or uniform (Lyndon) interpolation and obtained negative results of the form (III) for intermediate logics and various classes of (intuitionistic) modal logics. In Sect. 8, we presented positive results showing that various non-normal modal and conditional logics have uniform (Lyndon) interpolation. The proofs of these facts are based on the techniques developed for (I), still with $\mathcal{P}$ being uniform (Lyndon) interpolation.

This choice for $\mathcal{P}$ is convenient because uniform interpolation seems to be a rare property so that one can conclude results of the form (III) for broad classes of logics. However, there is no reason to believe that the method only works for interpolation properties. And in fact, it does not, as was shown by Akbar Tabatabai and Jalali (2025a). The class of logics they consider is that of intuitionistic modal logics in the language $\mathcal{L} = \{\wedge, \vee, \to, \top, \bot, \Box, \Diamond\}$. For the property $\mathcal{P}$ we turn to the realm of admissible rules.

Given an intuitionistic modal logic L in $\mathcal{L}$, a rule $\varphi \;/\; \psi$ is *admissible* in L, denoted $\varphi \mathrel{|\!\sim}_{\mathsf{L}} \psi$, if for every $\mathcal{L}$-substitution σ: $\vdash_{\mathsf{L}} \sigma\varphi$ implies $\vdash_{\mathsf{L}} \sigma\psi$. Here $\sigma : \mathcal{F}_{\mathcal{L}} \to \mathcal{F}_{\mathcal{L}}$ is an $\mathcal{L}$-substitution if it commutes with the connectives and modal operators. Clearly, if $\vdash_{\mathsf{L}} \varphi \to \psi$, then $\varphi \mathrel{|\!\sim}_{\mathsf{L}} \psi$. But the converse is not always the case. For example, in IPC and $\mathsf{iK}_{\Box}$ the Kreisel–Putnam rule

$$\frac{\neg\varphi \to \psi \vee \psi'}{(\neg\varphi \to \psi) \vee (\neg\varphi \to \psi')}$$

is admissible, but not derivable. For more on the notion of admissibility, see Iemhoff (2015), Iemhoff (2016), and Rybakov (1997).

Here we consider the notion of admissibility in the setting of sequent calculi and combine it with the notion of feasibility in the following definition. In it we use the notion of a Harrop formula (Troelstra and van Dalen 1988) but then extended to our modal language: the set of *Harrop formulas* is inductively defined as the smallest class of formulas containing the atoms and constants of $\mathcal{L}$, is closed under $\wedge$, $\Box$ and implications $\varphi \to \psi$ in which ψ is a Harrop formula.

Definition 46 Consider a sequent calculus G. We denote with $\vdash^{\pi} S$ that π is a proof in G of S. G has the *feasible Visser–Harrop property*, if there is a polynomial time algorithm f such that for all $\Gamma, \varphi_i, \psi_i, \chi_j$, where Γ consists of Harrop formulas, and any proof π in G:

$$\begin{array}{ll} \text{if} & \vdash^{\pi} \Gamma, \{\varphi_i \to \psi_i\}_{i\in I} \Rightarrow \chi_1 \vee \chi_2, \\ \text{then} & \vdash^{f(\pi)} \Gamma, \{\varphi_i \to \psi_i\}_{i\in I} \Rightarrow \chi_j \text{ for some } j \in \{1,2\} \\ \text{or} & \vdash^{f(\pi)} \Gamma, \{\varphi_i \to \psi_i\}_{i\in I} \Rightarrow \varphi_i \text{ for some } i \in I. \end{array}$$

In case I is empty, it is called the *feasible disjunction property* of G.

Returning to our method, let property $\mathcal{P}$ be the feasible Visser–Harrop property, and let the class of logics $\mathcal{C}$ be the logics in language $\mathcal{L}$ that are equal to or extend CK, which extends IPC by Modus Ponens, Necessitation, and the axioms

$$\Box(\varphi \to \psi) \to (\Box\varphi \to \Box\psi) \qquad \Box(\varphi \to \psi) \to (\Diamond\varphi \to \Diamond\psi).$$

Define the sequent calculus CK as the calculus that extends LJ by the two modal rules that correspond to the axioms above:

$$\frac{\Gamma \Rightarrow p}{\Box\Gamma \Rightarrow \Box p} \qquad \frac{\Gamma, p \Rightarrow q}{\Box\Gamma, \Diamond p \Rightarrow \Diamond q}.$$

Note that the logic and sequent calculus CK defined here are different from the conditional logic CK defined and used in Sect. 8.2. However, this abuse of notation will cause no confusion, as from now on, we will only work with CK as defined here. The class of sequent calculi we are interested in extends CK with possibly a finite number of *constructive* axioms, defined below. We call this class of calculi $\mathcal{CK}$. Note that these calculi are allowed to contain the Cut rule.

Definition 47 We define the following three sets of formulas:

- (*Basic*) $B := p \mid \top \mid \bot \mid B \wedge B \mid B \vee B \mid \Diamond B$, with p an atom
- (*Almost positive*) $P := B \mid P \wedge P \mid P \vee P \mid \Box P \mid \Diamond P \mid B \to P$, with B a basic formula
- (*Constructive*) $C := B \mid C \wedge C \mid \Box C \mid P \to C$, with B a basic and P an almost positive formula.

An axiom is called *constructive* if it is of the form $\Gamma \Rightarrow \varphi$, where φ is a constructive formula.

The following are examples of formulas that are basic, almost positive, and constructive: $p, (p \wedge q), (p \vee q), (p \to q), \neg p, \Box p, \Diamond p$. The formula $p \vee \neg p$ is almost positive but not constructive. The formulas $\neg\neg p$ and $(p \to q) \to r$ are constructive but not almost positive.

Denote the Kripke frame that has only one reflexive node by $\mathcal{K}_r$ and the Kripke frame that has one irreflexive node by $\mathcal{K}_i$. If the class of calculi $\mathcal{CK}$ satisfies a mild technical condition, namely, being *T-free* or *T-full*, then it has the feasible disjunction property.

Definition 48 Let G be a sequent calculus such that $\mathsf{CK} \subseteq \mathsf{G}$ and L be a logic such that $\mathsf{CK} \subseteq \mathsf{L}$. We say G (resp. L) is *T-free* if it is valid in $\mathcal{K}_i$. We say G (resp. L) is *T-full* if it is valid in $\mathcal{K}_r$ and both $T_a = (\Box p \to p)$ and $T_b = (p \to \Diamond p)$ are provable in it (resp. $T_a, T_b \in \mathsf{L}$).

Theorem 49 (Akbar Tabatabai and Jalali (2025a)) *Any calculus* G *in* $\mathcal{CK}$ *that is either T-free or T-full has the feasible disjunction property.*

Recall that a logic L has the *disjunction property (DP)* if $A \vee B \in L$ implies either $A \in L$ or $B \in L$. Define a *Visser rules* as

$$V_n \quad \frac{\Rightarrow \left(\bigwedge_{i=1}^{n} (p_i \to q_i) \to p_{n+1} \vee p_{n+2}\right) \vee r}{\Rightarrow \bigvee_{j=1}^{n+2} \left(\bigwedge_{i=1}^{n} (p_i \to q_i) \to p_j\right) \vee r}$$

for $n \geq 1$. By abuse of notation, if the logic L has DP, we say L admits V_0. We say the logic L *admits all Visser's rules* when it admits all V_n for $n \geq 0$.

Corollary 50 *If* L *is a T-free or T-full logic in which at least one Visser rule is not admissible, then* L *does not have a sequent calculus in* $\mathcal{CK}$.

The paper by Akbar Tabatabai and Jalali (2025a) contains many further results that we cannot do justice here. But we hope that with this selection of results from it, we have illustrated a variant of the general method and shown how also, in this case, strong negative and positive results can be obtained from it.

10 The Proof Systems

10.1 Natural Deduction

10.1.1 Natural Deduction Systems ND and NDi

See Fig. 3.

$$\dfrac{\begin{array}{cc}\mathcal{D} & \mathcal{D}' \\ \varphi & \psi\end{array}}{\varphi \wedge \psi}\, I\wedge \qquad \dfrac{\begin{array}{c}\mathcal{D} \\ \varphi_0 \wedge \varphi_1\end{array}}{\varphi_i}\, E\wedge \text{ (i=0,1)}$$

$$\dfrac{\begin{array}{c}\mathcal{D} \\ \varphi_i\end{array}}{\varphi_0 \vee \varphi_1}\, I\vee \text{ (i=0,1)} \qquad a,b\ \dfrac{\begin{array}{ccc} & [\varphi]^a & [\psi]^b \\ \mathcal{D} & \mathcal{D}_1 & \mathcal{D}_2 \\ \varphi \vee \psi & \chi & \chi\end{array}}{\chi}\, E\vee$$

$$a\ \dfrac{\begin{array}{c}[\varphi]^a \\ \mathcal{D} \\ \psi\end{array}}{\varphi \rightarrow \psi}\, I\!\rightarrow \qquad \dfrac{\begin{array}{cc}\mathcal{D} & \mathcal{D}' \\ \varphi \rightarrow \psi & \varphi\end{array}}{\psi}\, E \rightarrow$$

$$\dfrac{\begin{array}{c}\mathcal{D} \\ \varphi(t)\end{array}}{\exists x \varphi(x)}\, I\exists \qquad a\ \dfrac{\begin{array}{cc} & [\varphi(y)]^a \\ \mathcal{D} & \mathcal{D}' \\ \exists x \varphi(x) & \chi\end{array}}{\chi}\, E\exists$$

$$\dfrac{\begin{array}{c}\mathcal{D} \\ \varphi(y)\end{array}}{\forall x \varphi(x)}\, I\forall \qquad \dfrac{\begin{array}{c}\mathcal{D} \\ \forall x \varphi(x)\end{array}}{\varphi(t)}\, E\forall$$

$$a\ \dfrac{\begin{array}{c}[\neg\varphi]^a \\ \mathcal{D} \\ \bot\end{array}}{\varphi}\, E_c\bot \qquad \dfrac{\begin{array}{c}\mathcal{D} \\ \bot\end{array}}{\varphi}\, E_i\bot$$

Fig. 3 The natural deduction system ND for CQC (resp. NDi for IQC) consists of the above rules (resp. the above rules except for $E_c\bot$). In $E\exists$ and $I\forall$, either $y = x$ or y is not free in χ and not free in φ and not free in any open assumption of $\mathcal{D}$ except in the indicated one, $[\varphi(y)]^a$. The variable y is called an *eigenvariable*. In $E\vee$, $I \rightarrow$ and $E\exists$ the assumptions labeled a and b are closed in a derivation once the conclusion of the inference is reached

10.2 Sequent Calculi

10.2.1 Sequent Calculus G1cp

The rules LW and RW are the *left* and *right weakening* rules, and the rules LC and RC are the *left* and *right contraction* rules.

10.2.2 Sequent Calculus G1ip

See Fig. 5.

Axioms

$$p \Rightarrow p \quad At \ (p \text{ an atom}) \qquad\qquad \bot \Rightarrow \quad L\bot$$

Structural rules

$$\frac{\Gamma \Rightarrow \Delta}{\Gamma, \varphi \Rightarrow \Delta}\ LW \qquad\qquad \frac{\Gamma \Rightarrow \Delta}{\Gamma \Rightarrow \varphi, \Delta}\ RW$$

$$\frac{\Gamma, \varphi, \varphi \Rightarrow \Delta}{\Gamma, \varphi \Rightarrow \Delta}\ LC \qquad\qquad \frac{\Gamma \Rightarrow \varphi, \varphi, \Delta}{\Gamma \Rightarrow \varphi, \Delta}\ RC$$

Logical rules

$$\frac{\Gamma, \varphi, \psi \Rightarrow \Delta}{\Gamma, \varphi \wedge \psi \Rightarrow \Delta}\ L\wedge \qquad\qquad \frac{\Gamma \Rightarrow \varphi, \Delta \qquad \Gamma \Rightarrow \psi, \Delta}{\Gamma \Rightarrow \varphi \wedge \psi, \Delta}\ R\wedge$$

$$\frac{\Gamma, \varphi \Rightarrow \Delta \qquad \Gamma, \psi \Rightarrow \Delta}{\Gamma, \varphi \vee \psi \Rightarrow \Delta}\ L\vee \quad \frac{\Gamma \Rightarrow \varphi, \psi, \Delta}{\Gamma \Rightarrow \varphi \vee \psi, \Delta}\ R\vee$$

$$\frac{\Gamma \Rightarrow \varphi, \Delta \qquad \Gamma, \psi \Rightarrow \Delta}{\Gamma, \varphi \rightarrow \psi \Rightarrow \Delta}\ L\rightarrow \quad \frac{\Gamma, \varphi \Rightarrow \psi, \Delta}{\Gamma \Rightarrow \varphi \rightarrow \psi, \Delta}\ R\rightarrow$$

Fig. 4 The sequent calculus G1cp for CPC

Axioms

$$p \Rightarrow p \quad At \ (p \text{ an atom}) \qquad\qquad \bot \Rightarrow \quad L\bot$$

Structural rules

$$\frac{\Gamma \Rightarrow \Delta}{\Gamma, \varphi \Rightarrow \Delta}\ LW \quad \frac{\Gamma \Rightarrow}{\Gamma \Rightarrow \varphi}\ RW \quad \frac{\Gamma, \varphi, \varphi \Rightarrow \Delta}{\Gamma, \varphi \Rightarrow \Delta}\ LC$$

Logical rules

$$\frac{\Gamma, \varphi, \psi \Rightarrow \Delta}{\Gamma, \varphi \wedge \psi \Rightarrow \Delta}\ L\wedge \qquad\qquad \frac{\Gamma \Rightarrow \varphi \qquad \Gamma \Rightarrow \psi}{\Gamma \Rightarrow \varphi \wedge \psi}\ R\wedge$$

$$\frac{\Gamma, \varphi \Rightarrow \Delta \qquad \Gamma, \psi \Rightarrow \Delta}{\Gamma, \varphi \vee \psi \Rightarrow \Delta}\ L\vee \ (i = 0, 1) \ \frac{\Gamma \Rightarrow \varphi_i}{\Gamma \Rightarrow \varphi_0 \vee \varphi_1}\ R\vee$$

$$\frac{\Gamma \Rightarrow \varphi \qquad \Gamma, \psi \Rightarrow \Delta}{\Gamma, \varphi \rightarrow \psi \Rightarrow \Delta}\ L\rightarrow \quad \frac{\Gamma, \varphi \Rightarrow \psi}{\Gamma \Rightarrow \varphi \rightarrow \psi}\ R\rightarrow$$

Fig. 5 The sequent calculus G1ip for IPC. Δ contains at most one formula

Axioms

$\Gamma, p \Rightarrow p, \Delta$ *At* (p an atom) $\qquad \Gamma, \bot \Rightarrow \Delta$ $L\bot$

Logical rules

$$\frac{\Gamma, \varphi, \psi \Rightarrow \Delta}{\Gamma, \varphi \wedge \psi \Rightarrow \Delta} L\wedge \qquad \frac{\Gamma \Rightarrow \varphi, \Delta \qquad \Gamma \Rightarrow \psi, \Delta}{\Gamma \Rightarrow \varphi \wedge \psi, \Delta} R\wedge$$

$$\frac{\Gamma, \varphi \Rightarrow \Delta \qquad \Gamma, \psi \Rightarrow \Delta}{\Gamma, \varphi \vee \psi \Rightarrow \Delta} L\vee \qquad \frac{\Gamma \Rightarrow \varphi, \psi, \Delta}{\Gamma \Rightarrow \varphi \vee \psi, \Delta} R\vee$$

$$\frac{\Gamma \Rightarrow \varphi, \Delta \qquad \Gamma, \psi \Rightarrow \Delta}{\Gamma, \varphi \to \psi \Rightarrow \Delta} L{\to} \qquad \frac{\Gamma, \varphi \Rightarrow \psi, \Delta}{\Gamma \Rightarrow \varphi \to \psi, \Delta} R{\to}$$

$$\frac{\Gamma, \varphi(y) \Rightarrow \Delta}{\Gamma, \exists x \varphi(x) \Rightarrow \Delta} L\exists \qquad \frac{\Gamma \Rightarrow \varphi(t), \exists x \varphi(x), \Delta}{\Gamma \Rightarrow \exists x \varphi(x), \Delta} R\exists$$

$$\frac{\Gamma, \forall x \varphi(x), \varphi(t) \Rightarrow \Delta}{\Gamma, \forall x \varphi(x) \Rightarrow \Delta} L\forall \qquad \frac{\Gamma \Rightarrow \varphi(y), \Delta}{\Gamma \Rightarrow \forall x \varphi(x), \Delta} R\forall$$

Fig. 6 The sequent calculus G3c for CQC. In $L\exists$ and $R\forall$, y is not free in Γ, Δ, $\varphi$and is called an *eigenvariable*

10.2.3 Sequent Calculus G3c

See Fig. 6.

10.2.4 Sequent Calculus G3i

See Fig. 7.

10.2.5 Sequent Calculus G4ip

See Fig. 8.

Axioms

$\Gamma, p \Rightarrow p \quad At \ (p \text{ an atom})$ $\qquad \Gamma, \bot \Rightarrow \Delta \quad L\bot$

Logical rules

$$\frac{\Gamma, \varphi, \psi \Rightarrow \Delta}{\Gamma, \varphi \wedge \psi \Rightarrow \Delta} L\wedge \qquad \frac{\Gamma \Rightarrow \varphi \qquad \Gamma \Rightarrow \psi}{\Gamma \Rightarrow \varphi \wedge \psi} R\wedge$$

$$\frac{\Gamma, \varphi \Rightarrow \Delta \qquad \Gamma, \psi \Rightarrow \Delta}{\Gamma, \varphi \vee \psi \Rightarrow \Delta} L\vee \qquad \frac{\Gamma \Rightarrow \varphi_i}{\Gamma \Rightarrow \varphi_0 \vee \varphi_1} R\vee \ (\text{i=0,1})$$

$$\frac{\Gamma, \varphi \to \psi \Rightarrow \varphi \qquad \Gamma, \psi \Rightarrow \Delta}{\Gamma, \varphi \to \psi \Rightarrow \Delta} L\to \qquad \frac{\Gamma, \varphi \Rightarrow \psi}{\Gamma \Rightarrow \varphi \to \psi} R\to$$

$$\frac{\Gamma, \varphi(y) \Rightarrow \Delta}{\Gamma, \exists x \varphi(x) \Rightarrow \Delta} L\exists \qquad \frac{\Gamma \Rightarrow \varphi(t)}{\Gamma \Rightarrow \exists x \varphi(x)} R\exists$$

$$\frac{\Gamma, \forall x \varphi(x), \varphi(t) \Rightarrow \Delta}{\Gamma, \forall x \varphi(x) \Rightarrow \Delta} L\forall \qquad \frac{\Gamma \Rightarrow \varphi(y)}{\Gamma \Rightarrow \forall x \varphi(x)} R\forall$$

Fig. 7 The sequent calculus G3i for IQC in which $|\Delta| \leq 1$. In $L\exists$ and $R\forall$, y is not free in Γ, Δ, φ and is called an *eigenvariable*

Axioms

$\Gamma, p \Rightarrow p \quad At \ (p \text{ an atom})$ $\qquad \Gamma, \bot \Rightarrow \Delta \quad L\bot$

Logical rules

$$\frac{\Gamma, \varphi, \psi \Rightarrow \Delta}{\Gamma, \varphi \wedge \psi \Rightarrow \Delta} L\wedge \qquad \frac{\Gamma \Rightarrow \varphi \qquad \Gamma \Rightarrow \psi}{\Gamma \Rightarrow \varphi \wedge \psi} R\wedge$$

$$\frac{\Gamma, \varphi \Rightarrow \Delta \qquad \Gamma, \psi \Rightarrow \Delta}{\Gamma, \varphi \vee \psi \Rightarrow \Delta} L\vee \qquad \frac{\Gamma \Rightarrow \varphi_i}{\Gamma \Rightarrow \varphi_0 \vee \varphi_1} R\vee \ (\text{i=0,1})$$

$$\frac{\Gamma, p, \varphi \Rightarrow \Delta}{\Gamma, p, p \to \varphi \Rightarrow \Delta} Lp\to \ (p \text{ an atom}) \qquad \frac{\Gamma, \varphi \Rightarrow \psi}{\Gamma \Rightarrow \varphi \to \psi} R\to$$

$$\frac{\Gamma, \varphi \to (\psi \to \gamma) \Rightarrow \Delta}{\Gamma, \varphi \wedge \psi \to \gamma \Rightarrow \Delta} L\wedge\to \qquad \frac{\Gamma, \varphi \to \gamma, \psi \to \gamma \Rightarrow \Delta}{\Gamma, \varphi \vee \psi \to \gamma \Rightarrow \Delta} L\vee\to$$

$$\frac{\Gamma, \psi \to \gamma \Rightarrow \varphi \to \psi \qquad \gamma, \Gamma \Rightarrow \Delta}{\Gamma, (\varphi \to \psi) \to \gamma \Rightarrow \Delta} L{\to}{\to}$$

Fig. 8 The sequent calculus G4ip for IQC in which $|\Delta| \leq 1$

Acknowledgements Support by the Netherlands Organization for Scientific Research under grant 639.073.807 as well as by the EU H2020-MSCA-RISE-2020 Project 101007627 is gratefully acknowledged.

References

Akbar Tabatabai A, Iemhoff R, Jalali R (2021) Uniform lyndon interpolation for basic non-normal modal logics. In: Silva A, Wassermann R, de Queiroz R (eds) Logic, language, information, and computation: 27th international workshop, WoLLIC 2021, virtual event, October 5–8, 2021, Lecture Notes in Computer Science, vol 13038. Springer International Publishing, Cham, pp 287–301. https://doi.org/10.1007/978-3-030-88853-4_18

Akbar Tabatabai A, Iemhoff R, Jalali R (2022) Uniform Lyndon interpolation for basic non-normal modal and conditional logics. arXiv:2208.05202

Akbar Tabatabai A, Jalali R (2018) Universal proof theory: semi-analytic rules and uniform interpolation. arXiv:1808.06258

Akbar Tabatabai A, Jalali R (2025) Universal proof theory: feasible admissibility in intuitionistic modal logics. Ann Pure Appl Log 176(2):40. Paper no. 103526. https://doi.org/10.1016/j.apal.2024.103526

Akbar Tabatabai A, Jalali R (2025) Universal proof theory: semi-analytic rules and Craig interpolation. Ann Pure Appl Log 176(1):26. Paper no. 103509. https://doi.org/10.1016/j.apal.2024.103509

Arana A, Detlefsen M (2011) Purity of methods. Philos Impr 11(2):1–20. https://quod.lib.umich.edu/cgi/p/pod/dod-idx/purity-of-methods.pdf?c=phimp;idno=3521354.0011.002;format=pdf

Benton PN, Bierman GM, de Paiva VCV (1998) Computational types from a logical perspective. J Funct Program 8(2):177–193. https://doi.org/10.1017/S0956796898002998

Bezhanishvili G, Bezhanishvili N, Ilin J (2019) Subframization and stabilization for superintuitionistic logics. J Log Comput 29(1):1–35. https://doi.org/10.1093/logcom/exy035

Bílková M (2006) Interpolation in modal logic. PhD thesis, Charles University Prague, Czech Republic

Bílková M (2007) Uniform interpolation and propositional quantifiers in modal logics. Stud Log 85(1):1–31. https://doi.org/10.1007/s11225-007-9021-5

Chellas BF (1980) Modal logic: an introduction. Cambridge University Press. https://doi.org/10.1017/CBO9780511621192

Cross C, Nute D (2002) Conditional logic. In: Gabbay DM, Guenthner F (eds) Handbook of philosophical logic, 2nd edn. Springer, Dordrecht, pp 1–98. https://doi.org/10.1007/978-94-017-0456-4_1. Revision of Nute's chapter from the first edition with a new 40+pp section on conditionals and the ramsey test written by cross

Curry HB (1966) A theory of formal deducibility. Notre dame mathematical lectures, vol 6, 2nd edn. University of Notre Dame Press, South Bend, Indiana

Dershowitz N, Manna Z (1979) Proving termination with multiset orderings. Commun ACM 22(8):465–476. https://doi.org/10.1145/359138.359142

Dyckhoff R (1992) Contraction-free sequent calculi for intuitionistic logic. J Symb Log 57(3):795–807. https://doi.org/10.2307/2275431

Fairtlough M, Mendler M (1995) An intuitionistic modal logic with applications to the formal verification of hardware. In: Pacholski L, Tiuryn J (eds) Computer science logic (CSL 1994). Lecture notes in computer science, vol 933. Springer, Berlin, pp 354–368. https://doi.org/10.1007/BFb0022268

Fairtlough M, Mendler M (1997) Propositional lax logic. Inf Comput 137(1):1–33. https://doi.org/10.1006/inco.1997.2627

Friedman N, Halpern J (1994) On the complexity of conditional logics. In: Doyle J, Sandewall E, Torasso P (eds) Principles of knowledge representation and reasoning: proceedings of the fourth international conference (KR '94). The Morgan Kaufmann series in representation and reasoning. Morgan Kaufmann, pp 202–213. https://www.sciencedirect.com/science/article/pii/B9781483214528501159

Galatos N, Jipsen P, Kowalski T, Ono H (2007) Residuated lattices. An algebraic glimpse at substructural logics. Studies in logic and the foundations of mathematics, vol 151. Elsevier, Amsterdam. https://www.sciencedirect.com/bookseries/studies-in-logic-and-the-foundations-of-mathematics/vol/151/

Gentzen,G (1935) Untersuchungen über das logische Schließen. I Math Z 39(1):176–210. https://doi.org/10.1007/BF01201353

Ghilardi S, Zawadowski M (1995) Undefinability of propositional quantifiers in the modal system $S4$. Stud Log 55(2):259–271. https://doi.org/10.1007/BF01061237

Ghilardi S, Zawadowski M (2002) Sheaves, games and model completions: a categorical approach to nonclassical propositional logics. Trends in logic—studia logica library, vol 14. Springer. https://doi.org/10.1007/978-94-015-9936-8

van der Giessen I (2022) Uniform interpolation and admissible rules: proof-theoretic investigations into (intuitionistic) modal logics. PhD thesis, Utrecht University

Hudelmaier J (1988) A prolog program for intuitionistic logic. Seminar für natüiirlich-sprachliche Systeme SNS-Bericht, vol 88–28, Universität Tübingen, Tübingen

Hudelmaier, J (1992) Bounds for cut elimination in intuitionistic propositional logic. Arch Math Log 31(5):331–353. https://doi.org/10.1007/BF01627506

Hudelmaier J (1993) An $O(n \log n)$-space decision procedure for intuitionistic propositional logic. J Log Comput 3(1):63–75. https://doi.org/10.1093/logcom/3.1.63

Iemhoff R (2015) On Rules. J Philos Logic 44(6):697–711. https://doi.org/10.1007/s10992-015-9354-x

Iemhoff R (2016) Consequence relations and admissible rules. J Philos Logic 45(3):327–348. https://doi.org/10.1007/s10992-015-9380-8

Iemhoff R (2017) Remarks on simple proofs. In: Kossak R, Ording P (eds) Simplicity: ideals of practice in mathematics and the arts. Mathematics, culture, and the arts. Springer, Cham, pp 145–151

Iemhoff R (2019) Uniform interpolation and sequent calculi in modal logic. Arch Math Log 58(1–2):155–181. https://doi.org/10.1007/s00153-018-0629-0

Iemhoff R (2019) Uniform interpolation and the existence of sequent calculi. Ann Pure Appl Log 170(11):102711, 33. https://doi.org/10.1016/j.apal.2019.05.008

Iemhoff R (2024) Proof theory for Lax Logic. In: Dick de Jongh on intuitionistic and provability logics. Outstanding contributions to logic, vol 28. Springer, Cham, pp 203–229. https://doi.org/10.1007/978-3-031-47921-2_8

Kahle R, Pulcini G (2018) Towards an operational view of purity. In: Arazim P, Pávička T (eds) The logica yearbook 2017. College Publications, London, pp 125–138

Kurahashi T (2020) Uniform Lyndon interpolation property in propositional modal logics. Arch Math Log 59(5–6):659–678. https://doi.org/10.1007/s00153-020-00713-y

Lewis D (1973) Counterfactuals found lang 13(1):145–151

Maksimova LL (1977) Craig's theorem in superintuitionistic logics and amalgamable varieties of pseudo-Boolean algebras. Algebr Log 16(6):643–681, 741. https://www.mathnet.ru/eng/al1584

Maksimova LL (1991)Amalgamation and interpolation in normal modal logic. Stud Log 50(3–4):457–471. https://doi.org/10.1007/BF00370682

Melzer M (2020) Canonical formulas for the lax logic. MSc thesis, University of Amsterdam

Mendler M (1993) A modal logic for handling behavioural constraints in formal hardware verification. PhD thesis, University of Edingburgh

Negri S (2011) Proof theory for modal logic. Philos Compass 6(8):523–538. https://doi.org/10.1111/j.1747-9991.2011.00418.x

Negri S, von Plato J (2001) Structural proof theory. Cambridge University Press, Cambridge. https://doi.org/10.1017/CBO9780511527340

Orlandelli E (2021) Sequent calculi and interpolation for non-normal modal and deontic logics. Log Log Philos 30(1):139–183. https://doi.org/10.12775/llp.2020.018

Pacuit E (2017) Neighborhood semantics for modal logic. Short textbooks in logic. Springer, Cham. https://doi.org/10.1007/978-3-319-67149-9

Pattinson D (2013) The logic of exact covers: completeness and uniform interpolation. In: 2013 28th annual ACM/IEEE symposium on logic in computer science, New Orleans, June 25–28, 2013. IEEE Computer Society, Los Alamitos, CA, pp 418–427. https://doi.org/10.1109/LICS.2013.48

Pitts A (1992) On an interpretation of second-order quantification in first-order intuitionistic propositional logic. J Symb Log 57(1):33–52. https://doi.org/10.2307/2275175

Prawitz D (1965) Natural deduction. A proof-theoretical study. Acta universitatis stockholmiensis. Stockholm studies in philosophy, vol 3. Almqvist & Wiksell, Stockholm

Prawitz D (1971) Ideas and results in proof theory. In: Fenstad JE (ed) Proceedings of the second Scandinavian logic symposium (University of Oslo, Oslo, 1970). Studies in logic foundations of mathematics, vol 63. Elsevier, Amsterdam, pp 235–307. https://doi.org/10.1016/S0049-237X(08)70849-8

Prawitz D (1974) On the idea of a general proof theory. Synthese 27:63–77. https://doi.org/10.1007/BF00660889

Rybakov V (1997) Admissibility of logical inference rules. Studies in logic and the foundations of mathematics, vol 136. Elsevier, Amsterdam

Santocanale L, Venema Y (2010) Uniform interpolation for monotone modal logic. In: Proceedings of the 8th conference (AiML 2010), Moscow, Russia, August 24–27, 2010. Advances in modal logic, vol 8. College Publications, London, pp 350–370

Shavrukov VY (1993)Subalgebras of diagonalizable algebras of theories containing arithmetic. Diss Math (Rozprawy Matematyczne) 323:1–82. http://matwbn.icm.edu.pl/ksiazki/rm/rm323/rm32301.pdf

Simpson AK (1994) The proof theory and semantics of intuitionistic modal logic. PhD thesis, University of Edingburgh

Sørensen MH, Urzyczyn P (2006) Lectures on the Curry-Howard isomorphism. Studies in logic and the foundations of mathematics, vol 149, Elsevier, Amsterdam

Stalnaker R (1968) A theory of conditionals. In: Rescher N (ed) Studies in logical theory: American philosophical quarterly supplementary monograph series. Blackwell, Oxford, pp 98–112

Troelstra AS, van Dalen D (1988) Constructivism in mathematics: an introduction. Volume I. Studies in logic and the foundations of mathematics, vol 121. Elsevier, Amsterdam

Troelstra AS, Schwichtenberg H (2000) Basic proof theory, Cambridge tracts in theoretical computer science vol 43, 2nd edn. Cambridge University Press, Cambridge. https://doi.org/10.14771017/CBO9781139168717

Visser A (1996a) Bisimulations, model descriptions and propositional quantifiers. Logic group preprint series, vol 161. Utrecht University

Visser A (1996b) Uniform interpolation and layered bisimulation. In: Hájek P (ed) Gödel '96 (Brno, 1996). Logical foundations of mathematics, computer science and computer science and physics-kurt Gödel's Legacy. Lecture notes in logic, vol 6. Springer, Berlin, pp 139–164

Pointfree Topology and Constructive Mathematics

Graham Manuell

Abstract Constructive mathematics replaces abstract existence claims with explicit constructions and allows for the broader application of mathematical results. Unfortunately, many standard theorems of analysis and topology fail in their usual form. Pointfree topology, which studies spaces via their lattices of opens, resolves these difficulties and allows us to recover constructive versions of classical results together with new consequences. We hope these notes will provide an accessible introduction to the subject.

1 Introduction

Constructive mathematics has many advantages over its classical counterpart: vague promises of existence are replaced with concrete examples and theorems can be unwound to obtain algorithms. Even better, constructive results can be interpreted in any topos, giving many different results for the price of one.

Unfortunately, there is also a cost. We are told that we must give up many familiar results: not only Tychonoff's theorem and the Hahn–Banach theorem, but the intermediate value theorem, the extreme value theorem and the fundamental theorem of algebra. We can sympathise with David Hilbert who asserted that asking a mathematician to avoid excluded middle is akin to 'proscribing the telescope to the astronomer or to the boxer the use of his fists'. However, despite what you might have heard, all the results listed above are still true constructively! They simply need to be formulated in the appropriate way.

The cost is still nonzero, since most mathematicians lack familiarity with constructive techniques, and it can take a little time to understand what is permitted and what is not. Nonetheless, the obstacles are not insurmountable. I hope that these notes will provide some assistance, as well as some motivation for further study.

G. Manuell (✉)
Mathematics Division, Department of Mathematical Sciences, Stellenbosch University, Private Bag X1, Matieland, 7602 Stellenbosch, Western Cape, South Africa
e-mail: graham@manuell.me

M. M. Clementino et al. (eds.), *Topology, Algebra and Categories in Logic*,
Coimbra Mathematical Texts 5, https://doi.org/10.1007/978-3-032-13760-9_3

The solution to the problems above lies in a reformulation of topology. This is perhaps the area where the apparent problems of constructive mathematics are most acute. It is true that point-set topology is very badly behaved constructively, but the problems disappear if we instead use the *pointfree* approach to topology. Pointfree topology studies spaces through their lattices of opens without reference to a predefined underlying set of points and can also be understood from a logical perspective. While such an algebraic/logical approach has some advantages even in classical mathematics, it really shines in its smooth generalisation to the constructive setting.

But as we will see, pointfree topology does more than just allow for the constructive development of topology; the topological perspective also sheds light on aspects of constructive mathematics that might at first appear to have little to do with topology. From this point of view even a version of the axiom of choice can be understood as a constructive theorem.

This chapter will provide a gentle introduction to the main aspects of constructive topology and some of its applications.

2 Motivation and Basic Concepts

2.1 Topology

In order to reformulate topology it is worthwhile to first understand what topology is actually studying. Let us forget any preconceived notions we might have for a moment and imagine a simple situation.

Suppose we have some physical object that we want to measure. For instance, we might want to determine the length ℓ of a steel rod. In the real world we cannot measure the length of the rod to infinite precision. Perhaps we use a ruler to find that it has a length of 10.1 ± 0.3 cm. If we wish to be more precise we could use a ruler with finer gradations and measure the rod to be 10.0 ± 0.1 cm long. We might suspect that the rod is exactly 10 cm in length, but note that only being able to measure to finite precision we will never be able to be certain that this is true.

On the other hand, suppose we are told that the rod is strictly less than 10 cm. This is something we will be able to verify for ourselves using arbitrarily, but not infinitely, precise physical measurements. If $\ell < 10$ cm then there is some $\varepsilon > 0$ such that $\ell + \varepsilon < 10$ cm and so if our measurement error is less than ε we can check that this is indeed the case. We do not know ε beforehand, but if we keep on taking more and more precise measurements, at some point we will be able to tell.

Do note, however, that if it is not true that $\ell < 10$ cm, then we might continue to make finer and finer measurements without ever being certain we have been misled (since perhaps $\ell = 10$ cm exactly). We say that the proposition $\ell < 10$ cm is *verifiable*, but the proposition $\ell \geq 10$ cm is not.

There is a space $\mathbb{R}^+$ of all possible lengths for the rod, and the verifiable properties correspond precisely to the open subsets of $\mathbb{R}^+$. We will now use verifiability as our *definition* of open sets, even in situations that are less familiar. For further discussion of this approach see Chap. 2 of Vickers (1989).

The power of this approach is that it immediately motivates the definition of a topological space. Notice that verifiable properties satisfy the following conditions:

(i) The property that is always true is verifiable.
(ii) If U and V are verifiable properties, then so is $U \wedge V$.
(iii) If $\mathcal{U}$ is a set of verifiable properties, then $\bigvee \mathcal{U}$ is verifiable.

The first condition holds since if the property is always true, then there is nothing to check. For the second, we can simply check U and V in turn in order to verify their conjunction. Observe there is no apparent way that we could verify an *infinite* conjunction of such properties, since this process would take infinitely long. On the other hand, even infinite disjunctions of verifiable properties are verifiable, since if $\bigvee \mathcal{U}$ holds, so must some $U \in \mathcal{U}$ and this U can then simply be verified in finite time. Note that the empty disjunction $\bot$ is vacuously verifiable, since we only need to be able to verify propositions when they are true.

We obtain the definition of a topological space by starting with a set X of things that might satisfy some properties and identifying verifiable properties on X with the subsets of things that satisfy them. Furthermore, if $f\colon X \to Y$ is a physically realisable function then for an open U of Y we can verify that $x \in f^{-1}(U)$ by checking that $f(x) \in U$. Thus, $f^{-1}(U)$ should be an open of X. This motivates the definition of a continuous map.

2.1.1 Frames and Geometric Logic

Instead of identifying verifiable properties with sets of points we can also treat them more abstractly. The logic of verifiable propositions is called *geometric logic*. A formula in geometric logic can be built up from 'basic propositions' using finite conjunctions or arbitrary disjunctions. There is no implication connective, since implications are not generally verifiable. However, we can describe top-level implications between formulae by *sequents* $\varphi \vdash \psi$, which are interpreted as meaning that ψ holds whenever φ does.

A *geometric theory* is specified by giving by a set of basic propositions (such as $\ell < 10\,\text{cm}$ in the example above) and a set of axioms (such as $\ell < 10\,\text{cm} \vdash \ell < 11\,\text{cm}$). From these we can derive additional sequents using intuitive logical rules. (For example, if we have $\varphi_\alpha \vdash \psi$ for all $\alpha \in I$, then we can conclude that $\bigvee_{\alpha\in I} \varphi_\alpha \vdash \psi$. For more details, see Sect. 4 of Pitts (1991), but ignore the rules concerning implication and negation, and change the finite disjunctions to infinite ones.)

Just as classical propositional logic can be studied algebraically using Boolean algebras, geometric logic can be studied using frames.

Definition 1 A *frame* is a poset with finite meets[1] (or infima) and arbitrary joins[2] (or suprema) satisfying the distributivity condition $a \wedge \bigvee_{\alpha \in I} b_\alpha = \bigvee_{\alpha \in I} a \wedge b_\alpha$, where $\wedge$ denotes a binary meet and $\bigvee$ denotes an arbitrary join.

Elements of a frame correspond to (classes of provably equivalent) geometric formulae. Meets and joins in the frame correspond to the conjunctions and disjunctions in geometric logic and the distributivity condition ensures that the conjunctions and disjunctions interact as we expect them to. A sequent is simply interpreted using the $\leq$ relation. We say that frames are the *Lindenbaum–Tarski* algebras for geometric logic. We will explore this relationship further in Sect. 3.1.

Note that the lattice of open sets of any topological space forms a frame. Indeed, a topology on X is nothing but a *subframe* of the powerset of X (i.e. a subset closed under finite meets and arbitrary joins). Pointfree topology is the study of topology using frames directly, eschewing the underlying sets of points.

However, while we do not start with a predefined set of points, it is still possible to define points of a frame. From a logical perspective these correspond to *models* of the geometric theory represented by the frame. Here a model is a consistent assignment of truth values to each verifiable property (saying which opens 'contain' the point). Explicitly, we have the following conditions:

- The top element 1 must hold in a model.
- If a holds and $a \leq b$ then b must hold too.
- If a and b both hold in the model, then so does $a \wedge b$.
- If $\bigvee_{\alpha \in I} a_\alpha$ holds, then a_α holds for some $\alpha \in I$.

In order-theoretic terms this defines what is known as a *completely prime filter* on the frame, but we will simply call these *points*.

If $\mathcal{O}X$ is the frame of opens of a topological space X, then every point of X gives a point of $\mathcal{O}X$, but they might not be in one-to-one correspondence. For example, if X contains two points that lie in precisely the same open sets, they will both correspond to the same point of $\mathcal{O}X$. However, in most cases of interest the points of $\mathcal{O}X$ can indeed be used to recover X completely. If this is the case, we say that the space X is *sober*. Sobriety is a rather weak condition; classically, every Hausdorff space and every finite T_0 space is sober.

Frames can be understood as algebraic structures with operations 1, $\wedge$ and a proper class of join operations of various arities. A *frame homomorphism* is simply a function that preserves these operations.

A continuous map of topological spaces induces a frame homomorphism between their frames of open sets *in the opposite direction* by taking preimages. This motivates the following definition.

---

[1] It suffices to ask for nullary meets (i.e. the top element 1) and binary meets.

[2] Any poset admitting arbitrary joins also has arbitrary meets by a standard argument and is hence a complete lattice. However, the infinite meets are merely incidental to the structure of a frame. They do not have a logical interpretation and will not be preserved by frame homomorphisms.

Definition 2 The category **Loc** of *locales* is the opposite of the category **Frm** of frames and frame homomorphisms. A locale is simply a frame, but viewed from a topological, instead of an algebraic, perspective.

To avoid confusion we will make a clear notational distinction between frames and locales: if X is a locale we write OX for its corresponding 'frame of opens' and if $f: X \to Y$ is a morphism of locales we write $f^*: OY \to OX$ for the corresponding frame homomorphism.

Taking open sets gives a functor from **Top** to **Loc**. This has a right adjoint, which sends a locale X to its set of points equipped with the topology consisting of the open sets of the form $O_a = \{P \mid a \in P\}$, and which sends a locale morphism $f: X \to Y$ to a map that acts on completely prime filters by taking preimages under f^*. This is an idempotent adjunction which restricts to an equivalence between sober topological spaces and *spatial* locales. A locale is spatial if its opens are completely determined by the points they contain.

Not all locales are spatial. In classical mathematics non-spatial locales might appear to be rather unusual, but constructively locales that cannot be proved to be spatial are ubiquitous.

2.2 Constructive Mathematics

The geometric logic introduced above is useful for reasoning about properties of points within a topological space or locale. In contrast, the classical logic that you are familiar with is more far-reaching, and can be used to reason about all of classical mathematics (which itself involves dealing with many topological spaces). *Intuitionistic logic* is what replaces classical logic in this role when doing constructive mathematics.

As we did with our discussion of topology above, before we take a look at intuitionistic logic, let us take a step back and consider what logic is about. This will be helpful since it can be difficult for those who are only used to classical mathematics to understand what it could mean to weaken some of the assumptions they feel obviously hold. The assumption that constructive mathematics weakens is the *law of the excluded middle*: $\varphi \vee \neg\varphi$, which is deeply ingrained in the way most mathematicians think to the extent that it can be very difficult for them to tell when they are using it.

One way to understand classical logic is as the logic describing Platonic truth. If mathematical objects truly exist, then how could they not have definite properties? Surely a proposition either holds of such an object or does not (though whether we can prove which is the case is another matter entirely). This is fine as far as it goes, but it is not the only thing we might want to use logic to do.

Non-classical logics can be understood as describing something different. We have already seen an example of this: geometric logic describes *verifiable truths*. There are also non-classical logics for knowledge, possibilities, fuzzy concepts and resources (see Bell et al. (2001) or Chap. 2 of Gabbay and Rodrigues (2016) for more details).

Constructive mathematics uses *intuitionistic logic*, which is often understood as describing things *we already know to be true*. From this perspective $\varphi \vee \neg\varphi$ would mean that either we know that φ holds, or we know that assuming φ leads to a contradiction (since $\neg\varphi$ means $\varphi \Rightarrow \bot$). Now it is not hard to see why it would not be reasonable to accept this as always being true. Do note that constructive mathematicians do *not* assert that excluded middle is false either, since that would mean that if we ever did find out that φ was true or φ was false we would have a contradiction. For further details on constructive mathematics see Bauer (2012, 2017) and van Dalen (2001).

We remark that the law of non-contradiction $\neg(\varphi \wedge \neg\varphi)$ is still valid constructively, since it just says that from φ and $\varphi \Rightarrow \bot$ we can conclude $\bot$. From a classical perspective one might feel that this is equivalent to excluded middle, but to derive this equivalence would require using either excluded middle itself or some other equivalent axiom.

One logical axiom that is equivalent to excluded middle is double negation elimination: $\neg\neg\varphi \implies \varphi$. This means that it is difficult to get rid of negations in constructive mathematics once they have been introduced and so it is important to define basic concepts without using negations if possible—they should be defined by *positive formulae*.

For example, asking that a set S is nonempty ($S \neq \emptyset$) is a rather weak condition constructively and is often not terribly useful. We usually instead ask that the set is *inhabited*—that is, $\exists x \in S$. Now we can use the element x to more easily prove further results. (On the other hand, if a set is not inhabited then it *is* empty.)

It is often said that proof by contradiction is not permitted in constructive mathematics, but this claim should be interpreted carefully. We can, of course, prove $\neg\varphi$ by assuming φ and deriving a contradiction, since this is the definition of $\neg\varphi$. On the other hand, assuming $\neg\varphi$ and deriving a contradiction only proves $\neg\neg\varphi$ and this is not generally enough to conclude φ.

The *axiom of choice* is another famously nonconstructive axiom due to its claim that choice functions exist without giving any indication of how to actually construct one. It can be shown that the axiom of choice implies excluded middle (for example, see Bauer (2017, Theorem 1.3)), though it is much stronger. On the other hand, countable choice and dependent choice can be assumed without forcing excluded middle to hold, but we do not consider them to be entirely constructive either. A version of *finite* choice can be proved constructively by induction.

2.2.1 Some Models of Constructive Logic

Another good way to understand intuitionistic logic is through its *models*. This can be formalised as interpreting the logic in different toposes, but we can simply understand them as different logical universes in which to do mathematics. In order to make sense of a statement in intuitionistic logic it is often helpful to consider what it might mean in one or more of these different models.

One rather intuitive model corresponds to computability by Turing machines. This is related to the Brouwer–Heyting–Kolmogorov interpretation, which asserts that a logical formula is true if it has a *realiser*, which can be defined recursively based on its syntactic form. For example, a realiser for a disjunction is simply a choice of one of the disjuncts together with a realiser for that. A realiser for an implication $\varphi \implies \psi$ is given by a procedure that converts a realiser for φ into a realiser for ψ. (See Bauer (2012) for more details.) Here we take these realisers be given by Turing machines. If $\varphi(n)$ means that the nth Turing machine halts then the formula $\forall n \in \mathbb{N}.\ \varphi(n) \vee \neg\varphi(n)$ would be interpreted as saying that the halting problem is decidable, and thus it is false in this model.[3]

Since intuitionistic logic is more general than classical logic our usual understanding of classical logic also gives a model of intuitionistic logic. It can be helpful to bear in mind that constructive theorems cannot contradict classical results. On the other hand, this also exemplifies the fact that one model might not necessarily be enough for a complete understanding.

Our final class of models arises from topology and corresponds to the concept of *local truth*, where the truth value of propositions is allowed to vary from place to place over some space. In this model truth values correspond to opens in some fixed locale X. The proposition $\top$ means true *everywhere*, the proposition $\bot$ means true *nowhere* and conjunctions and disjunctions correspond to meets and joins.

Implication is more subtle. Since operation $a \wedge (-)$ on the frame $\mathcal{O}X$ preserves joins, it has a right adjoint $a \Rightarrow (-)$ making $\mathcal{O}X$ into a Heyting algebra. We use this Heyting implication to interpret logical implication. If X is spatial then $U \Rightarrow V$ is given by the *interior* of the set $\{x \in X \mid x \in U \implies x \in V\}$, which is the closest approximation to what we might naively expect while still giving an open set. Note that $U \vee \neg U$ is only equal to $\top$ if U has a *complement*—that is, if U is clopen. Thus, excluded middle seldom holds in such models.

Remark 3 There are some definite similarities between intuitionistic logic as it is used to describe local truth and geometric logic. However, they are not the same. Most evidently intuitionistic logic involves an implication connective, but geometric logic has none. This is related to the fact that the Heyting implication in a frame is not preserved by frame homomorphisms. We will see that topology does have its own way to deal with negation, however. There the complement of an open set is not another open set, but something new: a *closed* set (corresponding to a *refutable* property). Higher-order intuitionistic logic also permits existential and universal quantification over any sets, while the status of quantifiers in geometric logic is more subtle. In summary, intuitionistic logic has greater expressive power at the expense of having a weaker link to topological notions.

---

[3] Note that for any *given* proposition ψ we still have $\neg\neg(\psi \vee \neg\psi)$ as we argued above. It is the universally quantified formula involving many different propositions at once that is false.

2.2.2 Truth Values

Classically there are precisely two truth values: $\top$ (true) and $\bot$ (false). Constructively, the situation is more subtle. We write Ω for the set of truth values. Certainly, we still have $\bot, \top \in \Omega$. If φ is a proposition with no free variables, we write $[\![\varphi]\!]$ for the truth value of φ. Then Ω is ordered by $[\![\varphi]\!] \leq [\![\psi]\!]$ if and only if $\varphi \implies \psi$ and the logical connectives give it the structure of a Heyting algebra. In particular, binary meet is given by logical conjunction and binary join is given by logical disjunction.

Note that since $\varphi \iff (\varphi \iff \top)$, we have $p = [\![p = \top]\!]$ for any $p \in \Omega$. So truth values are completely determined by 'whether they equal $\top$'. This can be used to show that there is still a sense in which Ω does not have more than two elements.

Lemma 4 *For $p \in \Omega$, if $p \neq \top$ then $p = \bot$.*

Proof Suppose $[\![\varphi]\!] \neq \top$. This means $\neg([\![\varphi]\!] = \top)$. Hence, $\neg\varphi$ holds and so $[\![\varphi]\!] \leq \bot$. Thus, $[\![\varphi]\!] = \bot$. □

On the other hand, if $p \neq \bot$ we can only show $\neg\neg(p = \top)$. We cannot prove that Ω only contains $\top$ and $\bot$.

Lemma 5 *The equality $\Omega = \{\bot, \top\}$ is equivalent to excluded middle.*

Proof Simply note that $[\![\varphi]\!] \in \{\bot, \top\} \iff \neg\varphi \vee \varphi$. □

Remark 6 A proof that a statement implies some nonconstructive principle is called a *Brouwerian counterexample*. While assuming such a statement holds might not lead to a contradiction, it cannot have a constructive proof and will fail in any model of mathematics where the nonconstructive principle in question does not hold.

A proposition φ for which $\varphi \vee \neg\varphi$ is true is said to be *decidable*. Constructively there is still some use for the set $2 = \{\bot, \top\}$ of decidable truth values.

The defining property of Ω is that it classifies subsets. If X is a set and $\chi : X \to \Omega$, then $\{x \in X \mid \chi(x) = \top\}$ defines a subset of X. Conversely, if $S \subseteq X$ then we can define the characteristic map $\chi_S : X \to \Omega$ by $x \mapsto [\![x \in S]\!]$. These correspondences are inverses and so Ω^X may be identified with the *powerset* of X.

Remark 7 In particular, note that functions into Ω are completely determined by which elements they send to $\top$. Furthermore, since the above correspondences are monotone, we have $f \leq g$ for two such maps if and only if $f(x) = \top \implies g(x) = \top$.

On the other hand, the exponential 2^X contains only the *decidable subsets of X*.

Definition 8 A subset $S \subseteq X$ is said to be *decidable* if $\forall x \in X.\, x \in S \vee x \notin S$.

Decidable subsets are analogous to clopen subsets in topology or to complemented elements in a frame. Indeed, they are precisely the complemented elements of Ω^X, with the complement being given by $\{x \in X \mid x \notin S\}$.

Definition 9 When the equality relation on a set X is a decidable subset of $X \times X$ we say that X has *decidable equality*.

The natural numbers $\mathbb{N}$ and the rationals $\mathbb{Q}$ both have decidable equality, but this is not true for Ω unless excluded middle holds.

The lattice Ω also plays an important role in pointfree topology.

Lemma 10 *The lattice Ω is frame, with the arbitrary joins given by $\bigvee S = [\![\top \in S]\!]$. Moreover, it is the* initial *frame: the unique frame map $!^* \colon \Omega \to OX$ sends $p \in \Omega$ to $\bigvee\{1 \mid p = \top\}$.*

Proof We already know Ω is a Heyting algebra. So $a \wedge (-)$ has a right adjoint and hence preserves existing suprema. Thus, it to show Ω is a frame, it suffices to show it admits arbitrary joins. We now prove that the join of $S \subseteq \Omega$ is given by $[\![\top \in S]\!]$.

To see that this is an upper bound for S we take $s \in S$ and show $s \leq [\![\top \in S]\!]$. If $s = \top$ then $\top = s \in S$, so that $s = \top \implies \top \in S$ and the desired inequality holds.

To see that $[\![\top \in S]\!]$ is the *least* upper bound we must show that for any upper bound $u \in \Omega$, we have $\top \in S \implies u = \top$. Suppose u is an upper bound of S and $\top \in S$. Then in particular, $\top \leq u$ and so $u = \top$, as required.

Now we show that Ω is the initial frame. Let OX be a frame. Any frame map from Ω to OX must send $\top$ to 1 and preserve joins. But now for any $p \in \Omega$, we have that $p = [\![p = \top]\!] = \bigvee\{\top \mid p = \top\}$[4] and so the only possible map is $!^* \colon p \mapsto \bigvee\{1 \mid p = \top\}$. Finally, it is not hard to check that this is indeed a frame homomorphism. □

Remark 11 Interestingly, the set $2 = \{0, 1\}$ cannot be proved to be a frame constructively. Joins of decidable subsets of 2 always exist, but joins of arbitrary subsets might not. Indeed, if 2 is complete then $\bigvee\{1 \mid \varphi\} = 0 \iff \bigvee\{1 \mid \varphi\} \leq 0 \iff (\varphi \implies 1 \leq 0) \iff \neg\varphi$. But 2 has decidable equality and so either $\bigvee\{1 \mid \varphi\} = 0$ or $\bigvee\{1 \mid \varphi\} \neq 0$. Hence, we conclude that $\neg\varphi \vee \neg\neg\varphi$. This logical principle is weaker than excluded middle, but it is nevertheless nonconstructive.

Recall that *points* of a frame OX are given by consistent assignments of its elements to truth values. We quickly see that these are simply frame homomorphisms from OX to Ω. Since Ω is the initial frame, it corresponds to the *terminal locale* 1 and so points of a locale X are simply morphisms from 1 to X. Thus, our definition coincides with the usual definition of point from category theory.

Note that 1 has a unique point by virtue of being terminal and Ω is isomorphic to the powerset of $1 = \{*\}$ and hence to the frame of opens of the unique one-point topological space. So we can identify the spatial locale 1 with the sober space 1. More generally, we can show that locales faithfully represent sets (viewed as discrete spaces). The frames of opens of such *discrete locales* are given by powerset lattices.

[4] The set $\{\top \mid p = \top\}$ might seem a little strange from the classical perspective. It is the subset of $\{\top\}$ corresponding to the constant map sending $\top$ to the element $[\![p = \top]\!]$ in Ω. Classically, we seldom consider subsets defined by propositions with no free variables since any such subset is either the entire set or empty.

2.2.3 Why Work Constructively?

Constructive proofs often require more care than classical ones would and so it is natural to ask why it is worth the trouble. Probably the most convincing argument is the much wider applicability of constructive proofs: they hold in any topos (with natural numbers object). Interpreting the results in these models in terms of external concepts then gives rise to new and different results depending on which topos we use.

We stress that it is *not* necessary to have a deep understanding of topos theory to do constructive mathematics, just as one can do classical mathematics without deep knowledge of set theory. When it comes to interpreting a result in a particular topos, some basic knowledge of that topos and of categorical logic in general is helpful, though even then, if one knows how to interpret basic structures, it is usually possible to fill in the details. This is the approach we will take in these notes. For further information see Johnstone (2002, Part D).

Let us consider an example of how interpreting results in a topos can be useful. Consider the topos $\mathrm{Sh}(B)$ of sheaves over a locale B. The internal logic of this topos corresponds to 'local truth' over B. It can be shown that a locale internal to $\mathrm{Sh}(B)$ corresponds externally to a locale morphism into B (see Johnstone (2002, Sect. C1.6)). Thus, if we interpret a constructive theorem about locales in this topos, we obtain an analogous result about locale morphisms into B. Classically, there is an entire field of *fibrewise topology* that studies such a situation (for topological spaces), but by working constructively we can deduce many of these results automatically. We will see a more specific example of this in Sect. 5.2.4.

Of course, interpreting the result in the category **Set** (in a classical metatheory) simply recovers the classical result. Some other important examples are shown in Table 1. If we are primarily interested in interpreting the results in a particular class of toposes, we might be able to make use of certain classical principles, but for the most general results we should not assume any of these.

Table 1 Interpreting constructive topology in different toposes gives results in a number of subfields for free. Here B is a locale (or sober space) and G is any étale-complete localic group (such as a discrete or profinite group, see Moerdijk (1988))

Topos	Interpretation	Permitted classical axioms
Set	Classical results	Axiom of choice
G-**Set**	G-equivariant topology	Excluded middle
Eff	Computable analysis	Dependent choice
$\mathrm{Sh}(\boldsymbol{B})$	Fibrewise topology over $\boldsymbol{B}$	—

3 Frame Presentations and Algebraic Techniques

A convenient aspect of pointfree topology is that frames are simply a kind of algebraic structure. There is a subtlety in that they have a proper class of operations, but we can nonetheless show that the forgetful functor from **Frm** to **Set** has a left adjoint and **Frm** has coequalisers.[5] Thus, most of the basic ideas from universal algebra are still applicable. In particular, **Frm** is complete and cocomplete and we can talk about subframes and quotient frames in the usual way.

A particularly important consequence is that we can present frames by generators and relations. A *presentation* has the form $\langle G \mid R\rangle$ where G is a set of generators and R is a set of equalities[6] between formal combinations of the generators, called relations. Formally, R can be defined to be a subset of $FG \times FG$ where FG is the free frame on G. The frame defined by a presentation can be constructed as the quotient of FG by the congruence generated by R.

The function from G to $\langle G \mid R\rangle$ sending generators to their image in the quotient of FG satisfies the following universal property: any function f from G to a frame L such that the images of the generators under f satisfy the relations in R induces a unique frame homomorphism $\overline{f}\colon \langle G \mid R\rangle \to L$ such that the following diagram commutes:

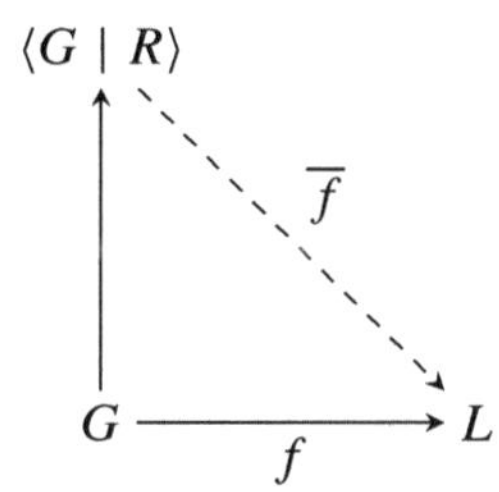

We will see that this provides a very convenient way to construct spaces which is completely unavailable in point-set topology.

3.1 *Classifying Locales*

Presentations can be understood as describing (propositional) geometric theories. The generators correspond to basic propositions and the relations correspond to axioms. A sequent $\varphi \vdash \psi$ corresponds to the relation $\varphi \le \psi$ (or as an equality $\varphi \wedge \psi = \psi$).

---

[5] We remark that the existence of a left adjoint is enough to conclude any equationally presented category is monadic over **Set**. However, constructively this alone is not sufficient to deduce cocompleteness. Nonetheless, cocompleteness does follow as long as we have sufficiently many coequalisers. For instance, see Riehl (2016, Proposition 5.6.11).

[6] We will also find it convenient to use *inequalities* as relations, but since $a \le b \iff a \wedge b = a$ (or alternatively, $a \le b \iff a \vee b = b$), these can be replaced by equivalent equalities.

We call the locale corresponding to this presentation the *classifying locale* of the geometric theory.

As mentioned before, points of the locale correspond to models of the geometric theory. The universal property of the presentation corresponds to the fact that we can show something is a model just by checking the axioms. A point of $\langle G \mid R \rangle$ is simply a frame homomorphism from $\langle G \mid R \rangle$ to Ω, which corresponds to an assignment of the generators to truth values that respects the relations.

As a very simple example, points of the free frame on one generator correspond to the elements of Ω. It turns out that this frame is spatial, and the topology on Ω is generated by the single open $\{\top\}$. This space (as well as the corresponding locale) is called *Sierpiński space*.

We usually conceptualise geometric theories as describing their hypothetical models—that is, the points of the locale. This is despite the fact that we do not have a completeness theorem for geometric logic in the usual sense and so nontrivial theories might even fail to have any models at all.[7] Moreover, the theory alone describes not only the points of the locale, but also the 'topology'.

It will be helpful to look at an example. We will describe a theory that gives the locale of real numbers. Recall that real numbers can be constructed using Dedekind cuts and so let us consider the theory of Dedekind cuts on the rationals. Such a *Dedekind cut* is a pair (L, U) of subsets of $\mathbb{Q}$ satisfying a number of conditions. The intuition is that we expect $L = \{q \in \mathbb{Q} \mid q < \rho\}$ and $U = \{q \in \mathbb{Q} \mid q > \rho\}$ for some real number ρ. The precise conditions are as follows:

(1) L is downward closed: if $q \in L$ and $p \leq q$ then $p \in L$.
(2) L is *rounded*: if $p \in L$ then $q \in L$ for some $q > p$.
(3) L is inhabited: there is some $q \in L$ (to rule out $\rho = -\infty$).
(4) U is upward closed: if $p \in U$ and $p \leq q$ then $q \in U$.
(5) U is rounded: if $q \in U$ then $p \in U$ for some $p < q$.
(6) U is inhabited: there is some $q \in U$ (to rule out $\rho = \infty$).
(7) L and U are disjoint: if $p \in L$ and $q \in U$ then $p < q$.
(8) The pair (L, U) is *located*: if $p < q$ then either $p \in L$ or $q \in U$.

These can be formulated as a geometric theory. We define for each $q \in \mathbb{Q}$ a basic proposition ℓ_q, which we interpret to mean '$q \in L$', and a basic proposition u_q, which we interpret to mean '$q \in U$'. In terms of these the conditions become the following axioms:

[7] The completeness theorem is restored if we look at models not only in **Set**, but in all Grothendieck toposes. Indeed, one way to understand locales is as a succinct way to encode what the points are from the perspective of every Grothendieck topos at once.

(1)	$\ell_q \vdash \ell_p$	for $p \leq q$	(L downward closed)
(2)	$\ell_p \vdash \bigvee_{q>p} \ell_q$	for $p \in \mathbb{Q}$	(L rounded)
(3)	$\top \vdash \bigvee_{q\in\mathbb{Q}} \ell_q$		(L inhabited)
(4)	$u_p \vdash u_q$	for $p \leq q$	(U upward closed)
(5)	$u_q \vdash \bigvee_{p<q} u_p$	for $q \in \mathbb{Q}$	(U rounded)
(6)	$\top \vdash \bigvee_{q\in\mathbb{Q}} u_q$		(U inhabited)
(7)	$\ell_p \wedge u_q \vdash p < q$	for $p, q \in \mathbb{Q}$	(L and U disjoint)
(8)	$\top \vdash \ell_p \vee u_q$	for $p < q$	(locatedness)

Note that the models of this theory are indeed the Dedekind cuts of $\mathbb{Q}$ (if for each model we take L to be the set of $q \in \mathbb{Q}$ such that ℓ_q holds and U to be the set of $q \in \mathbb{Q}$ such that u_q holds).

This (description of the) geometric theory can then be translated to give a frame presentation as above. The basic propositions ℓ_q and u_q become generators and the axioms become relations. For instance, $\top \vdash \bigvee_{q\in\mathbb{Q}} \ell_q$ gives the relation $1 \leq \bigvee_{q\in\mathbb{Q}} \ell_q$. Axiom schema (7) deserves particular mention; it involves the metatheoretical truth value $p < q$, which can be understood in geometric logic by interpreting it as a shorthand for the disjunction $\bigvee\{\top \mid p < q\}$. In this way we see the corresponding relations are $\ell_p \wedge u_q \leq \bigvee\{1 \mid p < q\}$. Note that in the resulting frame $p < q$ becomes the open $!^*([\![p < q]\!])$ where $!^*$ is the unique frame map from Ω.

These relations can be simplified a little. For example, (1) and (2) together are equivalent to the equality $\ell_p = \bigvee_{q>p} \ell_q$. Also, note that since inequality of rational numbers is decidable, we can equivalently express the relations for (7) as $\ell_p \wedge u_q = 0$ for $p \geq q$ (while the condition is trivial in the case $p < q$). We arrive at the following proposition.

Proposition 12 *The classifying locale for the theory of Dedekind cuts has underlying frame*

$$\begin{aligned} O\mathbb{R} = \langle \ell_q, u_q,\ q \in \mathbb{Q} \mid\ & \ell_p = \bigvee_{q>p} \ell_q,\ u_q = \bigvee_{p<q} u_p, \\ & \bigvee_{q\in\mathbb{Q}} \ell_q = 1,\ \bigvee_{q\in\mathbb{Q}} u_q = 1, \\ & \ell_p \wedge u_q = 0 \textit{ for } p \geq q, \\ & \ell_p \vee u_q = 1 \textit{ for } p < q \rangle. \end{aligned}$$

This is called the *locale of reals*. The elements ℓ_q and u_p can be interpreted as the subbasic open sets (q, ∞) and $(-\infty, p)$. The points of the locale are real numbers by construction: if $p\colon 1 \to \mathbb{R}$ is a locale map then $L = \{q \in \mathbb{Q} \mid p^*(\ell_q) = \top\}$ and $U = \{q \in \mathbb{Q} \mid p^*(u_q) = \top\}$ define a Dedekind cut. The induced topology on this set of points is the usual one. However, the locale of reals is not necessarily spatial constructively![8] The failure of such an important locale to be spatial is a big reason why the pointfree approach is so important in constructive mathematics.

Note that by describing the Dedekind cuts as a geometric theory and finding the classifying locale, we didn't just get the correct set of points; we also got the right topology 'for free'! Moreover, we obtained the full locale even though it is not necessarily determined by its (global) points.

We end this subsection with a few additional examples.

Example 13 (*Cantor space*) Cantor space is the locale of binary sequences. We can verify if the bit at any given index is 0 or 1, giving generators z_n and u_n, respectively. Exactly one of these two options holds, and so we have the presentation

$$O(2^{\mathbb{N}}) \cong \langle z_n, u_n,\ n \in \mathbb{N} \mid z_n \wedge u_n = 0, z_n \vee u_n = 1 \rangle.$$

As expected, the points of this locale correspond to decidable subsets of $\mathbb{N}$.

Example 14 (*The Stone spectrum*) Let L be a bounded distributive lattice. The Stone spectrum of L is the space of *prime filters* of L. A prime filter is a subset $F \subseteq L$ such that

- if $a \in F$ and $a \leq b$ then $b \in F$,
- $1 \in F$,
- if $a \in F$ and $b \in F$ then $a \wedge b \in F$,
- $0 \notin F$,
- if $a \vee b \in F$ then $a \in F$ or $b \in F$.

Writing a^f for a basic proposition meaning $a \in F$ this can be expressed as a geometric theory. The resulting frame presentation is then

$$\langle a^f,\ a \in L \mid 1^f = 1, (a \wedge b)^f = a^f \wedge b^f, 0^f = 0, (a \vee b)^f = a^f \vee b^f \rangle.$$

The Zariski spectrum of a commutative ring can be defined similarly. In classical mathematics the spatiality of all Stone or Zariski spectra is equivalent to the Boolean prime ideal theorem (a consequence of the axiom of choice), but we can always define the appropriate locale. With the pointfree approach Stone duality is constructively valid.

Remark 15 Analogously to the above example, if L is a frame then we can recover L as the (frame of opens of) classifying locale of the theory of completely prime ideals of L.

Example 16 (*Surjections from* $\mathbb{N}$ to X) Fix an inhabited set X and consider the following geometric theory. There is a basic proposition denoted by $[f(n) = x]$ for each $n \in \mathbb{N}$ and $x \in X$. The axioms are

---

[8] An example of a model of constructive mathematics where $\mathbb{R}$ fails to be spatial is the effective topos Eff from computability theory. See Remark 104 and Bauer (2006) for more information.

- $[f(n) = x] \wedge [f(n) = y] \vdash x = y$ for $n \in \mathbb{N}$ and $x, y \in X$,
- $\top \vdash \bigvee_{x \in X} [f(n) = x]$ for $n \in \mathbb{N}$,
- $\top \vdash \bigvee_{n \in \mathbb{N}} [f(n) = x]$ for $x \in X$.

The models of such a theory are given by surjective functions $f\colon \mathbb{N} \twoheadrightarrow X$ where the proposition $[f(n) = x]$ is interpreted to mean that $f(n) = x$. The first axiom schema ensures the relation is functional, the second that it is total, and the third imposes surjectivity.

This locale can be proved to be nontrivial (see Johnstone (2002, Example C1.2.8)). However, if we take $X = \Omega^{\mathbb{N}}$ then there are no surjections from $\mathbb{N}$ to X by Cantor's theorem and thus the locale has no points whatsoever! The nontriviality of this locale is related to the fact that such surjections can appear after passing to a forcing extension. This example is arguably rather pathological, but a similar locale has found use in topos theory (see Joyal and Tierney (1984)). It is remarkable that an intuitive axiomatisation of a theory that has no models has lead to a complete characterisation of a nontrivial locale.

3.2 *Free Frames*

To show that frames are algebraic, and in particular to be able to present them by generators and relations, we must prove the existence of free frames. This means constructing a left adjoint to the forgetful functor $U\colon \mathbf{Frm} \to \mathbf{Set}$.

It will be helpful to proceed in two steps. The forgetful functor U factors through the category $\wedge$-**SLat** of $\wedge$-semilattices.

Definition 17 Recall that a *semilattice* is simply a commutative idempotent monoid. Such a structure induces a natural order structure by $a \le b \iff ab = b$ and with respect to this order ab gives the greatest lower bound of $\{a, b\}$. To emphasise this viewpoint we write ab as $a \wedge b$ and call the structure a $\wedge$-semilattice. We could have equally defined the order in the reverse direction, in which case we write ab as $a \vee b$ and call it a $\vee$-semilattice.

Any frame can be viewed as a $\wedge$-semilattice by simply forgetting the join operations. Since taking adjoints respects composition, the adjoint to $U\colon \mathbf{Frm} \to \mathbf{Set}$ can be constructed as the composition of the adjoint $\mathcal{D}\colon \wedge\text{-}\mathbf{SLat} \to \mathbf{Frm}$ of this forgetful functor and the free semilattice functor, which can be constructed just as for any other finitary algebra. Thus, it just remains to construct the free frame on a $\wedge$-semilattice.

3.2.1 Frames of Downsets

We need to freely add joins to a $\wedge$-semilattice M. A formal join of elements of M will be specified by some subset $S \subseteq M$. By idempotence, a join of such formal joins is again one of these formal joins, and we do not need to iterate this procedure. But distinct subsets of M do not always give distinct formal joins. For instance, if $a \leq b$ then $\bigvee\{a, b\}$ should be the same as $\bigvee\{b\} = b$.

Definition 18 A subset S of a poset P is called a *downset* if it is downward closed. If $a \in P$ we write $\downarrow a$ for the principal downset generated by a, which is given by $\{b \in P \mid b \leq a\}$.

We expect that the join of a subset S should be completely specified by the downset generated by S. It turns out there are no additional conditions and we have the following result.

Proposition 19 *The free frame on a $\wedge$-semilattice M is given by the frame $\mathcal{D}M$ of downsets of M (ordered by subset inclusion).*

Proof We first note that $\mathcal{D}M$ is indeed a frame, since it is a subframe of the powerset of M. To show it is a free frame on M we must show there is a $\wedge$-semilattice homomorphism $\eta_M \colon M \to \mathcal{D}M$ such that for any $\wedge$-semilattice homomorphism f from M to a frame L there is a unique frame homomorphism $f^\flat \colon \mathcal{D}M \to L$ for which the following diagram commutes.

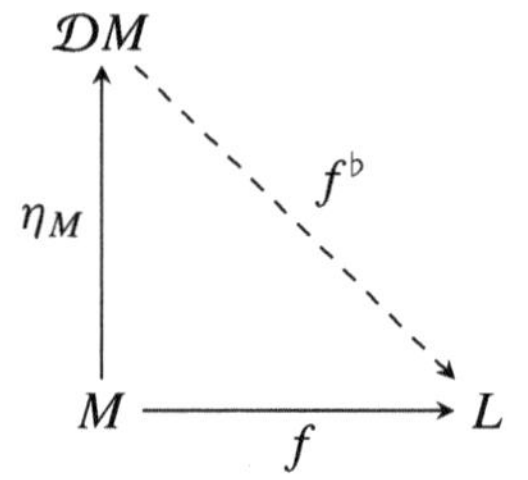

The map η_M sends m to $\downarrow m$, which is quickly seen to preserve finite meets. Any map $f^\flat$ as above must send $\downarrow m$ to $f(m)$ and so by join preservation we have

$$f(D) = f^\flat(\bigcup_{d \in D} \downarrow d) = \bigvee_{d \in D} f(d).$$

Thus, this condition specifies $f^\flat$ uniquely. It remains to show $f^\flat$ so defined is indeed a frame homomorphism.

It is easy to see it preserves joins and 1. Now consider $D, E \in \mathcal{D}M$. We have

$$\begin{aligned} f^\flat(D) \wedge f^\flat(E) &= \bigvee_{d \in D, e \in E} f(d) \wedge f(e) \\ &= \bigvee_{d \in D, e \in E} f(d \wedge e) \\ &\leq \bigvee_{a \in D \cap E} f(a) \\ &= f^\flat(D \cap E), \end{aligned}$$

where the second last inequality holds since $d \wedge e \leq d \in D$ and $d \wedge e \leq e \in E$ give that $d \wedge e \in D \cap E$. The reverse inequality holds by monotonicity and so $f^\flat$ preserves binary meets. □

3.2.2 Free Semilattices and Finiteness

The above argument suffices to prove that the free frame on a set exists. However, to better understand the structure of free frames it will be useful to have a more concrete description of the free $\wedge$-semilattice on a set. We will actually find it more convenient to work with $\vee$-semilattices, but of course, we obtain the corresponding $\wedge$-semilattices by simply reversing the order.

The free semilattice on a set G should be the quotient of the free commutative monoid on G by a congruence enforcing idempotence. Classically, it is easy to see that this is isomorphic to the finite powerset $\mathcal{P}_{\text{fin}}(G)$ under union. Constructively, finiteness can be quite subtle, bifurcating into a number of different inequivalent notions. Let us take this opportunity to discuss some of them.

We start with what is perhaps the most obvious notion of finiteness.

Definition 20 A *finite cardinal* is a set of the form $[n] := \{i \in \mathbb{N} \mid i < n\}$ for some $n \in \mathbb{N}$. A set is said to be *Bishop-finite* if it is isomorphic[9] to a finite cardinal.

Bishop-finite sets have decidable equality and are closed under binary (and even Bishop-finite) products and coproducts, as well as under taking decidable subsets.

However, they are *not* necessarily closed under arbitrary subsets. Indeed, a subsingleton $\{* \mid p = \top\}$ is only Bishop-finite if $p = \bot \vee p = \top$. This can be quite upsetting for classical mathematicians, though arguably their intuition only applies to decidable subsets. Perhaps it helps to consider a model: in the topos $\text{Sh}(B)$ of sheaves over a locale B the Bishop-finite sets correspond to finite covering spaces over B (i.e. covering spaces with finite fibres). Then the failure of closure under subsets simply means that an open subspace of a finite covering space can fail to be a finite covering space.

Surprising aspects of constructive mathematics can often be understood by generalising from sets to locales. We will see that finiteness is closely related to compactness, and it is no surprise that (open) subspaces of compact spaces might fail to be compact.

Most importantly for us, Bishop-finite subsets are not closed under taking unions. For example, we cannot decide if the union of $\{\top\}$ and $\{p\}$ in Ω has two elements or not unless $p = \top \vee p = \bot$. For a good notion of finite powerset we will need a different definition.

---

[9] Here by isomorphic, we mean that 'there exists an isomorphism' in the sense of higher-order intuitionistic logic. In type theory one can express a more restrictive definition of isomorphism, but this is beyond the scope of these notes.

Definition 21 A set X is said to be *Kuratowski-finite* or *finitely indexed* if it is a quotient of a finite cardinal—that is, if there is a natural number n and a surjection $e\colon [n] \twoheadrightarrow X$.

Kuratowski-finite sets have better closure properties than Bishop-finite sets. In addition to being closed under decidable subsets, Bishop-finite products and Bishop-finite coproducts, they are also closed under quotients (by construction) and under Kuratowski-finite unions. We give the proof of one these claims for illustrative purposes. The others are not difficult and make for good exercises.

Lemma 22 *A decidable subset of a Kuratowski-finite set is Kuratowski-finite.*

Proof Let us first show that decidable subsets of $[n]$ are Bishop-finite by induction on n. This is trivial for $n = 0$. Now suppose $n = k + 1$ and that S is a decidable subset of $[n]$. Note that the set $S \setminus \{k\}$ is a decidable subset of $[k]$. Thus, by the inductive hypothesis we have an isomorphism $\varphi\colon [m] \to S \setminus \{k\}$. Now by decidability, either $k \in S$ or $k \notin S$. If $k \notin S$ then $S = S \setminus \{k\}$ and we are done. Otherwise, we can construct a map $\varphi'\colon [m + 1] \to S$ by

$$\varphi'(x) = \begin{cases} \varphi(x) & \text{if } x < m \\ k & \text{if } x = m \end{cases},$$

which is well defined since $\mathbb{N}$ has decidable equality and is easily seen to be an isomorphism.

Now consider a decidable subset S of an arbitrary Kuratowski-finite set X. There is a surjection $e\colon [n] \twoheadrightarrow X$ for some $n \in \mathbb{N}$. Observe that $e^{-1}(S)$ is decidable subset of $[n]$ and hence Bishop-finite by the previous argument. Finally, we have $S = e(e^{-1}(S))$ by surjectivity. Hence, S is the image of a Bishop-finite set and is therefore Kuratowski-finite itself. □

Remark 23 In the above proof we defined the map φ by cases. There is some subtlety in ensuring such definitions are well defined in constructive mathematics. The branches must be exhaustive and agree on any overlap. In this case the branching condition is decidable and so there is no problem, but we could not case match on either an arbitrary proposition or its negation holding, since this is not necessarily exhaustive.

However, Kuratowski-finite sets are still not closed under arbitrary subsets. Indeed, the following result implies that if every subsingleton (i.e. set of the from $\{* \mid p = \top\}$) is Kuratowski-finite then we can prove excluded middle.

Lemma 24 *A Kuratowski-finite set X is either empty or inhabited.*

Proof We have a surjection $e\colon [n] \twoheadrightarrow X$ for some $n \in \mathbb{N}$. Either $n = 0$ or $n \geq 1$. In the former case, $[n] = [0]$ is empty and hence so is X. In the latter case, $e(0) \in X$ and so X is inhabited. □

Remark 25 Kuratowski-finite sets are 'finitely listable'. The difference between them and Bishop-finite sets is that Bishop-finite sets may be listed *without repeats*. The inability to resolve repeats indicates one disadvantage of Kuratowski-finite sets: they might not have decidable equality (as with the set $\{\top, p\}$ we considered before). Indeed, this idea can be used to show that a set is Bishop-finite if and only if it is Kuratowski-finite and has decidable equality.

We are now in position to link this discussion back to semilattices. We start by noting that $\vee$-semilattices admit all Kuratowski-finite joins.

Lemma 26 *Let L be a $\vee$-semilattice and let $S \subseteq L$ be a Kuratowski-finite subset. Then $\bigvee S$ exists and is equal to $\bigvee_{i=0}^{n-1} e(i)$ for any surjection $e\colon [n] \twoheadrightarrow S$.*

Proof For each $s \in S$ there is a $j \in [n]$ such that $s = e(j) \le \bigvee_{i=0}^{n-1} e(i)$. Hence $\bigvee_{i=0}^{n-1} e(i)$ is an upper bound of S.

Now suppose $u \in L$ is an upper bound of S. Then for each $j \in [n]$, we have $e(j) \le u$ and hence $\bigvee_{i=0}^{n-1} e(i) \le u$. Thus, this is the least upper bound of S. □

Corollary 27 *A $\vee$-semilattice homomorphism preserves Kuratowski-finite joins.*

Now as might be expected, we can show that a free semilattice is given by the set of Kuratowski-finite subsets.

Proposition 28 *The free semilattice on a set G is given by the set $\mathcal{P}_{\text{fin}}(G)$ of Kuratowski-finite subsets of G under union.*

Proof Certainly, $\mathcal{P}_{\text{fin}}(G)$ is a $\vee$-semilattice. Now let L be a $\vee$-semilattice and consider a function $f\colon G \to L$. This should correspond to a unique $\vee$-semilattice homomorphism $f^\flat\colon \mathcal{P}_{\text{fin}}(G) \to L$ such that $f^\flat(\{g\}) = f(g)$ for all $g \in G$.

By Corollary 27 such a map must satisfy $f^\flat(S) = \bigvee_{s\in S} f(s)$. It remains to show that this indeed defines a map which preserves 0 and $\vee$, but this easy to check. □

In summary, we obtain the following description of free frames.

Proposition 29 *The free frame on a set G is given by $\mathcal{D}(\mathcal{P}_{\text{fin}}(G)^{\text{op}})$.*

So an element of a free frame is a set of finite subsets of the generating set. Interpreting the inner sets as formal finite meets and the outer sets as formal joins, we see that general elements are given by formal joins of formal finite meets. Indeed, any more complicated expression using the frame operations can be brought into this form using the frame distributivity law.

3.3 Sublocales and Frame Quotients

To show that frame presentations exist, we require not only free frames, but also coequalisers.

Definition 30 A *congruence* on a frame L is an equivalence relation on L that is also a subframe of $L \times L$.

The kernel equivalence relation $\{(u, v) \in L \times L \mid h(u) = h(v)\}$ of any frame homomorphism $h: L \to M$ is a congruence relation. On the other hand, the quotient of a frame by a congruence is again a frame, and the quotient map is a frame homomorphism. The frame structure on the quotient requires some explanation. It is easy to define finitary operations on a quotient by picking representatives as necessary, but in general this does not work for infinitary operations since we would need to use the axiom of choice to choose representatives. However, every congruence class of a frame congruence has a largest element (by taking the join of all the elements) and hence in this case we can pick canonical representatives constructively.

Congruences are closed under intersections and hence form a complete lattice. We say the smallest congruence containing a relation R is the congruence generated by R and denote it by $\langle R \rangle$. These can be used to construct coequalisers in **Frm**. In particular, the frame presentation $\langle G \mid R \rangle$ can be constructed as the quotient $\mathcal{D}(\mathcal{P}_{\text{fin}}(G)^{\text{op}})/\langle R \rangle$. From presentations we can then construct arbitrary colimits in the usual way for algebraic structures.

3.3.1 Sublocales

Let us now consider how to understand quotient frames from a more geometric perspective. Frame surjections correspond to regular epimorphisms in **Frm** and hence regular monomorphisms in **Loc**. Thus, a quotient frame is a kind of sublocale. Indeed, subspaces are precisely the regular subobjects in **Top**. Furthermore, if S is a subspace of a topological space X, then restricting opens to S gives a surjective frame homomorphism from $\mathcal{O}X$ to $\mathcal{O}S$. Thus, it seems appropriate to call these *sublocales*.

From a logical point of view, quotient frames correspond to adding additional axioms to the geometric theory. These axioms are simply a generating set for the congruence associated to the quotient. They add additional restrictions and hence reduce the number of models (i.e. points), which helps explain why sublocales act like subspaces.

Remark 31 Instead of using frame congruences it is also common to represent sublocales using certain closure operators called *nuclei* (obtained by composing a quotient map with its right adjoint) or certain subsets of the frame called *sublocale sets* (which are the fixed points of nuclei). For more details on these alternative approaches see Picado and Pultr (2012).

Classical topology suggests some important classes of sublocales. Firstly, for each element $a \in \mathcal{O}X$ we expect there to be an *open sublocale*.

Definition 32 Let X be a locale and consider $a \in \mathcal{O}X$. The *open sublocale* corresponding to a is given by the frame quotient from $\mathcal{O}X$ to $\downarrow a$ sending u to $u \wedge a$.

Note that this accords with our topological intuition: taking the meet with a restricts the open to the open subspace and opens in the subspace correspond to the opens in X which are contained in a.

The kernel equivalence relation of such an open frame quotient is given by $\Delta_a = \{(u, v) \mid u \wedge a = v \wedge a\}$. Logically, in the open sublocale the proposition a becomes true. Thus, we should expect Δ_a to be generated by the pair $(a, 1)$. This is indeed the case.

Lemma 33 *The open congruence* $\Delta_a = \langle (a, 1) \rangle$.

Proof Certainly, $(a, 1) \in \Delta_a$ since $a \wedge a = a = 1 \wedge a$. Thus, $\langle (a, 1) \rangle \subseteq \Delta_a$.

Now suppose $u \wedge a = v \wedge a$. In the congruence $\langle (a, 1) \rangle$, we have $u \sim u$ and $a \sim 1$ so that $u \wedge a \sim u \wedge 1 = u$ by closure under binary meets. Similarly, $v \wedge a \sim v$. But $u \wedge a = v \wedge a$ by assumption and so $u \sim v$ by transitivity (and symmetry). Therefore, $\Delta_a \subseteq \langle (a, 1) \rangle$ and the claim follows. □

By considering the least element in the equivalence class of 1 with respect to an open congruence Δ_a we can recover the element a and hence the above assignment of opens to (open) sublocales is injective. Therefore, we can identify opens and open sublocales without causing confusion.

There is also a notion of closed sublocale. For each $a \in \mathcal{O}X$ the corresponding closed sublocale is given by setting the proposition a to be *false*.

Definition 34 Let X be a locale and consider $a \in \mathcal{O}X$. The *closed sublocale* corresponding to a is given by the frame quotient from $\mathcal{O}X$ to $\uparrow a$ sending u to $u \vee a$.

As above, we can prove that the kernel ∇_a of this map is indeed given by $\langle (0, a) \rangle$.

This definition of closed sublocales also makes sense intuitively. If we think of closed 'sets' as formal complements of opens, then $(\mathcal{O}X)^{\mathrm{op}}$ can be thought of as the lattice of closed sets of X. With respect to this reversed order we see that closed sets of X restrict to closed sublocales by taking an intersection, as we would expect.

Now let $\mathcal{S}X$ denote the lattice of all sublocales of X. This is isomorphic to the lattice of congruences on $\mathcal{O}X$ under the *reverse order*. We will not need this, but it is notable that $\mathcal{S}X$ is a distributive lattice. In fact, $(\mathcal{S}X)^{\mathrm{op}}$ is a frame! (See Joyal and Tierney (1984, Sect. IV.5) for details.) The open and closed sublocales behave as we might expect inside $\mathcal{S}X$. Most importantly, we have the following.

Lemma 35 *The open and closed sublocales induced by the element a are mutual complements in the lattice of sublocales.*[10]

Proof We will work in the lattice of congruences. Firstly, note that $\nabla_a \vee \Delta_a = \langle (0, a) \rangle \vee \langle (a, 1) \rangle \supseteq \langle (0, 1) \rangle$, which is easily seen to be the largest congruence $\mathcal{O}X \times \mathcal{O}X$.

[10] We remark that since $\mathcal{S}X$ is a distributive lattice, the complement of a sublocale is unique when it exists.

Now take $(u, v) \in \nabla_a \cap \Delta_a$. This means $u \vee a = v \vee a$ and $u \wedge a = v \wedge a$. We wish to show $u = v$ so that this intersection is the equality relation, the least congruence on OX. This is a standard result about distributive lattices, but we give the full proof for completeness.

Consider $(u \wedge v) \vee (u \wedge a) = u \wedge (v \vee a)$. Since $(u, v) \in \nabla_a$, we know $v \vee a = u \vee a$ and so $u \wedge (v \vee a) = u$. Similarly, $(u \wedge v) \vee (v \wedge a) = v$. But also $u \wedge a = v \wedge a$ since $(u, v) \in \Delta_a$ and hence these are all equal. □

Remark 36 If open sublocales correspond to verifiable properties, closed sublocales correspond to *refutable* properties. These are properties for which we can ascertain their falsehood whenever they are false.

We can also show that open sublocales are closed under finite meets and arbitrary intersections in $\mathcal{S}X$ (and dually for closed sublocales). This follows from the following result about closed congruences. (Recall that congruences have the reverse order to sublocales.)

Lemma 37 *The map $a \mapsto \nabla_a$ preserves finite meets and arbitrary joins.*

Proof The assignment is easily seen to be monotone. Thus, $\nabla_{a \wedge b} \subseteq \nabla_a \cap \nabla_b$. Conversely, suppose $(u, v) \in \nabla_a \cap \nabla_b$. Then $u \vee (a \wedge b) = (u \vee a) \wedge (u \vee b) = (v \vee a) \wedge (v \vee b) = v \vee (a \wedge b)$ and so $(u, v) \in \nabla_{a \wedge v}$, as required.

As for join preservation, monotonicity gives $\bigvee_{a \in A} \nabla_a \subseteq \nabla_{\bigvee A}$. On the other hand, we have $(0, \bigvee A) \in \bigvee_{a \in A} \langle (0, a) \rangle$ by closure under joins and hence $\nabla_{\bigvee A} = \langle (0, \bigvee A) \rangle \subseteq \bigvee_{a \in A} \nabla_a$. □

Corollary 38 *The open sublocale given by an element $a \in OX$ is closed if and only if a is complemented in OX.*

Thus, these 'clopen' sublocales correspond to decidable properties, in the sense that we can verify whether they are true or false.

It is enlightening to consider the case of open and closed sublocales of discrete locales. The open sublocales of a set X viewed as discrete locale simply correspond to subsets of X. Then by Corollary 38 the clopen sublocales are precisely the decidable subsets. On the other hand, if *every* sublocale of the terminal locale 1 is open, then in particular the closed complement of each open sublocale is open, so that all subsets are decidable and excluded middle holds.

General closed sublocales of discrete locales are more subtle. They can be understood as formal complements of subsets. This explains some strange definitions that sometimes appear in constructive mathematics. For example, consider the following definition from constructive algebra.

Definition 39 An *anti-ideal* of a commutative ring R is a subset $A \subseteq R$ such that

- $0 \notin A$,
- if $x + y \in A$ then $x \in A$ or $y \in A$,
- if $xy \in A$ then $y \in A$.

Classically such a set is simply the complement of an ideal, but this is not the case constructively since negation is badly behaved. The localic perspective clarifies things. The usual notion of ideal is an open sublocale satisfying certain conditions, while an anti-ideal is a *closed* sublocale satisfying the same conditions. When writing these conditions in terms of the complementary open sublocale (i.e. a subset of R) we recover the above definition (see Manuell (2019, Sect. 1.7 and Proposition 1.3.8)). One would not be surprised to see closed ideals appearing in topological algebra and so it is not actually too surprising to see them show up when studying discrete rings.

3.3.2 Images and Preimages

Since frames are well-behaved algebraic structures, **Frm** is a regular category and so has (regular epi, mono) factorisations (given by the usual factorisation of maps into a surjection followed by an injection). Hence, **Loc** has (epi, regular mono) factorisations and regular monomorphisms in **Loc** are stable under pullback. Therefore, we can use pullbacks to take preimages of sublocales along locale morphisms as shown in the following diagram. This makes the sublocale lattice into a functor $\mathcal{S}\colon \mathbf{Loc}^{\mathrm{op}} \to \mathbf{Pos}$.

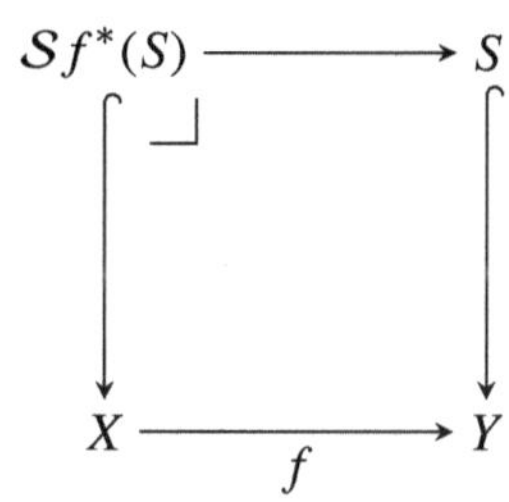

By standard algebraic arguments, if C_S is the frame congruence associated to S then the congruence associated to $\mathcal{S}f^*(S)$ is given by $\langle (f^* \times f^*)(C_S) \rangle$. Thus, preimages of open/closed sublocales are open/closed (by their descriptions in terms of generators).

On the other hand, the (epi, regular mono) factorisation can be used directly to define *images* of sublocales.

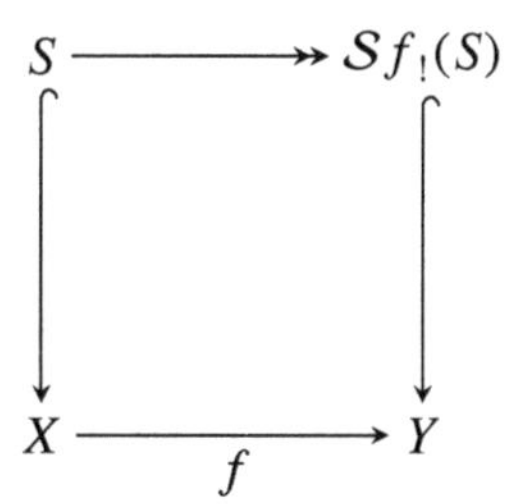

By calculating the image factorisation in **Frm** we easily find that the congruence corresponding to the image $\mathcal{S}f_!(S)$ is given by $(f^* \times f^*)^{-1}(C_S)$.

Using the similar adjunction between images and preimages of sets (and remembering that congruences are ordered in the reverse direction to sublocales) we can see that $\mathcal{S}f_!$ is left adjoint to $\mathcal{S}f^*$.

3.4 *Frame Coproducts and Suplattices*

As mentioned above, we can use frame presentations to construct arbitrary colimits. Let us look at the case of frame coproducts in more detail. These correspond to *products* of locales.

If $(X_\alpha)_{\alpha \in I}$ is a family of locales, then the frame of the product $\prod_{\alpha \in I} X_\alpha$ has a presentation with a generator $\iota_\alpha(u)$ for each $\alpha \in I$ and $u \in OX_\alpha$ and relations making the ι_α maps into frame homomorphisms. The maps $\iota_\beta \colon OX_\beta \to \coprod_{\alpha \in I} OX_\alpha$ are then the coproduct injections, though going forward we will instead refer to these as the frame maps associated to the product projections $\pi_\beta \colon \prod_{\alpha \in I} X_\alpha \to X_\beta$.

Furthermore, if each OX_α is itself given by a presentation $\langle G_\alpha \mid R_\alpha \rangle$, then the product locale has a presentation $\langle \bigsqcup_{\alpha \in I} G_\alpha \mid \bigsqcup_{\alpha \in I} R_\alpha \rangle$, where $\bigsqcup_{\alpha \in I} R_\alpha$ is understood in the obvious way.

Logically, we see that the theory given by the coproduct frame describes something with aspects that are modelled by each OX_α. Indeed, the points of product are simply families of points from each OX_α, as we would expect. The opens in the $\prod_{\alpha \in I} X_\alpha$ are given by joins of finite meets of elements of the form $\pi_\alpha^*(u)$. These finite meets are analogous to the basic open rectangles of the Tychonoff topology. From the perspective of verifiability, it is intuitive that we can only constrain finitely many indices with such a basic open, since verifying something in the product involves checking each component in turn, and we can only carry out finitely many such checks.

3.4.1 Finite Products

We now look at the case of finite products in more detail. Here by 'finite' we mean 'Bishop-finite', since this is the most natural version of finiteness to consider for non-idempotent operations. A general Bishop-finite product can be built up from empty products and binary products. The empty product is the terminal locale 1, whose frame of opens is given by Ω.

For a binary product $X \times Y$, we write the corresponding frame coproduct as $OX \oplus OY$. It will be convenient to define the notation $u \oplus v = \pi_1^*(u) \wedge \pi_2^*(v)$ for the basic rectangles. These are basic opens for the product in the following sense.

Definition 40 We say a subset B of a frame OX is a *base* if every element of OX is a join of elements from B. If G is a generating set, then the finite meets of elements of G form a base.

If $f: Z \to X$ and $g: Z \to Y$ are locale maps, the map $(f, g): Z \to X \times Y$ obtained from the universal property of the product is defined by $(f, g)^*(u \oplus v) = f^*(u) \wedge g^*(v)$.

We have argued that it can be useful to treat set-based structures as discrete locales (for example, in the discussion after Definition 39). The following result is important for allowing a seamless transition between these perspectives when only finite products are involved.

Proposition 41 *Bishop-finite products of discrete locales are discrete and agree with the products of the underlying sets.*

Proof Certainly the empty product 1 is discrete. The result will then follow if we prove it for binary products. As right adjoint, the functor from **Loc** to **Top** preserves products, and so it suffices to show that the product locale in question is spatial.

We must show that elements of $\Omega^X \oplus \Omega^Y$ are determined by the points they contain. A general such element is of the form $u = \bigvee_{\alpha \in I} S_\alpha \oplus T_\alpha$. We have $(x, y) \in u$ if and only if $(x, y) \in S_\alpha \oplus T_\alpha$ for some $\alpha \in I$ (where (x, y) corresponds to a locale map $(x, y): 1 \to X \times Y$ that decomposes into $x: 1 \to X$ and $y: 1 \to Y$). Note that $(x, y) \in S_\alpha \oplus T_\alpha = \pi_1^*(S_\alpha) \wedge \pi_2^*(T_\alpha) \iff x \in S_\alpha \wedge y \in T_\alpha$. But $x \in S_\alpha \iff \{x\} \subseteq S_\alpha$ and similarly for T_α. Hence, $(x, y) \in u \iff \{x\} \oplus \{y\} \leq u$. Finally, since singletons generate the frames Ω^X and Ω^Y under joins, we can conclude that u is determined by the points it contains. □

3.4.2 Suplattices

We can draw an analogy to binary coproducts of commutative rings. In that setting, one can first construct the underlying abelian group structure of the coproduct as a tensor product of the additive groups of the constituents, before defining the appropriate multiplication operation. Something similar happens with frame coproducts if we view the joins as a kind of addition and meets as a multiplication. The analogue of abelian groups is then played by *suplattices*.

Definition 42 A *suplattice* is a poset which has arbitrary joins.[11] A *suplattice homomorphism* is a join-preserving map between suplattices. We denote the category of suplattices and suplattice homomorphisms by **Sup**.

Like frames, suplattices are (large) algebraic structures.

Lemma 43 *The free suplattice of a set G is given by the powerset Ω^G.*

Proof If L is a suplattice and $f: G \to L$ is a function then we can define a suplattice homomorphism from Ω^G by sending $S \subseteq G$ to $\bigvee_{s \in S} f(s)$. It is easy to check that this is indeed a suplattice homomorphism, and, in fact, the unique one making the relevant diagram commute. □

[11] This is the same thing as a complete lattice, but we use the term *suplattice* to emphasise that it is the structure given by the joins which is relevant for this discussion.

Lemma 44 *The category* **Sup** *is complete and cocomplete.*

Proof Limits are computed as for any algebraic structure. Now note that reversing the order gives an equivalence between **Sup** and the category **Inf** of inflattices (which are defined dually in terms of meets). On the other hand, there is a *dual* equivalence between **Sup** and **Inf** given by leaving the objects alone and taking the right adjoints of the suplattice homomorphisms. Composing these gives an auto-duality of **Sup**. Thus, we can compute colimits in **Sup** by taking the right adjoints of all the morphisms involved, reversing the order, and finding the limit in the usual way. □

In particular, we can define suplattices by generators and relations. Also note that the adjunction between **Sup** and **Set** factors through the category of posets **Pos**, and the free suplattice on a poset P can easily be shown to be the lattice of downsets $\mathcal{D}P$. We can use this to define suplattice presentations given by a *poset* of generators. The convenient aspects of the category **Sup** allow us to give a rather explicit construction of the suplattice described by a presentation.

Proposition 45 *If G is a poset and R is a set of relations of the form $\bigvee S \leq \bigvee T$, then $\langle G \text{ poset} \mid R\rangle_{\mathbf{Sup}}$ is order-isomorphic to the subset of $\mathcal{D}G$ consisting of downsets D such that for each relation $\bigvee S \leq \bigvee T$ in R, we have $S \subseteq D$ whenever $T \subseteq D$.*

Proof It will be convenient to view R as indexing set with the functions $s, t \colon R \to \Omega^G$ picking out the sets S and T, respectively, for each relation. We now define suplattice homomorphisms $\overline{s}, \overline{t} \colon \Omega^R \to \mathcal{D}G$ so that $\overline{s}(\{r\})$ and $\overline{t}(\{r\})$ are the downsets generated by $s(r)$ and $t(r)$, respectively.

The presented suplattice can then be expressed by the coequaliser

$$\Omega^R \underset{\overline{s} \vee \overline{t}}{\overset{\overline{t}}{\rightrightarrows}} \mathcal{D}G \overset{q}{\twoheadrightarrow} \langle G \text{ poset} \mid R\rangle_{\mathbf{Sup}}$$

where the suplattice homomorphism $\overline{s} \vee \overline{t}$ is given by the pointwise join of $\overline{s}$ and $\overline{t}$.

The right adjoint $\overline{t}_*$ of $\overline{t}$ is given by

$$\begin{aligned} \overline{t}_*(D) &= \bigvee\{U \in \Omega^R \mid \overline{t}(U) \subseteq D\} \\ &= \{r \in R \mid t(r) \subseteq D\}. \end{aligned}$$

Similarly, $(\overline{s} \vee \overline{t})_*(D) = \{r \in R \mid s(r) \cup t(r) \subseteq D\}$. Taking the equaliser of these in **Inf** we obtain the set of $D \in \mathcal{D}G$ such that $\overline{t}_*(D) = (\overline{s} \vee \overline{t})_*(D)$, which is to say those D such that $t(r) \subseteq D \iff s(r) \cup t(r) \subseteq D$ for all $r \in R$. The result follows. □

There is a strong link between frame presentations and suplattice presentations. Just as we can present suplattices using a poset of generators, we can present frames with a $\wedge$-semilattice of generators. We then have the following theorem (see Abramsky and Vickers (1993)).

Theorem 46 (Coverage theorem) *Let G be a $\wedge$-semilattice and let R be a set of relations of the form $\bigvee S \le \bigvee T$ such that whenever $\bigvee S \le \bigvee T$ is in R, then so is $\bigvee_{s \in S} u \wedge s \le \bigvee_{t \in T} u \wedge t$ for each $u \in G$. Then there is an order isomorphism*

$$\langle G \wedge\text{-semilattice} \mid R \rangle_{\mathbf{Frm}} \cong \langle G \text{ poset} \mid R \rangle_{\mathbf{Sup}}.$$

Note that *every* frame presentation can be brought into the above form by taking formal finite meets of the generators and adding the appropriate relations.

Proof We first show that $\langle G \text{ poset} \mid R \rangle_{\mathbf{Sup}}$ is a frame. It suffices to prove it is a Heyting algebra. We use the construction from Proposition 45. Recall that $\mathcal{D}G$ is a frame and hence a Heyting algebra. In there we have $g \in D \Rightarrow E \iff {\downarrow} g \cap D \subseteq E$. We will show that if E satisfies the conditions of Proposition 45, then so does $D \Rightarrow E$ and hence this also gives the Heyting implication in the sub-inflattice.

Suppose $\bigvee S \le \bigvee T$ is a basic relation. Then by assumption, for each $d \in G$ we have a basic relation $\bigvee (S \cap {\downarrow} d) \le \bigvee (T \cap {\downarrow} d)$ and hence $T \cap {\downarrow} d \subseteq E$ implies $S \cap {\downarrow} d \subseteq E$. Taking the join over all $d \in D$ and applying the defining adjunction of Heyting implication we see that $T \subseteq D \Rightarrow E$ implies $S \subseteq D \Rightarrow E$, as required.

We now argue that the canonical map from G to $\langle G \text{ poset} \mid R \rangle_{\mathbf{Sup}}$ satisfies the necessary universal property for $\langle G \wedge\text{-semilattice} \mid R \rangle_{\mathbf{Frm}}$. Note that it preserves finite meets. Now let L be a frame and let $f : G \to L$ be a $\wedge$-semilattice homomorphism such that $\bigvee_{s \in S} f(s) \le \bigvee_{t \in T} f(t)$ for each basic relation $\bigvee S \le \bigvee T$. Since f is in particular monotone, there is a unique suplattice homomorphism $\overline{f} : \langle G \text{ poset} \mid R \rangle_{\mathbf{Sup}} \to L$ such that $\overline{f}$ agrees with f on generators. It just remains to show that $\overline{f}$ preserves finite meets. Since f is a $\wedge$-semilattice homomorphism, $\overline{f}$ preserves finite meets on a base. The result then follows by taking joins. □

Now we would like to define tensor products of suplattices. Recall that the tensor product $A \otimes B$ of abelian groups is given by the universal bilinear map in the sense that bilinear maps from $A \times B$ to C are in bijection with linear maps (i.e. group homomorphisms) from $A \otimes B$ to C, naturally in C.

Definition 47 If L, M and N are suplattices, we say a function $f : L \times M \to N$ is *bilinear* in L and M if $f(\bigvee_{s \in S} s, t) = \bigvee_{s \in S} f(s, t)$ and $f(s, \bigvee_{t \in T} t) = \bigvee_{t \in T} f(s, t)$.

The tensor product $L \otimes M$ is defined by the following universal property: there is a bilinear map $\beta : L \times M \to L \otimes M$ such that for every bilinear map $f : L \times M \to N$ there is a unique suplattice homomorphism $\widetilde{f} : L \otimes M \to N$ such that $f = \widetilde{f}\beta$.

The tensor product $L \otimes M$ is easily seen to be given by the free suplattice on the generators $\ell \otimes m$ for $\ell \in L$ and $m \in M$, subject to the relations that make the map $\beta : (\ell, m) \mapsto \ell \otimes m$ bilinear.

Remark 48 We note that as a corollary of Proposition 45, $L \otimes M$ can be expressed as the set of downsets D of $L \times M$ such that whenever $(\ell_\alpha, m) \in D$ for all $\alpha \in I$ we have $(\bigvee_{\alpha \in I} \ell_\alpha, m) \in D$ and whenever $(\ell, m_\beta) \in D$ for all $\beta \in J$ we have $(\ell, \bigvee_{\beta \in J} m_\beta) \in D$.

We can now finally show that frame coproducts can be expressed in terms of suplattice tensor products.

Proposition 49 *If X and Y are locales, then $\mathcal{O}X \oplus \mathcal{O}Y$ is order-isomorphic to $\mathcal{O}X \otimes \mathcal{O}Y$.*

Proof This is an immediate consequence of the coverage theorem applied to the above presentation. □

The tensor product can be extended to a functor $\otimes\colon \mathbf{Sup} \times \mathbf{Sup} \to \mathbf{Sup}$ by defining $f \otimes g\colon \ell \otimes m \mapsto f(\ell) \otimes g(m)$. We also note that the lattice of truth values acts as a unit with respect to tensor product. There is a canonical isomorphism $\lambda\colon \Omega \otimes L \cong L$ defined by $\lambda\colon p \otimes \ell \mapsto \bigvee\{\ell \mid p\}$, and the inverse is given by $\ell \mapsto \top \otimes \ell$.

The tensor product functor $\otimes$ and the unit Ω, together with the canonical unit, associativity and symmetry isomorphisms, give **Sup** the structure of a symmetric monoidal category. In fact, $L \otimes (-)$ is left adjoint to the functor $\hom(L, (-))$, which sends a suplattice M to the set $\mathrm{Hom}(L, M)$ of suplattice homomorphisms from L to M equipped with the suplattice structure coming from the pointwise ordering. This makes **Sup** into a symmetric monoidal closed category.

We now obtain a very easy alternative proof of Proposition 41: we have $\Omega^X \otimes \Omega^Y \cong (\coprod_{x \in X} \Omega) \otimes (\coprod_{y \in Y} \Omega) \cong (\coprod_{(x,y) \in X \times Y} \Omega) \cong \Omega^{X \times Y}$, since the left adjoint $L \otimes (-)$ preserves coproducts.

4 Topological Notions

We are now in a position to discuss standard topological concepts such as the Hausdorff separation axiom and compactness. We will also see some other topological properties that are most relevant in the constructive setting.

4.1 Diagonal Conditions

Recall that classically a space X is Hausdorff if and only if the diagonal of $X \times X$ is closed. We will define Hausdorff locales similarly, but let us first briefly consider the dual condition of having an *open* diagonal.

Lemma 50 *The diagonal of a discrete locale is open.*

Proof By Proposition 41 the diagonal of a discrete locale is computed as in **Set** and hence given by subset, which is, of course, an open set in the discrete topology. □

So in a discrete locale equality is verifiable. This is not quite a characterisation of discrete locales, since, for instance, it is preserved by arbitrary sublocales. We will give the precise characterisation in Proposition 78.

A Hausdorff locale has *refutable* equality—that is, verifiable *inequality*.

Definition 51 We say a locale X is *Hausdorff* if the diagonal of $X \times X$ is closed.

Note that a set is Hausdorff as a discrete locale precisely when it has decidable equality. So while decidable equality is not something we worry about classically, it is a manifestation of a familiar topological concept.

The diagonal map $\Delta \colon X \to X \times X$ (obtained from applying the universal property of the product to id_X and id_X) is given by the frame homomorphism $u \oplus v \mapsto u \wedge v$. This is clearly surjective, and it being closed means that $\Delta^*(a) = \Delta^*(b) \iff a \vee d = b \vee d$ for some $d \in \mathcal{O}X \oplus \mathcal{O}X$. Of course, here d is the open complement of the diagonal. The following technical lemma can be helpful for proving Hausdorffness.

Lemma 52 *A locale X is Hausdorff if and only if for all $u, v \in \mathcal{O}X$ we have $u \oplus v \leq (u \wedge v) \oplus (u \wedge v) \vee d$ for some $d \in \mathcal{O}X \oplus \mathcal{O}X$ such that $\Delta^*(d) = 0$.*

Proof Suppose X is Hausdorff and the diagonal sublocale is represented by the congruence ∇_d for some $d \in \mathcal{O}X \oplus \mathcal{O}X$. Then $(d, 0) \in \nabla_d$, so that $\Delta^*(d) = 0$. Now since $\Delta^*(u \oplus v) = u \wedge v = (u \wedge v) \wedge (u \wedge v) = \Delta^*((u \wedge v) \oplus (u \wedge v))$, we know $u \oplus v \leq u \oplus v \vee d = (u \wedge v) \oplus (u \wedge v) \vee d$, as required.

On the other hand, suppose we have the above inequality. We must show $\Delta^*(a) = \Delta^*(b) \iff a \vee d = b \vee d$. The reverse implication is immediate after applying Δ^* and using $\Delta^*(d) = 0$. For the forward implication, observe that if $u \wedge v \leq \Delta^*(b)$, then $u \oplus v \leq (u \wedge v) \oplus (u \wedge v) \vee d \leq \Delta^*(b) \oplus \Delta^*(b) \vee d$. Expressing b as $\bigvee_{\alpha \in I} u'_\alpha \oplus v'_\alpha$ we have $\Delta^*(b) = \bigvee_{\alpha \in I} u'_\alpha \wedge v'_\alpha$ and hence $\Delta^*(b) \oplus \Delta^*(b) = \bigvee_{\alpha,\beta \in I} (u'_\alpha \wedge v'_\alpha) \oplus (u'_\beta \wedge v'_\beta) \leq \bigvee_{\alpha,\beta \in I} (u'_\alpha \wedge v'_\alpha \wedge u'_\beta \wedge v'_\beta) \oplus (u'_\alpha \wedge v'_\alpha \wedge u'_\beta \wedge v'_\beta)$ $\vee\, d \leq b \vee d$, where the first inequality is by applying the hypothesis to each pair of $u'_\alpha \wedge v'_\alpha$ and $u'_\beta \wedge v'_\beta$. So $u \oplus v \leq b \vee d$. Now taking the join over all $u \oplus v \leq a$, we find that whenever $\Delta^*(a) \leq \Delta^*(b)$ we have $a \leq b \vee d$. So if $\Delta^*(a) = \Delta^*(b)$ then $a \leq b \vee d$ and $b \leq a \vee d$. Thus, $a \vee d = b \vee d$ and we have shown that X is Hausdorff. □

Remark 53 In fact, it suffices to only check the above for all u, v in some generating set for $\mathcal{O}X$, since then it follows for general elements by taking finite meets and arbitrary joins.

A very important example of a Hausdorff locale is given by the locale $\mathbb{R}$ of reals.

Proposition 54 *The locale $\mathbb{R}$ is Hausdorff.*

Proof Recall that $\mathcal{O}\mathbb{R}$ has the following presentation.

$$\begin{aligned}\mathcal{O}\mathbb{R} = \langle ((q, \infty)), ((-\infty, q)),\ q \in \mathbb{Q} \mid \\ ((p, \infty)) = \bigvee_{q>p} ((q, \infty)),\ ((-\infty, q)) = \bigvee_{p<q} ((-\infty, p)), \\ \bigvee_{q \in \mathbb{Q}} ((q, \infty)) = 1,\ \bigvee_{q \in \mathbb{Q}} ((-\infty, q)) = 1, \\ ((-\infty, q)) \wedge ((p, \infty)) = 0 \text{ for } p \geq q, \\ ((-\infty, q)) \vee ((p, \infty)) = 1 \text{ for } p < q \rangle\end{aligned}$$

For convenience we define $(\!(p, q)\!) = (\!(p, \infty)\!) \wedge (\!(-\infty, q)\!)$.

The putative diagonal complement is

$$d = \bigvee_{r \in \mathbb{Q}} (\!(-\infty, r)\!) \oplus (\!(r, \infty)\!) \vee \bigvee_{r \in \mathbb{Q}} (\!(r, \infty)\!) \oplus (\!(-\infty, r)\!).$$

To show Hausdorffness we have the following cases to prove:

- $(\!(p, \infty)\!) \oplus (\!(q, \infty)\!) \leq (\!(\max(p, q), \infty)\!) \oplus (\!(\max(p, q), \infty)\!) \vee d$,
- $(\!(-\infty, p)\!) \oplus (\!(-\infty, p)\!) \leq (\!(-\infty, \min(p, q))\!) \oplus (\!(-\infty, \min(p, q))\!) \vee d$,
- $(\!(p, \infty)\!) \oplus (\!(-\infty, q)\!) \leq (\!(p, q)\!) \oplus (\!(p, q)\!) \vee d$,
- $(\!(-\infty, q)\!) \oplus (\!(p, \infty)\!) \leq (\!(p, q)\!) \oplus (\!(p, q)\!) \vee d$.

We will show the third one as a representative example. The other cases are similar. Note that if $p \geq q$ then trivially $(\!(p, \infty)\!) \oplus (\!(-\infty, q)\!) \leq (\!(q, \infty)\!) \oplus (\!(-\infty, q)\!) \leq d$, and so we may assume $p < q$.

The following illustration suggests we try to show $(\!(p, \infty)\!) \oplus (\!(-\infty, q)\!) \leq (\!(p, q)\!) \oplus (\!(p, q)\!) \vee (\!(p, \infty)\!) \oplus (\!(-\infty, p)\!) \vee (\!(q, \infty)\!) \oplus (\!(-\infty, q)\!)$:

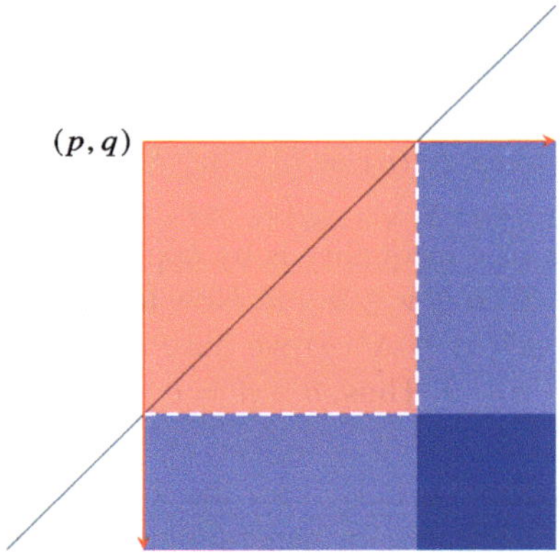

However, this is not quite right since the white dotted lines are missing from the right-hand side. But we can fix this if we shrink the left-hand side by a small amount to allow for an overlap and instead try showing $(\!(p + \varepsilon, \infty)\!) \oplus (\!(-\infty, q - \varepsilon)\!) \leq (\!(p, q)\!) \oplus (\!(p, q)\!) \vee (\!(p + \varepsilon, \infty)\!) \oplus (\!(-\infty, p + \varepsilon)\!) \vee (\!(q - \varepsilon, \infty)\!) \oplus (\!(-\infty, q - \varepsilon)\!)$ as in the illustration below:

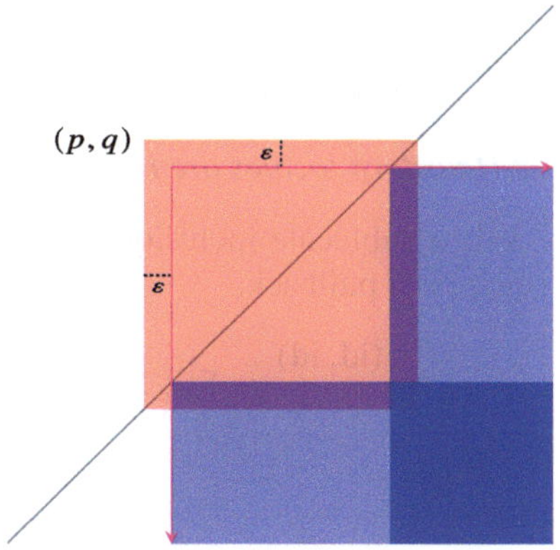

Let us prove this holds. First consider the join of the smaller red square and the part of the blue region below it:

$$\begin{aligned}&((p+\varepsilon,q))\oplus((p,q-\varepsilon))\vee((p+\varepsilon,q))\oplus((-\infty,p+\varepsilon))\\&=((p+\varepsilon,q))\oplus[((p,q-\varepsilon))\vee((-\infty,p+\varepsilon))]\\&=((p+\varepsilon,q))\oplus[((p,\infty))\wedge((-\infty,q-\varepsilon))\vee((-\infty,p+\varepsilon))]\\&=((p+\varepsilon,q))\oplus[(((p,\infty))\vee((-\infty,p+\varepsilon)))\wedge((-\infty,q-\varepsilon))]\\&=((p+\varepsilon,q))\oplus((-\infty,q-\varepsilon)).\end{aligned}$$

Now we connect this resulting region to the blue region beside it:

$$\begin{aligned}&((p+\varepsilon,q))\oplus((-\infty,q-\varepsilon))\vee((q-\varepsilon,\infty))\oplus((-\infty,q-\varepsilon))\\&=[((p+\varepsilon,q))\vee((q-\varepsilon,\infty))]\oplus((-\infty,q-\varepsilon))\\&=[((p+\varepsilon,\infty))\wedge((-\infty,q))\vee((q-\varepsilon,\infty))]\oplus((-\infty,q-\varepsilon))\\&=[((p+\varepsilon,\infty))\wedge(((-\infty,q))\vee((q-\varepsilon,\infty)))]\oplus((-\infty,q-\varepsilon))\\&=((p+\varepsilon,\infty))\oplus((-\infty,q-\varepsilon)).\end{aligned}$$

In summary, we have the join $((p+\varepsilon,q))\oplus((p,q-\varepsilon))\vee((p+\varepsilon,q))\oplus((-\infty,p+\varepsilon))\vee((q-\varepsilon,\infty))\oplus((-\infty,q-\varepsilon))=((p+\varepsilon,q))\oplus((-\infty,q-\varepsilon))\vee((q-\varepsilon,\infty))\oplus((-\infty,q-\varepsilon))=((p+\varepsilon,\infty))\oplus((-\infty,q-\varepsilon))$. Thus, $((p+\varepsilon,\infty))\oplus((-\infty,q-\varepsilon))\leq((p,q))\oplus((p,q))\vee d$.

But now taking the join over all sufficiently small $\varepsilon>0$, we have

$$\begin{aligned}((p,q))\oplus((p,q))\vee d&\geq\bigvee_{\varepsilon}((p+\varepsilon,\infty))\oplus((-\infty,q-\varepsilon))\\&\geq\bigvee_{\varepsilon',\varepsilon''}((p+\varepsilon',\infty))\oplus((-\infty,q-\varepsilon''))\\&=\bigvee_{\varepsilon'}\pi_1^*(((p+\varepsilon',\infty)))\wedge\bigvee_{\varepsilon''}\pi_2^*(((-\infty,q-\varepsilon'')))\\&=((p,\infty))\oplus((-\infty,q)),\end{aligned}$$

and this is precisely what we wanted to show. □

Hausdorffness for locales satisfies all the familiar closure properties.

Lemma 55 *Sublocales of Hausdorff locales are Hausdorff.*

Proof Suppose $i\colon S \hookrightarrow X$ is a sublocale inclusion. It is not difficult to show the following commutative square is a pullback.

$$\begin{array}{ccc} S & \xrightarrow{(\mathrm{id},\,\mathrm{id})} & S \times S \\ {\scriptstyle i}\downarrow & \lrcorner & \downarrow{\scriptstyle i \times i} \\ X & \xrightarrow[(\mathrm{id},\,\mathrm{id})]{} & X \times X \end{array}$$

So the diagonal map $(\mathrm{id}, \mathrm{id})\colon S \hookrightarrow S \times S$ is the restriction of the diagonal of X along the sublocale inclusion $i \times i$ and is therefore closed whenever the diagonal of X is closed. □

Proposition 56 *Products of Hausdorff locales are Hausdorff.*

Proof Let $(X_\alpha)_{\alpha \in I}$ be a family of Hausdorff locales. For each $\beta \in I$ we can form the pullback.

$$\begin{array}{ccc} E_\beta & \xrightarrow{e_\beta} & (\prod_{\alpha \in I} X_\alpha)^2 \\ {\scriptstyle p_\beta}\downarrow & \lrcorner & \downarrow{\scriptstyle (\pi_\beta \pi_1,\, \pi_\beta \pi_2)} \\ X_\beta & \xrightarrow[(\mathrm{id},\,\mathrm{id})]{} & X_\beta \times X_\beta \end{array}$$

Here E_β is intuitively the closed sublocale of $(\prod_{\alpha \in I} X_\alpha) \times (\prod_{\alpha \in I} X_\alpha)$ where the two β coordinates are equal. (We do not try to express this as $X_\beta \times \prod_{\alpha \neq \beta} X_\alpha$, since I might not have decidable equality.) The sublocale is closed, since it is the preimage of the closed sublocale given by the diagonal of X_β.

Now we claim the diagonal $(\mathrm{id}, \mathrm{id})\colon \prod_{\alpha \in I} X_\alpha \to (\prod_{\alpha \in I} X_\alpha) \times (\prod_{\alpha \in I} X_\alpha)$ is the intersection of all the sublocales E_β. From this diagonal map $(\mathrm{id}, \mathrm{id})$ and the projection $\pi_\beta\colon \prod_{\alpha \in I} X_\alpha \to X_\beta$ we obtain a map $d_\beta\colon \prod_{\alpha \in I} X_\alpha \to E_\beta$ for each $\beta \in I$ from the universal property of the pullback. We want to show these maps make $\prod_{\alpha \in I} X_\alpha$ the wide pullback of the e_β maps.

Suppose we have maps $h_\beta\colon Y \to E_\beta$ such that $e_\beta h_\beta$ is the same for each $\beta \in I$. Call this common composite $(f, g)\colon Y \to (\prod_{\alpha \in I} X_\alpha)^2$. Then by commutativity of the diagram above we have $(\pi_\beta \pi_1 f, \pi_\beta \pi_2 g) = (\pi_\beta \pi_1, \pi_\beta p_2) \circ e_\beta h_\beta =$

$(\mathrm{id}, \mathrm{id}) \circ p_\beta h_\beta = (p_\beta h_\beta, p_\beta h_\beta)$. So the composites of f and g with each projection are equal and hence $f = g$. Now from $(\mathrm{id}, \mathrm{id}) \circ f = (f, f) = (f, g)$ it follows that $f: Y \to \prod_{\alpha \in I} X_\alpha$ is a map making the following diagram commute. It is unique since $(\mathrm{id}, \mathrm{id}): \prod_{\alpha \in I} X_\alpha \to (\prod_{\alpha \in I} X_\alpha)^2$ is monic.

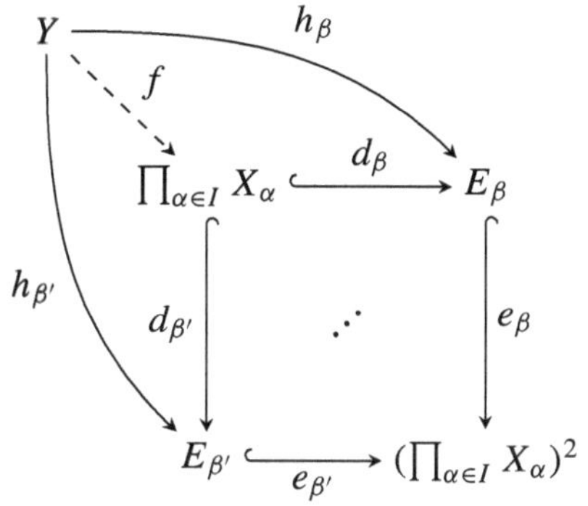

Thus, we have shown the diagonal is indeed the intersection of the E_β sublocales. Now since the E_β sublocales are closed and closed sublocales are closed under arbitrary intersections, we find that the diagonal of $\prod_{\alpha \in I} X_\alpha$ is closed, as required. □

Note that the previous two results are rather intuitive from the verifiability perspective: if we can refute equality in a larger space we can surely also do so in a subspace and to verify that two tuples in a product are different, we need to only verify that they differ at some index.

4.2 *Compactness and Overtness*

Compactness is a notion of central importance in classical topology, but it can take some time before students understand its significance. The intuition of compact spaces being 'like finite spaces' in some sense is good, but too vague. Yet again, the perspective of verifiability provides a very satisfactory answer: compact spaces are those which verifiable properties can be universally quantified over.

Given a verifiable predicate φ on the product $X \times Y$, when does $\forall x \in X.\, \varphi(x, y)$ give a verifiable predicate on Y? This is easy if X is (Kuratowski-)finite—just check that $\varphi(x, y)$ holds for each element of X in turn. For an infinite set this is impossible to verify in finite time, but it may come as a surprise that there are locales with infinitely many points for which this is indeed possible.[12]

[12] For an interpretation of this idea in a slightly different setting see Escardó (2003, Sect. 3.11), which gives an algorithm for universal quantification over Cantor space.

To describe what we mean by universal quantification formally without reference to points it is best to use hyperdoctrines.[13] However, in these notes we would like to avoid the apparatuses of categorical logic and so we will make do with an informal justification using sets of points and classical logic.

Suppose U is an open subset of $X \times Y$. We want to know if the set $\{y \in Y \mid \forall x \in X.\, (x, y) \in U\}$ is open. Taking complements, this is the same as the set $\{y \in Y \mid \exists x \in X.\; (x, y) \notin U\}$ being closed. But this is just the image of the closed set U^{c} under the projection $\pi_2\colon X \times Y \to Y$. Thus, at least intuitively, we have argued that the universal quantification over X of a verifiable predicate is always verifiable precisely if the images of closed sublocales under $\pi_2\colon X \times Y \to Y$ are always closed for all Y.

We will see later that this indeed characterises when X is compact. However, first let us consider the dual concept: what are the locales which we can *existentially* quantify verifiable propositions over? Analogously to the compact case we can argue these are the locales X such that the projections $\pi_2\colon X \times Y \to Y$ always send open sublocales to open sublocales. We call such locales *overt*.

You are probably wondering why, if compactness is such an important concept, you have not encountered the dual concept of overtness before. This is because classically *every* locale is overt! However, constructively it is nontrivial, and indeed, just as important as compactness. Overtness is technically a little easier to deal with than compactness, so we will take a look at it first.

4.2.1 Overtness

Before we can give a formal definition of overtness we must first define the notion of an open map.

Definition 57 A locale map $f\colon X \to Y$ is *open* if the image map $\mathcal{S}f_!\colon \mathcal{S}X \to \mathcal{S}Y$ sends open sublocales to open sublocales.

Remark 58 Note that a sublocale inclusion is open if and only if it is an open sublocale as defined in Definition 32.

It will be convenient to have an alternative description of open maps.

Lemma 59 *A locale map $f\colon X \to Y$ is open if and only if $f^*\colon \mathcal{O}Y \to \mathcal{O}X$ has a left adjoint $f_!\colon \mathcal{O}X \to \mathcal{O}Y$ satisfying $f_!(f^*(b) \wedge a) = b \wedge f_!(a)$.*

Proof If f is open then by the definition of $\mathcal{S}f_!$ we have $(f^* \times f^*)^{-1}(\Delta_a) = \Delta_{g(a)}$ for some map $g\colon \mathcal{O}X \to \mathcal{O}Y$. This means $f^*(u) \wedge a = f^*(v) \wedge a \iff u \wedge g(a) = v \wedge g(a)$.

[13] See Pitts (1991) for a good introduction to categorical logic and Manuell (2019) for some applications to topology. We can formally relate compactness to universal quantification if we use the hyperdoctrine of open sublocales, and dually we can relate it to *existential* quantification if we use the hyperdoctrine of closed sublocales. We note that in the former case we should assume an additional Frobenius reciprocity condition for universal quantification that is usually omitted outside of classical logic.

Now taking $u = 1$ we find $a \leq f^*(v) \iff g(a) \leq v$ and so g is left adjoint to f^*. On the other hand, letting $v = w \wedge u$, the right-hand side becomes $u \wedge g(a) \leq w$ and the left-hand side becomes $f^*(u) \wedge a \leq f^*(w)$, which is equivalent to $f_!(f^*(u) \wedge a) \leq w$. Thus, $f_!(f^*(u) \wedge a) = u \wedge g(a) = u \wedge f_!(a)$.

Conversely, suppose $f_!(f^*(b) \wedge a) = b \wedge f_!(a)$ for all $a \in OX$ and $b \in OY$. Then $f^*(b) \wedge a \leq f^*(w) \iff b \wedge f_!(a) \leq w$.

Now $f^*(b) \wedge a = f^*(w) \wedge a$ implies $f^*(b) \wedge a \leq f^*(w)$ and $f^*(w) \wedge a \leq f^*(b)$, yielding $b \wedge f_!(a) \leq w$ and $w \wedge f_!(a) \leq b$ by the above. Hence, it entails $b \wedge f_!(a) = w \wedge f_!(a)$. On the other hand, $b \wedge f_!(a) = w \wedge f_!(a)$ implies $b \wedge f_!(a) \leq w$ and $w \wedge f_!(a) \leq b$ and hence $f^*(b) \wedge a \leq f^*(w)$ and $f^*(w) \wedge a \leq f^*(b)$, giving $f^*(b) \wedge a = f^*(w) \wedge a$.

Thus, we have shown $f^*(b) \wedge a = f^*(w) \wedge a \iff b \wedge f_!(a) = w \wedge f_!(a)$, which is precisely what it means to have $(f^* \times f^*)^{-1}(\Delta_a) = \Delta_{f_!(a)}$ and so f is open. □

We can now define overtness.

Definition 60 A locale X is called *overt* if the unique map $!\colon X \to 1$ is open.

Some people call these *open* locales, but to avoid confusion we will not.

Remark 61 We argued above that a locale should be overt if and only if $\pi_2\colon X \times Y \to Y$ is open for all locales Y. Definition 60 only asserts this for $Y = 1$, but it actually follows for all Y, since we will see later (in Theorem 73) that open maps are stable under pullback.

Lemma 62 *A locale X is overt if and only if the unique frame map $!^*\colon \Omega \to OX$ has a left adjoint $\exists\colon OX \to \Omega$.*

Proof If X is overt then $!^*$ has a left adjoint by Lemma 59. Again by Lemma 59, for the other direction we need to show that $\exists(!^*(b) \wedge a) = b \wedge \exists(a)$. But $!^*(b) \wedge a = \bigvee\{a \mid b = \top\}$ and this join is preserved by the left adjoint $\exists$. Hence, the result follows. □

Let us look more closely at the map $\exists$. Since it maps into Ω it is determined by what it sends to $\top$. Suppose $\exists(u) = \top$. Then if $u \leq \bigvee S$ we have $u \leq \bigvee\{s \mid s \in S\} \leq \bigvee\{1 \mid s \in S\} = !^*([\![\exists s \in S]\!])$, so that $\top = \exists(u) \leq [\![\exists s \in S]\!]$ and hence S is inhabited. We say that such an element u is *positive*.

Definition 63 An open u in a locale X is said to be *positive* (written $u > 0$) if $u \leq \bigvee S$ implies S is inhabited. We say X is positive if $1 \in OX$ is positive.

This is a more useful way to formulate a frame element being nonzero constructively than simply saying $u \neq 0$. It is easy to see that an open of a discrete locale is positive if and only if the corresponding subset is inhabited. More generally, any open that contains a point is positive, but the converse does not hold.

Lemma 64 *A locale X is positive if and only if the unique map $!\colon X \to 1$ is epic.*

Proof Suppose X is positive. We must show $!^*\colon \Omega \to OX$ is injective. Suppose $\bigvee\{1 \mid p = \top\} \leq \bigvee\{1 \mid q = \top\}$. Then if $p = \top$, we have $\bigvee\{1 \mid q = \top\} \geq 1 > 0$ and hence $q = \top$. Thus, $!^*$ reflects order and is therefore injective.

Now suppose $!^*$ is injective and consider $\bigvee S = 1 \in \mathcal{O}X$. We must show S is inhabited. Note that $1 \leq \bigvee\{s \mid s \in S\} \leq \{1 \mid s \in S\} = !^*([\![\exists s \in S]\!])$. But then $!^*([\![\exists s \in S]\!]) = !^*(\top)$ and so $[\![\exists s \in S]\!] = \top$ by injectivity. □

Corollary 65 *If $f: X \to Y$ is a map of locales and X is positive, then so is Y. The converse holds if f is epic.*

Proof This follows from simple properties of composites of epimorphisms. □

As suggested above the map $\exists$ measures the positivity of opens.

Lemma 66 *In an overt locale $\exists(u) = [\![u > 0]\!]$.*

Proof We have already shown $\exists(u) = \top \implies u > 0$. For the converse, suppose $u > 0$. Then $u \leq !^*\exists(u) = \bigvee\{1 \mid \exists(u) = \top\}$ and hence $\exists(u) = \top$ by positivity. □

Remark 67 Note that if an open u in an overt locale X is *not* positive, then $\exists(u) \leq \bot$ and hence $u \leq !^*(\bot) = 0$.

We can characterise overt locales in terms of having sufficiently many positive elements. For this reason overt locales are sometimes called *locally positive* locales.

Proposition 68 *A locale X is overt if and only if $\mathcal{O}X$ has a base of positive elements.*

Proof Suppose X is overt. We will show the set of all positive elements forms a base. For $u \in \mathcal{O}X$ we have $u \leq !^*\exists(u) = \bigvee\{1 \mid u > 0\}$. Meeting with u then gives $u = \bigvee\{u \mid u > 0\}$, which is indeed a join of positive elements.

Now suppose B is a base of positive elements for $\mathcal{O}X$. We claim the map $\exists: u \mapsto [\![u > 0]\!]$ is left adjoint to $!^*$. It is clear that if $\bigvee\{1 \mid p = \top\} > 0$ then $p = \top$ and so $\exists \circ !^* \leq \mathrm{id}_\Omega$. Now take $a \in \mathcal{O}X$. We have $a = \bigvee\{b \in B \mid b \leq a\} \leq \bigvee\{1 \mid b \in B,\ b \leq a\} = !^*([\![\exists b \in B.\, b \leq a]\!])$. But if $a \geq b > 0$ then $a > 0$ and so $a \leq !^*([\![\exists b \in B.\, b \leq a]\!]) \leq !^*\exists(a)$. Thus, $\mathrm{id}_{\mathcal{O}X} \leq !^* \circ \exists$ and $\exists$ is left adjoint to $!^*$. □

Remark 69 If X is overt we can replace any join by one consisting only of positive elements, since as above $\bigvee A = \bigvee\{\bigvee\{a \mid a > 0\} \mid a \in A\} = \bigvee\{a \in A \mid a > 0\}$.

It is now clear that discrete locales are overt, since the singletons form a base of positive elements. Moreover, classically the set of nonzero elements always form a base and so *every* locale is overt!

Lemma 70 *Open sublocales of overt locales are overt.*

Proof Observe that the composition of open maps is open. Thus, if $!: X \to 1$ is open and $U \hookrightarrow X$ is open, then so is the composite map from U to 1. □

Remark 71 This result is in a sense dual to the more familiar result that closed subspaces of compact spaces are compact (see Lemma 95). It is also intuitively clear from the point of view of verifiability, since we can verify $\exists x \in U.\, \varphi(x)$ by verifying $\exists x \in X.\, x \in U \wedge \varphi(x)$.

Lemma 72 *Images of overt (sub)locales are overt.*

Proof It suffices to show subframes of overt frames are overt. Let $e^*: OY \hookrightarrow OX$ be a subframe inclusion and suppose $!^*: \Omega \to OX$ has a left adjoint $\exists: OX \to \Omega$. We claim that $\exists \circ e^*$ is the left adjoint of $!^*: OY \to \Omega$. Indeed, observe that $\exists(e^*(u)) \le p \iff e^*(u) \le !^*(p) = e^*(!^*(p)) \iff u \le !^*(p)$, since e^* is an order embedding. □

The following theorem is an important result about open maps.

Theorem 73 *Open maps are stable under pullback.*

Proof Consider the following pullback diagram in **Loc**.

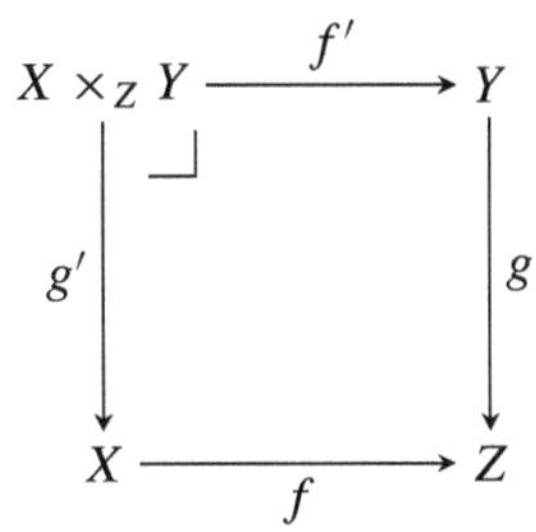

The frame $O(X \times_Z Y)$ can be found as a quotient of the frame coproduct $OX \oplus OY$ by the congruence generated by $\pi_1^* f^*(w) \sim \pi_2^* g^*(w)$ for $w \in OZ$. As in Proposition 49 we can show that $O(X \times_Z Y)$ is order-isomorphic to the quotient of $OX \otimes OY$ by the suplattice congruence generated by $(u \wedge f^*(w)) \otimes v \sim u \otimes (g^*(w) \wedge v)$. In terms of this description we have $g'^*(u) = [u \oplus 1]$ and $f'^*(v) = [1 \oplus v]$.

Now suppose g is an open map and consider the assignment $[u \otimes v] \mapsto u \wedge f^* g_!(v)$. Note that $(u \wedge f^*(w)) \otimes v \mapsto u \wedge f^*(w \wedge g_!(v))$ and $u \otimes (g^*(w) \wedge v) \mapsto u \wedge f^* g_!(g^*(w) \wedge v)$, which coincide by openness (and Lemma 59). Thus, using the suplattice presentation above, this defines a suplattice homomorphism $g'_!: O(X \times_Z Y) \to OX$.

As the notation suggests, $g'_!$ is left adjoint to g'^*. Indeed, $g'_! g'^*(u) = u \wedge f^* g_!(1) \le u$ and $g'^* g'_!([u \otimes v]) = [(u \wedge f^* g_!(v)) \otimes 1] = [u \otimes g^* g_!(v)] \ge [u \otimes v]$. Finally, we have $g'_!(g'^*(u') \wedge [u \otimes v]) = g'_!([(u' \wedge u) \otimes v]) = u' \wedge u \wedge f^* g_!(v) = u' \wedge g'_!([u \otimes v])$ and hence g' is open. □

We can now finally prove the characterisation of overt locales that we originally motivated intuitively.

Corollary 74 *A locale X is overt if and only if for all locales Y the product projection $\pi_2: X \times Y \to Y$ is open.*

Proof The projection condition gives our definition of overtness if we take $Y = 1$. On the other hand, if X is overt we can consider the following pullback diagram.

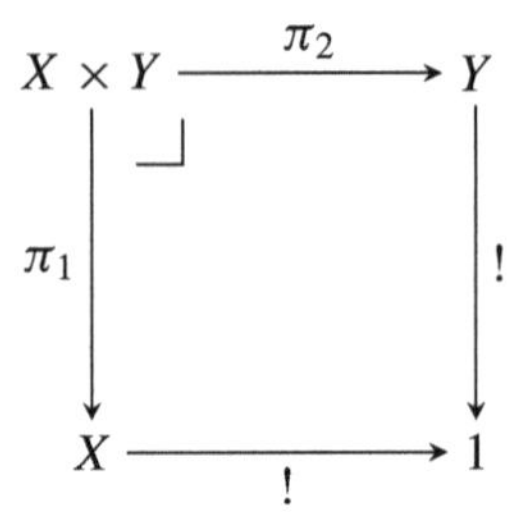

Theorem 73 gives that π_2 is open since $!\colon X \to 1$ is. Explicitly, the left adjoint of π_2^* is given by $(\pi_2)_!\colon u \otimes v \mapsto !^*\exists(u) \wedge v$. □

It is now easy to see that overt locales are closed under Bishop-finite products.

Lemma 75 *If X and Y are overt, then so is $X \times Y$.*

Proof The map $!\colon X \times Y \to 1$ can be expressed as the composite of $\pi_2\colon X \times Y \to Y$ and $!\colon Y \to 1$, which are both open. Thus, it is open too. □

This result can also be understood intuitively: we can existentially quantify over the product by quantifying over each factor in turn.

The following result will help give a Brouwerian counterexample to the claim that all locales are overt.

Lemma 76 *Every overt sublocale of a discrete locale is open.*

Proof Intuitively, if V is an overt sublocale of the discrete locale X, we can verify $x \in V$ by showing $\exists y \in V.\, x = y$. More formally, consider the following pullback.

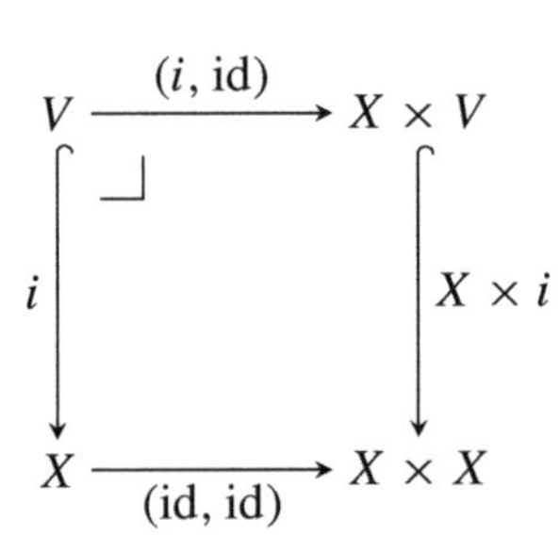

Since the diagonal $(\mathrm{id}, \mathrm{id})$ is open by Lemma 50, so is (i, id). Since V is overt, $\pi_1\colon X \times V \to X$ is open. Thus, the composite $i = \pi_1(i, \mathrm{id})$ is open. □

Corollary 77 *If every closed sublocale of* 1 *is overt, then excluded middle holds.*

Proof By the above this would mean that the closed complement of every truth value is open, and thus every truth value is decidable. □

In fact, overtness is the missing condition we needed to characterise discrete locales.

Proposition 78 *A locale X is discrete if and only if it is overt and has open diagonal.*

Proof We have already noted that discrete locales are overt and have open diagonal, so we only need to show the converse. Let d be the open corresponding to the diagonal inclusion $\Delta\colon X \hookrightarrow X \times X$. The idea is to isolate the opens which end up corresponding to singletons. We define S to be the set of elements $s \in OX$ such that $s \oplus s \leq d$ and $s > 0$.

We claim that $OX \cong \Omega^S$. We will show this by giving a bijective suplattice homomorphism $h\colon \Omega^S \to OX$. Since Ω^S is the free suplattice on S, a suplattice map from Ω^S to OX corresponds to a function from S to OX. We take this to be the obvious inclusion. It remains to show the map h is surjective and injective.

We start by observing that since Δ is open, $(u \oplus v) \wedge d = (u' \oplus v') \wedge d \iff \Delta^*(u \oplus v) = \Delta^*(u' \oplus v') \iff u \wedge v = u' \wedge v'$. Thus, if $u \oplus v \leq d$ then $u \oplus v = (u \wedge v) \oplus (u \wedge v)$.

Now since the elements of the form $u \oplus v$ give a base for $X \times X$ we have $d = \bigvee\{u \oplus v \mid u \oplus v \leq d\}$. By the previous argument we know each $u \oplus v = (u \wedge v) \oplus (u \wedge v)$ and so $d \leq \bigvee\{u \oplus u \mid u \oplus u \leq d\}$. We can project this down to X by applying the left adjoint $(\pi_2)_!$. Note that $\pi_2\Delta = \pi_2(\mathrm{id}, \mathrm{id}) = \mathrm{id}$ and so $(\mathcal{S}\pi_2)_!(\Delta\colon X \hookrightarrow X \times X) = X$. Thus, we find $1 \leq \bigvee\{!^*\exists(u) \wedge u \mid u \oplus u \leq d\} = \bigvee S$. Meeting with v and restricting the join to positive elements (using overtness) we obtain $v \leq \bigvee\{s \in S \mid s \leq v\}$. Thus, $v = h(S \cap {\downarrow}v)$ and so h is surjective.

We now show h is injective. First note that if $s, s' \in S$ we have $(s \wedge s') \oplus s = (s \wedge s') \oplus (s \wedge s')$ as above. So if $s \wedge s' > 0$, projecting onto the second factor we find $s = s \wedge s'$ and so $s \leq s'$. Similarly, $s' \leq s$ and hence $s = s'$. Now suppose $s \leq h(T)$. This means $s = \bigvee\{s \wedge t \mid t \in T\} = \bigvee\{s \wedge t \mid t \in T,\ s \wedge t > 0\}$ by overtness. Now $s \wedge t > 0$ means $t = s$ and so $s = \bigvee\{s \mid s \in T\}$ (where s is a fixed element and this denotes a join of a subsingleton, not a join over all of T). But this join is inhabited since s is positive and so $s \in T$. It follows that $s \in T \iff s \leq h(T)$. Hence, $h(T)$ determines T and h is injective. □

Another important example of an overt locale is given by the reals.

Proposition 79 *The locale $\mathbb{R}$ is overt.*

Proof Observe that the opens $((p, q)) = ((p, \infty)) \wedge ((-\infty, q))$ form a base for $O\mathbb{R}$. Furthermore, we have $((p, q)) = 0$ for $p \geq q$, and so, since the order on $\mathbb{Q}$ is decidable, we can restrict the base to those intervals with $p < q$. But it is easy to see that if $p < q$ then $\frac{p+q}{2}$ defines a point of $\mathbb{R}$ that lies in $((p, q))$. Hence, $((p, q)) > 0$ and $O\mathbb{R}$ has a base of positive elements. □

The following result provides a very general method of showing a locale is overt from a presentation.

Proposition 80 *Let X be a locale with a presentation $\langle G\ \wedge\text{-semilattice} \mid R\rangle_{\mathbf{Frm}}$ of the form used in the coverage theorem (Theorem 46). Then X is overt if and only if there is a set $P \subseteq G$ of candidate positive generators such that*

(i) P *is upward closed;*
(ii) *for each relation* $\bigvee S \leq \bigvee T$*, if* $\exists s \in S \cap P$ *then* $\exists t \in T \cap P$*;*
(iii) *for each generator* $g \in G$*, we have* $g \leq \bigvee\{1 \mid g \in P\}$ *in* OX*.*

Proof If X is overt, then the set of positive generators is easily seen to satisfy the hypotheses. To prove the converse we will construct a left adjoint to $!^*\colon \Omega \to OX$.

First define a map from G to Ω by $g \mapsto [\![g \in P]\!]$. This is monotone by condition (i) and respects the relations by condition (ii). Thus, by the coverage theorem it induces a suplattice homomorphism $\exists\colon OX \to \Omega$. We must show this is indeed a left adjoint to $!^*\colon \Omega \to OX$.

Note that for $g \in G$ we have $!^*\exists(g) = \bigvee\{1 \mid g \in P\} \geq g$ by condition (iii) and hence $!^* \circ \exists \geq \mathrm{id}_{OX}$. On the other hand, $\exists !^*(p) = \bigvee\{[\![1 \in P]\!] \mid p = \top\} = [\![1 \in P]\!] \wedge p \leq p$ and so $\exists \circ !^* \leq \mathrm{id}_\Omega$. Thus, $\exists$ is left adjoint to $!^*$ and X is overt. □

4.2.2 Compactness

We now return to the property of compactness. Things proceed similarly to before.

Definition 81 A locale map $f\colon X \to Y$ is *closed* if $\mathcal{S}f_!$ maps closed sublocales to closed sublocales.

Remark 82 Note that a sublocale inclusion is closed if and only if it is a closed sublocale as defined in Definition 34.

Since a frame map f^* preserves arbitrary joins, it always has a right adjoint f_*.

Lemma 83 *A locale map* $f\colon X \to Y$ *is closed if and only if the frame map* f^* *and its right adjoint* f_* *satisfy* $f_*(f^*(b) \vee a) = b \vee f_*(a)$*.*

Proof Omitted. (The proof is very similar to that of Lemma 59.) □

At this point our discussion diverges somewhat from the overt case, since unlike open maps, closed maps are not pullback stable. Before we deal with this let us define compactness in a more familiar way.

Definition 84 A locale X is *compact* if whenever $\bigvee S = 1$ in OX then there is some Kuratowski-finite subset $F \subseteq S$ such that $\bigvee F = 1$.

It will be useful to re-express this condition using the following concept from order theory.

Definition 85 A poset S is called *directed* if every Kuratowski-finite subset $F \subseteq S$ has an upper bound $b \in S$.

Remark 86 It can be shown by induction that S is directed if and only if it is inhabited and for each pair of elements $s, t \in S$ there is a $b \in S$ with $s, t \leq b$.

Lemma 87 *A locale* X *is compact if and only if whenever* $\bigvee S = 1$ *is a directed cover,* $1 \in S$*.*

Proof If X is compact and $\bigvee S = 1$ then there is a Kuratowski-finite subset $F \subseteq S$ such that $\bigvee F = 1$. But if S is directed, then there is a $b \in S$ such that $b \geq \bigvee F = 1$ and hence $b = 1$.

On the other hand, suppose ever directed cover of $\mathcal{O}X$ contains 1 and consider an arbitrary cover S of $\mathcal{O}X$. Set $S' = \{\bigvee F \mid F \in \mathcal{P}_{\text{fin}}(S)\}$. This set is directed since $0 = \bigvee \emptyset \in S'$ and if $\bigvee F_1, \bigvee F_2 \in S'$ then $\bigvee F_1 \vee \bigvee F_2 = \bigvee (F_1 \cup F_2) \in S'$, because Kuratowski-finite subsets are closed under binary joins. Also, note that $\bigvee S \leq \bigvee S'$ and so S' is a cover. Thus, by assumption $1 \in S'$, so $\bigvee F = 1$ for some $F \in \mathcal{P}_{\text{fin}}(S)$. Consequently, X is compact. □

Corollary 88 *A locale X is compact if and only if the right adjoint $!_*\colon \mathcal{O}X \to \Omega$ preserves directed suprema.*

Proof By the usual argument that maps into Ω are determined by what they send to $\top$ it suffices to show that $!_*$ preserves directed covers. Note that $1 = {!^*}(\top) \leq u \iff \top \leq {!_*}(u)$ and so ${!_*}(u) = \top \iff u = 1$. Thus, $!^*$ preserves directed joins if and only if for S directed, $\bigvee S = 1$ implies $\exists s \in S.\ s = 1$, which is simply the characterisation of compactness from Lemma 87. □

This motivates the following strengthening of closedness, which will behave more similar to openness.

Definition 89 A locale map $f\colon X \to Y$ is *proper* if it is closed and f_* preserves directed suprema.

Proposition 90 *A locale X is compact if and only if the unique map $!\colon X \to 1$ is proper.*

Proof By Corollary 88 it simply remains to show that if X is compact, then $!\colon X \to 1$ is closed.

We must show ${!_*}({!^*}(p) \vee a) \leq p \vee {!_*}(a)$. (The reverse inequality always holds.) Suppose ${!_*}({!^*}(p) \vee a) = \top$. Then ${!^*}(p) \vee a = 1$. Observe that ${!^*}(p) \vee a = \bigvee\{1 \mid p = \top\} \vee a = \bigvee(\{a\} \cup \{1 \mid p\})$. This join is directed and so by compactness, either $a = 1$ or $p = \top$. But this means $p \vee {!_*}(a) = \top$, as required. □

Remark 91 Inclusions of closed sublocales are another example of proper maps. In fact, in that case the right adjoints (which are of the form $u \mapsto u \vee a$) preserve all inhabited suprema.

Just like open maps, proper maps are pullback stable. In fact, we have the following characterisation, the proof of which is unfortunately beyond the scope of these notes. See Vermeulen (1994) for details.

Theorem 92 *Proper maps are precisely the pullback-stable closed maps.*

Remark 93 The proof that proper maps are pullback stable is similar to that for open maps, but makes use of 'preframes' instead of suplattices. A *preframe* is a poset with finite meets and directed joins in which finite meets distribute over directed joins. Schematically, we have frames = suplattices + finite meets = preframes + finite joins.

We can now give the characterisation of compactness we motivated in terms of universal quantification.

Corollary 94 *A locale X is compact if and only if for all locales Y the product projection $\pi_2 \colon X \times Y \to Y$ is closed.*

Proof This is simply Theorem 92 applied to $! \colon X \to 1$. □

Analogous preservation results to those we proved for overtness can now be given for compactness. They are proved in essentially the same way (and have similar intuitive explanations from the perspective of verifiability).

Lemma 95 *Closed sublocales of compact locales are compact.*

Lemma 96 *Images of compact (sub)locales are compact.*

Lemma 97 *Bishop-finite products of compact locales are compact.*

We also have an analogue of Lemma 76 (the proof of which did not actually use the overtness of the parent locale).

Lemma 98 *Every compact sublocale of a Hausdorff locale is closed.*

Let us consider some examples of compact locales.

Proposition 99 *A set X is compact as a locale if and only if it is Kuratowski-finite.*

Proof Suppose X is compact. We have $X = \bigcup_{x \in X} \{x\}$ and so by compactness, X is a Kuratowski-finite join of singletons. Hence, X is Kuratowski-finite.

Now suppose X is Kuratowski-finite. Then X is the image of some set of the form $[n] = \{m \in \mathbb{N} \mid m < n\}$. Thus, it suffices to show $[n]$ is compact.

We proceed by induction. Certainly, $[0]$ is compact. Suppose $[n]$ is compact and consider a union $\bigcup \mathcal{S} = [n] \cup \{n\}$. Then $[n] \subseteq \bigcup \mathcal{S}$ and so there is a Kuratowski-finite subset $\mathcal{F} \subseteq \mathcal{S}$ such that $[n] \subseteq \mathcal{F}$. Moreover, $n \in S$ for some $S \in \mathcal{S}$. Thus, $\mathcal{F} \cup \{S\} \subseteq \mathcal{S}$ is a Kuratowski-finite subcover. So $[n+1]$ is compact. □

Corollary 100 *A set X is compact Hausdorff if and only if it is Bishop-finite.*

Proof Recall that a set is Bishop-finite if and only if it is Kuratowski-finite and has decidable equality. □

Here we see that our different notions of finiteness again have familiar topological generalisations.

Remark 101 From this perspective it is perhaps less surprising that a subset of a Kuratowski-finite set can fail to be Kuratowski-finite. After all, we are familiar with the idea that an open sublocale of a compact locale need not be compact.

We can now discuss a slightly more complicated example. Recall that the Cantor space $2^{\mathbb{N}}$ has the presentation

$$O(2^{\mathbb{N}}) \cong \langle z_n, u_n,\ n \in \mathbb{N} \mid z_n \wedge u_n = 0, z_n \vee u_n = 1 \rangle.$$

If $\bigvee S = 1$ in $O(2^{\mathbb{N}})$ this should be forced by some relations and since every relation involves only finite joins, we might intuitively feel that we should be able to find a finite subset of S which already joins to give 1. This intuition can be made precise.

Proposition 102 *If every join in a frame presentation is finite, then the resulting locale is compact.*

Proof Since the presentation involves only finite joins and finite meets, it can also be viewed as a presentation for a distributive lattice L. The presented frame is then the free frame on this distributive lattice L. (This frame is also that obtained in Example 14.) It is not hard to show that this free frame is given explicitly by the frame $\mathscr{I}L$ of lattice ideals on L—that is, downsets on L which are closed under finite joins. Directed joins in $\mathscr{I}L$ are simply given by union and hence if $\bigvee_\alpha J_\alpha = {\downarrow}1$ is a directed join then we must have $1 \in J_\alpha$ for some α. Thus, $\mathscr{I}L$ is compact. □

Remark 103 In fact, analogously to Proposition 80 there is a more general result for showing a locale is compact from a presentation. See Theorem 10 of Vickers (2006).

We can now use the compactness of $2^{\mathbb{N}}$ to deduce that the closed real interval $[0, 1]$ is also compact by constructing an epimorphism $f : 2^{\mathbb{N}} \to [0, 1]$ which, intuitively, interprets the points of $2^{\mathbb{N}}$ as the binary expansions of real numbers. We omit the details.

Remark 104 You might have heard constructivists claim that Cantor space or the closed real interval might fail to be compact constructively. How can this be reconciled with the above results? The crucial point is that those constructivists are not referring to the true localic versions of $2^{\mathbb{N}}$ or $[0, 1]$, but instead to their spaces of points. The failure of these spaces to be compact is simply the failure of these locales to be spatial. This is an example of the localic approach being much better behaved than the point-set approach in the constructive setting. One topos where $2^{\mathbb{N}}$ and $[0, 1]$ fail to be spatial is the effective topos. See Bauer (2006) for details.

5 Applications

Now that we have seen the basic ideas behind constructive pointfree topology, we will take a look at some of the advantages of adopting a more topological approach to constructive mathematics. We will also see an example of how constructive topology can be useful even if you only care about proving classical theorems.

5.1 Ostensibly Classical Theorems That Always Hold for Locales

A number of topological results are famously nonconstructive and one might fear that this might make it difficult to do topology or analysis without them. However, we typically find that some formulation of the results do in fact hold for locales. Indeed, pointfree results tend to be remarkably robust with respect to precisely which foundational axioms are assumed.

5.1.1 Tychonoff's Theorem

Perhaps the most famous example of a nonconstructive theorem in topology is Tychonoff's theorem that an arbitrary product of compact spaces is compact. This is known to be equivalent to the axiom of choice. However, its localic variant is constructively valid. (The discrepancy is resolved by observing that the product might fail to be spatial even if all the factors are.)

Theorem 105 *Arbitrary products of compact locales are compact.*

The proof is somewhat too involved to give here, but see Vickers (2005) for details. A simple classical proof is also given in Johnstone and Vickers (1991), which can be made to work constructively if we restrict to indexing sets with decidable equality.

5.1.2 The Axiom of Choice

It is not only explicitly topological results that become constructively valid with the pointfree approach. Even classical set-theoretic theorems can often become constructive if we replace the sets with locales. The axiom of choice itself provides such an example.

Recall that the axiom of choice can be formulated as saying that a product of inhabited sets is inhabited. Constructively, we still have the following result from Henry (2023).

Theorem 106 *A product of positive overt locales indexed by a set with decidable equality is positive and overt.*

Proof Let $(X_\alpha)_{\alpha \in I}$ be a family of positive overt locales. Recall that $\mathcal{O}(\prod_{\alpha \in I} X_\alpha)$ has a presentation with a generator $\iota_\alpha(u)$ for each $\alpha \in I$ and $u \in \mathcal{O}X_\alpha$ and relations enforcing the relations holding in each $\mathcal{O}X_\alpha$. We can obtain a set of basic opens for the product by taking (Kuratowski-)finite meets of these generators. Explicitly, the basic opens are of the form $\bigwedge_{\alpha \in F} \iota_\alpha(u_\alpha)$ for $F \in \mathcal{P}_{\mathrm{fin}}(I)$.

If the indexing set I has decidable equality, then so does each Kuratowski-finite subset $F \subseteq I$ and hence such subsets are *Bishop*-finite. In other words, each such finite meet involves opens from *distinct* factors.

We now claim that $\bigwedge_{\alpha\in F}\iota_\alpha(u_\alpha)$ is positive if and only if u_α is positive for each $\alpha\in F$. (Note that it is important that each u_α comes from a different X_α, since if a and b lie in the same factor, their intersection might fail to be positive even if they both are.) We will show that the map $\bigwedge_{\alpha\in F}\iota_\alpha(u_\alpha)\mapsto\bigwedge_{\alpha\in F}[\![u_\alpha>0]\!]$ defines a suplattice homomorphism that is left adjoint to $!^*\colon\Omega\to\mathcal{O}(\prod_{\alpha\in I}X_\alpha)$ as in Proposition 80.

We would like to apply the coverage theorem and so we must find a presentation for $\mathcal{O}(\prod_{\alpha\in I}X_\alpha)$ that is of the appropriate form. The $\wedge$-semilattice of generators will consist of formal finite meets $\bigwedge_{\alpha\in F}\iota_\alpha(u_\alpha)$ for $F\in\mathcal{P}_{\text{fin}}(I)$. The meet operation is

$$\bigwedge_{\alpha\in F_1}\iota_\alpha(u_\alpha)\wedge\bigwedge_{\alpha\in F_2}\iota_\alpha(v_\alpha)=\bigwedge_{\alpha\in F_1\cup F_2}\iota_\alpha\Big(\bigwedge\big(\{u_\alpha\mid\alpha\in F_1\}\cup\{v_\alpha\mid\alpha\in F_2\}\big)\Big).$$

We should take a quotient so that $\iota_\beta(1)=1$ for each $\beta\in I$. That is, we have $\iota_\beta(1)\wedge\bigwedge_{\alpha\in F}\iota_\alpha(u_\alpha)=\bigwedge_{\alpha\in F}\iota_\alpha(u_\alpha)$ for all $F\in\mathcal{P}_{\text{fin}}(I)$. The relations in the frame presentation then consist of inequalities

$$\bigvee_{u\in U}\bigwedge_{\alpha\in F}\iota_\alpha(a_\alpha)\wedge\iota_\beta(u)\le\bigvee_{v\in V}\bigwedge_{\alpha\in F}\iota_\alpha(a_\alpha)\wedge\iota_\beta(v)$$

whenever $\bigvee U\le\bigvee V$ for $U,V\subseteq\mathcal{O}X_\beta$. It is not too difficult to show that this is indeed a presentation for the product locale.

Note that for the assignment $\bigwedge_{\alpha\in F}\iota_\alpha(u_\alpha)\mapsto\bigwedge_{\alpha\in F}\exists_{X_\alpha}(u_\alpha)$ to be well-defined map on the generators, we require that $\iota_\beta(1)=1$ for all $\beta\in I$, but this is true since each X_β is positive by assumption. Also note that this map is monotone. Now by the coverage theorem, this induces a suplattice homomorphism $\exists\colon\mathcal{O}(\prod_{\alpha\in I}X_\alpha)\to\Omega$ as long as it satisfies the necessary relations, which can easily be seen to hold.

Finally, we show that $\exists$ is left adjoint to $!^*\colon\Omega\to\mathcal{O}(\prod_{\alpha\in I}X_\alpha)$. We have

$$\begin{aligned}!^*\exists\Big(\bigwedge_{\alpha\in F}\iota_\alpha(u_\alpha)\Big)&=\bigwedge_{\alpha\in F}!^*\exists_{X_\alpha}(u_\alpha)\\&=\bigwedge_{\alpha\in F}\iota_\alpha!^*_{X_\alpha}\exists_{X_\alpha}(u_\alpha)\\&\ge\bigwedge_{\alpha\in F}\iota_\alpha(u_\alpha)\end{aligned}$$

and so $!^*\circ\exists\ge\text{id}$. On the other hand, $\exists!^*(p)=\bigvee\{\exists(1)\mid p\}=\bigvee\{\top\mid p\}=p$ and so $\exists\circ!^*=\text{id}$. Thus, $\exists$ is a left adjoint retraction of $!^*$ and hence $\prod_{\alpha\in I}X_\alpha$ is overt and positive, as required. □

Recall that sets are always overt, and a set is positive if and only if is inhabited. Furthermore, under the assumption of excluded middle, every set has decidable equality. So in this case a product of inhabited sets in **Loc** is always a positive locale. The set-theoretic axiom of choice is then equivalent to the claim that every such product locale is not just positive, but has a *point*.

5.1.3 Excluded Middle

Finally, there is a sense in which excluded middle itself holds from the localic perspective. Of course, a truth value $p \in \Omega$ is not guaranteed to have a true lattice-theoretic complement and so the closest approximation to such a complement $\neg p$ does not need to satisfy $p \vee \neg p = \top$ or $\neg\neg p = p$. On the other hand, we have already seen that the open sublocale P of 1 corresponding to the open $p \in \Omega = \mathcal{O}(1)$ always has a closed complement $P^{\mathsf{c}} \hookrightarrow 1$ in the lattice of all sublocales of 1.

It is not that the closed sublocale P^{c} fails to exist constructively, but that P^{c} is not necessarily an open sublocale (i.e. a subset) of 1. Indeed, the locale P^{c} cannot be shown to be a discrete locale at all. The open sublocale given by the intuitionistic negation of p is the *interior* of the closed sublocale P^{c} and the 'bad behaviour' of constructive negation is simply a reflection of the fact that we are forcing these locales to be sets.

A slightly different perspective on the same phenomenon is given in Vickers (2022). In **Set**, exponentials involving subsingletons are given by implication of truth values. In particular, $\emptyset^{\{* \mid p = \top\}}$ is isomorphic to $\{* \mid \neg p = \top\}$. In **Loc** on the other hand, the exponential object $0^P = \prod_p 0$ is isomorphic to the compact Hausdorff locale P^{c}. It can then be shown that $0^{0^P} \cong P$. This is another sense in which double negation elimination holds constructively for locales.

5.2 *An Extended Example: The Extreme Value Theorem*

Let us conclude this chapter with an example of a result from analysis proved using the constructive pointfree approach. We will then see how it might be applied to deduce a classical theorem in fibrewise topology.

We will prove a constructive version of the *extreme value theorem*. Classically, this states that if X is a compact topological space, a continuous function $f: X \to \mathbb{R}$ attains a maximum value. Our approach is based on that given in Taylor (2010).

5.2.1 One-Sided Real Numbers

We will need to understand suprema of real numbers. Classically, all inhabited, bounded sets of reals have suprema and infima, but constructively this is not the case. Suppose the set $\{0\} \cup \{1 \mid p = \top\}$ had a supremum s. By the locatedness property of Dedekind reals (see the start of Sect. 3.1) we have $s > 0 \vee s < 1$, but in the first case $p = \top$ and in the second $p \neq \top$, and hence excluded middle follows.

To discuss suprema and infima we must use slightly different variants of the reals that have a distinct constructive nature. Completing $\mathbb{Q}$ under inhabited suprema will give the *lower reals*. These are given by *one-sided* Dedekind cuts consisting only of the sets L of lower bounds instead of pairs (L, U). Similarly, completing $\mathbb{Q}$

under inhabited infima gives the *upper reals*, which are constructed from the sets of upper bounds U. Thus, lower reals are approximated from below and upper reals are approximated from above.

The formal axiomatisation of the lower reals $\overrightarrow{\mathbb{R}}$ involves a basic proposition ℓ_q for each $q \in \mathbb{Q}$ and the three families of axioms of Dedekind cuts that refer only to ℓ_q—namely, the following.

(1)	$\ell_q \vdash \ell_p$	for $p \leq q$	(L downward closed)
(2)	$\ell_p \vdash \bigvee_{q>p} \ell_q$	for $p \in \mathbb{Q}$	(L rounded)
(3)	$\top \vdash \bigvee_{q\in\mathbb{Q}} \ell_q$		(L inhabited)

This gives the presentation

$$\mathcal{O}\overrightarrow{\mathbb{R}} = \langle \ell_q,\ q \in \mathbb{Q} \mid \ell_p = \bigvee_{q>p} \ell_q,\ \bigvee_{q\in\mathbb{Q}} \ell_q = 1 \rangle.$$

Similarly, for the upper reals $\overleftarrow{\mathbb{R}}$ we have

$$\mathcal{O}\overleftarrow{\mathbb{R}} = \langle u_q,\ q \in \mathbb{Q} \mid u_q = \bigvee_{p<q} u_p,\ \bigvee_{q\in\mathbb{Q}} u_q = 1 \rangle.$$

Classically, $\overrightarrow{\mathbb{R}}$ is the space with underlying set $\mathbb{R} \sqcup \{\infty\}$ and the topology of lower semicontinuity. This has a base of opens of the form $(q, \infty]$ for $q \in \mathbb{Q}$. Similarly, $\overleftarrow{\mathbb{R}}$ is $\mathbb{R} \sqcup \{-\infty\}$ with the topology of upper semicontinuity.

Of course, the locales $\overrightarrow{\mathbb{R}}$ and $\overleftarrow{\mathbb{R}}$ are actually isomorphic (by negation), but it will be convenient to distinguish them. The familiar locale $\mathbb{R}$ of reals (also called *two-sided* or *Dedekind* reals in this context) is a sublocale of $\overrightarrow{\mathbb{R}} \times \overleftarrow{\mathbb{R}}$.

Since it is cumbersome to work with sets of rationals explicitly, in what follows we will simply use the letter r to denote a lower, upper or two-sided real and implicitly define the sets L or U (as appropriate) by saying when $q < r$ or $r < q$ for rational q.

5.2.2 Suprema over Sublocales

Note that classically the supremum r of a set $X \subseteq \mathbb{R}$ is characterised by $r \leq u$ if and only if $u \in \mathbb{R}$ is an upper bound of X (i.e. $\forall x \in X.\, x \leq u$). To define the supremum as a lower real we must specify for which $q \in \mathbb{Q}$ we have $q < \sup X$. We then have $q < \sup X \iff \neg(\sup X \leq q) \iff \neg(\forall x \in X.\, x \leq q) \iff \exists x \in X.\, q < x$.

Since we can existentially quantify over overt locales this suggests we can define the supremum of an overt sublocale V of $\mathbb{R}$ as a lower real r by

$$q < r \iff \exists x \in V.\, q < x.$$

The existential quantifier is interpreted by restricting the open $(\!(q, \infty)\!)$ to V (using the underlying frame homomorphism of the inclusion) and then applying $\exists\colon \mathcal{O}V \to \Omega$. If this gives $\top$ we write $V \between (\!(q, \infty)\!)$ which is understood to mean that V meets (i.e. intersects) $(\!(q, \infty)\!)$.

Lemma 107 *If V is a positive overt sublocale of $\mathbb{R}$, then $q < r \iff V \between (\!(q, \infty)\!)$ indeed defines a lower real r.*

Proof In $\mathcal{O}\mathbb{R}$ we have $(\!(p, \infty)\!) = \bigvee_{q>p} (\!(q, \infty)\!)$. Applying the join-preserving map $V \between (-)$ we obtain $V \between (\!(p, \infty)\!) \iff \exists q > p.\ V \between (\!(q, \infty)\!)$ and so we find $p < r \iff \exists q > p.\, q < r$. Thus, axioms (1) and (2) hold.

Similarly, $\bigvee_{q \in \mathbb{Q}} (\!(q, \infty)\!) = 1$ in $\mathcal{O}\mathbb{R}$ gives $\exists q \in \mathbb{Q}.\, q < r \iff V \between 1$. But $V \between 1$ is precisely the claim that V is positive and so axiom (3) follows and r is a lower real. □

Remark 108 Examining the proof it is apparent that we only used the axioms of $\mathbb{R}$ involving the generators of the form ℓ_q and so a similar result holds for positive overt sublocales of $\overrightarrow{\mathbb{R}}$.

Suprema of compact locales can also be defined, but this time we obtain an *upper* real. We define the supremum r of a compact sublocale K of $\mathbb{R}$ by

$$r < q \iff \forall x \in K.\, x < q.$$

Note that this is similar to the definition in our informal discussion above except that we restrict to rational q and replace $\leq$ with strict inequality.

Lemma 109 *If K is a compact sublocale of $\mathbb{R}$, then $r < q \iff K \leq (\!(-\infty, q)\!)$ indeed defines an upper real r.*

Proof Similarly to before we have $(\!(-\infty, q)\!) = \bigvee_{p<q} (\!(-\infty, p)\!)$ in $\mathcal{O}\mathbb{R}$. Thus, $K \leq (\!(-\infty, q)\!) \iff K \leq \bigvee_{p<q} (\!(-\infty, p)\!)$. Now note that the join is directed and so $K \leq \bigvee_{p<q} (\!(-\infty, p)\!)$ if and only if $K \leq (\!(-\infty, p)\!)$ for some $p < q$. Thus, we can conclude $r < q \iff \exists p < q.\, r < p$ and hence the upward closure and roundedness axioms for upper reals are satisfied.

Now consider $\bigvee_{q \in \mathbb{Q}} (\!(-\infty, q)\!) = 1$ in $\mathcal{O}\mathbb{R}$. We always have $K \leq 1$ and so by compactness $K \leq (\!(-\infty, q)\!)$ for some $q \in \mathbb{Q}$. Thus, $r < q$ for some $q \in \mathbb{Q}$ and the inhabitedness axiom holds. □

5.2.3 The Extreme Value Theorem

If K is a sublocale of $\mathbb{R}$ that is positive, overt *and* compact, then the above definitions define the supremum both as a lower real and as an upper real. As we might expect, these give the two sides of a single Dedekind cut.

To prove this makes the supremum r a Dedekind real it remains to prove the disjointness and locatedness axioms. Disjointness means that if $p < r$ and $r < q$ then $p < q$. By the definition of r, the assumption $p < r$ means $K \between (\!(p, \infty)\!)$ and $r < q$ means $K \leq (\!(-\infty, q)\!)$. Thus, $(\!(-\infty, q)\!) \between (\!(p, \infty)\!)$, or equivalently $(\!(-\infty, q)\!) \wedge (\!(p, \infty)\!) > 0$. But by the disjointness axiom of $\mathcal{O}\mathbb{R}$ this implies $p < q$, as required.

Locatedness means that if $p < q$ then either $p < r$ or $r < q$. Suppose $p < q$. We must show $K \between (\!(p, \infty)\!)$ or $K \leq (\!(-\infty, q)\!)$. By the locatedness axiom in $\mathcal{O}\mathbb{R}$ we know that $(\!(p, \infty)\!) \vee (\!(-\infty, q)\!) = 1$. Thus, the desired result follows from the following lemma.

Lemma 110 *Let K be a compact overt locale and suppose $u \vee v = 1$ in $\mathcal{O}K$. Then either $u > 0$ or $v = 1$.*

Proof By overtness we have $u = \bigvee\{u \mid u > 0\}$. Then by compactness $1 = u \vee v = \bigvee\{u \mid u > 0\} \vee v$ has a Kuratowski-finite subcover. So we have $c_1 \vee c_2 \vee \cdots \vee c_n = 1$ where for each i, either $c_i = v$ or $c_i = u > 0$. Since there are finitely many of these, we can conclude that either $c_i = v$ for all i or $u > 0$. In the former case $v = 1$ and in the latter case $u > 0$, as required. □

Finally, we note that r lies in the sublocale K and is thus a *maximum*. This is because by definition whether r lies in a subbasic open is defined purely in terms of that open's restriction to K and hence the map $r^*\colon \mathcal{O}\mathbb{R} \to \Omega$ factors through the frame quotient $\mathcal{O}\mathbb{R} \twoheadrightarrow \mathcal{O}K$.

Therefore, up to taking images, we have proved the following.

Theorem 111 (Extreme value theorem) *If X is any positive compact overt locale and $f\colon X \to \mathbb{R}$, then f has a supremum given by a Dedekind real number r defined by $q < r \iff f^*(\!(q, \infty)\!) > 0$ and $r < q \iff f^*(\!(-\infty, q)\!) = 1$. Furthermore, this supremum is a maximum in the sense that $\mathcal{S}f^*(\{r\})$ is (compact and) positive.*

Proof Simply take $K = \mathcal{S}f_!(X)$. The image of a positive compact overt sublocale is positive, compact and overt. The definition of r follows since $\mathcal{S}f_!(X) \leq u \iff X \leq f^*(u)$ by adjointness and $\mathcal{S}f_!(X) \between u \iff X \between f^*(u)$ by a simple extension of Corollary 65.

Finally, we show $\mathcal{S}f^*(\{r\})$ is positive. Since $\mathbb{R}$, and thus K, is Hausdorff, the inclusion $\{r\} \hookrightarrow K$ is closed. Now note that the map f is epic onto its image and consider the following pullback diagram.

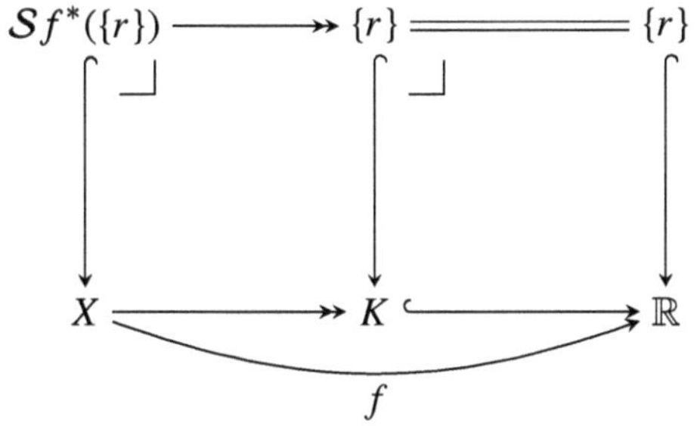

It is not hard to show that epimorphisms of locales are stable under pullback along closed inclusions, and hence the map $\mathcal{S}f^*(\{r\}) \to \{r\}$ is epic. It follows that $\mathcal{S}f^*(\{r\})$ is positive. (It is compact, because it is a closed sublocale of X.) □

Remark 112 As we might have come to expect by this point, the extreme value theorem is *not* a theorem of constructive *point-set* topology. In particular, the positive locale $\mathcal{S}f^*(\{r\})$ might fail to contain a point. There are still some constructive point-set results about the existence of suprema and infima of certain sets of reals (see Troelstra and van Dalen (1988, Sect. 6.1)), but these can be derived as consequences of our pointfree theorem.

Remark 113 It is also interesting to consider how the supremum of a positive compact overt sublocale of $\mathbb{R}$ changes as the sublocale is varied. We can understand this by using a localic version of the Vietoris hyperspace construction (see Johnstone (1985)). The points of the (positive) Vietoris hyperlocale $V_+(X)$ are the positive compact overt sublocales of X (see Vickers (1997)), and the supremum becomes a locale morphism from $V_+(\mathbb{R})$ to $\mathbb{R}$ (by essentially the same argument we gave above).

5.2.4 An Application to Fibrewise Topology

Now we can discuss an example of the pay-off we get from proving Theorem 111 constructively. As we mentioned in Sect. 2.2.3, constructive results can be interpreted in any topos. In particular if B is a locale, we can interpret the extreme value theorem in the topos $\mathrm{Sh}(B)$ of sheaves over B. Let us see what this gives us.

An in-depth discussion of how to interpret intuitionistic logic in $\mathrm{Sh}(B)$ is beyond the scope of these notes. See Johnstone (2002, Part C) for details. Instead, we provide the following dictionary to translate between internal constructive concepts and their externalisations. In particular, we will use that there is an equivalence of categories between locales internal to $\mathrm{Sh}(B)$ and the slice category $\mathbf{Loc}/B$.

Constructive concept	**Interpretation in** $\mathrm{Sh}(B)$
A locale X	A locale map $\chi\colon X \to B$
The locale $\mathbb{R}$	The projection $\pi_1\colon B \times \mathbb{R} \to B$
A locale morphism	A morphism in $\mathbf{Loc}/B$
A point of X	A section of χ
X is overt	χ is open
The map $\exists\colon \mathcal{O}X \to \Omega$ for X overt	The left adjoint $\chi_!\colon \mathcal{O}X \to \mathcal{O}B$
X is overt and positive	χ is open and epic
X is compact	χ is proper
The map $!_*\colon \mathcal{O}X \to \Omega$ for X compact	The right adjoint $\chi_*\colon \mathcal{O}X \to \mathcal{O}B$
X is compact and positive	χ is proper and epic

Using this we can now see that interpreting Theorem 111 in $\mathrm{Sh}(B)$ immediately gives the following result.

Theorem 114 *Let $\chi : X \to B$ be a locale map that is epic, proper and open and let $f : X \to \mathbb{R}$ be any locale map. Then there is a locale map $\rho : B \to \mathbb{R}$ defined by*

$$\rho^*((q, \infty)) = \chi_! f^*((q, \infty))$$
$$\rho^*((-\infty, q)) = \chi_* f^*((-\infty, q))$$

taking the fibrewise supremum[14] *of f.*

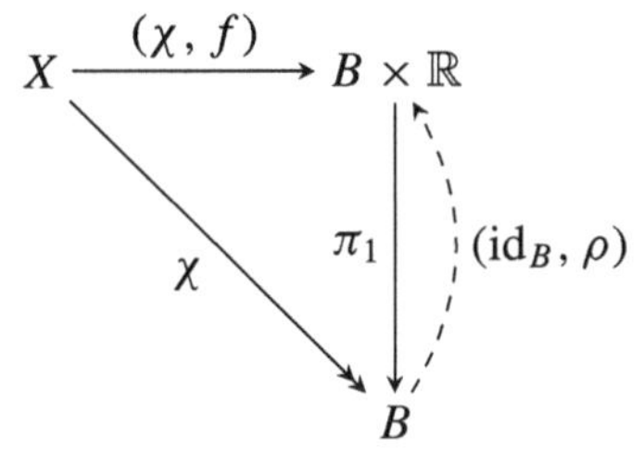

Moreover, this is a fibrewise maximum *in the sense the canonical morphism into B from the sublocale M of X where the maximum is attained is a (proper) epimorphism.*

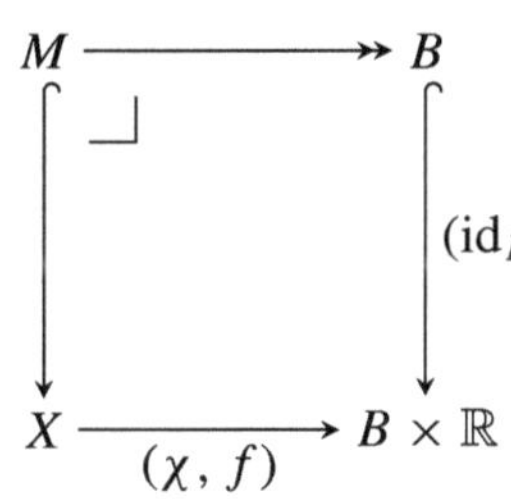

So we have obtained a 'fibrewise' extreme value theorem for free!

Remark 115 Suppose that contrary to Remark 112 the constructive extreme value theorem proved the existence of a *point* achieving the maximum. Then in the context of Theorem 114 we would find that the map χ always has a family of continuous local sections which pick out points that attain the maximum in each fibre. But note that if $X = B = \mathbb{C}$, $\chi(x) = x^2$ and $f(x) = |x|$, then no such section exists in the neighbourhood of 0, since there is no continuous choice of square roots.

While it is nice to obtain results for free, a sceptic might point out that Theorem 114 is still a result of constructive pointfree topology and so what use is it to classical mathematicians? Of course, every constructive result is also a classical result. Moreover, results in point-set topology often follow from their pointfree counterparts. In this case, we have the following corollary (assuming classical logic).

[14] The above specification of ρ is the *definition* of fibrewise supremum by the same argument we gave when defining suprema of real numbers above. See also the proof of Corollary 116.

Corollary 116 *Let $\chi \colon X \to B$ a proper and open quotient of topological spaces and let $f \colon X \to \mathbb{R}$ be a continuous function. Then the function $\rho \colon B \to \mathbb{R}$ defined by $\rho(b) = \max_{x \in \chi^{-1}(b)} f(x)$ is continuous.*

Proof It can be shown that if a continuous function between topological spaces is proper or open then the corresponding locale map is too. Thus, Theorem 114 applies and we obtain a map from the locale associated to B to $\mathbb{R}$, which in turn induces a continuous map $s \colon B \to \mathbb{R}$.

This map s is equal to ρ by construction, but let us spell out the formal details. In spatial terms the map $\chi_!$ is simply the image of χ restricted to open sets. Hence, we have $s^{-1}((q, \infty)) = \chi(f^{-1}(q, \infty))$ and so $s(b) > q \iff b \in \chi(f^{-1}(q, \infty)) \iff \exists x \in \chi^{-1}(b).\ f(x) > q \iff \sup_{x \in \chi^{-1}(b)} f(x) > q$. Thus, $s(b) = \sup_{x \in \chi^{-1}(b)} f(x)$, as required. (Note that this supremum is indeed a maximum by the extreme value theorem and because $\chi^{-1}(\{b\})$ is compact since χ is proper.) □

Remark 117 This result can be understood as a variant of Berge's maximum theorem (see Aliprantis and Border (2006, Sect. 17.5)) where the compact-valued correspondence is given by the preimage of a proper and open continuous map.

Therefore, the constructive approach has yielded a result that should be interesting even to an avowedly classical mathematician. In contrast, we would not have been able to obtain Corollary 116 from the classical extreme value theorem without additional work.

6 Further Reading

In these notes I have tried to give a taste of constructive pointfree topology, as well as some idea of its advantages over both classical topology and non-topological approaches to constructive mathematics. There is, of course, much that I have not been able to cover. If you are interested to learn more, I encourage you to look at the following resources:

- Picado and Pultr (2012) is a nice introduction to the classical theory of pointfree topology. It is fairly comprehensive and can be a good reference. Of the resources I recommend here, this demands the least of the reader. Unfortunately, the proofs given are usually nonconstructive, but once you become more comfortable with constructive reasoning it will generally not be too hard to constructivise them as necessary.
- Manuell (2019) is my PhD thesis. The Background chapter will be of most interest. It has some overlap with these notes, but covers certain topics, such as suplattices, in more detail. It also contains a crash course in categorical logic (though see Pitts (1991) for more details) and later gives examples of how it might be applied to the study of pointfree topology.

- Henry (2016) provides a constructive proof of the localic Gelfand duality, though probably most relevant is the helpful background material. It can be more difficult than the previous resources, but Henry works constructively throughout, and some of what is discussed cannot be found elsewhere.
- Johnstone (2002) is an enormous reference on topos theory, but Part C deals extensively with constructive pointfree topology. It is the most difficult reference of all, but it contains a huge amount of very useful information.

For further information on geometric logic, see Vickers (2014, 1997). For a different but related approach to constructive pointfree topology see Bauer and Taylor (2009) and Taylor (2010) on Abstract Stone Duality.

References

Abramsky S, Vickers S (1993) Quantales, observational logic and process semantics. Math Struct Comput Sci 3(2):161–227. https://doi.org/10.1017/S0960129500000189

Aliprantis CD, Border KC (2006) Infinite dimensional analysis—a hitchhiker's guide, 3rd edn. Springer, Heidelberg

Bauer A (2006) König's lemma and Kleene tree. Unpublished notes. http://math.andrej.com/wp-content/uploads/2006/05/kleene-tree.pdf

Bauer A (2012) Intuitionistic mathematics and realizability in the physical world. In: Zenil H (ed) A computable universe: understanding and exploring nature as computation. World Scientific, Hackensack, NJ, pp 143–157. https://doi.org/10.1142/9789814374309_0008

Bauer A (2017) Five stages of accepting constructive mathematics. Bull Am Math Soc (NS) 54(3):481–498. https://doi.org/10.1090/bull/1556

Bauer A, Taylor P (2009) The Dedekind reals in abstract stone duality. Math Struct Comput Sci 19(4):757–838. https://doi.org/10.1017/S0960129509007695

Bell JL, DeVidi D, Solomon G (2001) Logical options—an introduction to classical and alternative logics. Broadview Press, Peterborough, Ontario

Escardó M (2004) Synthetic topology: of data types and classical spaces. In: Proceedings of the workshop on domain theoretic methods for probabilistic processes, McGill University, Montreal, Canada, April 21–25, 2003. Electron Notes Theor Comput Sci 87:21–156. Elsevier, Amsterdam. www.sciencedirect.com/science/article/pii/S1571066104051357

Gabbay DM, Rodrigues OT (2016) Elementary logic with applications—a procedural perspective for computer scientists, studies in logic (London), vol 65. College Publications, London

Henry S (2016) Localic metric spaces and the localic Gelfand duality. Adv Math 294:634–688. https://doi.org/10.1016/j.aim.2016.03.006

Henry S (2023) Localic metric spaces and the localic Gelfand duality. arXiv:1411.0898v2

Johnstone PT (1985) Vietoris locales and localic semilattices. In: Hoffmann RE, Hofmann KH (eds) Continuous lattices and their applications (Bremen, 1982). Lecture notes in pure and applied mathematics, vol 101. Dekker, New York, pp 155–180

Johnstone PT (2002) Sketches of an elephant—a topos theory compendium, volume 2. Oxford logic guides, vol 44. Oxford University Press, Oxford

Johnstone P, Vickers S (1991) Preframe presentations present. In: Carboni A, Pedicchio MC, Rosolini G (eds) Category theory—proceedings of the international conference held in Como, 1990. Lecture notes in mathematics, vol 1488. Springer, Berlin, pp 193–212. https://doi.org/10.1007/BFb0084221

Joyal A, Tierney M (1984) An extension of the Galois theory of Grothendieck. Mem Am Math Soc 51(309). https://doi.org/10.1090/memo/0309

Manuell G (2019) Quantalic spectra of semirings. PhD thesis, University of Edinburgh
Moerdijk I (1988) The classifying topos of a continuous groupoid. I. Trans Am Math Soc 310(2):629–668. https://doi.org/10.2307/2000984
Picado J, Pultr A (2012) Frames and locales—topology without points. Frontiers in mathematics. Birkhäuser/Springer Basel AG, Basel. https://doi.org/10.1007/978-3-0348-0154-6
Pitts AM (1991) Notes on categorical logic. Unpublished lecture notes. https://www.cl.cam.ac.uk/~amp12/papers/notcl/notcl.pdf
Riehl E (2016) Category theory in context. Aurora dover modern math originals. Dover Publications Inc, Mineola, New York
Taylor P (2010) A lambda calculus for real analysis. J Log Anal 2(5):1–115. https://doi.org/10.4115/jla.2010.2.5
Troelstra AS, van Dalen D (1988) Constructivism in mathematics—an introduction, vol I. Studies in logic and the foundations of mathematics, vol 121. Elsevier, Amsterdam
van Dalen D (2001) Intuitionistic logic. In: Goble L (ed) The Blackwell guide to philosophical logic, Blackwell philosophy guides. Blackwell, Oxford, pp 224–257
Vermeulen JJC (1994) Proper maps of locales. J Pure Appl Algebra 92(1):79–107. https://doi.org/10.1016/0022-4049(94)90047-7
Vickers S (1989) Topology via logic. Cambridge tracts in theoretical computer science, vol 5. Cambridge University Press, Cambridge
Vickers S (1997) Constructive points of powerlocales. Math Proc Cambridge Philos Soc 122(2):207–222. https://doi.org/10.1017/S0305004196001636
Vickers S (2005) Some constructive roads to Tychonoff. In: Crosilla L, Schuster P (eds) From sets and types to topology and analysis: towards practicable foundations for constructive mathematics. Oxford logic guides, vol 48. Oxford University Press, Oxford, pp 223–238. https://doi.org/10.1093/acprof:oso/9780198566519.003.0014
Vickers S (2006) Compactness in locales and in formal topology. Ann Pure Appl Logic 137(1–3):413–438. https://doi.org/10.1016/j.apal.2005.05.028
Vickers S (2014) Continuity and geometric logic. J Appl Log 12(1):14–27. https://doi.org/10.1016/j.jal.2013.07.004
Vickers S (2022) Generalized point-free spaces, pointwise. arXiv:2206.01113

An Introduction to Categorical Proof Theory

Amirhossein Akbar Tabatabai

Abstract Categorical proof theory studies the structure of propositions and deductions by representing them as objects and morphisms within a category. In this gentle introduction to categorical proof theory, we begin with an extensive overview of category theory itself, motivating some of its basic notions through numerous examples and proof-theoretic intuitions. We then introduce three families of categories—bicartesian closed, almost bicartesian closed, and cartesian closed—to serve as categorical formalizations of three fragments of intuitionistic propositional deduction. We also present presheaf categories as the categorical formalization of the Brouwer-Heyting-Kolmogorov (BHK) interpretation and provide some representation theorems illustrating the generality of the BHK reading of intuitionistic deductions. In addition, we briefly discuss certain aspects of categorical proof theory for classical propositional logic. Subsequently, we employ the aforementioned three families of categories as semantics for formal intuitionistic propositional deductions and establish soundness and completeness results for three fundamental problems: the existence of proofs, the equality of proofs, and the identity of propositions. Finally, we extend the discussion from propositional to higher-order languages, introduce a form of realizability, and use it to construct computable (resp. continuous) worlds in which all functions are computable (resp. continuous). These constructions enable us to prove the consistency of certain anti-classical theories, such as finite-type Heyting arithmetic with the Church-Turing Thesis or with Brouwer's Continuity Principle. Moreover, realizability provides a means of extracting computational content from intuitionistic and classical arithmetical proofs. For instance, we show that any provably recursive function in Heyting or Peano arithmetic is a primitive recursive functional.

A. Akbar Tabatabai (✉)
Faculty of Science and Engineering (FSE), University of Groningen,
Groningen, The Netherlands
e-mail: amir.akbar@gmail.com

M. M. Clementino et al. (eds.), *Topology, Algebra and Categories in Logic*,
Coimbra Mathematical Texts 5, https://doi.org/10.1007/978-3-032-13760-9_4

1 A Historical Introduction

It is equally stupid and simple to consider mathematics to be just an axiom system as it is to see a tree as nothing but a quantity of planks van Stigt (1990). L. E. J. BROUWER

Proof theory was the child of a great panic. Attacked by a wild wave of paradoxes, the foundation of mathematics was in great danger, and to save it, Hilbert came up with a formalistic program to reduce mathematics (or at least its consistency) to its trusted finitary core. His idea was to first formalize mathematics as a purely syntactical theory and then show the *consistency* of that theory using what he vaguely called the *finitary methods*. This way, he hoped to bring peace back to the shaking foundation. From the very beginning, though, it was quite clear that for most interesting theories (e.g., Peano arithmetic), the usual strategy of proving consistency by model construction was not an option, simply because the models of such theories were usually infinite and hence beyond our trusted core. Therefore, the only remaining option was to work directly with the syntactical proofs, the *finitary* and tame objects that mathematicians use every day to communicate. As expected, Hilbert's strategy was to investigate the structure of *all* possible proofs of a proposition in a purely combinatorial way to hopefully rule out the contradiction as a provable proposition. Even more ambitious, through a pure formal investigation of the possible proofs of a theorem, he hoped to be able to prove the *conservativity* of mathematics over its finitary core by eliminating all possible ideal infinitary entities as helpful auxiliary tools that we use for our convenience, yet we know they have no real effect.

As Gödel showed in 1931, the program, as stated above, fails dramatically. In fact, there is no way to prove the consistency of mathematics (not even finitary mathematics) within finitary mathematics itself. This failure to prove consistency also shows the failure to prove conservativity, as the consistency of finitary mathematics is a simple finitary claim about the non-existence of a proof as a finitary object. This consistency statement, however, is provable in mathematics using the axiom of the existence of an infinite set, while it is unprovable in finitary mathematics based on what Gödel proved. As disappointing as it seems, the story, of course, has not ended with this huge negative impact of Gödel's theorems, and we are going to explain how. But before going any further, let us pause for a moment to consider the role of proofs in Hilbert's proof theory.

1.1 Proofs as Auxiliary Objects

As is clear even from our brief explanation of Hilbert's program, Hilbert employs proofs as a *tool* to learn something about the theorems of a given theory. For him, what has the central role in proof theory is not the proofs but the *provability*. Consistency is the unprovability of contradiction, and conservativity means that whatever is provable in a bigger theory is also provable in the smaller one. Even in many formalizations of constructive mathematics, where the constructions—and hence the proofs—must have the central role, people usually follow Hilbert-style proof theory and focus mostly on provability rather than on proofs. Such an interest in the shadows of an entity rather than the entity itself, such a move from the intensional to the extensional, from a notion to its extension, has similar incarnations in many mathematical fields. For instance, in the early days of the theory of computation, the central role was given to computability, or to a lesser extent to the functions that algorithms compute, rather than to the algorithms themselves. As is clear today, the crown of the central notion in the theory of computation must be awarded to computation itself, and not to mere computability. In a similar fashion, one can argue that proof theory must be elevated to a theory about proofs rather than merely provability.

To illustrate how auxiliary proofs are even within their own home of proof theory, let us consider the most basic questions a mathematician might ask herself when encountering a new mathematical concept: namely, what is the *definition* of the concept, and what is its corresponding *identity*? For the first question, the answer seems straightforward. A proof is a *syntactical* object constructed within a syntactical calculus, starting from certain axioms and following certain rules. This definition aligns with the aims of Hilbert's program, as it provides a sound and complete definition for the program's main interest: provability. However, it is fair to say that this definition is more of a practical first draft than a polished final answer. Here are two reasons to support this claim.

The first concern is the role of syntax in the definition. A proof, regardless of its nature, must be a *formal* entity distant from any sort of content. This is a philosophically acceptable assumption. However, the definition provided does not encompass form in its general sense. Instead, it relies on a specific kind of syntax as a substitute, and this reduction introduces its own problems. In Hilbert's program, equating the formal with the syntactical serves a technical purpose, as it allows for the use of combinatorial techniques to investigate the structure of proofs. However, one can argue that form is a much richer notion than any type of syntax can capture, and even if it could, the result would not meet modern standards of mathematics. To illustrate this, let us consider an abstract mathematical entity, such as a ring. A ring is indeed constructed from elements, but what a ring theorist is truly interested in is the abstract structure of the ring beyond the nature of its elements. This structure is what can be referred to as the form of the ring.

Is it possible to reduce this form to a kind of syntax? For rings—and let us emphasize that this is not necessarily true in general—the answer is affirmative. In fact, one can observe this possibility by examining the nineteenth-century approach to "ring theory", which employed polynomials as its syntax. The theory begins with a given set of variables as the generic, meaningless generators of the ring and then imposes certain equations as the expected relationships among the resulting polynomials. Although it is possible to develop ring theory in this manner, it is clear that the form of rings extends beyond their polynomial representations. Mathematically, it is fair to assert that the significant advances in ring theory during the twentieth century owe their success to liberation from such polynomial presentations or any other form of presentation, for that matter. Taking this analogy seriously, one might even contend that the overly tight interdependence of proofs and syntax is a primary reason behind the relative narrowness of proof theory:

> *There ought to be a vibrant specialty of "proof theory". There is a subject with this title, started by David Hilbert in his attempt to employ finitistic methods to prove the correctness of classical mathematics... In 1957, at a famous conference in Ithaca, proof theory was recognized as one of the four pillars of mathematical logic (along with model theory, recursion theory and set theory). But the resulting proof theory is far too narrow to be an adequate pillar..* Mac Lane (1997)

The form-to-syntax reduction is not merely an abstract philosophical problem; it also carries practical implications. For instance, consider the immaterial yet necessary choices that any use of syntax compels us to make. Is a proof a sequence of formulas in a given Hilbert-style system, or is it a tree of formulas in the sense of natural deduction? There are numerous proof systems that employ various syntactical choices, making it difficult to assert that any single system is the true representation of real abstract formal proofs. Most likely, all these proof systems offer different presentations of a single abstract notion, akin to the various presentations of a ring through different generators and relations. The ring exists beyond its presentations, and so does the proof

The second problem with the conventional definition of a proof is its *reductive* nature. It characterizes a proof as a sequence or a tree of propositions, thereby reducing proofs to formulas. This reductive approach is analogous to the usual interpretation of an interval as a collection of points, a function as a specific type of relation, and movement as a sequence of snapshots. From a philosophical standpoint, it is far from trivial to believe whether the former can be faithfully reduced to the latter. Even mathematically speaking, one can argue that, even if this reduction is indeed possible, the outcome may not be desirable, as it could restrict the independence of the former and thus inhibit some of its alternative yet interesting interpretations.

For instance, consider the following three examples. First, we have Brouwer's refusal to reduce the continuum to its points, which unearthed alternative notions of real numbers that later reemerged as natural constructions in classical geometry via Grothendieck toposes Mac Lane and Moerdijk (1994). These alternative notions compel all functions over the reals to be continuous or even smooth, as a geometer might expect. They also allow for the existence of a non-zero real number ϵ satisfying $\epsilon^2 = 0$, making the development of differential geometry easier and even synthetically possible Kock (2006).

Secondly, consider the point-free approach to real numbers, where subsets are not merely collections of points but specific equivalence relations over the open sets of the reals. This shift in perspective provides more refined subsets and addresses some anomalies, such as the existence of non-measurable subsets of real numbers Simpson (2012).

Finally, consider the conceptual leap in modern mathematics that places homomorphisms on equal footing with their algebraic structures. This leap has enabled the elegant progress of modern mathematics. Taking the analogy in these three examples seriously, one can argue that a proof, whatever it may be, is more than a mere collection of propositions. It must be an independent notion that interacts with the propositions but is not necessarily reducible to them. In short, it is a tree, not a quantity of planks.

So far, we have examined several arguments against the conventional definition of a proof, illustrating the satisfaction of Hilbert's proof theory with a minimal working definition of proofs as mere auxiliary tools within the theory. Now, let us continue this line of reasoning by addressing the second significant yet overlooked problem: the identity problem, which has two facets: *proof equivalence* and *propositional identity*. As noted, Hilbert's proof theory has largely ignored these issues. The reason is straightforward: these identities were unrelated to the provability that the program was originally concerned with. Consequently, the program had no impetus to investigate these problems seriously. This lack of interest further underscores the auxiliary role that proofs play in Hilbert's proof theory. Where else in mathematics, the central notions of a theory do not even have their own notion of identity?

As a concluding remark in this subsection, let us briefly elucidate the two identity problems mentioned and their interrelationship. The first is the *equivalence problem* for proofs. This problem is profound and intriguing, connecting programming languages on one side and *coherence problems* in category theory and homotopy theory Došen (2003) on the other. As we will explain later, proof equivalence is typically founded on *redundancy elimination*. For instance, consider the following two derivations:

$$\dfrac{\dfrac{\begin{matrix}\mathsf{D}_1 \\ A\end{matrix} \qquad \begin{matrix}\mathsf{D}_2 \\ B\end{matrix}}{A \wedge B}\wedge I}{A}\wedge E_1 \qquad\qquad \begin{matrix}\mathsf{D}_1 \\ A\end{matrix}$$

One might argue that these derivations are equivalent, as the left one is merely the derivation D_1 with a redundant step that does not alter the essence of the proof. However, when redundancies are absent, the meaning of equivalence becomes unclear. For instance, consider the following two derivations:

$$\dfrac{A \wedge A}{A}\wedge_1 E \qquad \dfrac{A \wedge A}{A}\wedge_2 E$$

It is not intuitively clear whether they are equivalent in any meaningful sense. Additionally, the problem becomes even more complex when we are tasked with comparing two proofs that exist within different systems.

For the identity problem concerning propositions, there is fortunately an intuitive definition: two propositions are considered identical if and only if they hold the same meaning, allowing one to be used in place of the other in any context. The challenge lies in formalizing this *propositional identity* in a formal and abstract manner without delving into the specifics of their meanings.

The most straightforward formalization that comes to mind is the logical equivalence relation, which asserts that two propositions logically imply each other. In other words, there exists a proof D_1 of B from A and a proof D_2 of A from B. However, this formalization proves to be inadequate. To illustrate this, let us represent the propositions A and B as sets of their respective proofs. In this context, logical equivalence merely indicates the existence of a function mapping from the set of proofs for A to the set of proofs for B, and vice versa. If A and B are truly identical propositions, one would expect their associated sets of proofs to also be identical. This necessitates that there is a bijection between the two sets, which can be induced by composing the two proofs, D_1 and D_2. Therefore, we require these proofs to be inverses of each other, such that $\mathsf{D}_1 \circ \mathsf{D}_2 = id$ and $\mathsf{D}_2 \circ \mathsf{D}_1 = id$, where $\circ$ denotes the composition of the two proofs and id represents the trivial proof of a proposition by itself.

It is evident that the formalization of propositional identity necessitates a formalization of equivalence between the proofs themselves. Furthermore, it is important to note that even when our primary concern is the propositions rather than the proofs or their provability, certain purely propositional issues may still require an examination of proofs and their equivalence.

1.2 Gentzen and His Consistency Proof

In the opening of this section, we explained how Hilbert's proof theory emerged as a natural response to the foundational crisis of the early twentieth century, and we deviated from the narrative to emphasize how auxiliary the notion of a proof is in Hilbert's proof theory. Now, it is time to return to the rest of the story. We stated that Gödel demonstrated the impossibility of proving the consistency of mathematics (even finitary mathematics) within finitary mathematics itself. This is actually not entirely accurate. The formal statement of Gödel's incompleteness theorem shows that no recursively axiomatizable theory that is strong enough can prove its own consistency, and this theorem applies to Hilbert's program if we assume that his finitary mathematics is recursively axiomatizable. Although such an assumption is quite reasonable, it can also be argued that there is no way to formalize all of finitary mathematics but only its proper parts. Therefore, for any consistent theory T, it may still be possible to find a strong enough finitary theory F_T that is sufficiently powerful to prove the consistency of T. The only thing that Gödel's incompleteness theorem establishes is that this finitary theory F_T cannot be a subtheory of the theory T. This is precisely how Gentzen began to investigate the situation. For him, the theory T was Peano Arithmetic, PA, for which he identified the corresponding finitary theory F_T as the theory PRA, a very basic quantifier-free theory, along with quantifier-free transfinite induction up to the ordinal ϵ_0. One might object that such a theory cannot be considered finitary, as it operates with infinite ordinals. However, it is important to recognize that what we are actually working with is merely a finitary representation of the ordinals below ϵ_0 as complex combinatorial objects, and the set-theoretical language of ordinals is employed solely to aid our understanding of this intricate combinatorial structure.

What we are truly interested in here is not Gentzen's proof of the consistency of arithmetic, but rather his strategy. We observed that to prove the consistency of a theory, one must investigate all possible proofs of a contradiction in a purely combinatorial manner. However, it should have been quite clear from the very beginning that proofs, as combinatorial objects, are extremely complex and challenging to analyze. To understand why, we do not need to delve into any specific proof theory; it is sufficient to consider how mathematicians grapple with their conjectures every day. If it were easy to determine what is provable and what is not, any conjecture could be resolved with a sufficient amount of combinatorial effort in proof theory.

However, this observation does not imply that there is absolutely no hope in investigating the proofs of simple statements in natural theories, such as the contradiction in PA. To systematically investigate all possible proofs, Gentzen devised an elegant strategy. The first step involved developing a well-behaved and easily controllable proof system. In fact, he created two elegant proof systems known as *natural deduction* and *sequent calculus*. The primary characteristic of these systems, aside from their natural usability, is the capacity to transform their proofs into simpler forms in a certain formal sense. This process is referred to as *normalization* in *natural deduction* and *cut elimination* in *sequent calculus*. By achieving the simplest possible form of a proof, one may gain the insight needed to discern its structure; using that structure, one can demonstrate that the contradiction does not have a simple proof, and thus not any proof, within the system. It is even possible to employ this framework to prove certain conservativity results by showing that the simplest proofs in a stronger theory actually reside within the weaker one. We hope the reader can recognize the irony here. To accomplish the primary objective of proof theory, namely proving consistency, even though we are primarily interested in provability, we are mathematically compelled to be meticulous about the formalization we choose for proofs, the structure of those proofs within that formalization, and the proof transformations that serve as witnesses for proof equivalence. In short, understanding the full structure of the proofs is essential for comprehending the seemingly simpler concept of provability!

Inspired by Gentzen's elegant work, interest in proof theory flourished. However, it primarily followed Gentzen's footsteps in investigating the form of proofs to gain insights into the associated theories, ranging from their consistency to the characterization of the functions that the theory can define. It also inherited the Hilbertian attitude, wherein proofs remained syntactical, mostly confined to one formalization and serving the theorems they prove. It is true that proof theorists became increasingly interested in the structure of proofs, but this was largely because they were compelled to do so.

1.3 Categorical Proof Theory

Gentzen's elegant work also gave birth to an alternative type of proof theory—a theory in which proofs play the main role, elevating their status from the planks of provability to a truly alive tree of proofs. The first such theory is Dag Prawitz's *general proof theory*, a term he coined to distinguish his approach from what he termed *reductive proof theory* (Prawitz 1970, 1974). The latter essentially represents the Hilbertian proof theory, using proofs as tools to reduce one theory to another to demonstrate consistency and conservativity. In contrast, general proof theory shifts the focus from provable propositions to the proofs themselves. Here, proofs are treated as elements of an *algebraic structure* where the rules function as its operations. The study of this structure is the primary goal of general proof theory.

However, this algebraic structure is presented to us in an indirect manner. The situation resembles the style of nineteenth-century matrix theory. While matrices can be defined as certain algebraic entities with corresponding operations, these arrays of numbers actually represent the linear transformations that are of primary interest. The challenge lies in the fact that a single linear transformation can be represented by multiple matrices, depending on the basis used for the underlying vector spaces. Since we are interested in the transformations rather than their representations, our initial task is to determine when two matrices represent the same transformation, or in other words, when they are *equivalent*. A similar scenario occurs in general proof theory. Proofs are represented within a proof system, specifically natural deduction, and the main challenge is to find an appropriate notion of equivalence between them.

Prawitz's solution was intriguing; he utilized proof transformations and demonstrated that any proof can be simplified to a unique simplest possible form known as the *normal form*. Thus, two proofs can be considered equivalent if they have the same normal form. This can be likened to algebraic simplifications. For example, two polynomials, $x(x + 1) + x$ and $x^2 + x + x$, are deemed equivalent when their simplest forms, namely $x^2 + 2x$, are equal. General proof theory, however, extends beyond this equivalence problem. It also seeks to understand the meanings of logical constants based on their roles within proofs. Nonetheless, it is fair to assert that the equivalence problem is central to the theory.

As intriguing as Prawitz's general proof theory is, it comes with its own limitations. It conflates form with syntax, reduces proofs to propositions, and confines itself to a single fixed system—namely, natural deduction.[1] At this juncture, a reader might argue that while the notion of liberation from syntax and reduction to propositions is appealing, it remains unclear whether a non-reductionist definition of proofs is even feasible. A similar skepticism can be applied to linear transformations: if we aim to avoid representations, how can we discuss linear transformations directly? The answer lies in understanding the two orthogonal mindsets typically involved in defining a mathematical notion.

The *internal* mindset favors constructing new mathematical entities from existing ones, using the latter as building blocks. However, this approach can inadvertently incorporate unintended aspects, such as construction methods, into the new mathe-

[1] It is important to clarify that general proof theory does not entirely restrict itself to syntax. Prawitz explicitly differentiates proofs from their syntactical representations. Moreover, for him, natural deduction is not merely one syntactical system among many; rather, it serves as a canonical framework for elucidating how proofs behave and what the logical constants signify. One can easily discern the early influences of the later categorical approach within these ostensibly syntactical aspects of general proof theory. That said, it is also notable that during the normalization process, proofs are identified through the syntactical derivations of propositions, making the process an algebraic simplification of syntax, where the aforementioned drawbacks emerge.

matical entity. To eliminate these extraneous elements, one must introduce an equivalence relation to collapse them, working with the entity only up to this relation to maintain statements about the entity rather than its representation. The process of constructing proofs from propositions and deriving linear transformations from real numbers exemplifies this internal mindset. Dually, the *external* mindset focuses on defining a mathematical entity based on its *relative* behavior rather than on its constituents or construction. The modern approach to linear algebra is a prime example of this external mindset, as it does not define vectors and linear transformations in terms of arrays of real numbers; instead, it characterizes them as abstract objects that behave as expected within their respective contexts.

Modern mathematics owes its external mindset to Hilbert's strong emphasis on form, positing that mathematics is concerned with form rather than content. However, Hilbert's proof theory is more of an internal endeavor; it seeks to define a proof using a more fundamental notion, which, in this context, is the proposition. To modernize Hilbert's proof theory in alignment with his vision, we must axiomatize the algebraic structure of a proof system, where both propositions and proofs reside. To achieve this, we first need a natural mathematical candidate for such two-sorted structures, which is identified as a category.

Categories are algebraic structures comprised of two largely independent entities: objects and morphisms that connect these objects. One can conceptualize a proof system as a category in which the objects represent propositions and the morphisms signify the deductions between them. This approach leads to an attempt to find an axiomatization of proof systems framed as *structured categories*. This perspective on proof theory is termed *categorical proof theory*. Inspired by Lawvere's thesis Lawvere (2004), the theory began with Lambek's comprehensive exploration of proof systems as presented in Lambek (1968a, b, 1971), and was further developed in Lambek and Scott (1988), Szabo (1978). For a succinct overview, see (Došen 2011, 2003).

Interestingly, the abstract structure of proof systems uncovered by categorical proof theory is not confined strictly to proof theory; it is a well-recognized construct across various branches of mathematics known as the *cartesian closed structure* and *bicartesian closed structure*, depending on connectives the language uses. Leveraging the richness of this structure allows one to hypothesize that many mathematical frameworks, ranging from computable functions to continuous maps, can be viewed as different forms of proof systems. By utilizing these mathematical structures as semantic interpretations of conventional syntactical proofs, we can gain intriguing insights into both the nature of proofs and the concept of provability.

As a concluding remark to this part, we will summarize the potential solutions that categorical proof theory offers for the problems previously discussed. Firstly, regarding the issue of form-to-syntax reduction, proof systems within categorical proof theory are viewed as algebraic structures, where their essence is captured in the abstract structure they embody. This perspective parallels how a ring can be regarded as a formal object independent of any syntactical representation. Consequently, in categorical proof theory, the specific representations of proof systems through various syntaxes become inconsequential. Two proof systems are deemed equivalent if their underlying structures align. Turning to the issue of reducing proofs to propositions, it is noteworthy that in a category, morphisms cannot be reduced to objects; they maintain an independence analogous to the relationship between deductions and proofs. In addressing the equivalence problem, it is important to recognize that a category inherently possesses its own notion of equality among morphisms. Thus, in categorical proof theory, there is no need to define the equivalence of proofs, as this concept emerges as a fundamental notion. However, this does not permit arbitrary definitions, as we expect our category to exhibit certain desirable properties. Finally, the identity of propositions is effectively captured by the established notion of isomorphic objects within categories.

1.4 Type Theory

Parallel to the categorical approach to proof theory, there is a type-theoretical approach that is more syntactic and thus closer to the usual practices of proof theory. Despite the primary focus of logical theories on formulas, the language of type theory incorporates two fundamental notions: types and terms. Over this syntax, a type theory consists of a series of type and term formers, along with rules that describe the expected behavior of these formers.

Type theories have many informal interpretations. Types and terms can be understood as propositions and proofs, sets and elements, or data types and programs. This versatility arises from an unexpected similarity between these concepts, known as the *Curry-Howard correspondence* or *propositions as types* dictum (Howard 1980; Sørensen and Urzyczyn 2006). For example, the simplest type theory is the typed lambda calculus, designed to capture the behavior of sets and functions in the presence of products and function spaces Lambek and Scott (1988). This type theory can be extended with sums as disjoint unions, resulting in a syntactical system for sets as well as propositional deductions. Additionally, it is possible to extend typed lambda calculus to dependent type theories, where types can depend on the elements of other types. This allows us to formalize families of sets, on the one hand and first-order deductions, on the other.

Type theories are typically motivated by their role in formalizing data types and programs rather than proofs and propositions. However, some type theories have foundational or even proof-theoretical motivations. For instance, the well-known Martin Löf type theory Martin-Löf (1984) is a fundamental dependent type theory influenced by Prawitz's work on the equivalence of proofs via normalization, aiming to formalize what constructions—and, consequently, proofs—mean in constructive mathematics. Due to this connection with data types, type theory garners significant interest from the theoretical computer science community, and it is fair to say that most developments in both categorical and type-theoretical approaches to proofs are inspired by these interests.

Finally, to connect categorical proof theory to type theory, one can interpret the former as the categorical semantics for the latter. For example, cartesian closed categories provide a semantics for the typed lambda calculus. In this chapter, we will not cover type theories; for more information on type theories, especially their connection with categories, see (Crole 1993; Jacobs 1999; Taylor 1999; Hofmann 1995b; Lambek and Scott 1988).

1.5 Landmarks in Categorical Proof Theory

As mentioned before, categorical proof theory essentially began with Lambek (1968a, b, 1971), who sought to formalize *propositional* deductive systems as structured categories. He starts with substructural proof systems and demonstrates that the required structure is the *monoidal structure*. Then, he moves to the intuitionistic deductive systems over the fragment $\{\top, \wedge, \rightarrow\}$ and shows that the structure is nothing other than the well-known *cartesian closed structure*. Later, Lambek shows that the categorical approach to propositional intuitionistic proofs and the type-theoretical one using typed lambda calculus coincide Lambek and Scott (1988). For a thorough investigation of the intuitionistic case, see (Szabo 1978; Lambek and Scott 1988). For the substructural setting, especially linear logic, see (Benton et al. 1992, 1993; Bierman 1994, 1995; Seely 1987; De Paiva 1987).

Lambek's use of categorical machinery is beneficial for both category theorists and proof/type theorists. For the first community, it provides a combinatorially manageable syntax to investigate the properties of categories. Specifically, it aids in axiomatizing equalities between morphisms in a given family of categories in terms of a few easy-to-check equalities. Such problems are referred to as *coherence problems*, with the primordial example being Mac Lane's work on monoidal and symmetric monoidal categories Mac Lane (1963). In fact, Lambek was the one who introduced Gentzen's proof-theoretical methods into category theory, which Mac Lane and Kelly later exploited to solve a major coherence problem for closed categories Kelly and Laplaza (1980). For more on these coherence problems and their connection to proof theory, see the comprehensive book by Došen and Petrić (2004). It is also worth consulting Mac Lane (1981).

For the proof/type theory community, category theory provides a rich semantics that is useful for providing counterexamples. Here, there are three main problems of interest. First, the *existence problem* concerns the existence of terms in a type or proofs of a proposition. The second is the *equivalence problem*, which addresses the equivalence of terms or proofs, and the third is the *identity problem*, asking whether two types or propositions are identical. In each of these problems, one can use categories to demonstrate that some types (resp. propositions) are empty (resp. unprovable), some terms (resp. proofs) are unequal, or some types (resp. propositions) are non-isomorphic; for more see Sect. 7. One can even go further and ask whether a given family of categories is *complete* for the problem, meaning that to solve the problem, it is sufficient to check only the interpretations in those categories.

For the existence problem, since we are only concerned with the existence of proofs, it is sufficient to relax the categorical structure to posets. Finding concrete yet complete posets is an interesting line of inquiry. For instance, regarding the existence of intuitionistic propositional proofs, it is enough to consider the bicartesian closed poset $\mathcal{O}(\mathbb{R})$, which consists of the open sets of the topological space $\mathbb{R}$ Bezhanishvili et al. (2018). Although it is not strictly necessary, using concrete non-posetal categories for the existence problem is also intriguing. The first example of such a category was $\mathbf{Set}^{\mathbb{Z}}$, which was studied by Laüchli (1968). His original paper does not employ categorical language; instead, it assigns a pair (X, σ_X) of a set and a bijection on it to propositions, and interprets deductions as usual functions that respect the bijection. The modern categorical presentation, along with its generalizations, can be found in Harnik and Makkai (1992). The categorical interpretation is also generalized to the first-order setting in Makkai (1993a, b).

For the equivalence problem, Friedman (1972) and, independently, Plotkin (1973, 1980) demonstrated that beta-eta conversion is complete for deriving all equalities between the (simply-typed) lambda-definable functionals in the category of sets. Building on this result, Čubrić showed that the category of sets is complete for the equivalence problem Čubrić (1998) for the fragment $\{\top, \wedge, \to\}$ of intuitionistic proofs. The result is further reduced to the category of finite sets by Solovev (1981). This reduction can be employed to prove the decidability of the equivalence of proofs in the mentioned fragment. Later, Simpson (1995a) established that a cartesian closed category is complete for the equivalence problem for the fragment $\{\top, \wedge, \to\}$ if and only if it is not a preordered set. This theorem has a significant philosophical consequence, as it states that the equivalence between proofs is the maximum congruence relation one can impose on proofs without collapsing the entire proof structure to mere provability. For further details, see Došen and Petrić (2000). Recently, Scherer (2017) demonstrated that the category of finite sets is complete for the equivalence problem of the full propositional intuitionistic language leading again to the decidability of the equivalence of proofs problem. Using the argument provided by Simpson (1995b), one can show that a bicartesian closed category is complete for the full language if and only if it is not a preordered set.

For the identity problem concerning the fragment $\{\wedge, \top, \to\}$, it is known that a finite set of simple identities (see Sect. 7.3) axiomatizes all identities between propositions. It has also been proven that the category of finite sets is complete for this problem, which leads to the decidability of identities between propositions in the fragment. For the fragment $\{\wedge, \vee, \top, \to\}$, it is known that the canonical finite set of axioms, referred to as *Tarski's high school identities*, is not complete. Furthermore, Fiore et al. (2006) showed that the problem is not finitely axiomatizable. For the full language, the situation is more complicated. It is still not finitely axiomatizable Fiore et al. (2006). We also know that the category of sets cannot be complete, as it considers the formulas $\neg p \vee \neg\neg p$ and $\top$ to be isomorphic, despite the fact that they are not even intuitionistically equivalent. To the best of our knowledge, the decidability of the propositional identity problem for the last two fragments remains open. For a survey of the results and applications of the propositional identity problem, see (Di Cosmo 2005, 1995).

So far, we have discussed the propositional language. To some extent, the theory has been generalized to more expressive languages by Lawvere (2006, 1968) using hyperdoctrines, and to first-order languages by Bénabou (1974), Seely (1977, 1983), and Makkai (1993a, b) through a special family of fibrations. In this chapter, we will not cover these investigations. However, to provide a glimpse into the use of categorical machinery in higher-order settings, we will restrict our focus to the special yet rich case of *realizability*. The main idea is to use maps in a structured category or terms in a type theory as constructions. These constructions are usually called *realizers*. Realizability originated with Kleene, who employed algorithms as realizers Kleene (1945). This approach proved useful for establishing meta-mathematical theorems about constructive arithmetic. For instance, one can prove the consistency of classically invalid Church-Turing thesis, which asserts that all functions on natural numbers are computable. It is also beneficial to extract computational information from constructive proofs. For example, if $\forall x \exists y A(x, y)$ is provable in constructive arithmetic, then there exists a computable function $f : \mathbb{N} \to \mathbb{N}$ such that $A(n, f(n))$ holds for any $n \in \mathbb{N}$. This machinery has been generalized in various ways Troelstra (1998). The categorical version we are interested in emerged in Hyland (1981), where Hyland used algorithms to realize higher-order theories. His framework was further generalized in Birkedal (2002), Bauer (2000) to allow for the use of structured categories or typed theories to formalize constructions. For a comprehensive explanation, see (Birkedal 1999; Bauer 2000, 2005; van Oosten 2008).

1.6 A Word to the Reader

The categorical language can pose a significant conceptual barrier for those wishing to understand the applications of categorical methods in proof theory. To mitigate this issue and reach a broader audience—particularly among proof theorists—we have decided to structure the material without assuming prior familiarity with category theory. We only require a very basic understanding of mathematical logic, including the fundamentals of natural deductions, the basics of order theory and topology, and a basic familiarity with the theory of computation.

As a consequence, the chapter begins with a slow and somewhat extensive introduction to category theory. In our explanation, we employ numerous examples and motivate the introduced categorical concepts using logical and, in particular, proof-theoretical ideas. To make the introduction easier to digest, we refrain from using any unnecessary categorical notions. For instance, we avoid discussing limits and colimits, representable functors, or adjunctions. This categorical part is addressed in Sects. 3 and 4, where we also introduce the basic structure of a propositional proof system. Readers with a solid background in category theory may skim through these sections to see the connection with proof theory and start reading from Sect. 5. In Sect. 5, we will explain and formalize the BHK interpretation and demonstrate how it fits into the broader context of categorical proof theory. In Sect. 6, we will discuss classical deductions. In Sect. 7, we will utilize the categorical framework established in the previous sections as a form of semantics for syntactical proofs. To transition from propositional to higher-order cases, in Sect. 8, we will introduce higher-order theories in general and some constructive theories of arithmetic in particular. In Sect. 9, we will apply categorical proof theory to gain insights into general higher-order theories using a semantical machinery called realizability. Finally, in Sect. 10, we will use realizability to prove some consistency results in arithmetic and also extract information from the arithmetical proofs. In each section, we will provide numerous philosophical discussions and examples to convey the fundamental ideas, the constructions, and, at times, sketches of the proofs. However, we will generally leave the computational details of the examples to the reader. We will also prove the simpler theorems to demonstrate how the concepts work in practice while referring readers to external sources for the proofs of more complex results.

2 Preliminaries

In this section, we will recall some basic notions from order theory, algebra, propositional proof theory, and computability theory, to refresh our memory, emphasize the less known points, and fix the notation. Many other notions such as spaces, compactness, and local compactness in topology, first-order theories in logic, and Turing machines in computability theory are assumed without any further explanation.

A *preordered set* is a pair $(P, \leq)$ of a set P and a preorder $\leq$, i.e., a reflexive and transitive binary relation $\leq \subseteq P \times P$, i.e., $x \leq x$, for any $x \in P$ and if $x \leq y$ and $y \leq z$ then $x \leq z$, for any $x, y, z \in P$. If this relation is also anti-symmetric, i.e., if $x \leq y$ and $y \leq x$ implies $x = y$, for any $x, y \in P$, then the preorder is called a *partial order* and $(P, \leq)$ is called a *poset*. A poset $(P, \leq)$ is called *linear*, if for any $x, y \in P$, either $x \leq y$ or $y \leq x$. A poset $(P, \leq)$ is called a *tree*, if for any element $x \in P$, the poset $\{y \in P \mid y \leq x\}$ with the induced order is finite and linear. A tree is called *rooted*, if it has the least element, i.e., an element $m \in P$ such that $m \leq x$, for any $x \in P$.

For any two preordered sets $(P, \leq_P)$ and $(Q, \leq_Q)$, an *order-preserving* map $f : (P, \leq_P) \to (Q, \leq_Q)$ is a function $f : P \to Q$ such that $f(x) \leq_Q f(y)$, if $x \leq_P y$. It is called an *order-embedding* if it is order-preserving and $f(x) \leq_Q f(x)$ implies $x \leq_P y$, for any $x, y \in P$. Two prototypical examples of posets that we will use in this chapter are the poset of the opens of a topological space X, denoted by $\mathcal{O}(X)$, and the poset of all the upward-closed subsets (upset, for short) of a preordered set $(P, \leq)$ with the inclusion, denoted by $U(P, \leq)$. Note that the upsets of $(P, \leq)$ also form a topology called the *upset topology* and any space with a topology induced in that way is called an *Alexandrov* space. For a poset $(P, \leq)$, define $\mathbf{2}^{(P, \leq)}$ as the poset of all order-preserving maps from $(P, \leq)$ into $\mathbf{2}$, where $\mathbf{2}$ is the poset $(\{0, 1\}, \leq)$ and $\leq$ is the usual order on numbers. Note that $U(P, \leq)$ is isomorphic to $\mathbf{2}^{(P, \leq)}$. Hence, we use them interchangeably.

A *monoid* is a tuple $(M, \cdot, e)$, where M is a set, $\cdot : M \times M \to M$ is a binary operation on M, and $e \in M$ is an element such that $e \cdot x = x \cdot e = x$, for any $x \in M$

and $x \cdot (y \cdot z) = (x \cdot y) \cdot z$, for any $x, y, z \in M$. *Groups* are the monoids where any element has an inverse, i.e., for any $x \in M$, there exists $y \in M$ such that $x \cdot y = y \cdot x = e$.

Let $\mathcal{L}_p = \{\top, \bot, \wedge, \vee, \to\}$ be the propositional language and recall that $\neg A$ is an abbreviation for $A \to \bot$. We define the *natural deduction* system **NJ** by its usual rules:

$$\frac{}{\top}\,(\top) \qquad \frac{\begin{array}{c}\mathsf{D}\\ \bot\end{array}}{A}\,(\bot)$$

$$\frac{\begin{array}{c}\mathsf{D}_1\\ A\end{array} \quad \begin{array}{c}\mathsf{D}_2\\ B\end{array}}{A \wedge B}\,(I\wedge) \quad \frac{\begin{array}{c}\mathsf{D}\\ A \wedge B\end{array}}{A}\,(E\wedge_1) \quad \frac{\begin{array}{c}\mathsf{D}\\ A \wedge B\end{array}}{B}\,(E\wedge_2)$$

$$\frac{\begin{array}{c}\mathsf{D}\\ A\end{array}}{A \vee B}\,(I\vee_1) \quad \frac{\begin{array}{c}\mathsf{D}\\ B\end{array}}{A \vee B}\,(I\vee_2) \quad \frac{\begin{array}{c}\mathsf{D}\\ A \vee B\end{array} \quad \begin{array}{c}[A]^i\\ \mathsf{D}_1\\ C\end{array} \quad \begin{array}{c}[B]^j\\ \mathsf{D}_2\\ C\end{array}}{C}\,(E\vee_{i,j})$$

$$\frac{\begin{array}{c}[A]^i\\ \mathsf{D}\\ B\end{array}}{A \to B}\,(I \to_i) \quad \frac{\begin{array}{c}\mathsf{D}_1\\ A\end{array} \quad \begin{array}{c}\mathsf{D}_2\\ A \to B\end{array}}{B}\,(E \to)$$

where i and j are natural numbers to refer to the assumptions. One can compose these rules to construct rooted trees that we will use to define derivations. Here are some conditions we impose on the use of superscripts in trees. First, all leaves are formulas with a superscript. Second, in any tree, any number i must be used for a unique formula. However, it is possible to have one formula with different superscripts. Third, in $(E\vee_{i,j})$, the indices i and j cannot occur in D, the index i cannot occur in D_2 and j cannot occur in D_1.

We define a *derivation* of A from the set $\Gamma = \{[\gamma_1]^{i_1}, [\gamma_2]^{i_2}, \ldots, [\gamma_n]^{i_n}\}$ of assumptions as a tree with the leaves inside Γ. Note that it is not necessarily that every formula in Γ appears in the tree. When the set of assumptions is clear from the context, especially when Γ is equal to the set of formulas in the leaves, we omit Γ and simply talk about the derivations of A. If D is a derivation of A from the assumptions Γ, a sub-derivation E of D is a full subtree of D. It proves the formula in its root and its assumptions can be canonically computed from Γ by adding the eliminated assumptions in the path from the root of E to the root of D. If D is a derivation of A from the assumptions in Γ and for any $[\gamma_i]^{n_i}$, we have a derivation D_i of γ_i from Σ, then by composition of D with the D_i's, we mean gluing D_i to the leaf $[\gamma_i]^{n_i}$ of D and eliminating the $[-]^{n_i}$'s. The resulting tree is a derivation of A from Σ. We identify two derivations if one is reachable by the other by changing the names of the superscripts in a bijective manner. For instance, the following two derivations of $B \to A \wedge B$ from the assumptions $\{[A]^3, [B]^5\}$ and $\{[A]^4, [B]^7\}$, respectively, are considered as equal:

$$\dfrac{\dfrac{[A]^3 \qquad [B]^5}{A \wedge B}\,(I\wedge)}{B \to A \wedge B}\,(I \to_5) \quad \equiv \quad \dfrac{\dfrac{[A]^4 \qquad [B]^7}{A \wedge B}\,(I\wedge)}{B \to A \wedge B}\,(I \to_7)$$

Consider the following family of *basic β-equivalences*:

$$\dfrac{\dfrac{\begin{matrix}\mathsf{D}_1\\ A\end{matrix} \qquad \begin{matrix}\mathsf{D}_2\\ B\end{matrix}}{A \wedge B}\,(I\wedge)}{A}\,(E\wedge) \quad \equiv \quad \begin{matrix}\mathsf{D}_1\\ A\end{matrix}$$

$$\dfrac{\dfrac{\begin{matrix}[A]^i\\ \mathsf{D}_2\\ B\end{matrix}}{A \to B}\,(I \to_i) \qquad \begin{matrix}\mathsf{D}_1\\ A\end{matrix}}{B}\,(E \to) \quad \equiv \quad \begin{matrix}\mathsf{D}_1\\ A\\ \mathsf{D}_2\\ B\end{matrix}$$

$$\dfrac{\dfrac{\begin{matrix}\mathsf{D}\\ A\end{matrix}}{A \vee B}\,(I\vee) \qquad \begin{matrix}[A]^i\\ \mathsf{D}_1\\ C\end{matrix} \qquad \begin{matrix}[B]^j\\ \mathsf{D}_2\\ C\end{matrix}}{C}\,(E\vee_{i,j}) \quad \equiv \quad \begin{matrix}\mathsf{D}\\ A\\ \mathsf{D}_1\\ C\end{matrix}$$

$$\dfrac{\dfrac{\begin{matrix}\mathsf{D}\\ B\end{matrix}}{A \vee B}\,(I\vee) \qquad \begin{matrix}[A]^i\\ \mathsf{D}_1\\ C\end{matrix} \qquad \begin{matrix}[B]^j\\ \mathsf{D}_2\\ C\end{matrix}}{C}\,(E\vee_{i,j}) \quad \equiv \quad \begin{matrix}\mathsf{D}\\ B\\ \mathsf{D}_2\\ C\end{matrix}$$

Moreover, consider the following *basic η-equivalences*:

$$\dfrac{\dfrac{\begin{matrix}\mathsf{D}_1\\ A \wedge B\end{matrix}}{A} \qquad \dfrac{\begin{matrix}\mathsf{D}_1\\ A \wedge B\end{matrix}}{B}}{A \wedge B} \quad \equiv \quad \mathsf{D}_1 \qquad\qquad \dfrac{\dfrac{\begin{matrix}\mathsf{D}_2\\ A \to B\end{matrix} \qquad [A]^1}{B}}{A \to B}\,1 \quad \equiv \quad \mathsf{D}_2$$

$$\dfrac{A \vee B \qquad \begin{matrix}\dfrac{[A]^1}{A \vee B}\\ \mathsf{D}_3\\ C\end{matrix} \qquad \begin{matrix}\dfrac{[B]^2}{A \vee B}\\ \mathsf{D}_3\\ C\end{matrix}}{C} \quad \equiv \quad \mathsf{D}_3$$

Note that in the right side, we always consider the used derivations from the left side as a sub-derivation. Hence, their assumptions are equal. For $\top$ and $\bot$, the basic η-equivalences are the identifications of any two derivations of a fix formula from $\bot$ and any two derivations of $\top$ from a fixed family of assumptions. These basic β- and

η-equivalences[2] give rise to an equivalence relation defined as the least equivalence relation on derivations extending the above-mentioned β- and η-equivalences and respecting the composition of derivations. We always consider derivations up to this equivalence relation.

The systems **NM** and **NN** are defined similarly for the fragments $\mathcal{L}_p - \{\bot\}$ and $\mathcal{L}_p - \{\bot, \vee\}$, respectively, by eliminating all the rules for the missing connectives. The *intuitionistic propositional logic*, IPC, is defined as the set of all formulas $A \in \mathcal{L}_p$ such that there exists a derivation of A in **NJ** from the empty set of assumptions. Similarly, one can define the *minimal propositional logic*, MPC, using the system **NM**.

The class of all *primitive recursive functions* are defined as the least class of functions over natural numbers, containing the basic functions $Z = 0$, $s(x) = x + 1$, and $I_k^i(x_1, x_2, \ldots, x_k) = x_i$ for $1 \leq i \leq k$ and closed under composition and *primitive recursion*, i.e., for any primitive recursive functions $g : \mathbb{N}^k \to \mathbb{N}$ and $h : \mathbb{N} \times \mathbb{N}^k \times \mathbb{N} \to \mathbb{N}$, the function $f : \mathbb{N} \times \mathbb{N}^k \to \mathbb{N}$ defined by

$$\begin{cases} f(0, \bar{x}) = g(\bar{x}) \\ f(y+1, \bar{x}) = h(y, \bar{x}, f(y, \bar{x})) \end{cases}$$

is also primitive recursive. Turing machines are defined in their usual way. It is possible to encode Turing machines by natural numbers. We call these numbers the *codes* of the algorithms or machines. Recall that there is a *universal Turing machine* U that reads the pair (e, n) of the code e of an algorithm and the input n and provides the output of the algorithm e on the input n. We denote $U(e, n)$ by $e \cdot n$. Note that U is a partial function. When $U(e, n)$ is (resp. is not) defined, i.e., when the algorithm e halts (resp. does not halt) on the input n, we write $e \cdot n \downarrow$ (resp. $e \cdot n \uparrow$). One interesting theorem that we will use later is Kleene's S_n^m *theorem*:

Theorem 1 *There exists a primitive recursive function S such that for any code e of a function $f(x, y)$ with two variables, we have $S(e, x) \cdot y = f(x, y)$, for any $x, y \in \mathbb{N}$.*

3 Categories, Functors, and Natural Transformations

In this section, we will introduce the basic notions of categories, functors, and natural transformations. We will also present many instances of these concepts to provide concrete toy examples to familiarize readers with such abstract notions. These examples will also be required later in the chapter when we aim to provide a categorical formalization for a proof system.

[2] These equivalences also appear in lambda calculus and more complex type theories as the basic computational rules on terms.

3.1 Categories

Let us start this subsection with the definition of a category.

Definition 2 A *category* C consists of the following data:

- a collection of *objects*, denoted by Ob(C),
- a collection of *morphisms* or *maps*, denoted by Mor(C),
- for any morphism $f \in \text{Mor}(C)$, an object $s(f)$ called the *source* of f,
- for any morphism $f \in \text{Mor}(C)$, an object $t(f)$ called the *target* of f,
- for any object $A \in \text{Ob}(C)$, a morphism id_A, called the *identity* on A,
- for any two morphisms $f, g \in \text{Mor}(C)$ such that $s(f) = t(g)$, a morphism $f \circ g$, called the *composition* of f and g,

satisfying the following properties:

- $s(id_A) = t(id_A) = A$,
- $s(f \circ g) = s(g)$ and $t(f \circ g) = t(f)$,
- $f \circ id_A = f = id_B \circ f$, if $s(f) = A$ and $t(f) = B$,
- $f \circ (g \circ h) = (f \circ g) \circ h$.

Since composition is a partial function, by the above equalities, we mean that if one side is defined, the other must also be defined, and they must be equal. For any $f \in \text{Mor}(C)$, we present the data $s(f) = A$ and $t(f) = B$ by $f : A \to B$. For any two objects $A, B \in \text{Ob}(C)$, by $C(A, B)$ or $\text{Hom}_C(A, B)$, we mean the collection of all morphisms $f : A \to B$. A category is called *small* if Mor(C) is a set. It is called *locally small* if $\text{Hom}_C(A, B)$ is a set, for any two objects A and B. Sometimes, for simplicity, we drop $\circ$ in $f \circ g$ and denote the composition by fg. A map $f : A \to B$ is called an *isomorphism* iff there exists $g : B \to A$ such that $gf = id_A$ and $fg = id_B$. If there is an isomorphism between A and B, we write $A \cong B$.

Philosophical Comment 3 To have an informal interpretation in mind, consider the objects of a category as the entities of a given discourse, the maps as the transformations between these entities, the composition as the composition of these transformations, and the identity as the do-nothing transformation. More specifically, objects in a category can be interpreted as *propositions*, and morphisms $f : A \to B$ as the *deductions* of B from the assumption A. Thus, the identity map $id_A : A \to A$ can be seen as the canonical deduction of A from itself, and the composition as the process of applying two deductions sequentially. In this sense, any category can be viewed as a kind of *proof system*.

In the following, we will present many examples of categories. These examples are intended to provide a proper sense of what categories are. In each example, we will also attempt to interpret the introduced category as a proof system (Table 1).

Table 1 A schematic diagram of a category

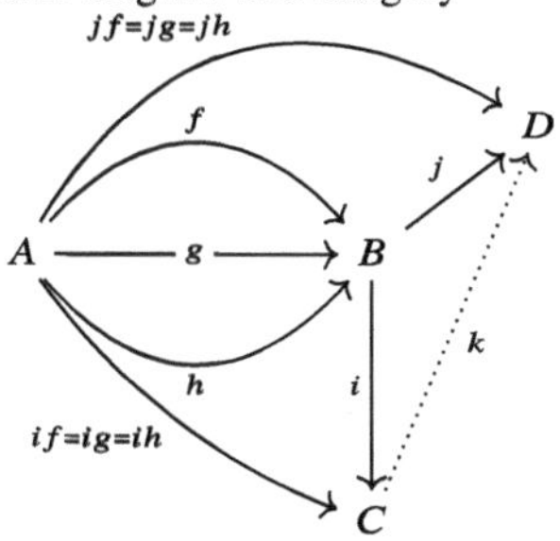

Example 4 The collection of all (resp. finite) sets as the objects and functions as the morphisms, with the usual composition and identity, constitutes a category denoted by **Set** (resp. **FinSet**). We can think of the set A as the set of all *proofs* of the proposition associated with A, denoted by ϕ_A. Then, the function $f : A \to B$ can be read as a *deduction* of ϕ_B from ϕ_A, transforming any proof of ϕ_A into a proof of ϕ_B.

Example 5 The collection of all topological spaces as the objects and continuous functions as the morphisms, with the usual composition and identity, constitutes a category, denoted by **Top**. We can think of the space X as the space of all proofs of the proposition associated to X, denoted by ϕ_X. These proofs can be infinitary. However, the topology of the space ensures that only a finite amount of information encoded in such infinitary proofs is used. More precisely, we can think of an open subset of X as a subset of all possible proofs of ϕ_X that extend a given finite amount of information. For instance, a proof of $\forall x \in \mathbb{N}, A(x)$ can be an infinite sequence $\{\pi_n\}_{n\in\mathbb{N}}$, where each π_n is a proof of $A(n)$, and the open sets are the subsets consisting of all sequences $\{\pi_n\}_{n\in\mathbb{N}}$ that extend a finite initial segment. In this interpretation, a deduction of Y from X is a *finitary* function transforming any proof of X into a proof of Y, where finitary means that it only uses a finite amount of information from its input to compute the output. The continuity condition simply encodes this finiteness condition. For example, in our case, a finitary deduction of B from $\forall x \in \mathbb{N}, A(x)$ must only use a finite initial segment of the proof $\{\pi_n\}_{n\in\mathbb{N}}$ of $\forall x \in \mathbb{N}, A(x)$.

Example 6 (*Discrete Categories*) A category C is called *discrete* if it has only identity maps. Therefore, any set can be considered as a small discrete category. For example, the set $\{A, B, C\}$ can be viewed as the following category:

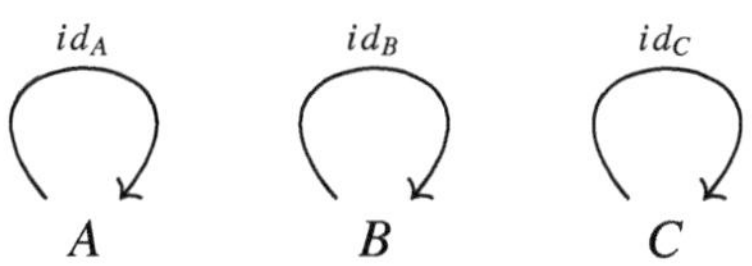

Example 7 (*Finite Categories*) These are some finite categories:

$$\mathbf{0}:$$

$$\mathbf{1}: \qquad \bullet$$

$$\mathbf{2}: \qquad \bullet \longrightarrow \bullet$$

The category **0** has no objects and no morphisms. The category **1** has one object and one morphism, namely, the identity of the object. We did not draw the identity map in the above diagram and will continue this practice of omitting identity maps in diagrams. The category **2** has two objects and one non-identity morphism. Throughout this chapter, we always use the notation **2** for this category, which should not be confused with the discrete category containing two elements.

Example 8 (*Preorders*) Any preordered set $(P, \leq)$ can be interpreted as a small category, where the objects are the elements of P, and there is a unique morphism from p to q iff $p \leq q$. Note that the identity and composition correspond to the reflexivity and transitivity of the preorder, respectively.

Philosophical Comment 9 It is useful to think of preordered sets as the shadows of usual categories, reducing all transformations between two objects to a single *transformability* between them. In the logical interpretation, this means that we collapse all deductions between two statements into one *deductibility* map.

Example 10 (*Monoids*) Any monoid $(M, \cdot, e)$ can be interpreted as a category with a single object $*$, where each element of M is a morphism from $*$ to $*$, e is the identity morphism id_*, and $\cdot$ represents the composition of morphisms.

Philosophical Comment 11 (*Groups and Identity*) Being monoids, groups can also be read as categories with one object $*$. One can read any element of the group or equivalently any morphism over $*$ as a proof of the equality $* = *$. With this interpretation, it is easy to see that the identity e is the *trivial* proof of $* = *$ and its existence is the proof-sensitive version of the reflexivity of the identity relation, while the composition and the inverse operation of the group are the proof-sensitive versions of its transitivity and symmetry, respectively.

Remark 12 A category combines the two extreme cases previously mentioned: a preordered set and a monoid. The first handles the existence of different objects in a category, while the second addresses the different morphisms between any two objects. From a proof-theoretic perspective, one can say that the first describes the propositions and the deductibility relation between them, whereas the second incorporates the actual deductions into the picture.

Example 13 (*Baby Variable Sets*) Consider the collection of functions

$$\begin{array}{c} A_1 \\ \uparrow{\scriptstyle f} \\ A_0 \end{array}$$

as the objects and define a morphism $\alpha : f \to g$ to be the pair of functions (α_0, α_1), where $\alpha_0 : A_0 \to B_0$ and $\alpha_1 : A_1 \to B_1$ such that $\alpha_1 f = g\alpha_0$:

$$\begin{array}{ccc} A_1 & \xrightarrow{\alpha_1} & B_1 \\ {\scriptstyle f}\uparrow & & \uparrow{\scriptstyle g} \\ A_0 & \xrightarrow[\alpha_0]{} & B_0 \end{array}$$

These collections, with the evident composition and identity, constitute a category denoted by $\mathbf{Set}^2$. Any object $f : A_0 \to A_1$ in $\mathbf{Set}^2$ can be interpreted as a *variable set*, varying over the discrete structure of time $\{0 \leq 1\}$. The set A_0 represents the elements of the variable set available at the moment $t = 0$, while the set A_1 represents its elements at the moment $t = 1$. Moving from $t = 0$ to $t = 1$, there are three main possibilities: either new elements are created, some elements remain unchanged (up to renaming), or some distinct elements in A_0 become equal in A_1. These scenarios are all formalized by the function f. The elements outside the range of f represent new elements at $t = 1$, while the elements within the range come from $t = 0$, reflecting the latter two possibilities. Any map between these variable sets naturally corresponds to a pair of maps, each representing the state of the sets at different moments in time, and must respect the changes that occur over time. From a proof-theoretical perspective, these variable sets can be interpreted as propositions whose proofs evolve over time. New proofs can be discovered, or two existing proofs may be found to be identical upon closer examination.

Remark 14 In the previous example, there is nothing special about the structure $\{0 \leq 1\}$ and it can be replaced by any other preordered set or even by any small category. We will see how later.

Example 15 (*Dynamical Systems*) A pair (A, σ_A) of a set A and a function $\sigma_A : A \to A$

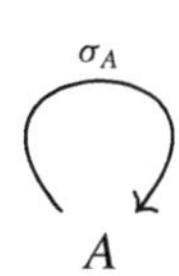

is called a *dynamic system*, where A is interpreted as the set of the states of the system and $\sigma_A : A \to A$ as its dynamism. Let (A, σ_A) and (B, σ_B) be two dynamic systems. A map $f : A \to B$ is called *equivariant* if $f \circ \sigma_A = \sigma_B \circ f$:

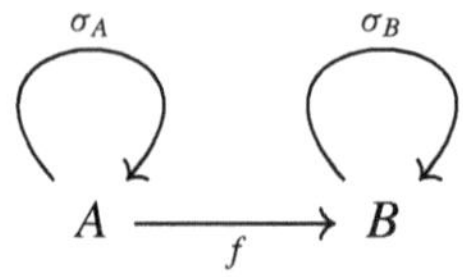

The equivariant maps are simply the functions that respect the dynamism, as we expect from any map between dynamic systems. The collection of all dynamic systems as the objects and equivariant maps as the morphisms with the evident composition and identity constitute a category, denoted by $\mathbf{Set}^{(\mathbb{N},+)}$ or $\mathbf{Set}^{\mathbb{N}}$, for short. (We will see the motivation of this rather strange denotation later). If we restrict the morphisms to equivariant maps that are also bijections, we reach the category of *reversible dynamic systems*, denoted by $\mathbf{Set}^{(\mathbb{Z},+)}$ or $\mathbf{Set}^{\mathbb{Z}}$, for short. It is possible to apply the same idea to define dynamic topological systems, dynamic monoids, and so on, using a topological space with one continuous map over it, a monoid with a homomorphism over it, and so on. Moreover, it is worth mentioning that a reversible dynamic system (A, σ_A) can be read as a proposition. The set A stores all possible proofs of that proposition, the bijection σ_A is a way to make the proof $a \in A$ identical to $\sigma_A(a)$ and we want to consider the proofs up to that identity. In this sense, an equivariant map $f : (A, \sigma_A) \to (B, \sigma_B)$ is just a function to transform the proofs of A to the proofs of B, respecting the newly put symmetry on the proofs.

Example 16 Let $\mathcal{L}_p = \{\top, \bot, \wedge, \vee, \to\}$ be the propositional language and A and B be two formulas in $\mathcal{L}_p$. By a morphism $f : A \to B$, we mean a derivation of B from the assumption A in **NJ**, up to $\beta\eta$-equivalence. The collection of all the formulas in $\mathcal{L}_p$ and morphisms with the identity on A as the trivial derivation of A from itself and composition as simply putting derivations one after another is a category. We denote this category by **NJ**. In a similar fashion, one can restrict oneself to the fragments $\mathcal{L}_p - \{\bot\}$ and $\mathcal{L}_p - \{\bot, \vee\}$ and define the categories **NM** and **NN**, respectively.

We conclude this subsection by presenting some simple methods for constructing new categories from existing ones.

Example 17 (*Opposite Category*) Let C be a category. By its dual (opposite), denoted by C^{op}, we mean a category with the same collections of objects and morphisms as of C with the source and the target assignments swapped and $f \circ' g = g \circ f$:

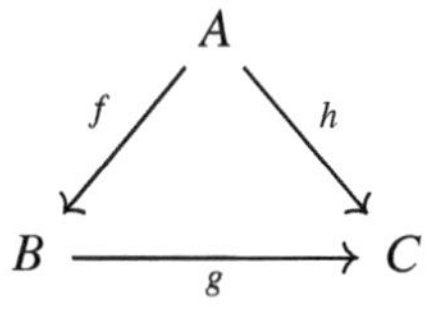

$\mathcal{C}$

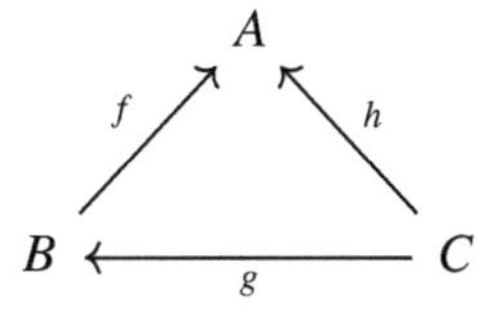

$\mathcal{C}^{op}$

Example 18 (*Product of Categories*) Let $\mathcal{C}$ and $\mathcal{D}$ be two categories. By $\mathcal{C} \times \mathcal{D}$, we mean the category with the pairs (C, D) as the objects, where C and D are the objects of $\mathcal{C}$ and $\mathcal{D}$, respectively, and the pairs $(f, g) : (C, D) \to (E, F)$ as the morphisms, where $f : C \to E$ and $g : D \to F$ are morphisms in $\mathcal{C}$ and $\mathcal{D}$, respectively. Note that this construction generalizes the product of monoids and posets, on the one hand, and the product of sets as discrete small categories, on the other.

Example 19 (*Preorder and Poset Reflection*) Let $\mathcal{C}$ be a small category. By its *preorder reflection*, we mean the preordered set $(\text{Ob}(\mathcal{C}), \leq)$ where $A \leq B$ iff $\text{Hom}(A, B)$ is non-empty. The *poset reflection* of $\mathcal{C}$, denoted by $\text{Po}(\mathcal{C})$, is the poset of the preorder reflection, i.e., the objects are the equivalence classes of the objects of the preorder reflection under the equivalence $A \sim B$ if $A \leq B$ and $B \leq A$ while the order is just the order induced by the preorder reflection.

3.2 Functors

Reading a category as a proof system, one naturally needs to formalize a transformation between proof systems as a map that respects the compositional structure of deductions.

Definition 20 (*Functors*) Let $\mathcal{C}$ and $\mathcal{D}$ be two categories. By a *functor* $F : \mathcal{C} \to \mathcal{D}$ we mean a pair of two assignments F_0 and F_1, such that F_0 maps any object A of $\mathcal{C}$ to an object of $\mathcal{D}$, denoted by $F_0(A)$, and F_1 maps any morphism f of $\mathcal{C}$ to a morphism in $\mathcal{D}$, denoted by $F_1(f)$, respecting the source, target, identity and composition operations as depicted in the following schematic diagram:

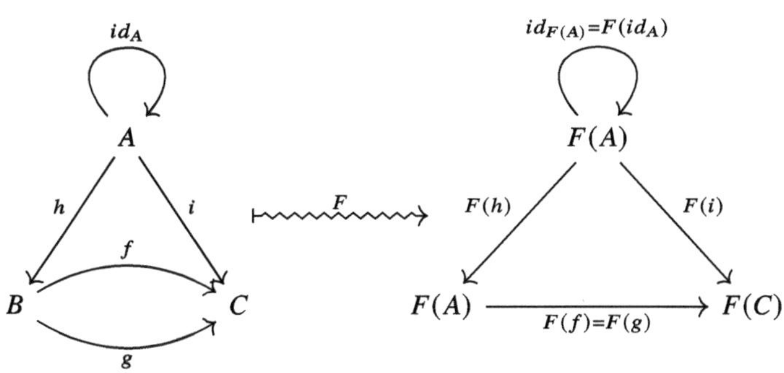

Usually, for simplicity, one drops the subscripts in F_0 and F_1 and denotes both of the assignments by F.

Philosophical Comment 21 It is possible to read a functor $F : \mathcal{C} \to \mathcal{D}$ as a way to interpret the discourse $\mathcal{C}$ into the discourse $\mathcal{D}$, as a way to realize $\mathcal{C}$ in $\mathcal{D}$, as a $\mathcal{C}$-indexed family in $\mathcal{D}$ or a $\mathcal{C}$-variable object in $\mathcal{D}$. When we interpret $\mathcal{C}$ and $\mathcal{D}$ as two proof systems, F may be read as an interpretation of the first proof system into the second.

Example 22 Homomorphisms between monoids and order-preserving maps between preordered sets are examples of functors. In fact, as categories are a common generalization of preordered sets and monoids, functors are also a common generalization of monoid homomorphisms and order-preserving maps.

Example 23 The assignment mapping any set A to its powerset $P(A)$ and any function $f : A \to B$ to the function $P(f) : P(A) \to P(B)$, defined by $P(f)(S) = f[S] = \{f(a) \mid a \in S\}$, is a functor from **Set** to itself. Similarly, the functor $P^\circ : \mathbf{Set}^{op} \to \mathbf{Set}$, mapping any set A to its powerset $P(A)$ and any map $f : B \to A$ in $\mathbf{Set}^{op}$ (i.e., a function $f : A \to B$) to the function $P^\circ(f) : P(B) \to P(A)$, defined by $P^\circ(f)(S) = f^{-1}(S) = \{a \in A \mid f(a) \in S\}$ is a functor.

Example 24 The assignment mapping any object (A, B) in $\mathbf{Set} \times \mathbf{Set}$ to the set $A \times B$ and any morphism $(f, g) : (A, B) \to (C, D)$ of $\mathbf{Set} \times \mathbf{Set}$ to the function $f \times g : A \times B \to C \times D$ defined by $[f \times g](a, b) = (f(a), g(b))$ is a functor, denoted by $(-) \times (-) : \mathbf{Set} \times \mathbf{Set} \to \mathbf{Set}$. Reading proof-theoretically, we can interpret this functor as the *conjunction operation* encoding the idea that a proof of a conjunction is a pair of the proofs of each component.

Example 25 The assignment mapping any object (A, B) in $\mathbf{Set} \times \mathbf{Set}$ to the set $A + B = \{(0, a) \mid a \in A\} \cup \{(1, b) \mid b \in B\}$ and any morphism $(f, g) : (A, B) \to (C, D)$ of $\mathbf{Set} \times \mathbf{Set}$ to the function $f + g : A + B \to C + D$ defined by $[f + g](0, a) = (0, f(a))$ and $[f + g](1, b) = (1, g(b))$ is a functor, denoted by $(-) + (-) : \mathbf{Set} \times \mathbf{Set} \to \mathbf{Set}$. Reading proof-theoretically, we can interpret this functor as the *disjunction operation* encoding the idea that a proof of a disjunction is a pair, where the first component states which of the disjuncts we are proving and the second component provides the proof.

Example 26 Let A be a fixed set. Define the assignment $(-)^A : \mathbf{Set} \to \mathbf{Set}$, mapping the set B to B^A, i.e., the set of all functions from A to B and mapping the function $f : B \to C$ to the function $f^A : B^A \to C^A$, defined by $f^A(g) = fg$. Then, $(-)^A : \mathbf{Set} \to \mathbf{Set}$ is a functor, generalizing the finite power functor $A \mapsto A^n$ generated by the iteration of the product functor. Similarly, it is possible to define the assignment $A^{(-)} : \mathbf{Set}^{op} \to \mathbf{Set}$, mapping the set B to the set $A^B = \{f : B \to A\}$ and the map $f : C \to B$ in $\mathbf{Set}^{op}$ (i.e., the function $f : B \to C$) to the function $A^f : A^C \to A^B$, defined by $A^f(g) = gf$. Then, $A^{(-)} : \mathbf{Set}^{op} \to \mathbf{Set}$ is a functor, generalizing the functor $P^\circ = 2^{(-)}$. It is possible to unify these two functors in one functor $(-)^{(-)} : \mathbf{Set}^{op} \times \mathbf{Set} \to \mathbf{Set}$ mapping (B, C) to C^B and (f, g) to the function

$g^f : C_1^{B_1} \to C_2^{B_2}$ defined by $g^f(h) = ghf$, where $f : B_2 \to B_1$ and $g : C_1 \to C_2$ are two functions. Reading proof-theoretically, we can interpret this functor as the *implication operation* encoding the idea that a proof of the implication $B \to C$ is a function mapping any proof of B to a proof of C.

Example 27 (*Forgetful Functors*) It is customary to have a category of some structures and then, for some reason, forget some of the structures the objects possess and the maps preserve. This way, one can think of the same data as it lives in a richer context. Functors can formalize such situations, and in these settings, they are called *forgetful functors* as they forget some parts of their input data. Let us provide three examples of such situations. First, the assignment U mapping any topological space X and any continuous map $f : X \to Y$ in **Top** to themselves in **Set** is a functor. The functor U forgets that there is a topology on the spaces that the continuous maps respect. For the second example, for any $i \in \{0, 1\}$, define the assignment $\pi_i : \mathbf{Set^2} \to \mathbf{Set}$, by mapping any $f : A_0 \to A_1$ to A_i and any $\alpha : f \to g$ to $\alpha_i : A_i \to B_i$, where $f : A_0 \to A_1$ and $g : B_0 \to B_1$. Both π_0 and π_1 are functors. By making two snapshots of a variable set in the two possible moments, π_0 and π_1 forget that the variable set actually varies. For the third, consider the assignment $U : \mathbf{Set}^{\mathbb{N}} \to \mathbf{Set}$ mapping any dynamic system to its set and any equivariant map to itself. The assignment U is a functor that forgets the dynamism of the dynamic space.

Conversely to the idea behind the forgetful functors, sometimes we want to add more structure to the existing object, and of course, there might be many ways to add these structures. In the following, we will see some examples of such a situation.

Example 28 Define the assignment $Disc : \mathbf{Set} \to \mathbf{Top}$ mapping the set X to the discrete space X (i.e., where all subsets are open) and the function $f : X \to Y$ to itself. This assignment is a functor. Similarly, if we define $InDisc : \mathbf{Set} \to \mathbf{Top}$, mapping the set X to the indiscrete space X (i.e., where the opens are $\varnothing$ and X) and the function $f : X \to Y$ to itself, we also obtain a functor.

Example 29 Define the assignment $St : \mathbf{Set} \to \mathbf{Set}^{\mathbb{Z}}$ mapping the set X to the *static* dynamic system (X, id_X) and the function $f : X \to Y$ to itself. This assignment is a functor.

There are several functors that are generally available over arbitrary categories. Here are a few examples.

Example 30 Let $\mathcal{C}$, $\mathcal{D}$, and $\mathcal{E}$ be some categories. Then, note that the identity functor $id_{\mathcal{C}} : \mathcal{C} \to \mathcal{C}$ mapping any object and morphism of $\mathcal{C}$ to itself is a functor. Moreover, if $F : \mathcal{D} \to \mathcal{E}$ and $G : \mathcal{C} \to \mathcal{D}$ are two functors, then the composition $FG : \mathcal{C} \to \mathcal{E}$ with its evident definition is also a functor. For any object D of $\mathcal{D}$, the assignment $\Delta_D : \mathcal{C} \to \mathcal{D}$ mapping all objects of $\mathcal{C}$ to D and all maps of $\mathcal{C}$ to id_D is a functor.

Example 31 Let C be a locally small category. The assignment $\mathrm{Hom}_C : C^{op} \times C \to \mathbf{Set}$, defined by $\mathrm{Hom}_C(A, B) = \{f : A \to B \mid f \in \mathrm{Mor}(C)\}$ and $\mathrm{Hom}_C(g, h) : \mathrm{Hom}_C(A, B) \to \mathrm{Hom}_C(C, D)$ as $\mathrm{Hom}_C(g, h)(f) = hfg$, for any maps $f : A \to B$, $g : C \to A$ and $h : B \to D$ in C, is a functor. This functor captures the whole structure of the category C and is called the Hom functor.

Example 32 Let C be a locally small category. For any object A in C, there is a canonical functor $y^A = \mathrm{Hom}_C(A, -) : C \to \mathbf{Set}$, capturing the behavior of the maps above A. It is defined by $y^A(B) = \mathrm{Hom}_C(A, B)$ and $y^A(f) : \mathrm{Hom}_C(A, B) \to \mathrm{Hom}_C(A, C)$ as $y^A(g) = fg$, for any $f : B \to C$. Similarly, there is a canonical functor $y_A = \mathrm{Hom}_C(-, A) : C^{op} \to \mathbf{Set}$, capturing the behavior of the maps below A. It is defined by $y_A(B) = \mathrm{Hom}_C(B, A)$ and $y_A(f) : \mathrm{Hom}_C(C, A) \to \mathrm{Hom}_C(B, A)$ as $y_A(f)(g) = gf$, for any $f : B \to C$. These functors are just the Hom functor where we fix one of the components.

Example 33 Let C be a small category. Define the assignment $\pi : C \to \mathrm{Po}(C)$, mapping an object A to its class $[A]$ and a map $f : A \to B$ to the unique map from $[A]$ to $[B]$. This assignment is a functor. From a proof-theoretical perspective, we can interpret this functor as the *collapsing operation*, which forgets all deductions between two propositions and retains only the record of their mere existence, encoded in the deductibility relation.

To conclude this subsection, we will introduce three families of functors that will be important later in this chapter.

Definition 34 A functor $F : C \to \mathcal{D}$ is called *faithful* if for any two morphisms $f, g : A \to B$ in C, if $F(f) = F(g)$, then $f = g$.

Example 35 Any functor with a preordered set as its domain is faithful, as in any preordered set, there is at most one map between any two objects. A homomorphism between two monoids is faithful iff it is injective. The forgetful functors from **Top** and $\mathbf{Set}^{\mathbb{N}}$ to **Set** are faithful, simply because they map the continuous functions and the equivariant maps to themselves. The functors $Disc, InDisc : \mathbf{Set} \to \mathbf{Top}$ and $St : \mathbf{Set} \to \mathbf{Set}^{\mathbb{Z}}$ are all faithful, as they also do not change the functions. The functor $\pi_0 : \mathbf{Set^2} \to \mathbf{Set}$ is not faithful, as it forgets the value of the variable set at the point $t = 1$. More precisely, consider the inclusion function $i : \{0\} \to \{0, 1\}$ as an object of $\mathbf{Set^2}$ and define the map $\alpha : i \to i$ by setting $\alpha_0 : \{0\} \to \{0\}$ as the identity function and $\alpha_1 : \{0, 1\} \to \{0, 1\}$ as the switch function, i.e., $\alpha_1(0) = 1$ and $\alpha_1(1) = 0$:

$$\begin{array}{ccc} \{0,1\} & \xrightarrow{\alpha_1} & \{0,1\} \\ \uparrow{\scriptstyle i} & & \uparrow{\scriptstyle i} \\ \{0\} & \xrightarrow[\alpha_0]{} & \{0\} \end{array}$$

It is clear that $\pi_0(\alpha) = \alpha_0 = \pi_0(id_i)$, but $\alpha \neq id_i$ as $\alpha_1 \neq id_{\{0,1\}}$.

Definition 36 A functor $F : \mathcal{C} \to \mathcal{D}$ is called *full* if for any morphism $g : F(A) \to F(B)$ in $\mathcal{D}$, there exists a map $f : A \to B$ such that $F(f) = g$. It is called *weakly-full* if the non-emptiness of $\mathrm{Hom}_{\mathcal{D}}(F(A), F(B))$ implies the non-emptiness of $\mathrm{Hom}_{\mathcal{C}}(A, B)$.

Example 37 An order-preserving map between two posets is weakly-full iff it is full iff it is an order-embedding. A homomorphism between two monoids is always weakly-full but it is full iff it is surjective. The forgetful functors from **Top** and $\mathbf{Set}^{\mathbb{N}}$ to **Set** are not full as there are functions that are not continuous or equivariant. However, these forgetful functors are both weakly-full. The forgetful functor $\pi : \mathcal{C} \to \mathrm{Po}(\mathcal{C})$ is full and hence weakly-full. The functors $Disc, Indisc : \mathbf{Set} \to \mathbf{Top}$ are trivially full. The functor $\pi_0 : \mathbf{Set^2} \to \mathbf{Set}$ is not full, because not all functions give rise to a map between variable sets. For instance, consider the inclusion function $! : \{0, 1\} \to \{0\}$ and the identity function $id_{\{0,1\}} : \{0, 1\} \to \{0, 1\}$ as the objects of $\mathbf{Set^2}$ and let $\alpha_0 = id_{\{0,1\}} : \{0, 1\} \to \{0, 1\}$ be the identity function:

$$\begin{array}{ccc} \{0\} & \xrightarrow{\alpha_1} & \{0,1\} \\ {\scriptstyle !}\uparrow & & \uparrow{\scriptstyle id_{\{0,1\}}} \\ \{0,1\} & \xrightarrow[\alpha_0 = id_{\{0,1\}}]{} & \{0,1\} \end{array}$$

It is clear that there is no function $\alpha_1 : \{0\} \to \{0, 1\}$ to make the diagram commutative and hence α_0 is not $\pi_0(\alpha)$, for any map $\alpha : ! \to id_{\{0,1\}}$.

Philosophical Comment 38 We noted that from a proof-theoretic perspective, functors should be understood as interpretations between proof systems. In this context, faithful functors can be viewed as faithful interpretations that preserve the structure of deductions. Full functors, on the other hand, are interpretations that do not introduce any new deductions between propositions. Weakly full functors may add some new deductions, but they do not render non-deductible propositions deductible. In other words, they do not alter the logic of the proof system.

3.3 Natural Transformations

Reading a category as a proof system and a functor as an interpretation, one can go one step further to think of the possible formalization of the transformations between the interpretations.

Definition 39 (*Natural transformations*) Let $\mathcal{C}$ and $\mathcal{D}$ be two categories and $F, G : \mathcal{C} \to \mathcal{D}$ be two functors. By a *natural transformation* $\alpha : F \Rightarrow G$, depicted as

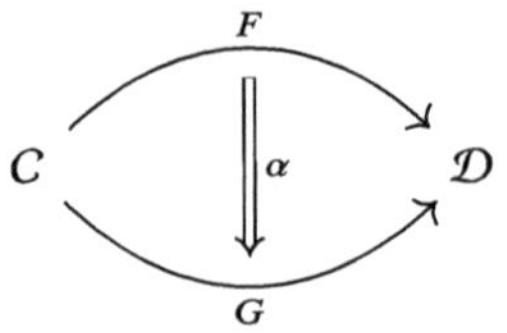

we mean an assignment mapping any object of $\mathcal{C}$ to a morphism $\alpha_C : F(C) \to G(C)$ in $\mathcal{D}$ such that for any morphism $f : A \to B$ in $\mathcal{C}$, the following diagram commutes

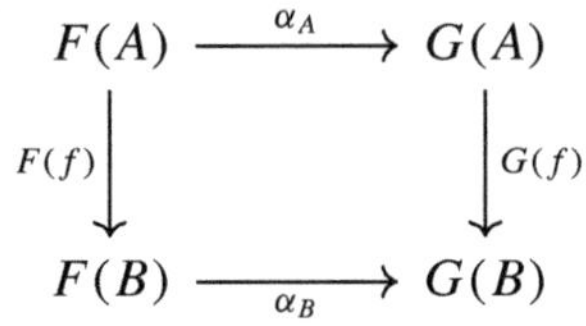

Philosophical Comment 40 Reading the functors $F, G : \mathcal{C} \to \mathcal{D}$ as $\mathcal{C}$-variable objects in $\mathcal{D}$, a natural transformation $\alpha : F \Rightarrow G$ serves as a natural candidate for a transformation between these variable objects. Naturally, any transformation between variable objects must specify the way we change the object $F(C)$ to the object $G(C)$ in $\mathcal{D}$, for each parameter $C \in \mathrm{Ob}(\mathcal{C})$. These changes cannot be arbitrary; they must respect the variation of the parameter C in $\mathcal{C}$, as illustrated in the following schematic diagram:

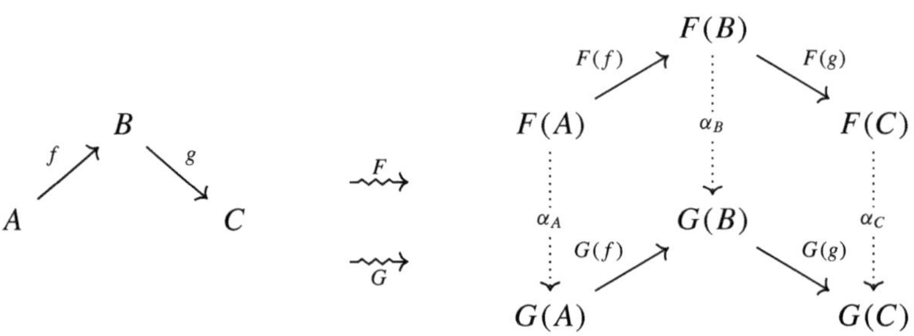

It is intuitively helpful to think of a natural transformation as a way of changing $F(C)$ to $G(C)$, *smoothly* in the parameter C.

It is also possible to interpret the functors $F, G : \mathcal{C} \to \mathcal{D}$ as two *construction methods* that take an object in $\mathcal{C}$ and transform it into an object in $\mathcal{D}$. When can we consider F and G to be "equal" as methods of construction? We certainly do not want to restrict ourselves to strict equality, which requires the functors to be the same on both objects and morphisms, as this is overly restrictive. For instance, consider $F, G : \mathbf{Set} \to \mathbf{Set}$ as $F(A) = A \times \{0\}$ and $G(A) = A \times \{1\}$. Although F and G are not strictly equal, they must be considered as the same methods of construction, as they are only different up to an isomorphism. Using this criterion, one natural candidate for the intended equality between F and G is the existence of an isomorphism between $F(A)$ and $G(A)$, for any object A in C. However, it is clear that any random assignment of isomorphisms between $F(A)$ and $G(A)$ does not suffice; the isomorphisms must be assigned in a *uniform* way, as we want F and G to be equal as two methods of construction not two mere structureless assignments. This uniformity requires that the isomorphisms be somewhat independent of the specific choice of the object A. Of course, one may object that the isomorphisms clearly depend on the object A (the source and the target of the isomorphism, for instance), but at the same time it is intuitively meaningful to talk about the constructions that apply the *same* method to *different* objects. An example may be more illuminating. Consider the canonical isomorphism $s_{A,B} : A \times B \to B \times A$ defined by $s_{A,B}(a, b) = (b, a)$ that shows the order in the product of two sets is not important. This map clearly depends on the choice of A and B, but at the same time it is defined in a uniform way of "swapping the elements in a pair" which does not use the sets in an essential way. Natural transformations are historically developed for the sole purpose of capturing this very intuition of uniformity.

Example 41 The assignment $s : id_{\mathbf{Set}} \Rightarrow P$ defined by $s_A : A \to P(A)$ as $s_A(a) = \{a\}$ is a natural transformation. It is natural, because for any map $f : A \to B$, if f maps $a \in A$ to $f(a) \in B$, then $P(f)$ maps $\{a\}$ to $f[\{a\}] = \{f(a)\}$:

$$\begin{array}{ccc} A & \xrightarrow{\{-\}} & P(A) \\ {\scriptstyle f}\downarrow & & \downarrow{\scriptstyle f[-]} \\ B & \xrightarrow[\{-\}]{} & P(B) \end{array} \qquad \begin{array}{ccc} a & \xmapsto{\{-\}} & \{a\} \\ {\scriptstyle f}\downarrow & & \downarrow{\scriptstyle f[-]} \\ f(a) & \xmapsto[\{-\}]{} & \{f(a)\} \end{array}$$

Example 42 Let $Ex : \mathbf{Set} \times \mathbf{Set} \to \mathbf{Set} \times \mathbf{Set}$ be the exchange functor, i.e, $Ex(A, B) = (B, A)$ and $Ex(f, g) = (g, f)$ and $(-) \times (-) : \mathbf{Set} \times \mathbf{Set} \to \mathbf{Set}$ be the product functor. Then, the assignment $s : (-) \times (-) \Rightarrow [(-) \times (-)] \circ Ex$ defined by $s_{(A,B)} : A \times B \to B \times A$ as $s_{A \times B}(a, b) = (b, a)$ is a natural transformation as for any map $(f, g) : (A, B) \to (C, D)$ in $\mathbf{Set} \times \mathbf{Set}$, we have

$$\begin{array}{ccc} A \times B & \xrightarrow{s_{(A,B)}} & B \times A \\ {\scriptstyle f \times g}\downarrow & & \downarrow{\scriptstyle g \times f} \\ C \times D & \xrightarrow[s_{(C,D)}]{} & D \times C \end{array} \qquad \begin{array}{ccc} (a, b) & \xmapsto{s_{(A,B)}} & (b, a) \\ {\scriptstyle f \times g}\downarrow & & \downarrow{\scriptstyle g \times f} \\ (f(a), g(b)) & \xmapsto[s_{(C,D)}]{} & (g(b), f(a)) \end{array}$$

Example 43 Let $(P, \leq_P)$ and $(Q, \leq_Q)$ be posets and $F, G : (P, \leq_P) \to (Q, \leq_Q)$ be two order-preserving maps. Then, there is at most one natural transformation $\alpha : F \Rightarrow G$, as there is at most one map from $F(p)$ to $G(p)$, for any $p \in P$. This unique natural transformation exists iff $F(p) \leq_Q G(p)$, for any $p \in P$:

$$\begin{array}{ccc} F(p) & \overset{\alpha_p}{\leq} & G(p) \\ \text{\rotatebox{90}{$\geq$}} & & \text{\rotatebox{90}{$\geq$}} \\ F(q) & \underset{\alpha_q}{\leq} & G(q) \end{array}$$

A similar phenomenon occurs when we replace $(P, \leq_P)$ with any other category. Specifically, for any category C and any functors $F, G : C \to (Q, \leq_Q)$, there is at most one natural transformation $\alpha : F \Rightarrow G$, and it exists if $F(A) \leq_Q G(A)$ for every object A in C.

Example 44 Let M and N be two monoids and $F, G : M \to N$ be two homomorphisms. Then, a natural transformation $\alpha : F \Rightarrow G$, by definition, is an element $\alpha_* = n \in N$ such that $nF(m) = G(m)n$, for any $m \in M$:

$$\begin{array}{ccc} F(*) & \xrightarrow{\alpha_*=n} & G(*) \\ {\scriptstyle F(m)}\downarrow & & \downarrow{\scriptstyle G(m)} \\ F(*) & \xrightarrow[\alpha_*=n]{} & G(*) \end{array}$$

Example 45 Let C be a category and $f : A \to B$ be a map. Then, the assignment $y_f : \mathrm{Hom}(-, A) \to \mathrm{Hom}(-, B)$ defined by $(y_f)_C = f \circ (-)$ is a natural transformation as for any map $g : C \to D$:

$$\begin{array}{ccc} \mathrm{Hom}(D, A) & \xrightarrow{f\circ(-)} & \mathrm{Hom}(D, B) \\ {\scriptstyle (-)\circ g}\downarrow & & \downarrow{\scriptstyle (-)\circ g} \\ \mathrm{Hom}(C, A) & \xrightarrow[f\circ(-)]{} & \mathrm{Hom}(C, B) \end{array}$$

Example 46 (*Non-natural Transformations*) Let α be an assignment choosing a function $\alpha_A : A \to A$ for any set A. Then $\alpha : id_{\mathbf{Set}} \Rightarrow id_{\mathbf{Set}}$ is a natural transformation iff $\alpha_A = id_A$, for any set A. It is clear that setting $\alpha_A = id_A$ for any set A makes α a natural transformation. For the converse, assume α is a natural transformation and consider the following commutative diagram:

$$\begin{array}{ccc} \{0\} & \xrightarrow{\alpha_{\{0\}}} & \{0\} \\ {\scriptstyle \hat{a}}\downarrow & & \downarrow{\scriptstyle \hat{a}} \\ A & \xrightarrow[\alpha_A]{} & A \end{array}$$

where $a \in A$ is an arbitrary element and $\hat{a}(0) = a$. It is clear that $\alpha_{\{0\}} = id_{\{0\}}$. Hence, $\alpha_A \circ \hat{a} = \hat{a}$. Applying both sides on 0, we have $\alpha_A(a) = a$. As $a \in A$ is arbitrary, we reach $\alpha_A = id_A$. Using the above characterization, it is enough to pick assignments that are not identity over at least one A to have non-natural transformations.

With functors and natural transformations at our disposal, we can generalize the notion of a variable set by generalizing the underlying temporal structure from the poset $\{0 \leq 1\}$ to any small category:

Example 47 (*Variable sets over* C) Let C be a small category. Then, a *variable set* over C is a functor $F : C \to \mathbf{Set}$. Variable sets over C with natural transformations and canonical identity and composition form a category denoted by $\mathbf{Set}^C$. The exponentiation-like notation suggests that $\mathbf{Set}^C$ collects all functors from C to $\mathbf{Set}$ similar to collecting all functions from A to B in B^A. It is evident that if C is the category $\mathbf{2}$, then a variable set over C corresponds to the notion of a variable set we had previously established. Additionally, when C is taken to be the monoid $(\mathbb{N}, +)$ or $(\mathbb{Z}, +)$, variable sets over C represent dynamic systems and reversible dynamic systems, respectively. This observation not only unifies these diverse categories under one notion of variable sets but also justifies the notation that we employ for them.

One can use the Hom functor to map any small category C to $\mathbf{Set}^{C^{op}}$. More precisely, define $y : C \to \mathbf{Set}^{C^{op}}$ by $y_A = \mathrm{Hom}(-, A)$ on objects and $y_f : \mathrm{Hom}(-, A) \to \mathrm{Hom}(-, B)$ as defined in Example 45 on morphisms. It is easy to check that y is actually a functor. This functor plays a crucial role in category theory and is known as the *Yoneda embedding*. It is known that y is a full and faithful functor. Therefore, one can always pretend that any small category is actually a category of variable sets.

Philosophical Comment 48 One of the prominent foundational paradigms in mathematics is Brouwerian intuitionism. Among many other things, the paradigm believes that mathematics is a mental narrative constructed by a creative subject for herself. Like any narrative, this mathematical story evolves over time as new constructions are added and new properties are proved. In this context, truth in mathematics is viewed as temporal and dynamic, which can be effectively represented by our variable sets in $\mathbf{Set}^C$, where C is a suitable category that captures the progression of time. There are many valid formalizations of this notion of time. For instance, the simplest formalization that comes to mind is the set of all natural numbers and its usual order encoding the instances and the arrow of time. However, in this example, we focus on Brouwer's own understanding of time:

> *This perception of a move of time may be described as the falling apart of a life moment into two distinct things, one of which gives way to the other, but is retained by memory. If the twoity thus born is divested of all quality, it passes into the empty form of the common substratum of all twoities. And it is this common substratum, this empty form, which is the basic intuition of mathematics* Brouwer (1981).

To formalize this notion of time, we use $[n] = \{0, 1, \ldots, n-1\}$, for $n \geq 0$, as the objects to represent the nth moment of time and, for any $n \leq m$, we define the morphisms from $[n]$ to $[m]$ as a function $f : [m] \to [n]$, where $f(i) = i$, for any $i < n$. The equation $f(i) = j$ represents the creation process of moments by encoding the fact that the moment i has been created from the moment j. Therefore, the condition $f(i) = i$ just says that when we are at the nth moment, the moment $i < n$ is fixed throughout the creation process, and only the moments greater than or equal to n are newly created. As it is expected, the category $\mathbf{Set}^C$ leads to an interesting intuitionistic dynamic version of sets. What is surprising, though, is the fact that some of these variable growing sets are in some sense *completed*, and the category of these completed sets satisfies all classical axioms of set theory except the axiom of choice. Hence, it can serve as a model to prove the unprovability of the axiom of choice from ZF. The details are, of course, beyond the scope of this chapter. For more, see (Mac Lane and Moerdijk 1994).

4 Some Universal Constructions

Category theory is renowned for its *relative approach*, in which the properties of an entity are defined by its relationships with other entities rather than by its internal components. For example, when we work with sets in a categorical context, we define the properties of a given set based on the morphisms (functions) that connect it to other sets, rather than inspecting the elements within the set itself.

As the relative approach completely ignores the internal structure of the entities, it is hardly clear whether it is actually capable of defining any non-trivial property. For instance, consider the task of defining the singleton set $\{0\}$ in **Set**. An analysis reveals that such a task is impossible as any two isomorphic objects in a category exhibit indistinguishable relative behavior. Therefore, a definition for $\{0\}$ would also apply to $\{1\}$ which is impossible. The first lesson from this observation is that any purely relative definition overlooks the differences between isomorphic objects. If we choose to pursue a relative perspective, we must accept this limitation. With that observation, let us change the task from defining the set $\{0\}$ to defining "being a singleton". Fortunately, we can see that this time, the definition is possible. A set A is a singleton iff for any set B, there exists exactly one map from B to A. To transform this modest observation to a general method of producing relative definitions, we must notice that our above definition of singletons is based on the informal intuition that singletons are the "*greatest*" objects with a certain property, i.e., the property of having a map from any object into them. Using this vague form of being greatest is not accidental. It is clear that if we must define everything relatively, then the only things we can define are the best objects satisfying a relative behavior with respect to some of the others. This being the best can take one of two forms: either it can be the least object or the greatest object with respect to some property. Such definitions are referred to as defining by a *universal property*, and this section is dedicated to exploring these definitions.

4.1 Cartesian Structure

As we have already seen, singleton sets are defined by the property of having exactly one map from any arbitrary source into them. This is a crucial notion that is useful in many categories, so let us give it a name.

Definition 49 (*Terminal Objects*) An object A is called *terminal* if for any object B, there exists a *unique* map from B to A. This unique map is denoted by $!_B : B \to A$ and if B is clear from the context, just by $!$.

Example 50 In **FinSet** and **Set**, an object is terminal iff it is a singleton. As we only care about the cardinality of the set and not its element, we usually denote these terminal objects by a common name $\{*\}$. In a poset $(P, \leq)$, a terminal object is by definition an element $a \in P$ such that for any $b \in P$, we have $b \leq a$. Hence, a terminal object is the greatest element of the poset if it exists. In $\mathbf{Set}^{\mathbb{N}}$ and $\mathbf{Set}^{\mathbb{Z}}$, a terminal object is a pair in the form $(\{*\}, id_{\{*\}})$. The same also holds for the category $\mathbf{Set}^{C}$, where a terminal object is a constant functor $\Delta_{\{*\}} : C \to \mathbf{Set}$. In **NJ**, the proposition $\top$ is a terminal object. The reason is that first, there is always a derivation from A to $\top$, for any proposition A. Second, for a fixed formula A, as we identified any two such derivations by the η-equivalence relation, the map from A to $\top$ is also unique. The same claim also holds for **NN** and **NM**.

Example 51 (*Non-existence of Terminal Objects*) Terminal objects do not necessarily exist. For instance, a poset without a greatest element such as $(\mathbb{N}, \leq)$ or a non-trivial monoid does not have a terminal object. As another example, the category of sets with at least two elements does not have a terminal object.

Remark 52 Define the order $\leq$ on the objects of C by setting $B \leq A$ if there is a morphism $f : B \to A$. In this sense, a terminal object A can be intuitively thought of as the "greatest" object in the category, as there is a map from any object B into A. However, we must use "greatest" with caution, as being the greatest in this order is not sufficient to be a terminal object. There is an additional condition requiring the uniqueness of morphisms from B to A, which is not captured by our *informal* order-theoretic description.

Remark 53 A category can have many terminal objects. For instance, all singletons in **Set** are terminal. However, up to an isomorphism, there is at most one terminal object. To prove this, let A and B be two terminal objects. Then, by the universal property of both A and B, there are maps $f : A \to B$ and $g : B \to A$. Now, consider the maps $fg : B \to B$ and $gf : A \to A$. By the uniqueness condition for both A and B, we have $fg = id_B$ and $gf = id_A$. Hence, $f : A \to B$ is an isomorphism. Now, as there is at most one terminal object up to an isomorphism and our relative approach sees everything up to isomorphisms, it is safe to denote all terminal objects by one reserved name, i.e., 1, and call it *the* terminal object.

Philosophical Comment 54 (*Terminal Objects as the Trivial Truth*) Reading any category as a proof system, a terminal object is a *trivially true proposition*, i.e., a proposition with a unique deduction from any given assumption. One may argue that although the intuition behind the uniqueness condition is clear for the singletons in **Set**, the same claim for $\top$ is a bit artificial and this is even evident in **NJ**, where we must use the apparently artificial η-equivalence to ensure this condition. To defend our interpretation, let us emphasize that the logical constant $\top$ is present in the language to formalize the *trivial* truth. Whatever it means, the trivial truth must be provable as it is true and have exactly one proof as it is trivial. Note that if we forget about the triviality part, then there is no need for $\top$ in the language as there are many other provable statements like $p \to p$ playing the same role as $\top$ does. In fact, in proof-irrelevant approaches, all provable statements are considered equivalent, while this should not hold in a proof-sensitive approach. For instance, the formula $p \to p$ and $\top$ must be considered as different formulas, because there are many deductions of p from p and hence many proofs of $p \to p$ while $\top$ has only one proof.

The second universal construction we are interested in is the binary product. Before going any further, it is useful to pause and consider cartesian products in **Set**, and to try to define them in a relative fashion without referring to their internal structure.

Definition 55 (*Binary Products*) Let A and B be two objects in a category $\mathcal{C}$. An object C together with two morphisms $p_0 : C \to A$ and $p_1 : C \to B$ is called a *binary product* of A and B if for any object D in $\mathcal{C}$ and any morphisms $f : D \to A$ and $g : D \to B$, there exists a *unique* map $h : D \to C$ such that

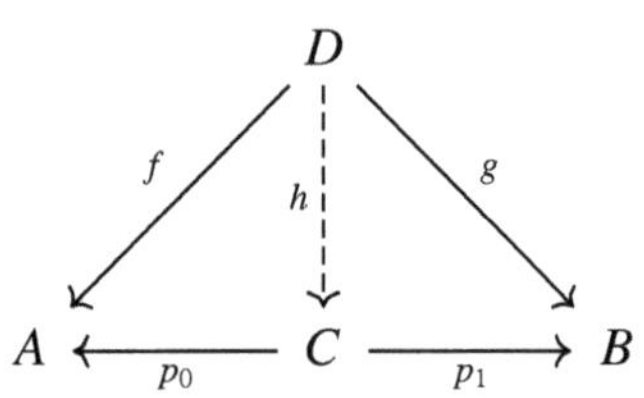

The maps p_0 and p_1 are called the *projections* of the binary product. The unique map h is denoted by $\langle f, g\rangle$. Note that $p_0 \circ \langle f, g\rangle = f$ and $p_1 \circ \langle f, g\rangle = g$. As we will mostly work with binary products in this chapter, whenever we say a product, unless specified otherwise, we always mean a binary product.

Remark 56 The uniqueness condition in Definition 55 can be equivalently phrased as the identity $\langle p_0 h, p_1 h\rangle = h$, for any $h : D \to C$. To show the equivalence, it is clear that having the identity, we can reach the uniqueness, because if $p_0 h = p_0 h' = f$ and $p_1 h = p_1 h' = g$, then $h = \langle p_0 h, p_1 h\rangle = \langle p_0 h', p_1 h'\rangle = h'$. Conversely, if we have the uniqueness condition, set $h' = \langle p_0 h, p_1 h\rangle$. As $p_0 h' = p_0 h = f$ and $p_1 h' = p_1 h = g$, then by uniqueness, we have $h = h'$. Writing universal conditions as equalities is useful in practice. However, it is also conceptually important as it helps to construct the free categories with that universal structure, by first producing all possible required objects and maps freely and then making the quotient over the equalities to ensure that they have the properties they should.

Remark 57 There is a high-level definition for binary products that is conceptually more elegant but computationally harder to work with. Here is the definition. A binary product of A and B is an object C together with an isomorphism between the functors $\mathrm{Hom}(-, C) : C^{op} \to \mathbf{Set}$ and $\mathrm{Hom}(-, A) \times \mathrm{Hom}(-, B) : C^{op} \to \mathbf{Set}$ in the category $\mathbf{Set}^{C^{op}}$. We do not prove the equivalence between the two definitions here. However, it is a good exercise in order to become fluent in the categorical language.

Remark 58 Consider all diagrams of the shape:

$$A \xleftarrow[f]{} D \xrightarrow[g]{} B$$

where A and B are fixed and we change D and the maps $f : D \to A$ and $g : D \to B$. Using maps between the middle objects, there is a natural order on these diagrams. In the following picture, we say that the lower diagram is greater than the upper one:

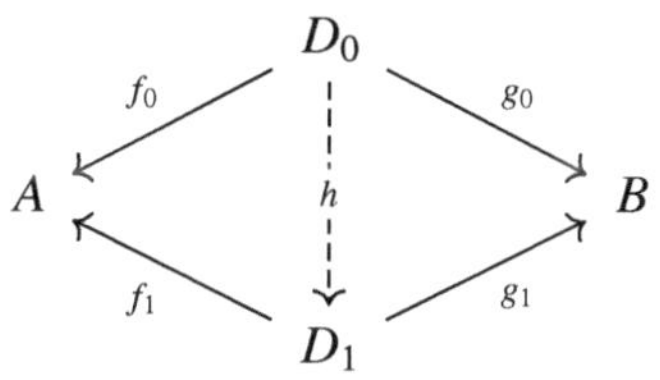

if there is $h : D_0 \to D_1$ making the whole diagram commutative. In this sense, the universal definition of binary product roughly states that the diagram

$$A \xleftarrow[p_0]{} C \xrightarrow[p_1]{} B$$

is the "greatest" among all the diagrams with the mentioned shape. Again, we put the "greatest" in quotation, as being the greatest in the above order is not enough to be a product due to the missing uniqueness condition.

Example 59 In **FinSet** and **Set**, the usual cartesian product $A \times B$ together with its projections $p_0 : A \times B \to A$ and $p_1 : A \times B \to B$ is a binary product of the sets A and B. In a poset $(P, \leq)$, a product of $a, b \in P$ is by definition the greatest lower bound of the subset $\{a, b\}$, i.e., an element c such that $c \leq a$ and $c \leq b$ and for any $d \in P$, if $d \leq a$ and $d \leq b$, then $d \leq c$. The product in a poset is usually called *meet*. For the typical poset $(P, \subseteq)$, where P is a set of subsets of a set X, if P is closed under intersection, then the product (meet) is the intersection itself. In $\mathbf{Set}^{\mathbb{N}}$, a product of (A, σ_A) and (B, σ_B) is the dynamic system $(A \times B, \sigma_A \times$

σ_B) together with the evident projections, where $[\sigma_A \times \sigma_B](a, b) = (\sigma_A(a), \sigma_B(b))$. Note that the projections respect the dynamisms and hence they live in $\mathbf{Set}^{\mathbb{N}}$. A similar situation happens in $\mathbf{Set}^{\mathbb{Z}}$. More generally, in $\mathbf{Set}^{C}$, a product of two functors $F, G : C \to \mathbf{Set}$ is the functor $H : C \to \mathbf{Set}$, together with the maps $p_0 : H \to F$ and $p_1 : H \to F$ such that H maps the object A to $F(A) \times G(A)$ and the morphism $f : A \to B$ to the function $H(f) : F(A) \times G(A) \to F(B) \times G(B)$, sending (x, y) to $(F(f)(x), G(f)(y))$. The projection $p_0 : H \to F$ is defined by setting $(p_0)_A : F(A) \times G(A) \to F(A)$ as the usual projection. The map $p_1 : H \to G$ is defined similarly. It is not hard to see that H is actually a functor, p_0 and p_1 are actually maps in $\mathbf{Set}^{C}$ and the data is actually a product.

Example 60 In **NJ**, the product of A and B is the conjunction $A \wedge B$ together with the following two derivations as $p_0 : A \wedge B \to A$ and $p_1 : A \wedge B \to B$:

$$\frac{A \wedge B}{A}\wedge E_1 \qquad \frac{A \wedge B}{B}\wedge E_2$$

For any two derivations D and D', take $\langle \mathsf{D}, \mathsf{D}' \rangle$ as the derivation:

$$\frac{\begin{array}{cc}\mathsf{D} & \mathsf{D}' \\ A & B\end{array}}{A \wedge B}\wedge I$$

To prove $p_0(\langle \mathsf{D}, \mathsf{D}' \rangle) = \mathsf{D}$, note that the left-hand side of the equality is the derivation:

$$\dfrac{\dfrac{\begin{array}{cc}\mathsf{D} & \mathsf{D}' \\ A & B\end{array}}{A \wedge B}\wedge I}{A}\wedge E_1$$

which is β-equivalent to D. As we consider the derivations up to β-equivalence, we have $p_0(\langle \mathsf{D}, \mathsf{D}' \rangle) = \mathsf{D}$. The proof for the other identity, i.e., $p_1(\langle \mathsf{D}, \mathsf{D}' \rangle) = \mathsf{D}'$ is similar. For the uniqueness, by Remark 56, it is enough to prove $\langle p_0 \mathsf{D}, p_1 \mathsf{D} \rangle = \mathsf{D}$, for any derivation D of $A \wedge B$. Notice that the left-hand side of the equality is

$$\dfrac{\dfrac{\begin{array}{c}\mathsf{D} \\ A \wedge B\end{array}}{A}\wedge E_1 \qquad \dfrac{\begin{array}{c}\mathsf{D} \\ A \wedge B\end{array}}{B}\wedge E_2}{A \wedge B}\wedge I$$

which is η-equivalent to D. Therefore, $\langle p_0 \mathsf{D}, p_1 \mathsf{D} \rangle = \mathsf{D}$. A similar construction is in place for the categories **NN** and **NM**. Note that conjunction, its rules, and their $\beta\eta$-equivalences form the minimal machinery required to make conjunction into the product in **NJ**. The only intuitively non-trivial aspect is the $\beta\eta$-equivalences, which might be seen as artificial conditions imposed on derivations to ensure the categorical interpretation works. However, in Philosophical Comment 62, we argue that these equivalences are also natural conditions to assume.

Example 61 (*Non-existence of Binary Products*) Consider the poset $(P, \subseteq)$ of all infinite subsets of $\mathbb{N}$. Then, the product (meet) of the set of even numbers and the set of odd numbers does not exist, as there is no infinite set below both of them, let alone the greatest of such lower bounds. For another example, consider a non-trivial group seen as a category. Then, the product of the category's only object $*$ with itself does not exist. Because, if it does, it must be $*$. Let us assume that $p_0, p_1 \in G$ are the projections of the product. Then, for any $g, h \in G$, there is $i \in G$ such that $p_0 i = g$ and $p_1 i = h$. Therefore, $p_1 p_0^{-1} g = h$, for any $g, h \in G$, which is impossible. Just put $h = p_1 p_0^{-1}$ and $g \neq e$.

Philosophical Comment 62 (*Product as the Conjunction*) Reading any category as a proof system, we can interpret a product of two propositions as their conjunction. However, looking deeply into the definition, one may also object that although the diagram commutativity and the uniqueness conditions for the cartesian product of sets is natural, the proof-theoretical counterpart, i.e., the β- and η-equivalences are not and hence the identification of the conjunction with the product may seem to be too demanding. To justify the choice, note that by Remark 57, a product of A and B is just an object C in $\mathcal{C}$ together with an isomorphism between the functors $\mathrm{Hom}(-, C)$ and $\mathrm{Hom}(-, A) \times \mathrm{Hom}(-, B)$ in the category $\mathbf{Set}^{\mathcal{C}^{op}}$. This just says that the deductions of $A \wedge B$ from D is in one-to-one and *uniform* correspondence with the pairs of deductions of A and B from D. This, of course, can be considered as a natural formalization for the conjunction operator. The projections, the diagram commutation and the uniqueness conditions, however, are inessential details to present this universal definition in a low level language of maps.

Lemma 63 *The product is unique up to a canonical isomorphism, i.e., for any two products (C, p_0, p_1) and (D, q_0, q_1) of A and B, the canonical map $i : C \to D$ induced by the universality of the product is an isomorphism.*

Proof Let (C, p_0, p_1) and (D, q_0, q_1) be both products of the two objects A and B. Therefore, there are maps $i : C \to D$ and $j : D \to C$ such that the diagram:

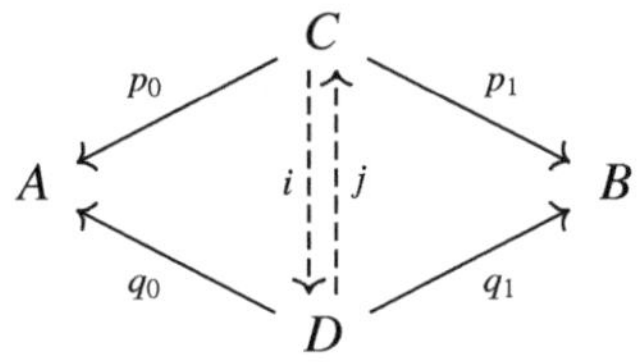

commutes. Therefore, the map $ji : C \to C$ makes the following diagram commute:

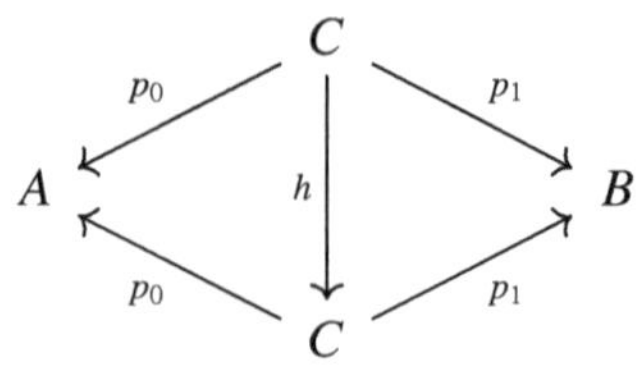

By the universality of the product (C, p_0, p_1), this map must be unique. As $id_C : C \to C$ does exactly the same thing, $ji = id_C$. Similarly, one can show that $ij = id_D$. Therefore, $i : C \to D$ is an isomorphism. □

Remark 64 Here are two remarks. First, note that the isomorphism $i : C \to D$ in Lemma 63 is not just an isomorphism between C and D. It also transforms the projections to each other, i.e., $q_0 i = p_0$ and $q_1 i = p_1$. This is what we should expect because a product is not just an object but the tuple of an object and two projections, and the isomorphism between the products must also interact with these projections. Secondly, using Lemma 63, we are allowed to denote a product of A and B and their projections with a reserved name $A \times B$ and $p_0^{A,B} : A \times B \to A$ and $p_1^{A,B} : A \times B \to B$. When there is no risk of confusion, we simply denote $p_0^{A,B}$ and $p_1^{A,B}$ by p_0 and p_1.

Definition 65 A category is called *cartesian* if it has a terminal object and all binary products.

Example 66 All categories **FinSet**, **Set**, **Top**, $\mathbf{Set}^{\mathcal{C}}$, **NN**, **NM**, and **NJ** are cartesian. A cartesian poset is usually called a *bounded meet-semilattice*.

Usually, universal constructions are functorial. Let us explain the phenomenon by an example. Assume that the product of A and B and the product of C and D exist. Now, assume that we are given two maps $f : A \to C$ and $g : B \to D$. We intend to come up with a *canonical map* $f \times g : A \times B \to C \times D$. For that purpose, first note that $p_0^{A,B} : A \times B \to A$ and $p_1^{A,B} : A \times B \to B$. Thus, $fp_0^{A,B} : A \times B \to C$ and $gp_1^{A,B} : A \times B \to D$. Therefore, $\langle fp_0^{A,B}, gp_1^{A,B} \rangle : A \times B \to C \times D$:

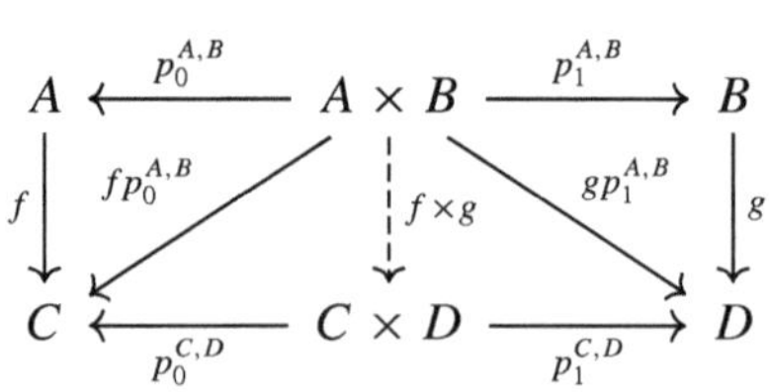

One can then define $f \times g : A \times B \to C \times D$ as $\langle fp_0^{A,B}, gp_1^{A,B} \rangle$. Using this definition, one can show that the assignment $(-) \times (-) : \mathcal{C} \times \mathcal{C} \to \mathcal{C}$ mapping the pair (A, B) to $A \times B$ and $(f, g) : (A, B) \to (C, D)$ to $f \times g : A \times B \to C \times D$ is a functor. We call this functor a *binary product* functor. Note that for each pair of objects, there are different choices for their binary product and hence there may be different product functors on a given category. However, as all those different choices are isomorphic, one can see that there is at most one binary product functor, up to an isomorphism. Using this observation and by a slight abuse of language, we denote all binary product functors with one reserved notation $(-) \times (-) : C \times C \to C$ and call it *the* binary product functor.

Using the universal property of the terminal object and the products, it is easy to prove the following list of isomorphisms:

Theorem 67 *Let C be a cartesian category. Then:*

- $A \times 1 \cong A$, *via the map* $p_0 : A \times 1 \to A$.
- $A \times B \cong B \times A$, *via the map* $\langle p_1, p_0 \rangle : A \times B \to B \times A$.
- $(A \times B) \times C \cong A \times (B \times C)$, *via the canonical map* $\langle r_0 q_0, \langle r_1 q_0, q_1 \rangle \rangle : (A \times B) \times C \to A \times (B \times C)$, *where* $q_i = p_i^{(A \times B), C}$ *and* $r_i = p_i^{A,B}$.

Proof We only prove the first claim and leave the rest to the reader. Consider the map $\langle id_A, ! \rangle : A \to A \times 1$. We claim that this map is the converse of p_0, i.e., $p_0 \langle id_A, ! \rangle = id_A$ and $\langle id_A, ! \rangle p_0 = id_{A \times 1}$. The first equality is clear. For the second, it is enough to show that both $\langle id_A, ! \rangle p_0$ and $id_{A \times 1}$ make the following diagram commute:

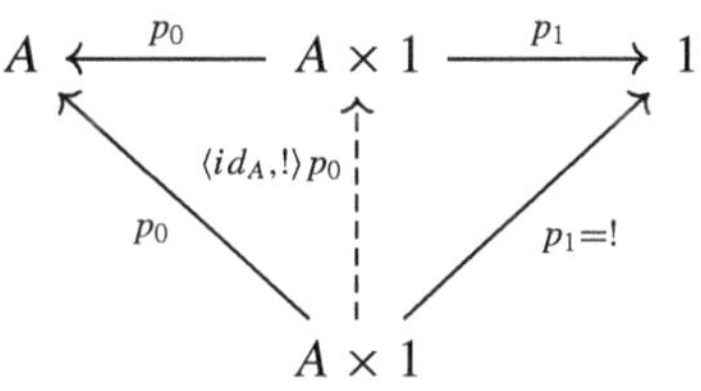

which is trivially true. Therefore, by the uniqueness condition in the definition of the product, we have $\langle id_A, ! \rangle p_0 = id_{A \times 1}$. □

Using binary products, it is possible to define the product of any finite family $A_0, A_1, \ldots, A_n$ of objects and denote it by $A_0 \times A_1 \times \ldots \times A_n$ or $\prod_{i=0}^n A_i$. Of course, we have to specify the way we split the product into binary products and by definition, we always start from the left, i.e., $\prod_{i=0}^n A_i = (\ldots(((A_0 \times A_1) \times A_2) \times \ldots) \times A_n)$. For the projections of this product, for any $0 \leq j \leq n$, there is a canonical combination of binary projections that goes from $\prod_{i=0}^n A_i$ to A_j. We denote this map by p_j. Moreover, for any families of maps $\{f_i : C \to A_i\}_{i=0}^n$, there is a canonical map from C to $\prod_{i=0}^n A_i$ that we denote by $\langle f_i \rangle_{i=0}^n$. Notice that using Theorem 67, it is clear that the way we split the product into binary products and the order of them is not important in the end.

4.2 Cocartesian Structure

The cartesian structure explained in the previous subsection has a dual that is also significant in categorical proof theory.

Definition 68 (*Initial Objects*) An object A is called *initial* if for any object B, there exists a unique map from A to B. This unique map is denoted by $!_B : A \to B$ and if B is clear from the context, just by $!$. Similar to terminal objects, there is at most one initial object up to an isomorphism. Therefore, we can denote all initial objects with a reserved name 0 and call it *the* initial object.

Example 69 In **FinSet** and **Set**, the initial object is the empty set $\varnothing$. In a poset $(P, \leq)$, the initial object is by definition an element $a \in P$ such that for any $b \in P$, we have $a \leq b$. Hence, the initial object is the least element of the poset. In $\mathbf{Set}^{\mathbb{N}}$ and $\mathbf{Set}^{\mathbb{Z}}$, the initial object is the pair $(\varnothing, id_{\varnothing})$. More generally, in $\mathbf{Set}^{\mathcal{C}}$, the initial object is the constant functor $\Delta_{\varnothing} : \mathcal{C} \to \mathbf{Set}$, mapping every object to the empty set and any morphism of $\mathcal{C}$ to the identity function over $\varnothing$. In **NJ**, the proposition $\bot$ is an initial object. The reasoning is similar to that of terminality of $\top$ presented in Example 50. Notice the role of the η-equivalence relation on derivations to ensure the uniqueness condition.

Philosophical Comment 70 (*Initial Objects as the Trivializing Inconsistency*) Reading any category as a proof system, we can interpret the initial object as the trivializing inconsistency, i.e., the proposition $\bot$ whose assumption makes all propositions deductible and all such deductions equivalent. Note that this type of definition is different from what we had for $\top$ and conjunction, as here, we identify a proposition by what and how it can prove the other propositions rather than by the form of its deductions. Trying to define $\bot$ in the latter way, one may argue that we must define $\bot$ as a proposition with no proof and then, as a derived property, show that such a proposition proves any other proposition in a trivial way. To see why, note that a deduction is a way to transform the proofs of the assumption to the proofs of the conclusion, and as $\bot$ has no proofs, there is only one unique way to map this empty set to the set of proofs of any other proposition. We will come back to this discussion when we reach the BHK interpretation in Sect. 5.

Example 71 (*Non-existence of Initial Objects*) For an easier example, consider the poset $(\mathbb{Z}, \leq)$. This poset has no least element and hence no initial object. For another, note that any non-trivial monoid seen as a category does not have an initial object. Moreover, the category of non-empty sets with functions as its morphisms does not have an initial object.

Definition 72 (*Binary Coproducts*) Let A and B be two objects. An object C together with two morphisms $i_0 : A \to C$ and $i_1 : B \to C$ is called a *coproduct* of A and B if for any object D and any morphisms $f : A \to D$ and $g : B \to D$, there exists a unique map $h : C \to D$ such that

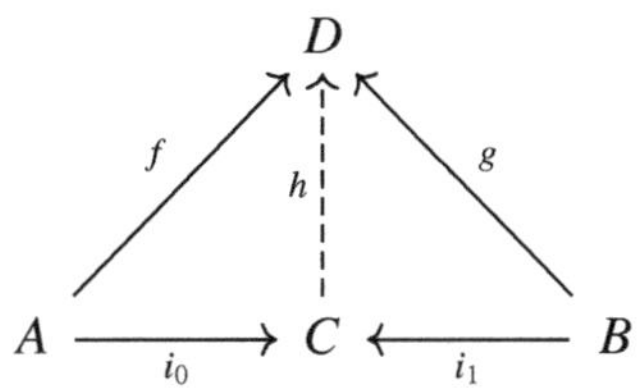

The maps i_0 and i_1 are called the *injections* of the coproduct. The unique h is denoted by (f, g). Similar to what we had for the other universal constructions, the coproduct of A and B is unique up to (a canonical) isomorphism and hence we can denote it and their injections with reserved names $A + B$ and $i_0^{A,B} : A \to A + B$ and $i_1^{A,B} : B \to A + B$, respectively. When there is no risk of confusion, we simply denote $i_0^{A,B}$ and $i_1^{A,B}$ by i_0 and i_1. Finally, as we will mostly work with binary coproducts in this chapter, whenever we say a coproduct, unless specified otherwise, we always mean a binary coproduct.

Remark 73 The uniqueness condition above can be equivalently phrased as the identity $(hi_0, hi_1) = h$, for any $h : C \to D$. The proof is similar to that of the product case.

Remark 74 Similar to products, there is also a high-level definition for coproducts. A coproduct of A and B is an object C together with an isomorphism between the functors $\mathrm{Hom}(C, -) : \mathcal{C} \to \mathbf{Set}$ and $\mathrm{Hom}(A, -) \times \mathrm{Hom}(B, -) : \mathcal{C} \to \mathbf{Set}$ in the category $\mathbf{Set}^{\mathcal{C}}$. We will not prove the equivalence between the two definitions.

Remark 75 Notice that the universal definition of the binary coproduct roughly states that the diagram

$$A \xrightarrow[i_0]{} C \xleftarrow[i_1]{} B$$

is the "least" between all such diagrams with fixed end nodes A and B.

Example 76 In **FinSet** and **Set**, the coproduct of A and B is the disjoint union $A + B = \{(0, a) \mid a \in A\} \cup \{(1, b) \mid b \in B\}$ together with the injections $i_0 : A \to A + B$ and $i_1 : B \to A + B$, defined by $i_0(a) = (0, a)$ and $i_1(b) = (1, b)$. In a poset $(P, \leq)$, the coproduct of $a, b \in P$ is by definition the least upper bound of the subset $\{a, b\}$, i.e., an element c such that $a \leq c$ and $b \leq c$ and for any $d \in P$, if $a \leq d$ and $b \leq d$, then $c \leq d$. Coproducts in posets are usually called *joins*. For the typical poset $(P, \subseteq)$, where P is a set of subsets of a set X, if P is closed under binary union, the coproduct (join) is the union, itself. In $\mathbf{Set}^{\mathbb{N}}$ and $\mathbf{Set}^{\mathbb{Z}}$, the coproduct of (A, σ_A) and (B, σ_B) is $(A + B, \sigma_{A+B})$, together with the usual injections, where $[\sigma_{A+B}](0, a) = (0, \sigma_A(a))$ and $[\sigma_{A+B}](1, b) = (1, \sigma_B(b))$. Note that the injections respect the dynamism. More generally, a coproduct of two objects $F : \mathcal{C} \to \mathbf{Set}$ and $G : \mathcal{C} \to \mathbf{Set}$ in $\mathbf{Set}^{\mathcal{C}}$ is the functor $H : \mathcal{C} \to \mathbf{Set}$ mapping the object A to $H(A) = F(A) + G(A)$ and the map $f : A \to B$ to the function $H(f) : F(A) + G(A) \to F(B) + G(B)$ defined by $H(f)(0, x) = (0, F(f)(x))$ and $H(f)(1, y) = (1, G(f)(y))$. The injections $i_0 : F \to H$ and $i_1 : G \to H$ are defined by $(i_0)_A : F(A) \to F(A) + G(A)$ and $(i_1)_A : G(A) \to F(A) + G(A)$ as the usual injections. It is easy to prove that $H : \mathcal{C} \to \mathbf{Set}$ is a functor, both i_0 and i_1 are natural transformations and the data is actually a coproduct.

Example 77 In **NJ**, the coproduct of A and B is the disjunction $A \vee B$ together with the two derivations:

$$\dfrac{A}{A \vee B}\,\vee I_1 \qquad\qquad \dfrac{B}{A \vee B}\,\vee I_2$$

as the injections. For any two derivations D and D', define the map $(\mathsf{D}, \mathsf{D}')$ as the derivation:

$$\dfrac{A \vee B \qquad \begin{array}{c}[A]\\ \mathsf{D}\\ C\end{array} \qquad \begin{array}{c}[B]\\ \mathsf{D}'\\ C\end{array}}{C}\,\vee E$$

To prove $(\mathsf{D}, \mathsf{D}')i_0 = \mathsf{D}$, note that the left-hand side is the derivation:

$$\dfrac{\dfrac{A}{A \vee B}\,\vee I_1 \qquad \begin{array}{c}[A]\\ \mathsf{D}\\ C\end{array} \qquad \begin{array}{c}[B]\\ \mathsf{D}'\\ C\end{array}}{C}\,\vee E$$

which is β-equivalent to D. As we consider the derivations up to β-equivalence, we have $(\mathsf{D}, \mathsf{D}')i_0 = \mathsf{D}$. The proof for the other identity, i.e., $(\mathsf{D}, \mathsf{D}')i_1 = \mathsf{D}'$ is similar. For the uniqueness, by Remark 73, it is enough to prove $(\mathsf{D}i_0, \mathsf{D}i_1) = \mathsf{D}$, for any derivation D with the assumption $A \vee B$. Notice that the left-hand side is

$$\dfrac{A \vee B \qquad \begin{array}{c}\dfrac{[A]^1}{A \vee B}\,\vee I_1\\ \mathsf{D}\\ C\end{array} \qquad \begin{array}{c}\dfrac{[B]^2}{A \vee B}\,\vee I_2\\ \mathsf{D}\\ C\end{array}}{C}\,\vee E$$

which is η-equivalent to D.

Philosophical Comment 78 (*Coproduct as the Disjunction*) Reading any category as a proof system, we can interpret the coproduct of two propositions as their disjunction. However, looking deeply into the definition, one may also object that although the diagram commutativity and the uniqueness conditions for the disjoint union of sets is natural, the proof-theoretical counterpart, i.e., the β- and η-equivalences are not and hence the identification of the disjunction with the coproduct may seem to be too demanding. To justify the choice, note that by Remark 74, a coproduct of A and B is just an object C in $\mathcal{C}$ together with an isomorphism between the functors $\mathrm{Hom}(C, -)$ and $\mathrm{Hom}(A, -) \times \mathrm{Hom}(B, -)$ in the category $\mathbf{Set}^{\mathcal{C}}$. This just says that the deductions of D from the assumption $A + B$ is in one-to-one and *uniform* correspondence with the pairs of deductions of D from A and B. This, of course, can be considered as a natural formalization for the disjunction operator. The projections, the diagram commutation, and the uniqueness conditions, however, are inessential details to present this universal definition in a low level language of maps. Note that this type of definition is again based on what and how a proposition can prove the other propositions rather than by the form of its deductions. Trying to define the disjunction in the latter way, one may argue that we must define $A \vee B$ as a proposition whose proofs are either the proofs of A or the proofs of B and we know which one is the case. Then again, as a derived property, we can show that the deductions of D from $A \vee B$ is in one-to-one correspondence with the pairs of the deductions of A and B from D. The reason simply is that to transform the proofs of $A \vee B$ to the proofs of D, as a proof of $A \vee B$ is either a proof of A or a proof of B and we know which case is happening, it is necessary and sufficient to transform the proofs of A and the proofs of B into the proofs of D. We will come back to this discussion when we reach the BHK interpretation in Sect. 5.

Definition 79 A category is called *cocartesian* if it has the initial object and all the binary coproducts.

Example 80 All categories **FinSet**, **Set**, **Top**, $\mathbf{Set}^{\mathcal{C}}$, and **NJ** are cocartesian. A cocartesian poset is usually called a *bounded join-semilattice*.

Example 81 (*Non-existence of Binary Coproducts*) Consider the poset $(P, \subseteq)$ of all subsets of $\mathbb{N}$ whose complement is infinite. Then, the coproduct (join) of the set E of even numbers and O of odd numbers does not exists, as the only subset above both of them is $\mathbb{N}$ whose complement is empty and as such finite. For another example, consider a non-trivial group considered as a category. Similar to the product case, it is easy to prove that the coproduct of $*$ with itself does not exist in this category.

Similar to what we had for the product, the coproduct can also rise to a functor. Assume that both the coproduct of A and B and the coproduct of C and D exist. Now, assume that we are given two morphisms $f : A \to C$ and $g : B \to D$. We intend to come up with a *canonical map* $f + g : A + B \to C + D$. For that purpose, first note that $i_0^{C,D} : C \to C + D$ and $i_1^{C,D} : D \to C + D$. Thus, $i_0^{C,D} f : A \to C + D$ and $i_1^{C,D} g : B \to C + D$. Therefore, $(i_0^{C,D} f, i_1^{C,D} g) : A + B \to C + D$:

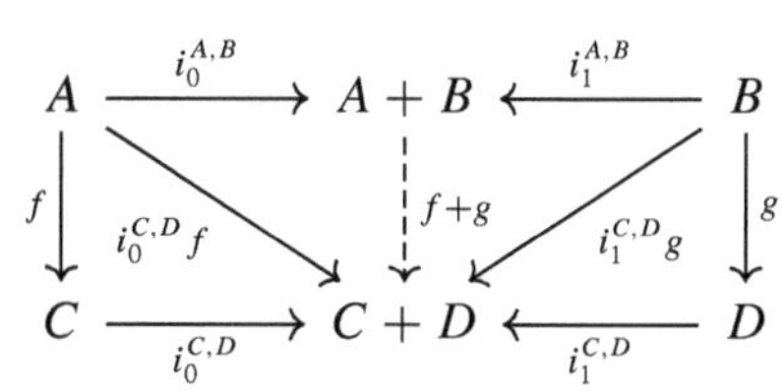

One can then define $f + g : A + B \to C + D$ as $(i_0^{C,D} f, i_1^{C,D} g)$. It is easy to see that this definition makes the assignment $(-) + (-) : \mathcal{C} \times \mathcal{C} \to \mathcal{C}$ mapping the pair

(A, B) to $A + B$ and the morphism $(f, g) : (A, B) \to (C, D)$ to $f + g : A + B \to C + D$ a functor. This functor is called a *coproduct functor*. Again, as all coproduct functors are isomorphic, we use one reserved notation $(-) + (-) : C \times C \to C$ for all of them and call it *the* coproduct functor.

Moreover, similar to products, we can define the coproduct of any finite family $A_0, \ldots, A_n$ of objects and denote it by $A_0 + \ldots + A_n$ or $\sum_{i=0}^{n} A_i$. Using the injections, one can come up with a canonical map $i_j : A_j \to \sum_{i=0}^{n} A_i$, for any $0 \leq j \leq n$. Moreover, for any family $\{f_i : A_i \to C\}_{i=0}^{n}$ of maps, there is a canonical map from $\sum_{i=0}^{n} A_i$ to C that we denote by $(f_i)_{i=0}^{n}$.

4.3 Exponential Objects

The third universal construction we are interested in is the exponential. Exponentials provide the machinery to internalize the structure of the category within itself.

Definition 82 (*Exponential object*) Let C be a cartesian category and A and B be two objects. An object C together with a morphism $ev : C \times A \to B$ is called an *exponential object* if for any $f : D \times A \to B$, there exists a unique $g : D \to C$ such that

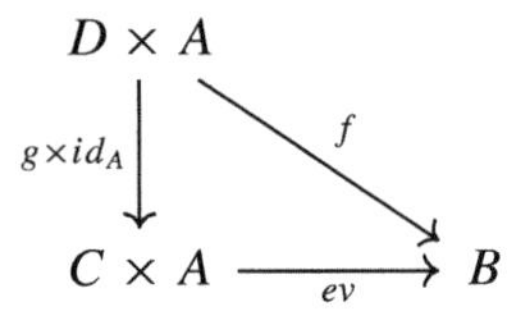

Again, it is easy to see that exponentials are unique up to a canonical isomorphism. Hence, we can safely refer to them as *the* exponential object of B by A and denote it and its evaluation map by the reserved names $[A, B]$ (or sometimes B^A) and $ev^{A,B} : [A, B] \times A \to B$, respectively. Moreover, for any $f : C \times A \to B$, we denote its unique g by $\lambda_A^{B,C} f : C \to [A, B]$. If there is no risk of confusion, we drop the superscripts in $ev^{A,B}$ and $\lambda_A^{B,C} f$ and write ev and $\lambda_A f$, respectively. Note that the above diagram's commutativity is simply the equality $ev(\lambda_A f \times id_A) = f$.

Remark 83 The uniqueness condition in Definition 82 can be equivalently phrased as the identity $\lambda_A ev(g \times id_A) = g$, for any $g : C \to [A, B]$.

Remark 84 Consider all diagrams of the shape:

$$D \times A \xrightarrow[f]{} B$$

where A and B are fixed and we change D and the map $f : D \times A \to B$. Using maps between D's, these diagrams have a natural order. In the following picture, we say that the lower diagram is greater than the upper one:

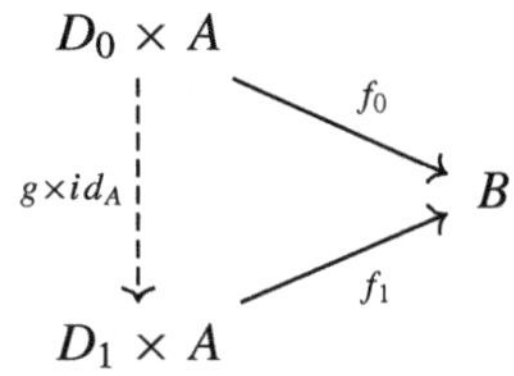

if there is $g : D_0 \to D_1$ making the whole diagram commutative. In this sense, the universal definition of the exponential object roughly states that the diagram

$$C \times A \xrightarrow{ev} B$$

is the "greatest" among all the diagrams with similar shapes. Note that the uniqueness condition is again missing in this informal discussion.

Remark 85 Similar to other universal constructions, there is a high-level formalization for the exponential object. An exponential object is an object C together with an isomorphism between the functors $\mathrm{Hom}(-, C)$ and $\mathrm{Hom}(- \times A, B)$ in the category $\mathbf{Set}^{\mathcal{C}^{op}}$. We do not prove the equivalence between the two definitions. However, one special instance of the isomorphism in the second definition, i.e., the one between $\mathrm{Hom}(1 \times A, B)$ or equivalently $\mathrm{Hom}(A, B)$ and $\mathrm{Hom}(1, [A, B])$ is worth explaining. Here is the description of the isomorphism. For any $f : A \to B$, using $p_1 : 1 \times A \to A$, we have the map $fp_1 : 1 \times A \to B$. By the universal property of the exponential $[A, B]$, we have the map $\lambda_A(fp_1) : 1 \to [A, B]$ such that

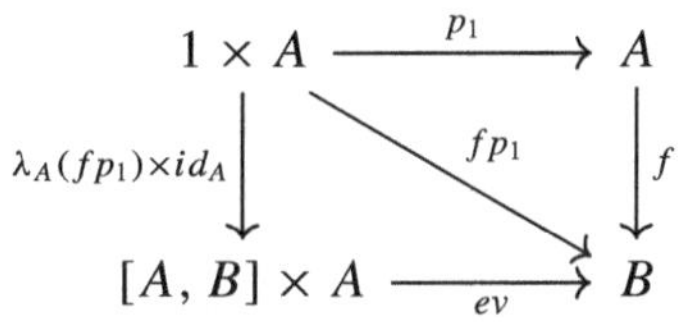

Conversely, having a map $g : 1 \to [A, B]$, we have

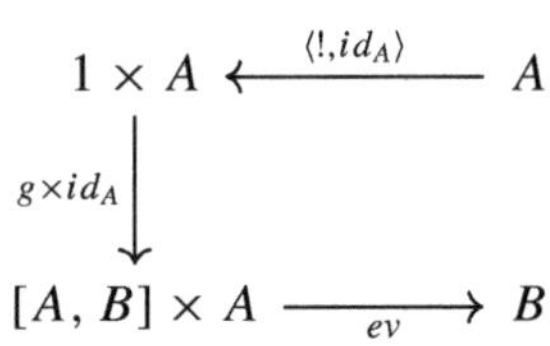

as a map from A to B. It is easy to see that these two operations are the converse to each other. For any $f : A \to B$, we denote $\lambda_A(fp_1)$ by λf and for any $g : 1 \to [A, B]$ and $a : X \to A$, we denote the map $ev(g \times id_A)\langle !, id_A\rangle a$, by $g \cdot a : X \to B$. It is easy to see that $(\lambda f) \cdot a = f \circ a$ and $\lambda(g \cdot id_A) = g$. Therefore, we can read λ as the usual lambda operator and $\cdot$ as the application operation.

Example 86 In the categories **FinSet** and **Set**, the exponential $[A, B]$ is the set B^A consisting of all the functions from A to B with the morphism $ev : B^A \times A \to B$ defined by $ev(f, a) = f(a)$. Note that for any map $f : C \times A \to B$, the map $\lambda_A f : C \to B^A$ is defined by $(\lambda_A f)(c)(a) = f(c, a)$. In the poset $(P, \leq)$, the exponential is by definition the greatest element c such that $c \wedge a \leq b$, i.e., an element c such that $c \wedge a \leq b$ and for any $d \in P$ if $d \wedge a \leq b$ then $d \leq c$. Exponential objects in posets are called *Heyting implications* and denoted by $\to$. In the poset $(\mathbf{2}^{(P,\leq)}, \subseteq)$, the exponential $U \to V$ is $\{x \in P \mid \forall y \geq x\, (y \in U \to y \in V)\}$. For the poset $\mathcal{O}(X)$, where X is a topological space, the exponential $U \to V$ is $int(U^c \cup V)$, where int is the interior operation.

Example 87 In $\mathbf{Set}^{\mathbb{Z}}$, the exponential object $[(A, \sigma_A), (B, \sigma_B)]$ is the dynamical system (C, σ_C) where C is the set of *all* functions from A to B and $\sigma_C(f) = \sigma_B f \sigma_A^{-1}$. The evaluation map $ev : (C, \sigma_C) \times (A, \sigma_A) \to (B, \sigma_B)$ is just the usual set-theoretical evaluation map. Notice that ev respects the dynamism as $ev(\sigma_C(f), \sigma_A(x)) = \sigma_C(f)(\sigma_A(x)) = \sigma_B f \sigma_A^{-1}(\sigma_A(x)) = \sigma_B(f(x)) = \sigma_B(ev(f, x))$. For any equivariant map $f : (D, \sigma_D) \times (A, \sigma_A) \to (B, \sigma_B)$, the map $\lambda_A f : (D, \sigma_D) \to [(A, \sigma_A), (B, \sigma_B)]$ is defined by $(\lambda_A f)(d)(a) = f(d, a)$ and it is equivariant. The reason is that for any $d \in D$ and any $a \in A$, we have $(\lambda_A f)(\sigma_D(d))(a) = f(\sigma_D(d), a)$, by definition. Moreover, as f is equivariant, we have

$$f(\sigma_D(d), a) = \sigma_B(f(d, \sigma_A^{-1}(a))) = \sigma_B[(\lambda_A f)(d)(\sigma_A^{-1}(a))].$$

Therefore, $(\lambda_A f)(\sigma_D(d)) = \sigma_B((\lambda_A f)(d))\sigma_A^{-1}$.

Example 88 Let C be a small category. Then, in $\mathbf{Set}^C$, the exponential object $[E, F]$ is the functor mapping the object A to the set $\mathrm{Hom}_{\mathbf{set}^C}(y^A \times E, F)$. The evaluation map is the natural transformation $ev : [E, F] \times E \to F$ defined by $ev_A : [E, F](A) \times E(A) \to F(A)$, mapping the pair (α, x) to $\alpha_A(id_A, x)$, for any $\alpha : y^A \times E \to F$ and $x \in E(A)$. It is an interesting exercise to set C as $\mathbf{1}$ or $(\mathbb{Z}, +)$, compute the exponential object described here, and show that it is exactly what was described in Examples 86 and 87.

Example 89 In **Top**, not all exponential objects exist. For instance, there is no topological space and no evaluation map that can act as $[\mathbb{Q}, I]$, where $\mathbb{Q}$ is the subspace of rational numbers in $\mathbb{R}$ and $I = [0, 1]$ is the unit interval Borceux (1994). In fact, for any topological space X, a necessary and sufficient condition for X is known that guarantees the existence of the exponential $[X, Y]$ for *all* spaces Y. This condition is called *core-compactness*. To define this notion, let $U, V \subseteq X$ be two open subsets of X. The open V is called *way below* U, denoted by $V \ll U$, if any open cover of U has a finite subcover for V. A space X is called core-compact if for every open neighborhood U of a point $x \in X$, there exists an open neighborhood V of x with $V \ll U$. If the space X is Hausdorff, it is core-compact iff it is locally compact. If X is core-compact, then the space $[X, Y]$ is simply the set of all continuous functions from X to Y with the topology with the subbasis

$$O_{U,V} = \{f \in [X, Y] \mid U \ll f^{-1}(V)\},$$

for any opens $U \subseteq X$ and $V \subseteq Y$. The evaluation map is the usual set-theoretical evaluation map. If both X and Y are Hausdorff, then this topology is just the usual compact-open topology that has the subbasis

$$O'_{K,V} = \{f \in [X, Y] \mid f[K] \subseteq V\},$$

for any compact $K \subseteq X$ and any open $V \subseteq Y$. For instance, as the discrete space $\mathbb{N}$ is locally compact and Hausdorff, the exponential $\mathbb{N}^{\mathbb{N}}$ exists and it is just the set of *all* functions over $\mathbb{N}$ together with the compact-open topology which is nothing but the usual product topology. Another example that we will use later is the exponential $[\{0, 1\}^{\mathbb{N}}, \mathbb{N}]$ that also exists, as $\{0, 1\}^{\mathbb{N}}$ is a compact and Hausdorff space and its topology is the compact-open topology.

Example 90 In **NJ**, the exponential $[A, B]$ is the implication $A \to B$ together with the derivation:

$$\dfrac{\dfrac{(A \to B) \wedge A}{A}\,\wedge E_2 \qquad \dfrac{(A \to B) \wedge A}{A \to B}\,\wedge E_1}{B}\,\to E$$

as the evaluation map. For any derivation D for B from $C \wedge A$, define $\lambda_A \mathsf{D}$ as

$$\begin{array}{c} \dfrac{C \qquad [A]}{C \wedge A}\,\wedge I \\ \mathsf{D} \\ \dfrac{B}{A \to B}\,\to I \end{array}$$

To show that $ev(\lambda_A \mathsf{D} \times id_A) = \mathsf{D}$, note that the right-hand side is

$$\begin{array}{c} \dfrac{\begin{array}{c} \dfrac{C \wedge A}{C} \\ \lambda_A \mathsf{D} \\ A \to B \end{array} \qquad \dfrac{C \wedge A}{A}}{(A \to B) \wedge A} \\ ev \\ B \end{array}$$

Substituting $\lambda_A \mathsf{D}$ and ev in this derivation, we have

$$\dfrac{\dfrac{\dfrac{\dfrac{\begin{array}{c}\dfrac{\dfrac{C \wedge A}{C} \qquad [A]}{C \wedge A}\\ \mathsf{D}\\ B\end{array}}{A \to B}\to I \qquad \dfrac{C \wedge A}{A}}{(A \to B) \wedge A}}{A} \qquad \dfrac{\dfrac{\dfrac{\begin{array}{c}\dfrac{\dfrac{C \wedge A}{C} \qquad [A]}{C \wedge A}\\ \mathsf{D}\\ B\end{array}}{A \to B}\to I \qquad \dfrac{C \wedge A}{A}}{(A \to B) \wedge A}}{A \to B}}{B}$$

which is $\beta\eta$-equivalent to D. The proof of the uniqueness part is similar.

Philosophical Comment 91 (*Exponentials as the Implication*) Reading any category as a proof system, we can interpret the exponential as the implication. To justify, note that the exponential $[A, B]$ is just an object together with a natural isomorphism between $\mathrm{Hom}(- \times A, B)$ and $\mathrm{Hom}(-, [A, B])$. This just says that the deduction of B from $D \wedge A$ are in one-to-one and uniform correspondence with the deductions of $A \to B$ from D, i.e., proving $A \to B$ is just proving B with the addition of A to the assumptions.

Definition 92 A cartesian category where all exponentials exist is called a *cartesian closed category* (CCC, for short). A CCC which has all the binary coproducts is called an *almost bicartesian closed category* (ABC, for short), and if it also has the initial object, a *bicartesian closed category* (BCC, for short). A BCC (resp. an ABC) that is also a poset is called a *Heyting algebra* (resp. an *almost Heyting algebra*). By the BC-structure, we mean the terminal and initial objects and the product, coproduct and exponential of a BCC. This is an informal notion, putting a name on the structure that makes a category a BCC. The CC- and AB-structures are defined accordingly.

Example 93 All categories **FinSet**, **Set**, $\mathbf{Set}^C$, and **NJ** are BCC's. The categories **NN** and **NM** are CCC and ABC, respectively.

Philosophical Comment 94 A BCC can be read as a proof system over the language $\mathcal{L}_p = \{\wedge, \vee, \to, \top, \bot\}$. Similarly, an ABC (resp. a CCC) is a proof system over the fragment $\mathcal{L}_p - \{\bot\} = \{\wedge, \vee, \to, \top\}$ (resp. $\mathcal{L}_p - \{\vee, \bot\} = \{\wedge, \to, \top\}$). With this interpretation, Heyting algebras are trivialized proof systems over $\mathcal{L}_p$ where all deductions are collapsed to one deductibility order. The same also holds for the poset ABC and the poset CCC. Moreover, it is worth mentioning that reading connectives as universal constructions naturally leads to intuitionistic proof systems. For classical proof systems, see Sect. 6.

Example 95 (*Non-existence of Exponential Objects*) Consider the set P consisting of all finite subsets of $\mathbb{N}$ together with the whole set $\mathbb{N}$. Note that the poset $(P, \subseteq)$ is cartesian with the terminal object $\mathbb{N}$ and the intersection as the product. We claim that the exponential $W = \{0\} \to \{1\}$ in the poset $(P, \subseteq)$ does not exist. If it does, it must have the property $S \cap \{0\} \subseteq \{1\}$ iff $S \subseteq W$, for any $S \in P$. As $S \cap \{0\} \subseteq \{1\}$ holds for any $S \subseteq \mathbb{N} - \{0\}$, we must have $\mathbb{N} - \{0\} \subseteq W$. Therefore, W is not finite and hence $W = \mathbb{N}$. But then as $\mathbb{N} \subseteq W = \mathbb{N}$, we must have $\mathbb{N} \cap \{0\} \subseteq \{1\}$ which is impossible.

Example 96 Let C be a non-preorder category with the initial and terminal objects and $0 \cong 1$. Then, C cannot have all the exponentials because if it can, then we must have

$$\mathrm{Hom}(A, B) \cong \mathrm{Hom}(1 \times A, B) \cong \mathrm{Hom}(1, [A, B]) \cong \mathrm{Hom}(0, [A, B]).$$

for any objects A and B. As 0 is the initial object, the rightmost set has exactly one element. Hence, $\mathrm{Hom}(A, B)$ must also be a singleton. This contradicts the assumption that C is not a preordered set. As a consequence, many algebraic categories, like the category of all groups and homomorphisms, do not have all exponential objects.

Example 97 Let X be a fixed infinite set. Define C as the category of sets in the form X^k for some $k \geq 0$ with the usual functions as the morphisms. This category is cartesian with the terminal object X^0 and the usual set-theoretical product as the product. We claim that the exponential $[X, X]$ does not exist in C. Because if it does, it must be in the form X^k, for some $k \geq 0$. Therefore, $\mathrm{Hom}(X^0, [X, X]) = \mathrm{Hom}(X^0, X^k)$ must be in one to one correspondence with $\mathrm{Hom}(X^0 \times X, X) \cong \mathrm{Hom}(X, X)$. However, the cardinality of the former equals the cardinality of X, while the cardinality of the latter is strictly greater than the cardinality of X. This phenomenon that there are more functions over an infinite set than the cardinality of the set itself is blocking the desirable construction of a cartesian closed category of *sets* with a non-terminal object X such that $[X, X] \cong X$. Such a situation can be useful to interpret lambda calculus where each element of the domain X can also be read as a function on X at the same time. Having such a dual nature is a usual phenomenon in computability theory where numbers are both the inputs and the codes of the algorithms. Such a CCC (using ordered/topological structures and not pure sets) was later discovered by Dana Scott (1988).

Similar to what we had for the product and coproduct, the exponential can also rise to a functor. Assume that both the exponentials $[A, B]$ and $[C, D]$ exist. Now, assume that we are given $f : C \to A$ and $g : B \to D$. We intend to come up with a *canonical map* $[f, g] : [A, B] \to [C, D]$. For that purpose, first note that $id_{[A,B]} \times f : [A, B] \times C \to [A, B] \times A$. Composing with $ev^{A,B} : [A, B] \times A \to B$ and $g : B \to D$, we will have the map $h : [A, B] \times C \to D$:

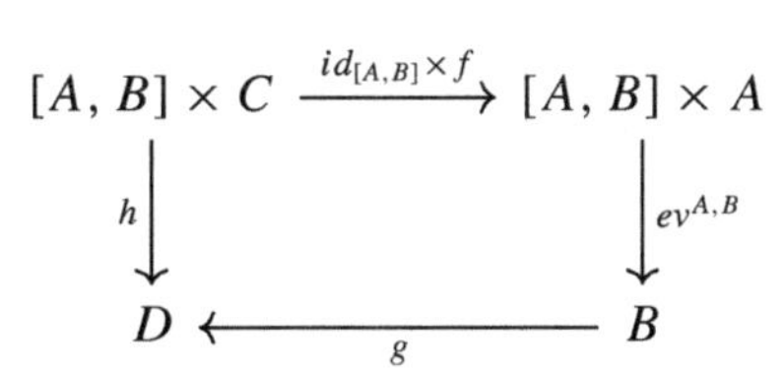

One can then define $[f, g] : [A, B] \to [C, D]$ as $\lambda_C h$. Now, define the assignment $[-, -] : \mathcal{C}^{op} \times \mathcal{C} \to \mathcal{C}$ by mapping the pair (A, B) to $[A, B]$ and $(f, g) : (A, B) \to (C, D)$ to $[f, g] : [A, B] \to [C, D]$. This assignment is a functor called the *exponentiation functor*.

4.4 Structure-Preserving Functors

So far, we have introduced terminal and initial objects, binary products and coproducts, and exponentials as proof-sensitive versions of logical constants. We also claimed that any BCC can be interpreted as a proof system over the language $\mathcal{L}_p$. The next natural step is to compare these proof systems using the natural machinery of functors. We expect these functors to preserve the logical constants of the systems. In this subsection, we will introduce these structure-preserving functors and then see some examples.

Definition 98 Let $\mathcal{C}$ and $\mathcal{D}$ be two BCC's. A functor $F : \mathcal{C} \to \mathcal{D}$ is said to preserve

- the terminal (initial) object, if $F(A)$ is a terminal (initial) object in $\mathcal{D}$, for any terminal (initial) object in $\mathcal{C}$,
- the product, if the diagram

$$F(A) \xleftarrow{F(p_0)} F(C) \xrightarrow{F(p_1)} F(B)$$

  is a product of $F(A)$ and $F(B)$ in $\mathcal{D}$, for any product

$$A \xleftarrow{p_0} C \xrightarrow{p_1} B$$

  in the category $\mathcal{C}$,
- the coproduct, if the diagram

$$F(A) \xrightarrow{F(i_0)} F(C) \xleftarrow{F(i_1)} F(B)$$

  is a coproduct of $F(A)$ and $F(B)$ in $\mathcal{D}$, for any coproduct

$$A \xrightarrow{i_0} C \xleftarrow{i_1} B$$

  in the category $\mathcal{C}$.

A product-preserving functor $F : \mathcal{C} \to \mathcal{D}$ preserves the exponentials, if the object $F(C)$ together with the evaluation map

$$F(C \times A) \xrightarrow{F(ev)} F(B)$$

is an exponential in $\mathcal{D}$, for any exponential

$$C \times A \xrightarrow{ev} B$$

in the category C. Note that as F preserves the product, $F(C \times A)$ is actually a product of $F(C)$ and $F(A)$.

Remark 99 Fixing a BC-structure for C and $\mathcal{D}$, one can define the preservation equivalently in the following way. We say that F preserves

- the terminal object iff the canonical map $! : F(1_C) \to 1_{\mathcal{D}}$ is an isomorphism,
- the initial object iff the canonical map $! : 0_C \to F(0_{\mathcal{D}})$ is an isomorphism,
- the product iff the canonical map $\langle F(p_0), F(p_1)\rangle : F(A \times B) \to F(A) \times F(B)$ is an isomorphism,
- the coproduct iff the canonical map $(F(i_0), F(i_1)) : F(A) + F(B) \to F(A + B)$ is an isomorphism,
- the exponential iff the canonical map $\lambda_{F(A)}(F(ev) \circ g) : F([A, B]) \to [F(A), F(B)]$ is an isomorphism, where g is the inverse of the canonical map $\langle F(p_0), F(p_1)\rangle : F([A, B] \times A) \to F([A, B]) \times F(A)$.

Note that the second definition of preservation demands the canonical morphisms (i.e., the ones induced by the universality of the structure) to be an isomorphism. As we will see later, this is not equivalent to simply having an isomorphism (except for the initial and terminal objects). Even if we have a natural isomorphism, it is still different from having the canonical map as an isomorphism. By these weaker versions, what we really do is preserving the object of the structure (product, coproduct, exponential), while we must also preserve its involved maps (projections, injections, evaluation) as they are also a part of the structure.

Definition 100 Let C and $\mathcal{D}$ be two BCC's. A functor $F : C \to \mathcal{D}$ is called a *BC-functor* if it preserves the BC-structure. The AB- and CC-functors are defined similarly, demanding the preservation of the structures available in the ABC's and CCC's, respectively.

Example 101 The inclusion functor i : **FinSet** $\to$ **Set**, mapping the finite sets and the functions between them to themselves is a BC-functor. The reason is that the construction of the BC-structure on finite sets is exactly the same as the one on all sets.

Example 102 The forgetful functor $U : \mathbf{Set}^{\mathbb{Z}} \to \mathbf{Set}$ mapping (A, σ_A) to A and equivariant maps to themselves is a BC-functor. To see why, it is enough to go back and check the construction of the BC-structure in the two categories. The only non-trivial thing to mention is that for the set of the exponential object $[(A, \sigma_A), (B, \sigma_B)]$, we used the set B^A of *all* functions from A to B and not just the equivariant functions. This is consistent with our claim that U preserves all exponentials.

Example 103 Let X and Y be two topological spaces and $f : X \to Y$ be an *open map*, i.e., a continuous map, where $f[U]$ is open in Y, for any open $U \subseteq X$. Then, the functor $f^{-1} : \mathcal{O}(Y) \to \mathcal{O}(X)$ is a BC-functor. It is clear that it preserves all finite meets and finite joins, simply because the inverse image function always preserves finite intersections and finite unions. For the exponentiation, first notice that $f[W \cap f^{-1}(U)] = f[W] \cap U$, for any subsets $W \subseteq X$ and $U \subseteq Y$. Then, consider the following series of equivalences:

$$W \subseteq f^{-1}(U) \to_X f^{-1}(V) \ \text{ iff } \ W \cap f^{-1}(U) \subseteq f^{-1}(V) \ \text{ iff}$$

$$f[W \cap f^{-1}(U)] \subseteq V \ \text{ iff } \ f[W] \cap U \subseteq V.$$

As f is an open map, $f[W]$ is open, for any open $W \in \mathcal{O}(X)$. Hence, $f[W] \cap U \subseteq V$ is equivalent to $f[W] \subseteq U \to_Y V$ which is also equivalent to $W \subseteq f^{-1}(U \to_Y V)$. Therefore, as W is an arbitrary open subset of X, we reach $f^{-1}(U \to_Y V) = f^{-1}(U) \to_X f^{-1}(V)$.

Example 104 Let $(P, \leq_P)$ and $(Q, \leq_Q)$ be two posets and $f : (P, \leq_P) \to (Q, \leq_Q)$ be a *p-morphism*, i.e., an order-preserving map such that for any $x \in P$ and $y \in Q$, if $f(x) \leq_Q y$, then there exists $z \in X$ such that $z \geq_P x$ and $f(z) = y$. Then, the functor $f^{-1} : U(Q, \leq_Q) \to U(P, \leq_P)$ is a BC-functor. To prove, it is enough to show that putting the upset topologies on P and Q makes the map $f : P \to Q$ an open map. Let $U \subseteq P$ be an upset. Then, to show that $f[U]$ is also an upset, assume that $y \in f[U]$ and $y \leq_Q w$. Therefore, there is $x \in U$ such that $y = f(x)$. Hence, by the property of the function f, there exists $z \in P$ such that $z \geq_P x$ and $w = f(z)$. As U is an upset and $x \in U$, we have $z \in U$. Hence, $w \in f[U]$. Therefore, $f[U]$ is an upset, which means that f is an open map.

Example 105 Let $\mathcal{C}$ be a BCC. Then, the forgetful functor $\pi : \mathcal{C} \to \mathrm{Po}(\mathcal{C})$ is a BC-functor. A similar claim holds for the ABC's and CCC's.

Example 106 Let $\mathcal{C}$ be a small CCC. Then, the functor $y : \mathcal{C} \to \mathbf{Set}^{\mathcal{C}^{op}}$ is a CC-functor. We only mention the main points in this example and leave the rest to the reader. For the terminal object, it is enough to note that $y(1) = \mathrm{Hom}(-, 1)$ is actually the constant functor $\Delta_{\{*\}}$ which is the terminal object in $\mathbf{Set}^{\mathcal{C}^{op}}$. For the product, note that the canonical map $y(A \times B) \to y(A) \times y(B)$ is the map $\mathrm{Hom}(-, A \times B) \to \mathrm{Hom}(-, A) \times \mathrm{Hom}(-, B)$ induced by a composition with projections. It is not hard to see that this map is actually an isomorphism. For the exponential, the canonical map $y([A, B]) \to [y(A), y(B)]$ is the map $\mathrm{Hom}(-, [A, B]) \to \mathrm{Hom}(y_- \times y_A, y_B)$ induced by combining the evaluation and the Yoneda functor. Using the Yoneda embedding, it is enough to show that the canonical map $\mathrm{Hom}(-, [A, B]) \to \mathrm{Hom}(- \times A, B)$ induced by the evaluation map is an isomorphism. The latter is mentioned in Remark 85.

Example 107 (*Non-preservation of Terminal Objects and Products*) Let C be a CCC and X be a set with at least two elements. Then, the constant functor $\Delta_X : C \to \mathbf{Set}$ is not a CC-functor. It does not preserve the terminal object as otherwise, X must have been a singleton. It does not preserve the product as otherwise, the canonical map $\Delta_X(A \times B) \to \Delta_X(A) \times \Delta_X(B)$ must be an isomorphism. Computing this map, we can see that it is actually the function $X \to X \times X$ mapping x to (x, x). However, as X has more than one element, the map $x \mapsto (x, x)$ cannot be surjective. Notice that if X is infinite, the objects $\Delta_X(A \times B)$ and $\Delta_X(A) \times \Delta_X(B)$, i.e., X and $X \times X$ are isomorphic in **Set**. This isomorphism also makes the functors $\Delta_X(- \times -) : C \times C \to C$ and $\Delta_X(-) \times \Delta_X(-) : C \times C \to C$ isomorphic. However, none of these facts imply that Δ_X is preserving the product. As we emphasized before, to preserve the product, it is important to have the *canonical morphism* as an isomorphism.

Example 108 (*Non-preservation of Exponentials*) Let X and Y be two topological spaces. Then, the inverse image of a continuous map $f : X \to Y$ is not necessarily a CC-functor as it may not preserve the exponentials. For instance, take the continuous map $f : \mathbb{R} \to \mathbb{R}$ defined by $f(x) = x^2$. Let $U = (-\infty, 0)$ and note that $U \to \varnothing = int[0, +\infty) = (0, +\infty)$. Then, $f^{-1}(U \to \varnothing) = f^{-1}((0, +\infty)) = \mathbb{R} - \{0\}$. However, $f^{-1}(U) \to f^{-1}(\varnothing) = \varnothing \to \varnothing = \mathbb{R}$ and hence $f^{-1}(U \to \varnothing) \neq f^{-1}(U) \to f^{-1}(\varnothing)$. Notice that $f : \mathbb{R} \to \mathbb{R}$ is not an open map, as $f[\mathbb{R}] = [0, +\infty)$ is not open.

Example 109 (*Non-preservation of Initial Objects and Coproducts*) Let C be a small cocartesian category. Then, the functor $y : C \to \mathbf{Set}^{C^{op}}$ does not preserve the initial object or *any* of the coproducts in C. For the initial object, note that $y_0(C) = \mathrm{Hom}(C, 0)$ and this set is not empty for $C = 0$. Therefore, y_0 cannot be the initial object of $\mathbf{Set}^{C^{op}}$. Similarly, for the coproducts, notice that $y_{A+B}(C) = \mathrm{Hom}(C, A + B)$ which is not isomorphic to $\mathrm{Hom}(C, A) + \mathrm{Hom}(C, B)$ for $C = 0$. Because the first set has exactly one element while the second has exactly two elements. Therefore, y_{A+B} and $y_A + y_B$ are not isomorphic as objects of $\mathbf{Set}^{C^{op}}$.

Remark 110 Note that in the previous example, we seriously used the existence of the initial object to prove that $y : C \to \mathbf{Set}^{C^{op}}$ does not preserve any existing coproducts. It is actually possible to prove it even without assuming that C is cocartesian.

4.5 Free Structured Categories

In the previous subsections, we identified the discourse of BCC's and BC-functors as a natural categorical framework for formalizing proof systems over $\mathcal{L}_p$ and their interpretations. With this framework established, one might take a conceptual leap to seek a universal property that redefines the syntactical BCC **NJ** in a relative manner, thereby providing a definition for syntax in a syntax-free way.

Theorem 111 *The category* **NJ** *is the free* BCC *generated by the infinite set* $\{p_0, p_1, \ldots\}$ *of objects, i.e., for any* BCC C *and any assignment* f *of the objects of* C *to the atoms* $\{p_0, p_1, \ldots\}$*, there is a unique (up to isomorphism)* BC*-functor* $F : \mathbf{NJ} \to C$ *extending* f*, i.e.,* $F(p_i) = f(p_i)$*, for any* $i \in \mathbb{N}$*. The same also holds for* **NM** *and* **NN***, as the free* ABC *and* CCC *generated by the set of objects* $\{p_0, p_1, \ldots\}$*, respectively.*

Proof We only explain the case **NJ**. The rest are similar. First, let us start with an informal explanation of why we expect **NJ** to be the free BCC generated by the set $\{p_0, p_1, \ldots\}$. The reason is simple. Reading category-theoretically, one can see that the propositions in the language $\mathcal{L}_p$ and the rules of **NJ** are designed to define a BC-structure on **NJ** in a minimal manner. For instance, the connective $\wedge$ defines the product object $A \wedge B$ of the objects A and B. The rules $(E\wedge_1)$ and $(E\wedge_2)$ act as the projection maps for the product $A \wedge B$, and $(I\wedge)$ corresponds to the operation $\langle -, - \rangle$ on the maps, which is required for the products. Similarly, there is a correspondence between the disjunction and its rules with coproducts, the implication and its rules with exponentiation, $\top$ and its axiom with the terminal object, and $\bot$ and its axiom with the initial object. Of course, there are some issues with the commutativity and the uniqueness conditions in the universal definition of the BC-structure. These are resolved with the minimal equalities that $\beta\eta$-equivalences put on the derivations. Note that these equivalences are nothing but what one needs to ensure the commutativity and the uniqueness conditions that the BC-structure of **NJ** demands. Therefore, we naturally expect **NJ** to be the "least" possible BCC generated by the set $\{p_0, p_1, \ldots\}$ of objects.

To formally prove the freeness, let f be an assignment of the objects of C to the atoms in $\{p_0, p_1, \ldots\}$. Our task is defining the BC-functor $F : \mathbf{NJ} \to C$ and proving its uniqueness up to an isomorphism. We will explain the main ingredients of the proof and leave the rest to the reader. First, fix a BC-structure on C, i.e., fix an initial and a terminal object and for any two objects, fix a product, a coproduct and an exponential for them. We will define the assignment F on propositions and derivations in an inductive way. For propositions, set $F(p_i) = f(p_i)$, for any $i \in \mathbb{N}$, $F(\top) = 1$, $F(\bot) = 0$, $F(A \wedge B) = F(A) \times F(B)$, $F(A \vee B) = F(A) + F(B)$ and $F(A \to B) = [F(A), F(B)]$.

For derivations, for any finite set of assumptions in the form $[A]^i$, define $F(\Gamma)$ as $\prod_{\gamma \in \Gamma} F(\gamma)$, using the order on Γ induced by the superscripts in $[A]^i$'s. If $\Gamma = \varnothing$, define $F(\Gamma) = 1$. Then, for any derivation D of A with the set of assumptions Γ, we want to assign a map $F(\mathsf{D}) : F(\Gamma) \to F(A)$. As the derivation D is a tree constructed from the rules of **NJ**, it is enough to define $F(\mathsf{D})$ recursively on the sub-derivations of D. If D is a one-node tree proving $[A]^i$, define $F(\mathsf{D})$ as the projection on the ith component of $F(\Gamma)$. For the other rules, use Table 2.

To read the table, we need the following notational conventions. First, for the map $F(\mathsf{D}) : F(\Gamma \cup [A]^i) \to F(B)$, we can use the commutativity of the product to transform $F(\Gamma \cup [A]^i)$ to $F(\Gamma) \times F(A)$, and then using λ_A, construct a map from

Table 2 Inductive definition of $F(\mathsf{D})$

Rules	Interpretations
$\dfrac{}{\top}\ (\top)$	$!_{F(\Gamma)}$
$\dfrac{\begin{matrix}\mathsf{D}\\ \bot\end{matrix}}{A}\ (\bot)$	$!_A \circ F(\mathsf{D})$
$\dfrac{\begin{matrix}\mathsf{D}_1\\ A\end{matrix}\quad \begin{matrix}\mathsf{D}_2\\ B\end{matrix}}{A \wedge B}\ (I\wedge)$	$\langle F(\mathsf{D}_1), F(\mathsf{D}_2)\rangle$
$\dfrac{\begin{matrix}\mathsf{D}\\ A\wedge B\end{matrix}}{A}\ (E\wedge_1)\ ,\quad \dfrac{\begin{matrix}\mathsf{D}\\ A\wedge B\end{matrix}}{B}\ (E\wedge_2)$	$p_0 \circ F(\mathsf{D})\ ,\quad p_1 \circ F(\mathsf{D})$
$\dfrac{\begin{matrix}\mathsf{D}\\ A\end{matrix}}{A \vee B}\ (I\vee_1)\ ,\quad \dfrac{\begin{matrix}\mathsf{D}\\ B\end{matrix}}{A \vee B}\ (I\vee_2)$	$i_0 \circ F(\mathsf{D})\ ,\quad i_1 \circ F(\mathsf{D})$
$\dfrac{\begin{matrix}\\ \mathsf{D}\\ A\vee B\end{matrix}\quad \begin{matrix}[A]^i\\ \mathsf{D}_1\\ C\end{matrix}\quad \begin{matrix}[B]^j\\ \mathsf{D}_2\\ C\end{matrix}}{C}\ (E\vee_{i,j})$	$ev\langle I\langle \lambda_i F(\mathsf{D}_1), \lambda_j F(\mathsf{D}_2)\rangle F(\mathsf{D})\rangle$
$\dfrac{\begin{matrix}[A]^i\\ \mathsf{D}\\ B\end{matrix}}{A \to B}\ (I \to_i)$	$\lambda_i F(\mathsf{D})$
$\dfrac{\begin{matrix}\mathsf{D}_1\\ A\end{matrix}\quad \begin{matrix}\mathsf{D}_2\\ A\to B\end{matrix}}{B}\ (E \to)$	$ev \circ \langle F(\mathsf{D}_2), F(\mathsf{D}_1)\rangle$

$F(\Gamma)$ to $[F(A), F(B)]$. We call this map $\lambda_i F(\mathsf{D})$. Second, notice that in any BCC, there is a canonical map $I : [X, Z] \times [Y, Z] \to [X + Y, Z]$. We use this map to interpret the rule $(E\vee)$.

The definition of $F(\mathsf{D})$ does not respect the renaming of superscripts as any such renaming may change the order of $F(\gamma)$'s in the product $F(\Gamma) = \Pi_{\gamma\in\Gamma} F(\gamma)$. However, as we will finally restrict ourselves to the derivations with one assumption, the map F becomes well-defined at the end. The definition of $F(\mathsf{D})$ respects the $\beta\eta$-equivalences on D. This holds due to the universal properties of the BC-structure on C. We leave its detailed proof to the reader. Then, to get a functor, one must restrict F to the maps in **NJ**, i.e., to the derivations with one assumption. It is easy to see that this restriction is a functor as F is defined on a one-node tree with one assumption as the identity map and we defined F by composing the values of F on sub-derivations. The functor F respects the BC-structure by design. For instance, we defined $F(A \wedge B) = F(A) \times F(B)$, and the projections in **NJ**, i.e., the rules $(E\wedge_1)$ and $(E\wedge_2)$ are mapped to the projections of C. As F clearly extends f, we constructed the BC-functor we sought.

To prove the uniqueness, note that every proposition is constructed from the atoms and $\{\top, \bot\}$ using $\{\wedge, \vee, \rightarrow\}$ in a unique way, and any derivation is constructed from the rules. If F is to be a BC-functor, it must preserve the BC-structure, and hence the definition of F is essentially forced upon us. For instance, we must define $F(A \wedge B)$ as the product $F(A) \times F(B)$, and the application of F on the derivation:

$$\frac{A \wedge B}{A}\ (E\wedge_1)$$

must be the projection $p_0 : F(A) \times F(B) \rightarrow F(A)$. Therefore, the only choice we have is the choice of the BC-structure on C which is unique up to isomorphism. Hence, F must be unique up to an isomorphism. □

Theorem 111 characterizes the BCC **NJ** as the "least" BCC containing the given set $\{p_0, p_1, \ldots\}$ of objects. This characterization, like those for products, coproducts, and other universal constructions, is a relative one based on universality and does not rely on the internal structure of the category **NJ**. The method of constructing **NJ** is not crucial; there may be alternative syntactical methods to construct the same category, which can be seen as different *presentations* of the same mathematical entity. In this sense, the definition of **NJ** provided by Theorem 111 serves as a conceptual characterization and offers insight into how to define syntax in a *presentation-free* manner.

5 BHK Interpretation and Variable Sets

In the previous section, we introduced some universal constructions to interpret the logical constants in the propositional language. For this purpose, we employed categorical language, where propositions are treated as objects and *deductions* as morphisms. This framework effectively formalized the expected behavior of logical constants. However, some anomalies emerged, such as defining the constants $\bot$ and $\vee$ by their behavior as assumptions, based on how they prove other propositions rather than how they are proved themselves. One could argue that these anomalies result from our choice of deduction as the primitive notion, while the primitive notion should actually be the *proof*, i.e., a deduction with no assumption. Once proofs are specified, a deduction of B from A can always be defined as a *construction* that maps proofs of A to proofs of B. But if proofs are taken as primitives, how do we define the logical connectives? The well-known Brouwer-Heyting-Kolmogorov

(BHK) interpretation attempts to answer this question. It defines the proofs of a proposition recursively as follows:

- there is a canonical proof for $\top$,
- there is no proof for $\bot$,
- a proof of $A \wedge B$ is a pair of a proof of A and a proof of B,
- a proof of $A \vee B$ is either a proof of A or a proof of B,
- a proof of $A \to B$ is a *construction* that transforms any proof of A to a proof of B.

The BHK interpretation is clearly based on a notion of *construction*, as the implication clause clearly demonstrates. Even for the other clauses, we must specify a world of constructions from which the clauses select the ones considered as proofs. To formalize this notion of construction, a natural approach is to employ a category with sufficient structure as the world of constructions. Here, we interpret the objects as types and the morphism $f : A \to B$ as a construction of type B based on a given input of type A.

There are many such categories, ranging from categories of recursive functions or continuous maps to the usual category of sets. In this section, we restrict ourselves to the classical category of sets to simplify the exposition. However, in Sects. 9 and 10, we will explore some alternative constructions to illustrate their usefulness and interesting properties. These alternative constructions are typically referred to as the *realizability* interpretation.

Fixing constructions, we also need to incorporate the notion of *time*, an important factor that is often ignored in the usual presentation of the BHK interpretation. To understand the role of time, consider the way a constructive mathematician rejects the axiom of the excluded middle. For her, $A \vee \neg A$ is not acceptable as a *general law* because it asserts that for any proposition A, there is either a proof of A or a proof that A leads to a contradiction. This is not true in her view, as there are propositions that are potentially provable but have not been proved *so far*, making them neither provable nor disprovable at this point in time.

However, in the static and timeless picture that the BHK interpretation presents, without any epistemological change, either A is provable and we have its proof, or we do not have any proof because it is impossible to have one. In the latter case, since there is no proof for A, one can claim that any proof of A leads to a proof of the contradiction, which provides a proof for $A \to \bot = \neg A$ by the BHK definition of the proofs of implication. Therefore, the BHK interpretation, with classical sets and functions as its world of constructions, validates the axiom of the excluded middle if we ignore the temporal aspect. Recalling that time is the seed of *Kripke semantics* for constructive theories, it is insightful to consider what Goodman interestingly highlighted:

> *Recursive realizability emphasizes the active aspect of constructive mathematics. However, Kleene's notion has the weakness that it disregards that aspect of constructive mathematics which concern epistemological change. Precisely that aspect of constructive mathematics which Kleene's notion neglects is emphasized by Kripke's semantics for intuitionistic logic. However, Kripke's notion makes it appear that the constructive mathematician is a passive observer of a structure which gradually reveals itself. What is lacking is the emphasis on the mathematician as active which Kleene's notion provides* Goodman (1978).

Using the temporal structure, the informal BHK interpretation can be updated as follows:

- there is *always* a canonical proof for $\top$,
- there is *never* a proof for $\bot$,
- a proof of $A \wedge B$ *at some point* is a pair of a proof of A and a proof of B *at the same point*,
- a proof of $A \vee B$ *at some point* is either a proof of A or a proof of B *at the same point*,
- a proof of $A \to B$ *at some point* is a construction that transforms any proof of A *at any point in the future* to a proof of B *at that point*.

There is another necessary update to the BHK interpretation concerning the clauses for $\bot$ and $\vee$. We will explain the latter, as the former is similar. In the traditional form of the BHK interpretation, a proof of $A \vee B$ at some point is either a proof of A or a proof of B at that same point. However, this definition can be too stringent for some practices. For instance, if we are generating a sequence of zeros and ones one bit at a time, at stage n, we know that the next bit a_{n+1} is either zero or one. Yet, we do not have a proof at that moment to determine which it will be.

This observation suggests weakening the clause for disjunction in the BHK interpretation to state that a proof of $A \vee B$ at time n is a construction that will eventually become a proof of A or B as time progresses, regardless of how it progresses. To formalize this concept, we need a notion called *coverage* or *Grothendieck topology*, which defines how future possibilities can cover the present moment. This chapter will not delve into this concept, as it extends far beyond the scope of our already extensive survey. However, you can explore a basic version of it in the renowned *Beth semantics* for intuitionistic logic Troelstra and Van Dalen (1988a, b).

Using the two ingredients of construction and time, and setting aside the role of coverages, we will provide a formalization for a weaker form of the BHK interpretation in the following two subsections.

5.1 Provability Via the BHK Interpretation

Let us start with the simpler task of formalizing the BHK interpretation considered as the inductive definition of the provability relation rather than the actual proofs. As we saw before, this means restricting our structures from categories to posets, where all the morphisms are collapsed into an order relation.

First, let $\mathcal{T} = (T, \leq)$ be a poset encoding the time structure. Note that time is not linear, as each point in time may branch into different possible futures. Ignoring explicit proofs and focusing solely on provability, a proposition is represented by an order-preserving map from $\mathcal{T}$ to $\mathbf{2}$, indicating whether the proposition is provable at each point in time. It is order-preserving because if a proposition is provable at some point $w \in T$, it remains provable as time progresses. Therefore, the set of all propositions is $\mathbf{2}^{\mathcal{T}}$. For the deductibility order between propositions, we define $A \leq B$ if and only if $A(w) \leq B(w)$ for every $w \in T$. This means that A proves B if the provability of A at any point in time implies the provability of B at that point.

For any proposition A and any point $w \in T$, we write $w \Vdash A$ to denote $A(w) = 1$, meaning that A is provable at w. Now, following the BHK interpretation and substituting proofs with provability, we define the provability of complex propositions as follows:

- $w \nVdash \bot$, meaning that $\bot$ is never provable.
- $w \Vdash \top$, meaning that $\top$ is always provable.
- $w \Vdash A \wedge B$ iff $w \Vdash A$ and $w \Vdash B$, meaning that $A \wedge B$ is provable at some point iff both A and B are provable at that point. This is the shadow of the definition of the proofs of a conjunction as the pairs of the proofs of the conjuncts.
- $w \Vdash A \vee B$ iff either $w \Vdash A$ or $w \Vdash B$, meaning that $A \vee B$ is provable at some point iff either A or B is provable at that point. This is the shadow of the definition of the proofs of a disjunction as the proofs of either of the disjuncts.
- $w \Vdash A \to B$ iff for any $u \geq w$ if $u \Vdash A$ then $u \Vdash B$, meaning that $A \to B$ is provable at some point iff at any point in the future, when A becomes provable, B also becomes provable at that point. This is the shadow of the definition of the proofs of an implication as a construction transforming a proof of A to a proof of B. When we have such a construction at w, we can apply it at all points in the future to any future proof of A to get a proof of B.

It is straightforward to verify that the operations $\{\bot, \top, \wedge, \vee, \to\}$ on the poset $\mathbf{2}^{\mathcal{T}}$ align with its BC-structure. For instance, for disjunction, we have $A \vee B \leq C$ if and only if $A \leq C$ and $B \leq C$, for any propositions A, B, and C. Given this BC-structure, we can conclude that the BHK interpretation yields a poset proof system, specifically a Heyting algebra, as discussed in the previous section.

Two points are worth noting here. First, the BHK interpretation defines provability rather than deductibility. Deductibility is simply the natural order on $\mathbf{2}^{\mathcal{T}}$ induced by the temporal structure of $\mathcal{T}$. This approach allows for a clearer definition of $\bot$ and $\vee$, as it directly addresses their provability rather than their role as assumptions for proving other propositions. Second, the BHK interpretation concretely formalizes poset proof systems, with propositions represented as order-preserving functions rather than abstract elements in a Heyting algebra.

Given these observations, one might wonder if the poset proof system we introduced earlier is more general than what the BHK interpretation provides. Fortunately, this is not the case:

Theorem 112 (Kripke representation for Heyting algebras Troelstra and Van Dalen (1988a)) *Let $\mathcal{H}$ be a Heyting algebra. Then, the canonical map $e : \mathcal{H} \to \mathbf{2}^{[\mathcal{H},\mathbf{2}]}$ defined by $e(a)(f) = f(a)$ is a full and faithful* BC*-functor, where $[\mathcal{H}, \mathbf{2}]$ is the poset of all* BC*-functors from $\mathcal{H}$ to* **2** *with the pointwise order.*

Therefore, any poset proof system can be viewed as a subsystem of the proof system derived from the BHK interpretation. In essence, the BHK interpretation offers a *concrete representation* for all poset proof systems.

5.2 Proofs via the BHK Interpretation

Having investigated the easier case of provability, we are now ready to focus on the original BHK interpretation. The first point is generalizing the structure of time from a poset to a small category $\mathcal{T}$. The idea is that not only the results of the growth of time are important, but also we care about *how* this expansion happens. Therefore, in the categorical formalization of time, the objects of $\mathcal{T}$ are the points on the timeline, while its morphisms, such as $f : w \to u$, explain how w grows to u.

Now, having the category $\mathcal{T}$ to formalize the temporal structure, a proposition is simply an assignment of sets to the objects of $\mathcal{T}$, specifying the proofs of the proposition at each point w, together with an assignment of functions to the maps of $\mathcal{T}$, specifying how proofs change through time. The natural way to formalize such data is by variable sets over $\mathcal{T}$, i.e., the functors from $\mathcal{T}$ to **Set**. Therefore, we assume that $\mathbf{Set}^{\mathcal{T}}$ is the category of propositions. Note that a deduction of the proposition B from the proposition A in this category is defined as a natural transformation that is simply a way to map the proofs of A to the proofs of B at each point w of $\mathcal{T}$ in a uniform and coherent way. This is compatible with our previous discussion of defining deductions as constructions that transform proofs and not as a primitive notion.

For any proposition A, write $w, x \Vdash A$ to denote $x \in A(w)$, meaning that x is a proof of A at the point w. Then, the BHK interpretation simply states

- there is no x such that $w, x \Vdash \bot$,
- $w, * \Vdash \top$, where $*$ is the unique x for which $w, x \Vdash \top$,
- $w, \langle x, y\rangle \Vdash A \wedge B$ iff $w, x \Vdash A$ and $w, y \Vdash B$,
- $w, \langle i, x\rangle \Vdash A \vee B$ iff either $i = 0$ and $w, x \Vdash A$ or $i = 1$ and $w, x \Vdash B$,
- $w, \{\alpha_u\}_{u \in \mathcal{T}} \Vdash A \to B$ iff for any expansion $f : w \to u$ and any x such that $u, x \Vdash A$, we have $u, \alpha_u(f, x) \Vdash B$.

In the last part, the sequence $\{\alpha_u\}_{u \in \mathcal{T}}$ must be coherent. To explain, note that for any map $g : u \to v$, there are two natural candidates to map the pair of $f : w \to u$ and x as a proof of A at u to a proof of B at v. One is using α_u to provide a proof of B at u and then apply $B(g)$ to reach a proof of B at v. The other is first changing (f, x) to $(gf, A(g)(x))$ and then apply α_v. The coherency simply says that these two ways are equal:

$$\begin{array}{ccc} \mathrm{Hom}(w,u)\times A(u) & \xrightarrow{\alpha_u} & B(u) \\ {\scriptstyle (g\circ -)\times A(g)}\downarrow & & \downarrow{\scriptstyle B(g)} \\ \mathrm{Hom}(w,v)\times A(v) & \xrightarrow[\alpha_v]{} & B(v) \end{array}$$

In other words, we are simply saying that the assignment $\{\alpha_u : \mathrm{Hom}(w,u)\times A(u) \to B(u)\}_{u\in\mathcal{T}}$ is a natural transformation.

The astute reader will soon realize that these clauses cannot fully define complex propositions such as $A \wedge B$, $A \vee B$, and $A \to B$, since any proposition is represented as a functor from $\mathcal{T}$ into **Set**, and what we've defined so far only covers the object part of the functor. However, there is a canonical method to extend this assignment to maps of $\mathcal{T}$, thereby obtaining a functor for each complex proposition. By doing so, it becomes clear that the operations $\{\bot, \top, \wedge, \vee, \to\}$ correspond precisely to the BC-structure of the category $\mathbf{Set}^{\mathcal{T}}$. Therefore, it follows that the BHK interpretation yields a proof system in the sense described in the previous sections, specifically a BCC.

Again, there are a few key points to consider. First, the BHK interpretation focuses on defining proofs rather than deductions. This approach resolves the peculiarities associated with $\bot$ and $\vee$, as they are now characterized by their proofs rather than by their deductive capacities. Second, it's important to emphasize that in this framework, propositions and deductions are represented concretely as variable sets and natural transformations, respectively, rather than as abstract objects and morphisms in a general BCC.

One might wonder whether there is any advantage to using this abstract setting, or if an abstract BCC can be represented by a category of variable sets. To address this, note that if we shift from BCCs to CCCs, such a representation becomes feasible. Specifically, the Yoneda embedding provides a CC-functor that maps the category $\mathcal{C}$ into $\mathbf{Set}^{\mathcal{C}^{op}}$ in a faithful and full manner. Leveraging a significant representation theorem from topos theory Joyal and Moerdijk (1990), we can even opt for a temporal structure that is a poset, providing a robust connection between abstract and concrete settings.

Theorem 113 (Kripke representation for small CCC'ss) *Let $\mathcal{C}$ be a small* CCC. *Then, there exist a poset $(P, \leq)$ and a full and faithful* CC*-functor $F : \mathcal{C} \to \mathbf{Set}^{(P,\leq)}$.*

Now, let us return to the general setting of BCCs. Contrary to what one might expect, it is not always possible to embed a BCC into a BCC of the form $\mathbf{Set}^{\mathcal{T}}$ via a full and faithful BC-functor. In fact, some BCCs cannot be mapped by any BC-functor into categories of variable sets.

The reason for this limitation is straightforward. In the category $\mathbf{Set}^{\mathcal{T}}$, the object $1+1$ is not isomorphic to 1 because $1+1$ has one element at each $w \in \mathcal{T}$, while 1 always has a single element. However, in certain BCCs, such as Heyting algebras, we have $1 \cong 1+1$, and this property is preserved by any BC-functor, whether or not it is full or faithful. Philosophically, this distinction highlights that the BHK interpretation imposes a specific structure that ensures the proofs of A and B within $A \vee B$ are disjoint. In contrast, arbitrary BCCs allow for a more flexible integration of proofs.

Remark 114 If the reader is familiar with Grothendieck toposes, which incorporate the coverage structures we omitted, they might suspect that using sheaves instead of presheaves could address this issue in the BHK interpretation. Unfortunately, this is not the case. The issue persists because, even in non-trivial Grothendieck toposes, we still have $1 + 1 \ncong 1$.

As it may be clear now, the issue with the representation claim lies in the cocartesian structure of a BCC. To provide stronger evidence, we demonstrate that some cocartesian categories cannot be mapped into a category of variable sets via a weakly-full functor that preserves the cocartesian structure. To prove this, we define the property $(*)$ in a cocartesian category as follows: For any diagram of the form:

$$\begin{array}{ccc} C & \longrightarrow & A \\ \big\downarrow & & \big\downarrow{\scriptstyle i_0} \\ B & \underset{i_1}{\longrightarrow} & A+B \end{array}$$

where i_0 and i_1 are the injections, there is a map in $\mathrm{Hom}(C, 0)$. Note that the property $(*)$ is somehow stating that the coproduct is disjoint.

Lemma 115 *Let $\mathcal{C}$ and $\mathcal{D}$ be two cocartesian categories such that there is a weakly-full functor $F : \mathcal{C} \to \mathcal{D}$ preserving the initial object and the binary coproducts. Then, if $\mathcal{D}$ has the property $(*)$, then so does $\mathcal{C}$.*

Proof Let the following diagram commute in $\mathcal{C}$:

$$\begin{array}{ccc} C & \longrightarrow & A \\ \big\downarrow & & \big\downarrow{\scriptstyle i_0} \\ B & \underset{i_1}{\longrightarrow} & A+B \end{array}$$

Then, applying the functor F and using the fact that F preserves the coproducts, we can use $(*)$ to have a map in $\mathrm{Hom}(F(C), 0)$. Since $F(0)$ is an initial object, there is a map in $\mathrm{Hom}(F(C), F(0))$. As F is weakly-full, there is a map in $\mathrm{Hom}(C, 0)$. $\square$

Now, observe that for any category of variable sets (and all Grothendieck toposes), the property $(*)$ holds, as $A + B$ is the pointwise *disjoint* union of sets. However, $(*)$ does not hold for non-trivial cocartesian posets. For example, consider a poset with $a \neq 0$. In such a poset, we have

$$\begin{array}{ccc} a & \leq & a \\ \rotatebox{-90}{$\leq$} & & \rotatebox{-90}{$\leq$} \\ a & \leq & a \vee a = a \end{array}$$

However, since $a \nleq 0$, if such posets could be mapped via a weakly-full functor that preserves the cocartesian structure into a category of the form $\mathbf{Set}^{\mathcal{T}}$, then $\mathbf{Set}^{\mathcal{T}}$, having the property $(*)$, would force the poset to also possess this property. This leads to a contradiction. It is crucial to note that the weakly-fullness condition cannot be omitted. Without this condition, any cocartesian category C could be trivially mapped into the initial object 0 of a category of variable sets using the constant functor Δ_0. While this functor is cocartesian, it is not necessarily weakly-full.

6 Classical Deductions

We have examined how BCC's offer a categorical formalization of intuitionistic proof systems and explored their interconnections, including the syntactical systems as free categories and the BHK interpretation as a source of a special family of proof systems. Given this comprehensive framework for intuitionistic proofs, it is natural to seek a comparable formalization for classical proof systems. Unfortunately, the situation is less satisfactory. To proceed, we begin with a useful lemma:

Lemma 116 *In a* BCC, *all maps in* $\mathrm{Hom}(A, 0)$ *are isomorphisms, for any object A. As a consequence, there is at most one map in* $\mathrm{Hom}(B, \neg C)$, *for any objects B and C.*

Proof First, observe that $0 \times A$ is an initial object. To prove, note that $\mathrm{Hom}(0 \times A, B) \cong \mathrm{Hom}(0, B^A)$ and as 0 is initial, there is exactly one map in $\mathrm{Hom}(0, B^A)$ which also means there is exactly one map from $0 \times A$ to B, for any arbitrary object B. Hence, $0 \times A$ is initial. Now, let $f : A \to 0$ be a map. We claim that the map $! : 0 \to A$ is the inverse of $f : A \to 0$. As 0 is initial, it is clear that $f! = id_0$. Therefore, we have to show that $!f = id_A$. For that purpose, consider the following diagram:

$$\begin{array}{ccccc} A & \xrightarrow{\langle f, id_A \rangle} & 0 \times A & \xrightarrow{p_1} & A \\ {\scriptstyle id_A}\downarrow & & {\scriptstyle p_0}\downarrow & & \downarrow{\scriptstyle id_A} \\ A & \xrightarrow[f]{} & 0 & \xrightarrow[!]{} & A \end{array}$$

The left square trivially commutes. The right square commutes as $0 \times A$ is an initial object. Therefore, the whole rectangular commutes. As the upper row is id_A, the lower row must also be id_A. Therefore, $!f = id_A$ and hence $f : A \to 0$ is an isomorphism. For the second part, first notice that there is at most one map in $\mathrm{Hom}(A, 0)$, for any object A. The reason is that if $\mathrm{Hom}(A, 0)$ is non-empty, by the first part, $A \cong 0$ which implies the initiality of A. Hence, $\mathrm{Hom}(A, 0)$ must be a singleton. Using this fact, as $\mathrm{Hom}(B, \neg C) \cong \mathrm{Hom}(B \times C, 0)$, it is trivial that $\mathrm{Hom}(B, \neg C)$ has at most one element. □

Philosophical Comment 117 Reading proof theoretically, Lemma 116 simply states that the negation of a formula has at most one proof because any such proof is just a deduction of $\bot$ from the formula itself. This result is quite counter-intuitive for some logicians, as there are apparently some negative formulas with different proofs in mathematics. For these logicians, what Lemma 116 actually implies is that identifying $\bot$ with the initial object is too demanding and must be relaxed. There are many ways to solve this issue. For instance, one can identify $\bot$ as a *weak initial object*, i.e., an object that has a (not necessarily unique) map to any other object. As another way, one can simply move to the setting of minimal logic, omitting $\bot$ completely from the language. For other logicians, the uniqueness of the proofs of the negative propositions is not only intuitive and justifiable but also has some explanatory power. For these logicians, all proofs of a negative formula are equivalent because the proofs of the negative statements carry no information other than their mere truth or provability. This is a known intuition in constructive mathematics that negative formulas behave more like classical propositions where only truth plays a role and not the actual proofs. One can even say that Lemma 116 actually provides a formal proof for this intuition.

Now, it is time to return to our task of providing a categorical formalization for classical proof systems. There are different interpretations of how classicality must be captured in the categorical setting. In the most obvious approach, one can argue that classical proofs, whatever they are, generalize intuitionistic proofs. Therefore, the category of classical deductions must at least be a BCC. To incorporate the essence of classical proofs, we have the following four proposals:

- **Logical Equivalence:** There is a map from $\neg\neg A$ to A, for any object A.
- **Isomorphism:** $A \cong \neg\neg A$, for any object A.
- **Natural Isomorphism:** $A \cong \neg\neg A$, natural in A, i.e., there is a natural transformation $\alpha : id_{\mathcal{C}} \to \neg\neg(-)$ such that $\alpha_A : A \to \neg\neg A$ is an isomorphisms, for any object A.
- **Canonical Isomorphism:** The canonical map $\lambda_{\neg A} ev\langle p_1, p_0\rangle$ from A to $\neg\neg A$ is an isomorphism.

Using Lemma 116, there is at most one map in $\mathrm{Hom}(A, \neg\neg A)$. Thus, the last three options are all equivalent. However, the weakest form is different from the others. It only ensures the existence of a deduction of A from $\neg\neg A$ for any proposition A. This cannot fully capture the essence of classical proofs, as classical mathematicians typically believe that the propositions A and $\neg\neg A$ are *identical*, not merely logically equivalent. Therefore, we are led to define a category of classical proofs as a BCC where $A \cong \neg\neg A$, for any object A.

Theorem 118 (Joyal) *A* BCC *in which* $A \cong \neg\neg A$, *for any object* A *is a preordered set.*

Proof By Lemma 116, there is at most one map in $\mathrm{Hom}(A, \neg\neg B)$. As $B \cong \neg\neg B$, there is at most one map in $\mathrm{Hom}(A, B)$, for any two objects A and B. Hence, the category is a preordered set. □

Similar to how we interpreted negations, Theorem 118 can be read in two distinct ways. The first interpretation suggests that there are no non-trivial classical deductions, reflecting the idea that classical mathematics is grounded in truth or, equivalently, in deductibility where deductions have no substantive role beyond their mere existence. In this view, classical proofs are seen as fundamentally trivial or non-informative.

Conversely, the second interpretation offers a different perspective: while classical proofs may not be as informative as constructive proofs, they still contain some amount of information and are not trivially unique. Thus, Theorem 118 demonstrates that the combination of the BC-structure and the isomorphism between A and $\neg\neg A$ might be too restrictive. Consequently, alternative formalizations are needed to capture classical proof systems. Finding a satisfactory formalization has been a longstanding open problem with many proposed solutions.

There are primarily two approaches to developing such alternatives. The first approach maintains the CC-structure but relaxes the cocartesian structure. The second approach acknowledges the importance of preserving the symmetry between conjunctive and disjunctive operations by weakening both cartesian and cocartesian structures while ensuring they remain dual to each other. Both of these approaches have many instances. To convey their taste, we will present one representative of each of these approaches: *control categories* and *classical categories*.

6.1 Control Categories

The first approach involves using the categorical reading of constructive proofs combined with a way to reduce classical deductions to the constructive ones. There are different ways to implement such a combination (Hofmann and Streicher 2002; Hofmann 1995a; Parigot 1992; Ong 1996; Pym and Ritter 2001; Selinger 2001). Here, we focus on a route leading to one specific example of *control categories*. In this case, the reduction of classical deductions to the constructive ones is via *negative translation*. Therefore, we must first generalize this translation from logics, i.e., the preorders, to categories. Then, the translation helps to identify classical deductions within constructive deductions and aims to axiomatize the structure of these deductions universally.

To present a generalization of the negative translation, recall that a *boolean algebra* is a Heyting algebra $\mathcal{H}$ where $a = \neg\neg a$ for any $a \in \mathcal{H}$. Then:

Theorem 119 *Let $\mathcal{H}$ be an almost Heyting algebra and $r \in \mathcal{H}$ be an arbitrary element. Then, the subset $\mathcal{H}_r = \{x \to r \mid x \in \mathcal{H}\}$ of $\mathcal{H}$ together with the order of $\mathcal{H}$ is a boolean algebra.*

Proof First, note that the following data actually describes the BC-structure of $\mathcal{H}_r$:

- $0_r = (1 \to r) = r$ and $1_r = (0 \to r) = 1$,
- $(x \to r) \wedge_r (y \to r) = (x \vee y \to r)$,
- $(x \to r) \to_r (y \to r) = ((x \to r) \wedge y) \to r$,
- $(x \to r) \vee_r (y \to r) = (x \wedge y \to r)$.

To prove that it is boolean, note that $\neg_r(x \to r) = (x \to r) \to r$ and hence $\neg_r\neg_r(x \to r) = [((x \to r) \to r)] \to r$. It is easy to prove that the latter is actually $x \to r$. Therefore, $\neg_r\neg_r(x \to r) = (x \to r)$. □

One can lift the above technique from the posetal ABC's, i.e., the almost Heyting algebras to all ABC's (Hofmann and Streicher 2002; Selinger 2001). Let C be an ABC and R be an arbitrary object such that the canonical map $\lambda_{R^A} ev\langle p_1, p_0\rangle : A \to R^{R^A}$ is monic, i.e., for any two maps $f, g : X \to A$, if $\lambda_{R^A} ev\langle p_1, p_0\rangle f = \lambda_{R^A} ev\langle p_1, p_0\rangle g$, then $f = g$. Consider the category C_R consisting of all the objects of the form R^A in C as its objects and *all* maps between these objects as its morphisms. The first observation is that the category C_R is actually a CCC. Its terminal object is simply 1_C^R, which is also a terminal object in C. For the product of R^A and R^B, first consider the maps:

$$R^{A+B} \times A \xrightarrow[id_{R^{A+B}} \times i_0]{} R^{A+B} \times (A + B) \xrightarrow[ev]{} R$$

$$R^{A+B} \times B \xrightarrow[id_{R^{A+B}} \times i_1]{} R^{A+B} \times (A + B) \xrightarrow[ev]{} R$$

Then, it is enough to set R^{A+B} as the product together with $\lambda_A ev(id_{R^{A+B}} \times i_0) : R^{A+B} \to R^A$ and $\lambda_B ev(id_{R^{A+B}} \times i_1) : R^{A+B} \to R^B$ as the projections. For the exponentiation of R^A and R^B, first consider the map:

$$R^{R^A \times B} \times R^A \times B \xrightarrow[ev]{} R$$

Then, it is enough to set $R^{R^A \times B}$ as the exponential object together with the evaluation map $\lambda_B ev : R^{R^A \times B} \times R^A \to R^B$. This completes the presentation of the CC-structure of the category C_R.

For the cocartesian structure, unlike the posetal case, the object R is not necessarily the initial object of C_R. For instance, set $C = \mathbf{Set}$ and $R = \{0, 1\}$. Then, C_R is the category of all powersets together with all the functions between them. It is easy to see that $\{0, 1\}$ is not an initial object in C_R. In fact, this category has no initial object. For the coproducts, the situation is worse. Our candidate for the coproduct, i.e., the operation mapping R^A and R^B to $R^{A \times B}$ is defined on the objects. However, there is no natural way to extend this assignment to the morphisms and gain a functor on

$C_R \times C_R$. In fact, it is possible to prove that this operation on the objects gives rise to a functor iff the category C_R is a preordered set Selinger (2001). Having said that, it is still possible to sacrifice the cocartesian structure and interpret $\bot$ and the disjunction as R and the operation $(R^A, R^B) \mapsto R^{A\times B}$, respectively. Of course, the lack of the functorial behavior of the latter blocks our disjunction to operate on the deductions properly. Finally, for the classicality, one can see that C_R is classical in the sense of the logical equivalence we had above, i.e., there is a map $f : \neg\neg A \to A$, for any object A. In fact, a stronger property holds, i.e., the map $f : \neg\neg A \to A$ is the left inverse of the canonical map $\lambda_{\neg A} ev\langle p_1, p_0\rangle$. One can axiomatize C_R's with their explained structures as the classical proof systems Selinger (2001). Those categories are called *control categories*.

6.2 A General Categorical Interpretation

We observed that employing the BC-structure, along with the assumption $A \cong \neg\neg A$, leads to a complete collapse of classical proofs into mere provability. Since $A \cong \neg\neg A$ lies at the core of classical reasoning, one way to prevent this collapse is to weaken the BC-structure. This requires revisiting our prior discussions to identify a weaker alternative for interpreting the connectives $\{\top, \bot, \wedge, \vee, \to\}$.

As mentioned earlier, in the second approach to handling the collapse, we aim to preserve the full duality between the conjunctive structure $\{\top, \wedge\}$ and the disjunctive structure $\{\bot, \vee\}$. To simplify our presentation, we therefore restrict ourselves to the complete and economical fragment $\{\top, \wedge, \to, \bot\}$, defining the remaining connectives by duality.[3]

To proceed, we begin with a category C. We interpret $\{\top, \bot\}$ as objects of C, $\wedge$ as a functor from $C \times C$ to C, and $\to$ as a functor from $C^{op} \times C$ to C, subject to properties that reflect the expected behavior of these connectives. Naturally, we cannot approach this blindly; an intuitive understanding of these connectives is essential to guide the formulation of a coherent family of axioms. In the next subsubsection, we will outline this intuitive framework and the fundamental properties it suggests. Such an interpretation and its resulting formalism are not unique (Straßburger 2011; Hyland 2002, 2004; Bellin et al. 2006; Došen n.d.; Došen and Petrić 2004). Here, we follow a path that leads to the specific example of *classical categories*, introduced by Fuhrmann and Pym (2004), Führmann and Pym (2006, 2007). Classical categories elegantly capture the inherent categorical structure of classical deductions and offer an insightful solution to the longstanding open problem of providing a natural categorical semantics for classical logic that does not reduce merely to provability. Moreover, the interpretation is *sound and complete* with respect to a natural form

[3] Alternative fragments, such as $\{\top, \wedge, \neg\}$ or $\{\top, \bot, \wedge, \vee, \neg\}$, are also viable and suited to specific purposes. For instance, the latter simplifies computation and facilitates the study of conjunction and disjunction, even in non-classical settings (Cockett and Seely 1997; Führmann and Pym 2006, 2007).

of classical proofs, known as **LK**-proofs, making classical categories a solid and well-founded framework for categorical interpretations of classical deductions.

6.2.1 ∗-Autonomous Categories

Let us begin with an intuitive understanding of the conjunctive structure. Previously, we characterized the conjunctive structure as the product structure by interpreting the deductions of $A \wedge B$ from C in terms of pairs of deductions of A and B from C. To develop a more general interpretation of the conjunctive structure, we must shift our focus from the *deductions of* a conjunction to the *deductions from* it. In other words, we must think of a conjunction as an *assumption* rather than a *conclusion*.

Following this intuition, $\top$ should be understood as the "empty set of assumptions", since adding $\top$ to the assumptions has no effect on deductions. Similarly, $\wedge$ should be seen as a "combining the assumptions" operation, as assuming $A \wedge B$ has the same effect as assuming both A and B. Guided by this intuition, several natural properties of $\top$ and $\wedge$ emerge.

For example, since "combining the assumptions" is inherently symmetric and associative, we expect $A \wedge B \cong B \wedge A$ and $(A \wedge B) \wedge C \cong A \wedge (B \wedge C)$. Additionally, since adding nothing to A should leave it unchanged, we anticipate $\top \wedge A \cong A \wedge \top \cong A$. This generalization of the monoid structure to categories, where we must also account for morphisms, is called a *symmetric monoidal structure*. Hence, the conjunctive structure must at least be symmetric monoidal.

However, to define a symmetric monoidal structure formally, the above isomorphisms alone are insufficient. The reason is that, although these isomorphisms are *not* equalities, we expect them to behave *as if* they were. This means that any two ways of composing these isomorphisms to move from one object to another must yield the same result.[4] For instance, let $\sigma_{A,B}$ denote the isomorphism from $A \wedge B$ to $B \wedge A$. There are two distinct ways to derive $A \wedge B$ from itself: one is by using the identity map, and the other is by composing $\sigma_{B,A}$ with $\sigma_{A,B}$. This situation can be visualized in the following diagram:

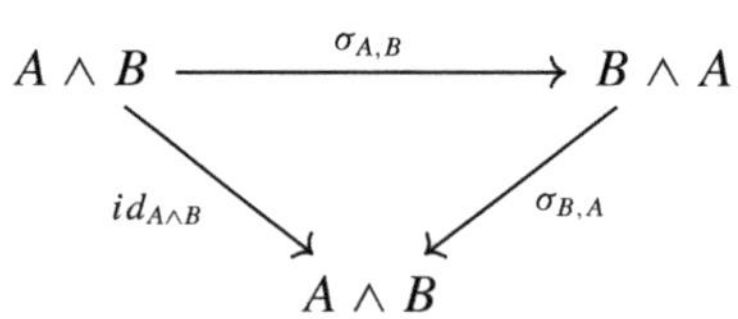

Naturally, we expect this diagram to commute, as $\sigma_{A,B}$ and $\sigma_{B,A}$ are expected to act as identities.

[4] This issue is known as the *coherency issue*, and its precise formal definition is rather subtle, varying depending on the context. Here, we use it only in an informal sense.

Given this discussion, completing the axiomatization of a symmetric monoidal structure requires enforcing all equalities between morphisms in the category that are induced by the above isomorphisms. Since there are numerous ways to compose these isomorphisms, one might find it challenging—or even seemingly impossible—to provide a clean and comprehensive axiomatization. Fortunately, a few fundamental equalities, listed in the next definition, exist that, once established, suffice to derive all other required identities within the structure Mac Lane (1998, 1963).

Definition 120 By a *symmetric monoidal structure* on C, we mean a tuple $(\otimes, I, \alpha, \lambda, \rho, \sigma)$, where

- $\otimes : C \times C \to C$ is a functor called the *tensor product* and $I \in C$ is an object called the *unit object*,
- $\alpha_{A,B,C} : (A \otimes B) \otimes C \to A \otimes (B \otimes C)$ is a natural isomorphism called the *associator*,
- $\lambda_A : I \otimes A \to A$ is a natural isomorphism called the *left unitor*,
- $\rho_A : A \otimes I \to A$ is a natural isomorphism called the *right unitor*, and
- $\sigma_{A,B} : A \otimes B \to B \otimes A$ is a natural isomorphism called the *braiding*

such that the following diagrams commute

- *Triangle Identity:*

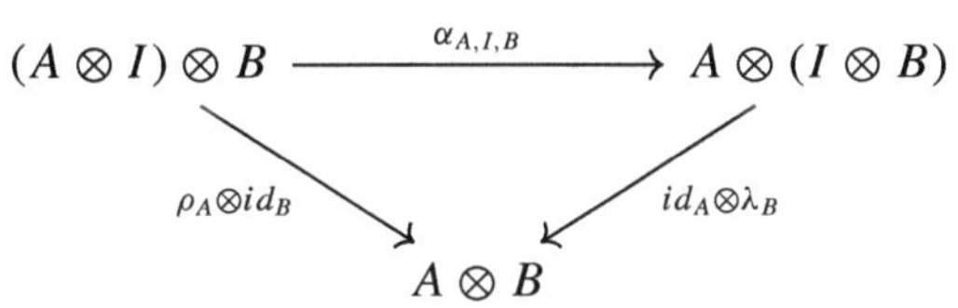

- *Pentagon Identity:*

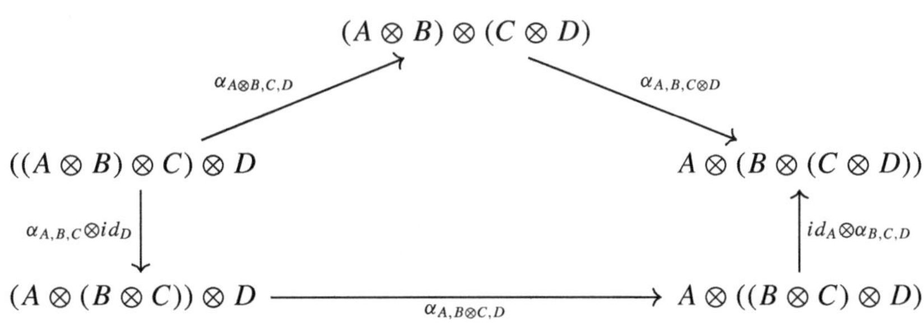

- *Hexagon Identity:*

$$\begin{array}{ccccc} (A \otimes B) \otimes C & \xrightarrow{\alpha_{A,B,C}} & A \otimes (B \otimes C) & \xrightarrow{\sigma_{A,B\otimes C}} & (B \otimes C) \otimes A \\ {\scriptstyle \sigma_{A,B}\otimes id_C}\downarrow & & & & \downarrow{\scriptstyle \alpha_{B,C,A}} \\ (B \otimes A) \otimes C & \xrightarrow[\alpha_{B,A,C}]{} & B \otimes (A \otimes C) & \xrightarrow[id_B \otimes \sigma_{A,C}]{} & B \otimes (C \otimes A) \end{array}$$

and $\sigma_{A,B}\sigma_{B,A} = id_{A\otimes B}$. By a slight abuse of notation, we occasionally denote the symmetric monoidal structure $(\otimes, I, \alpha, \lambda, \rho, \sigma)$ simply by $(\otimes, I)$. Moreover, we call the tuple $(C, \otimes, I)$ a *symmetric monoidal category*.

Philosophical Comment 121 The triangle identity states that the two canonical ways of absorbing the empty set between two assumptions are equal. The pentagon identity asserts that all canonical ways of rearranging parentheses—from left-to-right to right-to-left are equal. Lastly, the hexagon identity ensures that repeated swapping is equivalent to swapping with the combined set of assumptions.

This completes our first attempt at axiomatizing the conjunctive structure. However, some properties of the conjunctive structure remain unaddressed. Before proceeding further, let us incorporate the necessary structure for $\to$ and $\bot$. For implication, we follow the earlier idea: deductions of $A \to B$ from C correspond bijectively and uniformly to deductions of B from $C \wedge A$. Formally, this means:

Definition 122 Let $(\otimes, I)$ be a symmetric monoidal structure on C. The tuple $(\otimes, I, [-,-], \gamma)$ is called a *symmetric monoidal closed* structure if $[-,-] : C^{op} \times C \to C$ is a functor and

$$\gamma_{A,B,C} : \mathrm{Hom}_C(C \otimes A, B) \to \mathrm{Hom}_C(C, [A, B])$$

is a natural isomorphism. By abuse of notation, we often omit γ and refer to the symmetric monoidal closed structure simply as $(\otimes, I, [-,-])$. Moreover, we call the tuple $(C, \otimes, I, [-,-])$ a *symmetric monoidal closed category*.

Example 123 Let C be a cartesian category. Then, by interpreting $\otimes$ as the binary product, I as the terminal object, and taking α, λ, ρ, and σ in their canonical forms, we obtain a symmetric monoidal structure on C. If C is also a CCC, defining $[-,-] : C^{op} \times C \to C$ as exponentiation with the canonical choice of γ leads to a symmetric monoidal closed structure on C.

Example 124 Let $(M, \cdot, e)$ be a commutative monoid, and let C be the discrete category over the set M. Define $(\otimes, I)$ as $(\cdot, e)$ and set all the required natural isomorphisms as identity maps. It is straightforward to verify that $(\otimes, I)$ forms a symmetric monoidal structure on C. If we also assume that $(M, \cdot, e)$ is a group, by defining $[a, b] = a^{-1}b$, we can make the symmetric monoidal structure closed.

Philosophical Comment 125 By Example 124, a symmetric monoidal structure is a generalization of the monoid structure, extending it from small discrete categories to arbitrary categories.

Example 126 Let $C = \mathbf{Rel}$ be the category with sets as objects and relations $R \subseteq A \times B$ as morphisms $R : A \to B$. The identity morphism on A is given by the identity relation $\{(a, a) \mid a \in A\}$, and composition is the usual composition of relations. Define $\otimes : \mathbf{Rel} \times \mathbf{Rel} \to \mathbf{Rel}$ as the cartesian product functor, where $A \otimes B = A \times B$ for sets A and B, and for relations $R \subseteq A \times B$ and $S \subseteq C \times D$, we define $R \otimes S \subseteq (A \times C) \times (B \times D)$ by

$$((a, c), (b, d)) \in R \otimes S \iff (a, b) \in R \text{ and } (c, d) \in S.$$

Let $I = \{*\}$. With the canonical choices for α, λ, ρ, and σ, this defines a symmetric monoidal structure on **Rel**. It is worth noting that this cartesian product functor is not the binary product of **Rel**. The product and the coproduct functors are both the disjoint union functor on **Rel**.

Next, define the functor $(-)^{\perp} : \mathbf{Rel}^{op} \to \mathbf{Rel}$ by

$$A^{\perp} = A, \quad R^{\perp} = R^{op},$$

where $R^{op} \subseteq B \times A$ is the converse relation, i.e., $(b, a) \in R^{op}$ if and only if $(a, b) \in R$, for any $a \in A$ and $b \in B$. Now, define $[A, B] = A^{\perp} \otimes B$ with the canonical choice of the natural isomorphism γ. It is straightforward to verify that this setup satisfies the axioms of a symmetric monoidal closed structure. Throughout this section, whenever we refer to **Rel**, we always mean this category equipped with this symmetric monoidal closed structure.

Example 127 Let K be a fixed field, and let $C = \mathbf{Vect}_K$ be the category of all K-vector spaces with K-linear maps as morphisms. Define $\otimes$ as the usual tensor product functor and set $I = K$. With the canonical choices for α, λ, ρ, and σ, this gives a symmetric monoidal structure on $\mathbf{Vect}_K$.

Next, define the functor $[-, -] : \mathbf{Vect}_K^{op} \times \mathbf{Vect}_K \to \mathbf{Vect}_K$ by $[V, W] = \mathrm{Hom}_K(V, W)$ as the K-vector space of all K-linear maps from V to W. With the canonical choice for the natural isomorphism γ, we obtain a symmetric monoidal closed structure on $\mathbf{Vect}_K$. Restricting this structure to finite-dimensional K-vector spaces also yields a symmetric monoidal closed category, denoted by $\mathbf{FdVect}_K$. Throughout this section, whenever we refer to $\mathbf{Vect}_K$ or $\mathbf{FdVect}_K$, we always mean these categories equipped with the symmetric monoidal closed structure described above.

Finally, to formalize $\perp$, we need an object such that the functor $\neg : C^{op} \to C$ defined by $\neg A = [A, \perp]$ becomes an involution. To explain, note that for any object B, we have the functor $\neg_B : C^{op} \to C$ defined by $\neg_B A = [A, B]$. Using the natural isomorphism

$$\gamma_{X,Y,Z} : \mathrm{Hom}(X \otimes Y, Z) \to \mathrm{Hom}(X, [Y, Z])$$

and the identity map $id_{[Y,Z]} : [Y, Z] \to [Y, Z]$, there is a canonical map $[Y, Z] \otimes Y \to Z$. Therefore, we obtain a map $\neg_B A \otimes A \to B$. Using the symmetry of $\otimes$, we get a map $A \otimes \neg_B A \to B$. Again, applying γ, we derive a canonical map $A \to \neg_B \neg_B A$. An object B is called *dualizing* if the canonical map $A \to \neg_B \neg_B A$ is an isomorphism, for any object A. To reflect the classical behavior of negation, we need to assume that $\bot$ is dualizing:

Definition 128 Let $(\otimes, I, [-, -])$ be a symmetric monoidal closed structure on C. The tuple $(\otimes, I, [-, -], \bot)$ is called a $*$-*autonomous* structure on C if $\bot$ is a dualizing object of C, i.e., the canonical map $A \to \neg_\bot \neg_\bot A$ is an isomorphism for any object A of C. We denote $\neg_\bot$ by $\neg$. A $*$-autonomous structure is called *compact closed* if $[A, B] \cong \neg A \otimes B$, naturally in A and B. We call the tuple $(C, \otimes, I, [-, -], \bot)$ a $*$-*autonomous category* (resp. *compact closed category*) if $(\otimes, I, [-, -], \bot)$ is a $*$-*autonomous structure* (resp. *compact closed structure*) on C.

Example 129 Using Theorem 118, a BCC with its CC-structure as the symmetric monoidal closed structure and $\bot$ as its initial object is $*$-autonomous if and only if it is a boolean algebra. Note that in a non-trivial boolean algebra, there is no reason to have $a \to b = \neg a \wedge b$. Because, this equality would imply that $1 = (0 \to 0) = (\neg 0 \wedge 0) = 0$, which is a contradiction. Therefore, non-trivial boolean algebras provide an example of a $*$-autonomous category that is not compact closed.

The category **Rel** is compact closed if we define $\bot = \{*\}$. It is $*$-autonomous because the canonical map $A \to \neg\neg A$ is the identity. It is compact closed because $[A, B]$ is defined as $A^\perp \otimes B$, and $\neg A \cong A^\perp$ by definition. A similar result holds for $\mathbf{FdVect}_K$, if we set $\bot = K$. Since V is finite-dimensional and $\neg V = \mathrm{Hom}_K(V, K) = V^*$, the canonical map

$$V \to \mathrm{Hom}_K(\mathrm{Hom}_K(V, K), K) = V^{**}$$

is an isomorphism. Hence, $\mathbf{FdVect}_K$ is $*$-autonomous. It is also compact closed because $\mathrm{Hom}_K(V, W) \cong V^* \otimes W$, naturally in V and W.

Remark 130 Here are some remarks. First, $*$-autonomous categories are used to provide a natural categorical semantics for a fragment of classical linear logic Barr (1979, 1991). Adding more structure, we will use them to interpret deductions in classical logic as will be explained in the rest of this subsection. However, it is worth mentioning that as a special case of $*$-autonomous categories, compact closed categories (with additional structures) have also been proposed to formalize classical deductions Hyland (2004, 2002). Second, compact closed categories, when equipped with additional structures, can be used to formalize some fundamental and interesting aspects of quantum mechanics (Abramsky and Coecke 2004, 2009; Heunen and Vicary 2019). To have an intuition why, recall that the usual formalization of quantum mechanics is based on $\mathbb{C}$-vector spaces. Then, one can interpret compact closed categories as a categorical abstraction of the category $\mathbf{FdVect}_\mathbb{C}$.

Remark 131 In any $*$-autonomous category, we have $\neg I \cong \bot$ and $\neg\bot \cong I$. For the first isomorphism, observe that since $X \otimes I \cong X$ naturally in X, we obtain the following sequence of natural isomorphisms:

$$\mathrm{Hom}(X, Y) \cong \mathrm{Hom}(X \otimes I, Y) \cong \mathrm{Hom}(X, [I, Y]).$$

It follows that $[I, Y] \cong Y$. In particular, for $Y = \bot$, we get $\neg I = [I, \bot] \cong \bot$. For the second isomorphism, it suffices to use the fact that $\neg\neg I \cong I$. Additionally, it is worth noting that in a compact closed category, we have $\bot \cong I$. To prove this, first notice that by the above observation $[I, I] \cong I$. Then, recall that in a compact closed category, $[I, I] \cong \neg I \otimes I \cong \neg I$. Therefore, we obtain $\neg I \cong I$. Given that $\neg I \cong \bot$, we conclude that $I \cong \bot$.

6.2.2 Cloning and Deleting

In the previous subsubsection, we argued that the conjunctive part of the language possesses at least the structure of a symmetric monoidal category. However, this structure alone does not fully capture all the expected properties of conjunction. The reason is that two key properties of conjunction have not yet been addressed: *cloning* and *deleting*.[5] In proofs, an assumption can be used multiple times or discarded entirely. Formally, for any proposition A, we expect to have deductions $A \to A \wedge A$ and $A \to \top$ to *clone* the assumption A and *delete* it, respectively.

In a general symmetric monoidal category, there is no reason for these cloning and deleting morphisms to exist.

Example 132 Let $\mathcal{M} = (\mathbb{N}, +, 0)$ be the usual monoid of natural numbers, and let $a \neq 0$. Following Example 124, we can regard $\mathcal{M}$ as a symmetric monoidal category. Since morphisms in this setting are only identities, it is clear that no map $a \to 0$ or $a \to 2a$ exists, as either would imply $a = 0$, leading to a contradiction.

To formalize the cloning and deleting processes, we assume the existence of two morphisms, $A \to A \wedge A$ and $A \to \top$, for every object A. However, as with the structural isomorphisms in a symmetric monoidal category, these maps must satisfy certain coherence conditions. To begin, let us examine the relationship between these two maps for a fixed object A.

Definition 133 Let $(\mathcal{C}, \otimes, I)$ be a symmetric monoidal category. A *commutative comonoid structure* on an object A in $(\mathcal{C}, \otimes, I)$ is a pair $(\langle\rangle_A, \Delta_A)$ of morphisms, where $\langle\rangle_A : A \to I$ represents deletion and $\Delta_A : A \to A \otimes A$ represents cloning, satisfying the following commutative diagrams:

---

[5] The terms "cloning" and "deleting" have an information-theoretic flavor, referring to the ability to duplicate or discard bits of information. In proof theory, these operations are known as contraction and weakening, respectively.

- *Co-associativity:*

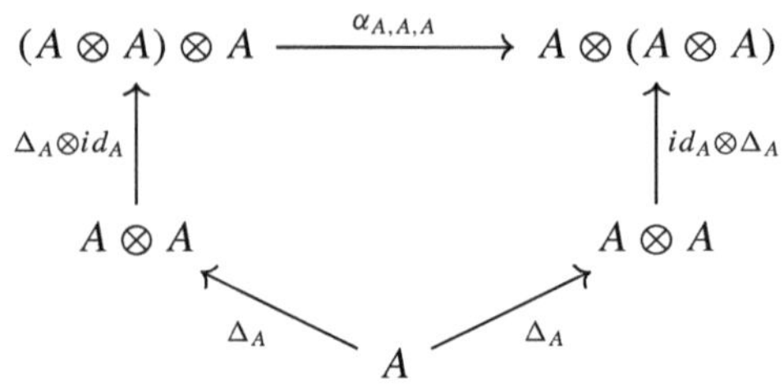

- *Co-unitality:*

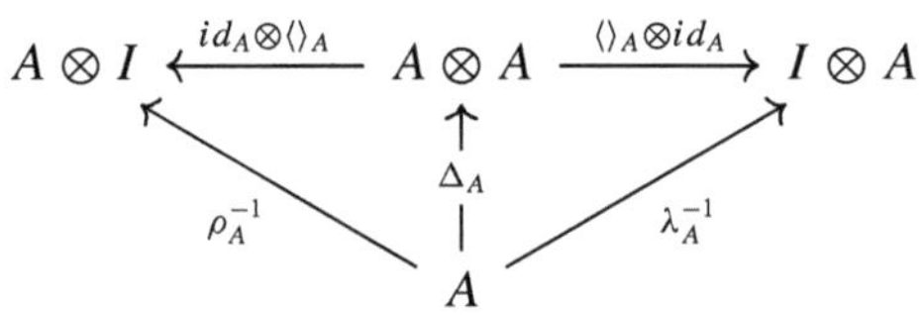

- *Co-symmetry:*

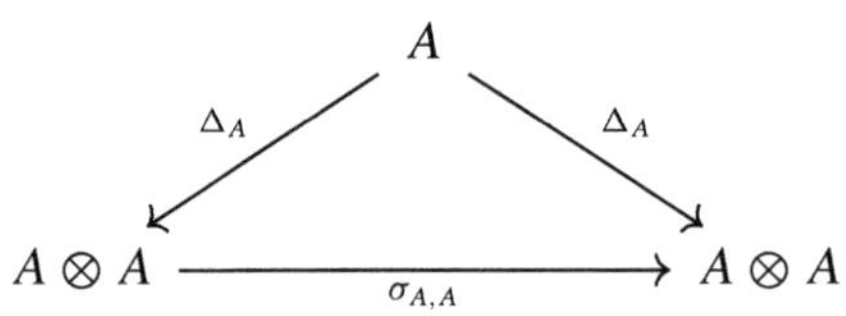

A *commutative comonoid* in $(C, \otimes, I)$ is a tuple $(A, \langle\rangle_A, \Delta_A)$, where A is an object in C and $(\langle\rangle_A, \Delta_A)$ is a commutative comonoid structure on A in $(C, \otimes, I)$.

Philosophical Comment 134 The co-associativity diagram formalizes the intuition that "the two canonical ways to obtain three copies of A are equal." The co-unitality diagram expresses that "cloning A and then deleting one copy has no effect." Finally, the co-symmetry diagram ensures that "the order of cloned copies is immaterial."

Example 135 In a cartesian category with its symmetric monoidal structure as its cartesian structure, there is a canonical commutative comonoid structure on any object A. Specifically, one can define $\langle\rangle_A$ as the unique morphism $!_A : A \to 1$ and set Δ_A as the diagonal morphism $\langle id_A, id_A\rangle : A \to A \times A$. Co-associativity, co-unitality, and co-symmetry can be easily verified using the properties of the binary product and terminal object.

Similarly, in the symmetric monoidal category **Rel**, any set A can be equipped with a commutative comonoid structure using the relations $\langle\rangle_A : A \to \{*\}$ and $\Delta_A : A \to A \times A$, defined by $\langle\rangle_A = A \times \{*\}$ and $\Delta_A = \{(a, (a, a)) \mid a \in A\}$. The commutativity of the diagrams is straightforward to verify.

Remark 136 The dual of a commutative comonoid is a *commutative monoid*. It consists of an object A together with maps $[]_A : I \to A$ and $\nabla_A : A \otimes A \to A$, which satisfy the duals of the equalities in Definition 133, namely *associativity*, *unitality*, and *symmetry*. Commutative monoids in a symmetric monoidal category generalize the usual commutative monoids in algebra. One can interpret $\nabla_A : A \otimes A \to A$ as the product of the monoid and $[]_A : I \to A$ as its neutral element, while associativity, unitality, and symmetry formalize the expected properties of a commutative monoid. In fact, a commutative monoid in **Set**, with its cartesian structure, is simply a commutative monoid in the traditional sense. Commutative monoids are abundant in mathematics. For another example, a commutative monoid in $\mathbf{Vect}_K$ corresponds to a commutative K-algebra. As in classical reasoning, the dual of the conjunctive structure is the disjunctive structure. A commutative monoid structure can also serve to witness the deductions $\bot \to A$ and $A \vee A \to A$.

To handle the relationship between cloning and deleting different assumptions, we can introduce additional coherency equalities. These equalities should reflect how different assumptions can be cloned or deleted in relation to each other.

Definition 137 Let $(C, \otimes, I)$ be a symmetric monoidal category. A *commutative comonoid structure on* $(C, \otimes, I)$ is an assignment of commutative comonoids $(A, \langle\rangle_A, \Delta_A)$ to any object A such that $\langle\rangle_I = id_I : I \to I$, $\Delta_I = \lambda_I^{-1} : I \to I \otimes I$ and

$$\begin{array}{ccc} (A \otimes B) \otimes (A \otimes B) & \xrightarrow{\quad\cong\quad} & (A \otimes A) \otimes (B \otimes B) \\ & {}_{\Delta_{A\otimes B}}\nwarrow \qquad \nearrow_{\Delta_A \otimes \Delta_B} & \\ & A \otimes B & \end{array}$$

$$\begin{array}{ccc} I \otimes I & \xrightarrow{\quad\lambda_I\quad} & I \\ & {}_{\langle\rangle_A \otimes \langle\rangle_B}\nwarrow \qquad \nearrow_{\langle\rangle_{A\otimes B}} & \\ & A \otimes B & \end{array}$$

where the map $\cong$ is the canonical map constructed by the associators and braiding. A comonoid structure is called *uniform* if both $\langle\rangle_A : A \to I$ and $\Delta_A : A \to A \otimes A$ are natural in A.

Philosophical Comment 138 The equalities in Definition 137 formalize the intuition that "cloning $\{A, B\}$ is equivalent to cloning A and cloning B", "deleting $\{A, B\}$ is the same as deleting A and B" and "deleting and cloning the empty set has no effect". The uniformity condition simply states that the processes of cloning and deleting are applied uniformly, rather than being arbitrarily chosen for different assumptions.

Example 139 Both assignments of commutative comonoids presented in Example 135 are commutative comonoid structures on their coressponding categories.

This seemingly completes our proposal for the conjunctive structure. We start with a symmetric monoidal category C and then introduce a uniform commutative comonoid structure on C to simulate the missing processes of cloning and deleting. Using this proposal, to formalize classical deduction, one might expect to employ a $*$-autonomous category C equipped with a uniform commutative comonoid structure. However, we are once again confronted with a collapse theorem, Corollary 141, which states that such a $*$-autonomous category must necessarily be a boolean algebra. To prove this theorem, we first show that the existence of a uniform comonoid structure forces the symmetric monoidal structure to collapse to the usual cartesian structure. This result is well-known among category theorists, although its proof is, to our knowledge, rarely found in print Heunen and Vicary (2019). Consequently, we provide the proof here.

Theorem 140 *Let $(C, \otimes, I)$ be a symmetric monoidal category. If there is a uniform commutative comonoid structure on $(C, \otimes, I)$, then the monoidal structure $(\otimes, I)$ must necessarily be the cartesian structure* $(\times, 1)$.

Proof Let $(\langle\rangle_A, \Delta_A)$ be the comonoid structure on A. Since $\langle\rangle_A : A \to I$ and $\Delta_A : A \to A \otimes A$ are natural in A, for any map $f : A \to B$, the following diagram commutes:

$$\begin{array}{ccccc} A \otimes A & \xleftarrow{\Delta_A} & A & \xrightarrow{\langle\rangle_A} & I \\ {\scriptstyle f\otimes f}\downarrow & & {\scriptstyle f}\downarrow & & \downarrow{\scriptstyle id_I} \\ B \otimes B & \xleftarrow[\Delta_B]{} & B & \xrightarrow[\langle\rangle_B]{} & I \end{array}$$

In words, we have $\langle\rangle_B \circ f = \langle\rangle_A$ and $\Delta_B \circ f = (f \otimes f) \circ \Delta_A$ for any map $f : A \to B$.

First, we prove that I is a terminal object. Since there is always the map $\langle\rangle_A : A \to I$, it suffices to show that this map is unique. Let $g : A \to I$. Since $\langle\rangle_I = \mathrm{id}_I$, we have $g = \mathrm{id}_I \circ g = \langle\rangle_I \circ g = \langle\rangle_A$. Therefore, $g = \langle\rangle_A$, which implies that I is a terminal object. Second, we prove that $A \otimes B$ is the binary product of A and B. Consider the maps $p_0 : A \otimes B \to A$ and $p_1 : A \otimes B \to B$ defined as follows:

$$A \otimes B \xrightarrow{id_A \otimes \langle\rangle_B} A \otimes I \xrightarrow{\rho_A} A$$

$$A \otimes B \xrightarrow[\langle\rangle_A \otimes id_B]{} I \otimes B \xrightarrow[\lambda_B]{} B$$

These maps are intended to serve as projections. Moreover, for any two morphisms $f : C \to A$ and $g : C \to B$, define $\langle f, g\rangle : C \to A \otimes B$ by

$$C \xrightarrow{\Delta_C} C \otimes C \xrightarrow{f \otimes g} A \otimes B$$

We prove that $p_0 \circ \langle f, g \rangle = f$ and $p_1 \circ \langle f, g \rangle = g$ for any $f : C \to A$ and $g : C \to B$. We will only prove the first equality; the second is similar. Consider the following diagram:

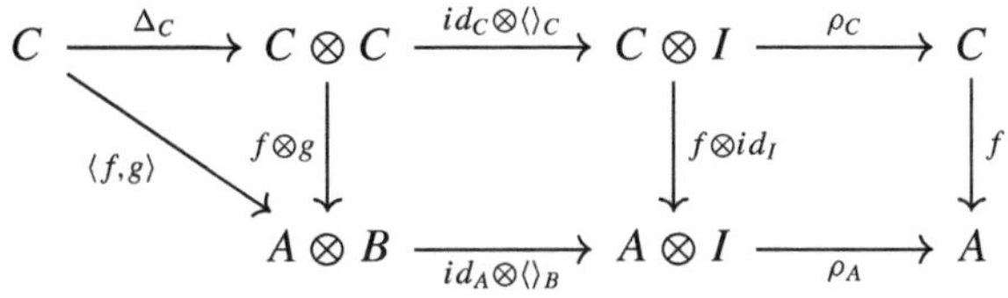

The right square commutes by the naturality of ρ. The left square commutes by the equality $\langle\rangle_B \circ g = \langle\rangle_C$. The commutativity of the triangle follows directly from the definition of $\langle f, g \rangle$. Now, by co-unitality, the upper row equals id_C and the lower row equals p_0, by definition. Hence, we obtain $p_0 \circ \langle f, g \rangle = f$.

Now, to complete the proof of the second part, it is enough to prove $\langle p_0 h, p_1 h \rangle = h$, for any $h : C \to A \otimes B$. For that purpose, consider the following diagram:

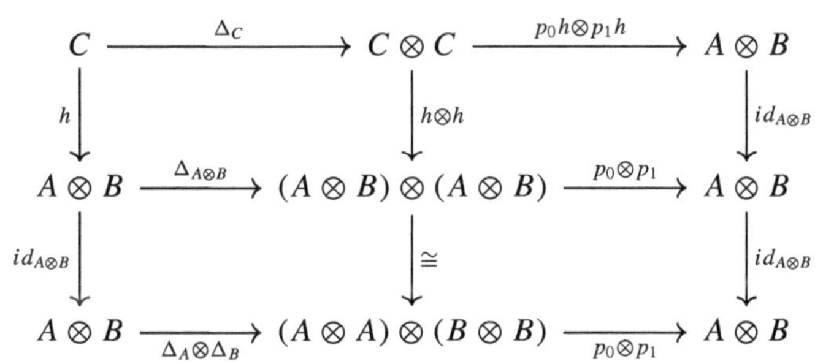

The upper left square commutes by the naturality of Δ. The commutativity of the upper right square is clear. The lower left square commutes by the definition of a commutative comonoid structure on $(\mathcal{C}, \otimes, I)$. The lower right square commutes by straightforward applications of the identities for the associator and braiding. Now, notice that the upper row is $\langle p_0 h, p_1 h \rangle$, by definition. The lower row is the identity, by co-unitality. Therefore, we have $\langle p_0 h, p_1 h \rangle = h$. This completes the proof that $A \otimes B$ is the binary product of A and B.

To show that the tensor product is the binary product as a functor, it suffices to prove that $f \otimes g = f \times g$ for any $f : A \to C$ and $g : B \to D$. By definition, $(f \times g) : A \otimes B \to C \otimes D$ is the unique map satisfying $p_0 \circ (f \times g) = f \circ p_0$ and $p_1 \circ (f \times g) = f \circ p_1$. Therefore, it is enough to show that $f \otimes g : A \otimes B \to C \otimes D$ also satisfies these equalities. We will prove the case for p_0; the other case is similar. Consider the following diagram:

$$
\begin{array}{ccccc}
A \otimes B & \xrightarrow{id_A \otimes \langle\rangle_B} & A \otimes I & \xrightarrow{\rho_A} & A \\
\Big\downarrow{\scriptstyle f \otimes g} & & \Big\downarrow{\scriptstyle f \otimes id_I} & & \Big\downarrow{\scriptstyle f} \\
C \otimes D & \xrightarrow[id_C \otimes \langle\rangle_D]{} & C \otimes I & \xrightarrow[\rho_C]{} & C
\end{array}
$$

The right square commutes by the naturality of ρ. The left square commutes by the equality $\langle\rangle_B = \langle\rangle_D \circ g$. Since the rows are both p_0, we obtain $p_0 \circ (f \otimes g) = f \circ p_0$, which completes the proof. □

Corollary 141 *Let $(C, \otimes, I, [-,-], \bot)$ be a $*$-autonomous category. If there is a uniform commutative comonoid structure on $(C, \otimes, I)$, then C must necessarily be a boolean algebra.*

Proof By Theorem 140, the monoidal structure must be the cartesian structure. Consequently, I is the terminal object, and $\otimes$ corresponds to the binary product. First, note that the collapse of $\otimes$ to the product forces the functor $[-,-] : C^{op} \times C \to C$ to be the exponentiation. Hence, C is a CCC. Second, using the isomorphism $A \cong \neg\neg A$, we can easily construct a natural isomorphism

$$\mathrm{Hom}(\neg A, \neg B) \cong \mathrm{Hom}(B, A).$$

Applying this isomorphism, along with the fact that $\bot \cong \neg I$ from Remark 131, we have

$$\mathrm{Hom}(\bot, A) \cong \mathrm{Hom}(\neg I, \neg\neg A) \cong \mathrm{Hom}(\neg A, I).$$

Since I is terminal, the last set is a singleton. Consequently, $\mathrm{Hom}(\bot, A)$ is also a singleton, which implies that $\bot$ is an initial object. Similarly, one can show that the functor $\oplus : C \times C \to C$, defined by $A \oplus B = \neg(\neg A \otimes \neg B)$ is the binary coproduct. This establishes that C is a BCC satisfying $A \cong \neg\neg A$ for all A. Finally, by Theorem 118, it follows that C must be a boolean algebra. □

Philosophical Comment 142 (*Lafont Example*) For readers familiar with sequent calculi, there is a stronger argument showing that not only classical reasoning but also multi-conclusion reasoning leads to a collapse. This argument is usually called *Lafont example* Fuhrmann and Pym (2004), Führmann and Pym (2006, 2007) as it is a variant of an example credited to Lafont Girard et al. (1989). To explain, consider a multi-conclusion sequent calculus with the weakening rule (denoted by W) and the contraction rule (denoted by C). Then, the following two proofs must be considered equivalent:

$$
\dfrac{\begin{array}{c}\mathcal{D}_1 \\ \Gamma_1 \Rightarrow \Delta_1, C\end{array} \qquad \dfrac{\begin{array}{c}\mathcal{D}_2 \\ \Gamma_2 \Rightarrow \Delta_2\end{array}}{\Gamma_2, C \Rightarrow \Delta_2}\,W}{\Gamma_1, \Gamma_2 \Rightarrow \Delta_1, \Delta_2}\,cut
\qquad\qquad
\dfrac{\begin{array}{c}\mathcal{D}_2 \\ \Gamma_2 \Rightarrow \Delta_2\end{array}}{\Gamma_1, \Gamma_2 \Rightarrow \Delta_1, \Delta_2}\,W
$$

The reason is that the formula C is added to the assumptions of the right-hand-side deduction in a dummy way. Therefore, using $\mathcal{D}_1$ to provide a proof for it must have no effect on the result, apart from adding other dummy formulas, i.e., Γ_1 and Δ_1 instead of C. Note that this is a proof-theoretic version of the naturality of the deletion map. Dually, we also have the following equality:

$$\dfrac{\dfrac{\overset{\mathcal{D}_1}{\Gamma_1 \Rightarrow \Delta_1}}{\Gamma_1 \Rightarrow \Delta_1, C}\,W \qquad \overset{\mathcal{D}_2}{\Gamma_2, C \Rightarrow \Delta_2}}{\Gamma_1, \Gamma_2 \Rightarrow \Delta_1, \Delta_2}\,cut \qquad \dfrac{\overset{\mathcal{D}_1}{\Gamma_1 \Rightarrow \Delta_1}}{\Gamma_1, \Gamma_2 \Rightarrow \Delta_1, \Delta_2}\,W$$

Now, using the above equalities, let us simplify the following proof tree:

$$\dfrac{\dfrac{\dfrac{\overset{\mathcal{D}_1}{\Gamma \Rightarrow \Delta}}{\Gamma \Rightarrow \Delta, C}\,W \qquad \dfrac{\overset{\mathcal{D}_2}{\Gamma \Rightarrow \Delta}}{\Gamma, C \Rightarrow \Delta}\,W}{\Gamma, \Gamma \Rightarrow \Delta, \Delta}}{\Gamma \Rightarrow \Delta}\,C$$

Using the above equalities, this proof tree is equal to both of the following proof trees:

$$\dfrac{\dfrac{\overset{\mathcal{D}_1}{\Gamma \Rightarrow \Delta}}{\Gamma, \Gamma \Rightarrow \Delta, \Delta}\,W}{\Gamma \Rightarrow \Delta}\,C \qquad \dfrac{\dfrac{\overset{\mathcal{D}_2}{\Gamma \Rightarrow \Delta}}{\Gamma, \Gamma \Rightarrow \Delta, \Delta}\,W}{\Gamma \Rightarrow \Delta}\,C$$

Hence, the latter two proof-trees are equal. Now, observe that applying contraction to weakening must lead to identity. This is the proof-theoretic presentation of the co-unitality condition of a commutative comonoid. Therefore, using the equality established above, we reach the equality $\mathcal{D}_1 = \mathcal{D}_2$, which proves the collapse of deductions of any sequent $\Gamma \Rightarrow \Delta$. Note that the argument does not use any negation or classicality. However, it critically depends on multi-conclusion type reasoning.

As mentioned earlier, compact closed categories (equipped with additional structures) can be used to formalize certain aspects of quantum mechanics. Since compact closed categories are a special case of $*$-autonomous categories, one can prove a stronger collapse theorem than Corollary 141, commonly known as the *no-deleting theorem,* which states that there is no uniform way to forget information in the quantum world.

Theorem 143 (No-Deleting) *Let* $(\mathcal{C}, \otimes, I, [-,-], \bot)$ *be a compact closed category. If there is a natural transformation* $\langle\rangle_A : A \to I$ *such that* $\langle\rangle_I = id_I$*, then* $\mathrm{Hom}_{\mathcal{C}}(A, B)$ *is a singleton, for any objects A and B.*

Proof First, we can use a similar argument to the one in the first part of the proof of Theorem 140 to show that I is terminal. Then, as $\bot \cong I$ by Remark 131, and using the isomorphism $B \cong \neg\neg B$, we obtain the following series of isomorphisms:

$$\mathrm{Hom}_C(A, B) \cong \mathrm{Hom}_C(A, \neg\neg B) \cong \mathrm{Hom}_C(A \otimes \neg B, \bot) \cong \mathrm{Hom}_C(A \otimes \neg B, I).$$

Since I is terminal, the last set is a singleton. Therefore, $\mathrm{Hom}_C(A, B)$ is also a singleton. □

6.2.3 Classical Categories

By Corollary 141, it appears that imposing a uniform commutative comonoid structure on a $*$-autonomous category is too strong a requirement. Since all the other structures are naturally desirable to retain, we must weaken the uniformity condition, i.e., the following commutativity:

$$\begin{array}{ccccc} A \otimes A & \xleftarrow{\Delta_A} & A & \xrightarrow{\langle\rangle_A} & I \\ \downarrow{\scriptstyle f \otimes f} & & \downarrow{\scriptstyle f} & & \downarrow{\scriptstyle id_I} \\ B \otimes B & \xleftarrow[\Delta_B]{} & B & \xrightarrow[\langle\rangle_B]{} & I \end{array}$$

For this purpose, we need to introduce a new structure into our categorical framework: *simplifications*. Intuitively, for any fixed A and B, there is a partial order between deductions from A to B that encodes the possibility of simplifying one deduction into another. Formally, we write $f_1 \leq f_2$ if f_1 can be simplified to f_2, where $f_1, f_2 : A \to B$. For example, let $f : A \to B$ be a deduction and let $\langle\rangle_B : B \to I$ be the deletion map for B. Then, the deduction $\langle\rangle_B \circ f : A \to I$ can be simplified to the deduction $\langle\rangle_A : A \to I$. The reasoning behind this simplification is that instead of first using A to prove B and then forgetting B, we can simply forget A from the outset.

The crucial observation of Fuhrmann and Pym (2004), Führmann and Pym (2006, 2007) is that this process of simplification is not necessarily symmetric and must therefore be formalized using a *partial order* on deductions rather than *equality*. Accordingly, we should have $\langle\rangle_B \circ f \leq \langle\rangle_A$, but there is no need to assume the equality $\langle\rangle_B \circ f = \langle\rangle_A$, which corresponds to the naturality condition that leads to a collapse.

Putting this in proof-theoretic terms, using this ordering on Lafont example, the proof tree:

$$\dfrac{\dfrac{\dfrac{\mathcal{D}_1}{\Gamma \Rightarrow \Delta}}{\Gamma \Rightarrow \Delta, C}\,W \qquad \dfrac{\dfrac{\mathcal{D}_2}{\Gamma \Rightarrow \Delta}}{\Gamma, C \Rightarrow \Delta}\,W}{\dfrac{\Gamma, \Gamma \Rightarrow \Delta, \Delta}{\Gamma \Rightarrow \Delta}\,C}$$

is less than or equal to both $\mathcal{D}_1$ and $\mathcal{D}_2$, as it can be simplified to both of these proof trees. However, this does not mean that $\mathcal{D}_1$ and $\mathcal{D}_2$ are equal.

To formalize this intuitive notion of simplification, one should start with a $*$-autonomous category and introduce a partial order on each $\mathrm{Hom}_{\mathcal{C}}(A, B)$ in a coherent manner.

Definition 144 A $*$-autonomous category $(\mathcal{C}, \otimes, I, [-, -], \gamma, \bot)$ equipped with a partial order on each $\mathrm{Hom}_{\mathcal{C}}(A, B)$ is called *order-enriched* if composition and the functors $\otimes$ and $[-, -]$ are monotone with respect to the partial order, and if the function $\gamma_{A,B,C} : \mathrm{Hom}_{\mathcal{C}}(C \otimes A, B) \cong \mathrm{Hom}_{\mathcal{C}}(C, [A, B])$ is an order-isomorphism for all A, B, and C.

Example 145 Any $*$-autonomous category, when equipped with equality as the partial order, is order-enriched. As another example, consider the $*$-autonomous category **Rel**. For any two relations $R, S \subseteq A \times B$ in $\mathrm{Hom}_{\mathbf{Rel}}(A, B)$, define $R \leq S$ if and only if $R \subseteq S$. Since both the cartesian product and the operation $(-)^{\perp} : \mathbf{Rel}^{op} \to \mathbf{Rel}$ are monotone, it is straightforward to verify that composition and the functors $\otimes$ and $[-, -]$ are monotone, and that the bijection $\gamma_{A,B,C}$ is an order-isomorphism.

Now, we can weaken the naturality of cloning and deleting from equality to inequality, thereby formalizing the idea that one deduction can be simplified into another.

Definition 146 An order-enriched $*$-autonomous category, equipped with a commutative comonoid structure, is called *classical* if it satisfies the following relaxed version of naturality:

$$\begin{array}{ccc} A & \xrightarrow{\langle\rangle_A} & I \\ {\scriptstyle f}\downarrow & \leq & \downarrow{\scriptstyle id_I} \\ B & \xrightarrow[\langle\rangle_B]{} & I \end{array} \qquad\qquad \begin{array}{ccc} A & \xrightarrow{\Delta_A} & A \otimes A \\ {\scriptstyle f}\downarrow & \leq & \downarrow{\scriptstyle f \otimes f} \\ B & \xrightarrow[\Delta_B]{} & B \otimes B \end{array}$$

In words, for any $f : A \to B$, we have $\langle\rangle_B \circ f \leq \langle\rangle_A$ and $\Delta_B \circ f \leq (f \otimes f) \circ \Delta_A$.

Example 147 The trivial example of a classical category is a boolean algebra, where the partial order is simply equality. For a non-trivial example, consider the order-enriched $*$-autonomous category **Rel** with the comonoid structure presented in Example 135, where the partial order is given by inclusion. To prove that **Rel** is a classical category, observe the following: for any $R \subseteq A \times B$, we have

$$\langle\rangle_B \circ R = \{(a, *) \mid \exists b \in B, (a, b) \in R\} \subseteq A \times \{*\} = \langle\rangle_A.$$

This represents an inequality, which becomes an equality if and only if R is a total relation. For the other condition, we note that

$$\Delta_B \circ R = \{(a, (b, b)) \mid (a, b) \in R\},$$

while

$$(R \otimes R) \circ \Delta_A = \{(a, (b, c)) \mid (a, b) \in R \text{ and } (a, c) \in R\}.$$

Therefore, we have $\Delta_B \circ R \subseteq (R \otimes R) \circ \Delta_A$. This inequality becomes an equality if and only if R is functional.

Remark 148 The category **Rel** in Example 147 illustrates that the notion of a classical category does not collapse to a boolean algebra. To expand on this observation, it is easy to see that the canonical structure induced from classical categories $\mathcal{C}$ and $\mathcal{D}$ to their product $\mathcal{C} \times \mathcal{D}$ forms a classical category itself. By using this closure under the product, one can take a non-trivial boolean algebra $\mathcal{B}$ and obtain the classical category **Rel** $\times\ \mathcal{B}$. This classical category is neither compact closed nor a preordered set. Hence, classical categories that are not compact closed are not restricted to preordered sets.

Remark 149 (*Dummett Categories*) The original definition of classical categories Fuhrmann and Pym (2004), Führmann and Pym (2006, 2007) differs, yet is equivalent, to the one we present here Straßburger (2011). It begins with a *symmetric linearly distributive category*, which is a category equipped with two symmetric monoidal structures: the conjunctive $(\otimes, I)$ and the disjunctive $(\oplus, \bot)$. It also includes a natural transformation $\delta : A \otimes (B \oplus C) \to (A \otimes B) \oplus C$, subject to certain coherency identities, ensuring *distributivity*. Subsequently, an order-enrichment is added coherently, along with a commutative comonoid structure for $(\otimes, I)$ and, dually, a commutative monoid structure for $(\oplus, \bot)$, subject to additional coherency equalities and inequalities.These categories are called *Dummett categories*. They capture *multi-conclusion* deductions in a propositional language without negation and implication, thereby addressing the issue raised by Lafont example and providing a non-trivial categorical interpretation of such deductions. Classical categories are then defined by adding an involutive negation functor that appropriately maps the conjunctive and disjunctive structures to each other. In this chapter, we have chosen to work with $*$-autonomous categories, as they are simpler to introduce, requiring fewer structures and thus involving fewer coherence issues. Moreover, they are more accessible, aligning better with both the spirit and structure of this chapter. However, it is worth noting that the route via Dummett categories is more general and provides a richer perspective than the one we have taken here.

Although the idea of simplification and its non-symmetric behavior is natural, the inequalities we used to weaken naturality might seem a bit arbitrary to the reader. However, this is not the case. To explain, first note that, similar to natural deduction proofs for intuitionistic logic, there is a syntactical calculus for classical proofs called *sequent calculus* **LK**. Moreover, similar to the $\beta\eta$-reductions in natural deduction, there is a natural process of simplification for **LK**-proofs called *cut reduction* Troelstra and Schwichtenberg (2000). The inequalities in the definition of a classical category are two instances of cut reduction, and they are actually enough to axiomatize all such cut reductions. In a more formal form:

Theorem 150 (Soundness-Completeness) *Let T be a set of inequalities between* **LK**-*proofs of the same sequent, and let $\mathcal{D}_1$ and $\mathcal{D}_2$ be two such* **LK**-*proofs. Then, $\mathcal{D}_1 \preceq \mathcal{D}_2$ is derivable from T by cut-reductions if and only if $F(\mathcal{D}_1) \leq F(\mathcal{D}_2)$ for any classical category C and any interpretation F of* **LK**-*proofs into C that satisfies the inequalities in T.*

As we have not formally defined **LK**-proofs and the cut-reduction process, notions such as "interpretation" and "derivable by cut reductions" have been used informally in the previous theorem. However, the theorem is stated and proved in its precise form in Fuhrmann and Pym (2004), Führmann and Pym (2006, 2007). This result demonstrates that classical categories, along with their order-enrichment, faithfully capture the categorical structure underlying sequent-style classical deductions and their cut reduction, providing further evidence for the naturality of these categories.

The connection between **LK**-proofs and classical categories can also be expressed in a more categorical form. Recall that for BCCs, we saw that natural deduction proofs, up to $\beta\eta$-equivalence, form a BCC that is also the free BCC generated by the set $\{p_0, p_1, \ldots\}$. In a similar vein, **LK**-proofs[6] and their simplifications via cut reduction form a classical category. This category is the *free classical category* generated by the set of atoms $\{p_0, p_1, \ldots\}$ as proved in Führmann and Pym (2006). Thus, classical categories capture the *exact structure and coherency identities* underlying classical deductions. Similar to the intuitionistic case, this freeness provides a syntax-free presentation of classical deductions.

The categorical structures of both Dummett and classical categories are studied in detail in Fuhrmann and Pym (2004), Führmann and Pym (2006, 2007). In addition to mathematical and syntactical examples of these categories, there is also a *combinatorial* construction based on the *classical propositional matrix method* presented in Pym et al. (2014). It is also worth noting that by strengthening classical categories, one can develop a categorical treatment of first-order classical deductions, as explored in Mckinley (2006).

---

[6] Technically, one should use proof nets for classical logic, introduced in Robinson (2003), which provide a less bureaucratic alternative to sequent-style proofs.

7 Categories as Semantics

In the previous sections, we introduced BCC's (resp. ABC's and CCC's) as the proof systems for the language $\mathcal{L}_p$ (resp. $\mathcal{L}_p - \{\bot\}$ and $\mathcal{L}_p - \{\bot, \vee\}$).[7] One can visualize the class of all BCC's and the BC-functors between them as the *models* for the abstract proofs. The free BCC, i.e., the category **NJ**, is living high above and captures the pure notion of proof, where no unnecessary structure is present. Then, below **NJ**, we have other BCC's, the shadows of the free BCC. As much as we go down in the picture, the BCC's become more concrete and hence easier to work with. Finally, on the ground level, we have the Heyting algebras, where all the deductions are collapsed to one map encoding the mere *deductibility*. One can use this picture as some sort of semantics for the deductions, where the free BCC plays the role of the *syntax*, all the others are the models of that syntax and the BC-functors from **NJ** into a BCC C are the *interpretations* of the syntax inside the model C.

To see how such a semantical setting works, first, let us recall the three problems we explained in Sect. 1:

- *The Existence Problem*: given two propositions A and B, if there is a map from A to B in **NJ**.
- *The Equivalence Problem*: given two maps $f, g : A \to B$ in **NJ**, if they are equal in **NJ**. Note that the problem is not just the pure equality of two syntactical entities as the equality of the derivations in **NJ** is considered up to the $\beta\eta$-equivalence.
- *The Identity Problem*: given two propositions A and B, if there is an isomorphism between them in **NJ**.

For any of these problems, if the answer is positive—i.e., if there exists a deduction or the deductions are equal or the propositions are identical—we must provide the evidence in the free BCC. However, for the negative claims, when there is no deduction, the deductions are unequal, or the propositions are non-identical, as any positive claim is inherited from the free category to all its shadows, it is enough to prove the claim for a shadow category, where hopefully its concrete nature makes the computation easier. Here, we use some examples to show how BCC's can be used to prove such negative claims.

Example 151 (*The Existence Problem*) We want to show that the proposition $p \vee \neg p$ is not provable, i.e., there is no map in $\mathrm{Hom}_{\mathbf{NJ}}(\top, p \vee \neg p)$. For the sake of contradiction, assume $\mathrm{Hom}_{\mathbf{NJ}}(\top, p \vee \neg p) \neq \varnothing$. Then, for any BCC C and any BC-functor $F : \mathbf{NJ} \to C$, there must be a map in $\mathrm{Hom}_C(1, F(p) + 0^{F(p)})$. To reach a contradiction, it is enough to find suitable BCC and BC-functor $F : \mathbf{NJ} \to C$ such that $\mathrm{Hom}_C(1, F(p) + 0^{F(p)})$ is empty. We provide two instances of such a suitable choice. First, as we only care about the existence of maps, one may aim for the

[7] In the following discussion, we will only focus on the full language and the BCC's. However, everything we say also holds for the two fragments.

lowest possible level of the BCC's, i.e., the Heyting algebras. For instance, take C as the Heyting algebra $\mathcal{O}(\mathbb{R})$ and set $f(p) = (0, \infty)$. Then, by the freeness of **NJ**, there is a BC-functor $F : \mathbf{NJ} \to \mathcal{O}(\mathbb{R})$ such that $F(p) = f(p)$. Now, note that $0^{F(p)} = (-\infty, 0)$ and hence $F(p) + 0^{F(p)} = \mathbb{R} - \{0\}$. Therefore, there is no map in $\mathrm{Hom}_C(1, F(p) + 0^{F(p)})$ as $\mathbb{R} \nsubseteq \mathbb{R} - \{0\}$.

As another instance and to have a more BHK-style argument, let $C = \mathbf{Set}^{\mathbb{Z}}$ and set $f(p) = (\{0, 1\}, \sigma)$, where $\sigma(0) = 1$ and $\sigma(1) = 0$. Again, f gives rise to a BC-functor $F : \mathbf{NJ} \to \mathbf{Set}^{\mathbb{Z}}$. Note that $0^{F(p)}$ is empty, as there is no function from $\{0, 1\}$ to $\varnothing$. Now, if there is a map in $\mathrm{Hom}(1, F(p) + 0^{F(p)})$, as $1 = (\{*\}, id_{\{*\}})$, there is an invariant element in $F(p) + 0^{F(p)}$. As $0^{F(p)}$ is empty, there must be an invariant element in $\{0, 1\}$ which is impossible.

Example 152 (*The Equivalence Problem*) We prove that the following two derivations:

$$\mathsf{D}_1 : \frac{p \wedge p}{p} \wedge E_1 \qquad \mathsf{D}_2 : \frac{p \wedge p}{p} \wedge E_2$$

are not equal in **NJ**. Assume $\mathsf{D}_1 = \mathsf{D}_2$. Then, for any BCC C and any BC-functor $F : \mathbf{NJ} \to C$, the maps $F(\mathsf{D}_1)$ and $F(\mathsf{D}_2)$ must be equal. Therefore, to prove the inequality, it is again enough to find suitable BCC and BC-functor $F : \mathbf{NJ} \to C$ such that $F(\mathsf{D}_1) \neq F(\mathsf{D}_2)$ in C. Notice that for this problem, we cannot use the preordered sets, as any two morphisms there are equal. Set $C = \mathbf{Set}$ and $f(p) = \{0, 1\}$. Then, the resulting BC-functor $F : \mathbf{NJ} \to \mathbf{Set}$ maps D_1 and D_2 to the projections $p_0 : \{0, 1\} \times \{0, 1\} \to \{0, 1\}$ and $p_1 : \{0, 1\} \times \{0, 1\} \to \{0, 1\}$, respectively. These two functions are not equal in **Set**. Therefore, D_1 and D_2 cannot be equal in **NJ**.

Example 153 (*The Identity Problem*) We show that the propositions p and $p \wedge p$ are not identical in **NJ**. Assume $p \wedge p \cong p$. Then, for any BCC C and any BC-functor $F : \mathbf{NJ} \to C$, the objects $F(p)$ and $F(p) \times F(p)$ must be isomorphic in C. Therefore, to prove the claim, it is again enough to find suitable BCC and BC-functor $F : \mathbf{NJ} \to C$ such that $F(p) \ncong F(p) \times F(p)$. Again, we cannot use preordered sets here, as for them, the logical equivalence and the propositional identity are the same and p and $p \wedge p$ are logically equivalent. Set $C = \mathbf{Set}$ and $f(p) = \{0, 1\}$. Then, the resulting BC-functor $F : \mathbf{NJ} \to \mathbf{Set}$ maps $p \wedge p$ and p to $\{0, 1\} \times \{0, 1\}$ and $\{0, 1\}$, respectively and $\{0, 1\} \times \{0, 1\}$ and $\{0, 1\}$ are clearly not isomorphic in **Set**.

These examples demonstrate how BCC's are useful in providing counter-models for negative claims. However, even for positive claims, sometimes working with the shadow of the syntax can be helpful, as it might be easier to handle compared to the pure syntax. To be more precise, for any of the three mentioned problems, it is clear that if the positive claim (the existence of deductions, the equivalence of derivations, or the existence of isomorphisms) holds for all BCC's, it also holds for **NJ**, simply because **NJ** is itself a BCC. This completeness-style result is not as informative as one might expect. To make it more informative, one may want to restrict the family of all BCC's to a complete subfamily such that working within that family is considerably easier than dealing with the raw syntax. This is the task of the following subsections, which will address the three problems discussed above.

7.1 The Existence Problem

For the existence problem, let us first define the suitable notion of completeness for a family of BCC's.

Definition 154 A family $\mathfrak{F}$ of BCC's is called *existence-complete* for **NJ** if for any two propositions A and B satisfying $\mathrm{Hom}_{\mathbf{NJ}}(A, B) = \varnothing$, there exists $C \in \mathfrak{F}$ and a BC-functor $F : \mathbf{NJ} \to C$ such that $\mathrm{Hom}_C(F(A), F(B)) = \varnothing$. A BCC is called *existence-complete* for **NJ**, if it is existence-complete as a singleton family. For **NM** and **NN**, a similar definition is in place, replacing the BC-structure by AB- or CC-structures, respectively.

Philosophical Comment 155 Reading any BCC C a proof system and any BC-functor $F : \mathbf{NJ} \to C$ as an interpretation of the pure deductions by the deductions living in C, the existence-completeness simply states that if B is not intuitionistically deductible from A, then there is an interpretation F into a BCC in $\mathfrak{F}$, where $F(A)$ does not prove $F(B)$.

Remark 156 There is a stronger version of the existence-completeness of C for **NJ** that simply demands the existence of a weakly-full BC-functor $F : \mathbf{NJ} \to C$. This is stronger as it is the uniform version of the existence-completeness, where one BC-functor F works for all the formulas A and B.

As the existence problem is just about the existence of morphisms, it is reasonable to think that C and Po(C) have the same behavior in this respect. The next corollary provides a proof for this intuition.

Lemma 157 *(i) The existence of a weakly-full* BC*-functor* $F : \mathbf{NJ} \to C$ *is equivalent to the existence of a weakly-full* BC*-functor* $F : \mathbf{NJ} \to \mathrm{Po}(C)$.
(ii) The existence-completeness of a family $\mathfrak{F}$ *of* BCC*'s for* **NJ** *is equivalent to that of the family* $\{\mathrm{Po}(C) \mid C \in \mathfrak{F}\}$.
A similar claim also holds for **NM** *and* **NN**.

Proof For (i), recall from Examples 37 and 105 that the poset reflection of a BCC is a Heyting algebra and the canonical forgetful map $\pi : C \to \mathrm{Po}(C)$ is a full and hence weakly-full BC-functor. As the composition of weakly-full functors is weakly-full, one direction of (i) is clear. For the other direction, let $G : \mathbf{NJ} \to \mathrm{Po}(C)$ be a weakly-full BC-functor. Then, as π is surjective on the objects, it is possible to find the objects $h(p_i)$ in C such that $\pi h(p_i) = G(p_i)$. Then, by the freeness of **NJ**, Theorem 111, there is a BC-functor $H : \mathbf{NJ} \to C$ such that $H(p_i) = h(p_i)$ and hence $\pi H(p_i) = G(p_i)$. We claim that $\pi H \cong G$:

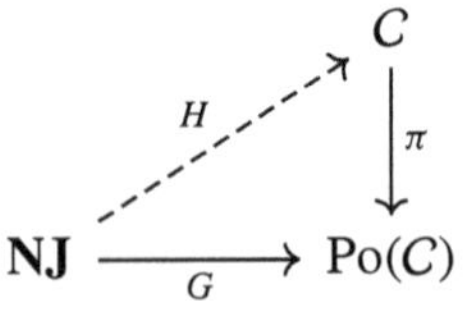

The reason simply is that $\pi H(p_i) = G(p_i)$, for any $i \geq 0$ and hence by the universality of the freeness of **NJ**, the two functors πH and G must be isomorphic. Using this fact, it is easy to show that H is weakly-full, because if $\mathrm{Hom}_{\mathbf{NJ}}(A, B) = \varnothing$, then $\mathrm{Hom}_{\mathrm{Po}(C)}(G(A), G(B)) = \varnothing$, as G is weakly-full. As $G(A) \cong \pi H(A)$ and $G(B) \cong \pi H(B)$, we have $\mathrm{Hom}_{\mathrm{Po}(C)}(\pi H(A), \pi H(B)) = \varnothing$. Therefore, $\mathrm{Hom}_C(H(A), H(B)) = \varnothing$ which completes the proof of weakly-fullness of H. The proof for (ii) is similar. □

The immediate consequence of Lemma 157 is the existence-completeness of one Heyting algebra for **NJ**:

Theorem 158 (Algebraic Completeness) *The Heyting algebra* Po(**NJ**) *and hence the family of all Heyting algebras are existence-complete for* **NJ**.

Proof As the identity functor $id : \mathbf{NJ} \to \mathbf{NJ}$ is weakly-full, by Lemma 157, we can prove the claim. □

To push the completeness further, one may want to restrict the family of Heyting algebra to Heyting algebras of some specific form. The first immediate restriction is to Heyting algebras in the form $\mathbf{2}^{(P, \leq)}$, where $(P, \leq)$ is a poset. To prove, it is enough to use Theorem 112 to embed the Heyting algebra $\mathcal{H}$ into $\mathbf{2}^{[\mathcal{H}, \mathbf{2}]}$, where $[\mathcal{H}, \mathbf{2}]$ is the poset of all BC-functors from $\mathcal{H}$ to $\mathbf{2}$:

Theorem 159 (Kripke Completeness)

(i) *There is a poset $(P, \leq)$ (i.e.,* [Po(**NJ**), **2**]*) and a weakly-full* BC-*functor* $\mathbf{NJ} \to \mathbf{2}^{(P, \leq)}$. *Therefore, the poset $\mathbf{2}^{(P, \leq)}$ and consequently the family of all Alexandrov spaces is existence-complete for* **NJ**.

(ii) *Let $FinTree$ be the family of all Heyting algebras in the form $\mathbf{2}^{(T, \leq)}$, where $(T, \leq)$ is a finite rooted tree. Then, $FinTree$ is existence-complete for* **NJ**.

Proof The proof of (i) is explained above. For (ii), a technique called *filtration* is required to show that for given formulas A and B, finite posets and then finite rooted trees are sufficient for the existence-completeness. We will not explain the details here. The reader must consult Chagrov and Zakharyaschev (1997). □

Using finite rooted trees, one can strengthen the existence-completeness even further. Recall that a space X is called *metrizble* if there is a metric on X inducing its topology. It is called *dense-in-itself* if $Cl(X - \{x\}) = X$, for any $x \in X$. The spaces $\mathbb{Q}$, $\mathbb{R}$ and $\{0, 1\}^{\mathbb{N}}$, where $\{0, 1\}$ is the discrete space with two elements are all dense-in-themselves.

Theorem 160 (McKinsey-Tarski (1944), Rasiowa and Sikorski (1963)) *Let X be a dense-in-itself metrizable space. Then, $O(X)$ is existence-complete for* **NJ**.

Proof To prove, it is enough to show that for any finite rooted tree $(T, \leq)$, there is a weakly-full BC-functor from $\mathbf{2}^{(T,\leq)}$ into $O(X)$. For that purpose and using Example 103, it is enough to provide an open surjection $f : X \to T$, where T is equipped with the upset topology. Note that the surjectivitiy of f implies that $f^{-1} : \mathbf{2}^{(T,\leq)} \to O(X)$ is weakly-full. This is not an easy task and one can find its proof in Bezhanishvili et al. (2018). □

Corollary 161 *Each of the Heyting algebras $O(\mathbb{Q})$, $O(\mathbb{R})$ and $O(\{0, 1\}^{\mathbb{N}})$ are existence-complete for* **NJ**.

Of course, any existence-complete family for **NJ** can also be used as an existence-complete family for **NM** and **NN**. The reason is the well-known conservativity of **NJ** over **NM** and **NN**. However, there are also some interesting existence-completeness results specific to the fragments. We present one of these results as we will need it in the next subsection.

Theorem 162 (Harnik and Makkai (1992)) *There is a weakly-full* AB*-functor from* **NM** *to $\mathbf{2}^{(\mathbb{N},|)}$, where $|$ is the divisibility order on $\mathbb{N}$. Therefore, $\mathbf{2}^{(\mathbb{N},|)}$ is existence-complete for* **NM**.

Proof The first part of the proof is similar to what we explained for Heyting algebras. First, notice that Po(**NM**) is an almost Heyting algebra and it is enough to provide a weakly-full AB-functor from Po(**NM**) to $\mathbf{2}^{(\mathbb{N},|)}$. For that purpose, we first embed Po(**NM**) into $\mathbf{2}^{(P,\leq)}$, where $(P, \leq)$ is the poset [Po(**NM**), **2**] of all AB-functors ordered by the pointwise order. One crucial point here is that the poset $(P, \leq)$ has a terminal object, which simply is the constant map 1. Notice that this map is an AB-functor but not a BC-functor, as it does not preserve the initial element 0. Having this terminal element is the main reason to deviate from **NJ** and Heyting algebras. Then, there is a technical step changing $(P, \leq)$ into a countable poset again with a terminal element Harnik and Makkai (1992). Finally, to provide a weakly-full AB-functor from $\mathbf{2}^{(P,\leq)}$ to $\mathbf{2}^{(\mathbb{N},|)}$, using Example 104, it is enough to provide a surjective p-morphism from $(\mathbb{N}, |)$ to $(P, \leq)$. This is the main part of the proof and can be found in Harnik and Makkai (1992). □

Note that the poset $\mathbf{2}^{(\mathbb{N},|)}$ or even more generally, the poset $\mathbf{2}^{(Q,\leq)}$, where $(Q, \leq)$ is a poset with the top element t is not existence-complete for **NJ**. The reason is that the formula $\neg p \vee \neg\neg p$ is not provable in intuitionistic logic, i.e., there is no map in $\mathrm{Hom}_{\mathbf{NJ}}(\top, \neg p \vee \neg\neg p)$, while for any order-preserving map $f : (Q, \leq) \to \mathbf{2}$ in $\mathbf{2}^{(Q,\leq)}$, we have $\neg f \vee \neg\neg f = 1$. The reason for the latter is that either $\neg f = 0$ or $\neg f = 1$. To prove, assume $(\neg f)(q) = 1$ and $(\neg f)(r) = 0$, for some $q, r \in Q$. Then, as $q \leq t$, we have $(\neg f)(t) = 1$. However, $(\neg f)(r) = 0$ means that there is $s \geq r$ such that $f(s) = 1$. Hence, $f(t) = 1$ which is impossible as $(\neg f)(t) = 1$.

7.1.1 BHK Existence-Completeness

One interesting family whose existence-completeness is worth investigating is the family of categories of the form $\mathbf{Set}^C$. The reason for this interest is their role in formalizing the BHK interpretation. For that purpose, and thanks to Lemma 157, it is enough to compute the poset reflection of $\mathbf{Set}^C$ and check if a family of these posets is existence-complete or not. To start, it is natural to consider the canonical choice among all categories of variable sets, i.e., the category of sets itself. However, this category is not existence-complete for **NJ** as it captures the whole classical logic.

Theorem 163 $\mathrm{Po}(\mathbf{Set}) \cong \mathbf{2}$. *Therefore, for any two propositions A and B,* $\mathsf{CPC} \vdash A \to B$ *iff* $\mathrm{Hom}_{\mathbf{Set}}(F(A), F(B)) \neq \varnothing$, *for any weakly-full* BC*-functor* $F : \mathbf{NJ} \to \mathbf{Set}$.

Proof First, note that in the preorder reflection of **Set**, all non-empty sets are equivalent to each other as there is a function between any two non-empty sets. As the empty set has a function into any set and there is no function from a non-empty set into the empty set, the poset reflection will be the poset **2**, where 0 represents the class of the empty set and 1 is the class of all non-empty sets. The second part is an easy consequence of the first part. □

The first non-trivial family to investigate is the family of variable sets over trees.

Theorem 164 *For any tree $(T, \leq)$, we have* $\mathrm{Po}(\mathbf{Set}^{(T,\leq)}) \cong \mathbf{2}^{(T,\leq)}$. *Therefore, the family of* BCC*'s in the form* $\mathbf{Set}^{(T,\leq)}$, *where $(T, \leq)$ is a finite rooted tree is existence-complete for* **NJ**.

Proof First, for any object E of $\mathbf{Set}^{(T,\leq)}$, define $C(E) = \{u \in T \mid E(u) \neq \varnothing\}$ as a subset of T. It is easy to see that $C(E)$ is actually an upset and hence lives in $\mathbf{2}^{(T,\leq)}$. Moreover, it is clear that if there is a map $\alpha : E \to F$, then $C(E) \subseteq C(F)$. Therefore, C induces an order-preserving map from $\mathrm{Po}(\mathbf{Set}^{(T,\leq)})$ to $\mathbf{2}^{(T,\leq)}$. We claim that this map is an isomorphism. First, it is clear that it is surjective, as for any upset U of $(T, \leq)$, it is enough to define $E_U(u) = \{*\}$ if $u \in U$ and otherwise $E_U(u) = \varnothing$ and define $E(i_{uv})$ as the only possible map from $E_U(u)$ to $E_U(v)$, where $i_{uv} : u \to v$ is the unique map in the poset $(T, \leq)$ from u to v. This map exists, as if $u \leq v$, either $u \notin U$ which implies $E_U(u) = \varnothing$ or $u \in U$ which implies $v \in U$ and hence $E_U(u) = E_U(v) = \{*\}$. It is also easy to check that $E_U : (T, \leq) \to \mathbf{Set}$ is a functor and $C(E) = U$.

Secondly, $C : \mathrm{Po}(\mathbf{Set}^{(T,\leq)}) \to \mathbf{2}^{(T,\leq)}$ is an order-embedding. To prove, assume that $C([E]) \subseteq C([F])$. We must construct a map $\alpha : E \to F$. As $C([E]) \subseteq C([F])$, either $F(u)$ is non-empty or $E(u) = F(u) = \varnothing$, for any $u \in T$. Define $M \subseteq T$ as the set of all minimal elements in T such that $F(u) \neq \varnothing$. For any $m \in M$, choose one $y_m \in F(m)$ in an arbitrary fashion. We can extend the assignment y to all u's for which $F(u) \neq \varnothing$, because for such u, there is a unique $m_u \in M$ such that $m_u \leq u$. The reason simply is that the segment below u is well-ordered and hence has the least element in M. Define y_u as $F(i_{m_u u})(y_{m_u})$. It is easy to see that if $u \leq v$ and $F(u)$ and

$F(v)$ are non-empty, then $m_u = m_v$ and $F(i_{uv})(y_u) = y_v$. Now, define $\alpha : E \to F$ in the following way. For any $u \in T$, if $F(u) = \varnothing$, then $E(u) = \varnothing$, and hence we can define $\alpha_u : E(u) \to F(u)$ as the unique function over the empty set. If $F(u) \neq \varnothing$, then define α_u as the constant function, mapping everything to y_u. Checking that $\alpha : E \to F$ is a natural transformation is straightforward. □

The next simple case whose poset reflection is easy to compute is the category $\mathbf{Set}^G$, where G is a group.

Theorem 165 *Let G be a group and $Sub(G)$ be the preordered set of its subgroups with the ordering $K \leq H$ defined by the existence of a $g \in G$ such that $g^{-1}Kg \subseteq H$. Then,* $\mathrm{Po}(\mathbf{Set}^G) = \mathbf{2}^{Sub(G)^{op}}$.

Proof To any object E in $\mathbf{Set}^G$, assign a subset of $Sub(G)$ by $\delta(E) = \{H \leq G \mid \exists x \in E(*)\ \forall h \in H, E(h)(x) = x\}$. Notice that $\delta(E)$ is downward-closed, because if $g^{-1}Kg \subseteq H$ and $H \in \delta(E)$, then there exists $x \in E(*)$ such that $E(h)(x) = x$, for any $h \in H$. Therefore, $E(g^{-1}kg)(x) = x$, for any $k \in K$ which implies that $E(k)(E(g)(x)) = E(g)(x)$. Hence, $K \in \delta(E)$. Now, we show that if there is a natural transformation $\alpha : E \to F$, then $\delta(E) \subseteq \delta(F)$. Assume that $H \in \delta(E)$. Hence, there exists $x \in E(*)$ such that $E(h)(x) = x$, for any $h \in H$. As α is a natural transformation, we have $F(h)(\alpha_*(x)) = \alpha_*(E(h)(x)) = \alpha_*(x)$. Therefore, $H \in \delta(F)$ and hence $\delta(E) \subseteq \delta(F)$. Using this observation, it is clear that δ induces a map from $\mathrm{Po}(\mathbf{Set}^G)$ to $\mathbf{2}^{Sub(G)^{op}}$. We claim that this map is an isomorphism.

To prove that it is an order-embedding, assume that $\delta(E) \subseteq \delta(F)$. We must construct a natural transformation $\alpha : E \to F$. Define the relation $\sim$ on the set $E(*)$ by $x \sim y$, if there is $g \in G$ such that $E(g)(x) = y$. It is easy to use the fact that G is a group and E is a functor to prove that the relation $\sim$ is an equivalence relation. Take the partition of $E(*)$ into the equivalence classes of $\sim$. To define $\alpha_* : E(*) \to F(*)$, we define α_* on any equivalence class, separately. Take a class O and pick an arbitrary element $x_O \in O$. Define $S_{x_O} = \{g \in G \mid E(g)(x_O) = x_O\}$. It is clear that $S_{x_O} \in \delta(E)$ and hence in $\delta(F)$. Therefore, there exists $y \in F(*)$ such that for any $g \in S_{x_O}$, we have $F(g)(y) = y$. This means that if $E(g)(x_O) = x_O$ then $F(g)(y) = y$, for any $g \in G$. Define $\alpha_*(x_O) = y$ and for any $z \in O$, if $z = E(g)(x_O)$, for some $g \in G$, define $\alpha_*(z) = F(g)(y)$. Note that the definition of $\alpha_*(z)$ is independent of the choice of g, because if $E(g)(x_O) = E(h)(x_O) = z$, then $E(g^{-1}h)(x_O) = x_O$ and hence $F(g^{-1}h)(y) = y$ which implies $F(g)(y) = F(h)(y)$. This completes the definition of α_* on O, but as O is an arbitrary class, we have a function $\alpha_* : E(*) \to F(*)$. To show that $\alpha : E \to F$ is a natural transformation, one must prove $F(g)(\alpha_*(z)) = \alpha_*(E(g)(z))$, for any $g \in G$ and any $z \in E(*)$. Assume that $z \in O$, for some class O and $z = E(h)(x_O)$. Notice that $E(g)(z) = E(g)E(h)(x_O) = E(gh)(x_O)$. Then, by definition, $\alpha_*(z) = F(h)(x_O)$. Hence, $F(g)(\alpha_*(z)) = F(g)(F(h)(x_O)) = F(gh)(x_O)$. The last is equal to $\alpha_*(E(g)(z))$, because $E(g)(z) = E(gh)(x_O)$ and $E(g)(z)$ belongs to the class O.

Finally, we have to show that δ is surjective. For any downward-closed set $I \subseteq Sub(G)$, define $E_I(*) = \Sigma_{H \in I} G/H$ and $E(g)(H, [r]_H) = (H, [gr]_H)$. Then, it is easy to see that $E_I : G \to \mathbf{Set}$ is actually a functor. Moreover, we have $\delta(E_I) = I$.

The reason is that for any subgroup K, we have $K \in \delta(E_I)$ iff there is $H \in I$ and $r \in G$ such that for any $k \in K$, we have $[r]_H = [kr]_H$. This is simply equivalent to the existence of $H \in I$ such that $K \leq H$. As I is downward closed, the last is equivalent to $K \in I$. □

Remark 166 Notice that if the group G is abelian, then the preorder $\leq$ in Theorem 165 becomes the usual inclusion.

Corollary 167 (Harnik and Makkai (1992)) $\text{Po}(\mathbf{Set}^{\mathbb{Z}}) \cong \mathbf{2}^{(\mathbb{N},|)}$, *where* $|$ *is the divisibility relation.*

Proof For $G = (\mathbb{Z}, +)$, the poset $Sub(G)$ is just $(\mathbb{N}, |)^{op}$, as the subgroups of G are of the form $n\mathbb{Z}$ and $m\mathbb{Z} \subseteq n\mathbb{Z}$ iff $n \mid m$. □

Now, having $\text{Po}(\mathbf{Set}^G)$ computed, we can combine it with Theorem 162 to prove some existence-completeness results for **NM**. For that purpose, we must connect $Sub(G)^{op}$ to $(\mathbb{N}, |)$. Let G be a group. An element $x \in G$ is called of an *infinite non-commutative order* if x^m is not in the subgroup generated by $\{g^{-1}x^n g \mid g \in G\}$, for any $n > m \geq 0$. If G is abelian, then x is of infinite non-commutative order iff $x^n \neq e$, for any $n > 0$.

Lemma 168 *If G has an element of an infinite non-commutative order, then there exists a surjective p-morphism* $f : Sub(G)^{op} \to (\mathbb{N}, |)$.

Proof Let x be an element of an infinite non-commutative order. For any subgroup H, if there is a positive number n such that $\{g^{-1}x^n g \mid g \in G\} \subseteq H$, define $f(H)$ as the least of such n's and if there is no n with that property, define $f(H) = 0$. It is easy to prove that for any such n, we have $f(H) \mid n$. We first show that f is order-preserving. Let $H \leq K$. If $f(H) = 0$, we have $f(K) \mid f(H)$. Therefore, w.l.o.g., we can assume that $f(H) \neq 0$. As $H \leq K$, there is some $g \in G$ such that $g^{-1}Hg \subseteq K$. We claim that $r^{-1}x^{f(H)}r \in K$, for any $r \in G$. To prove, set $s = rg^{-1}$. Then, $r^{-1}x^{f(H)}r = g^{-1}s^{-1}x^{f(H)}sg$. By definition of $f(H)$, we have $s^{-1}x^{f(H)}s \in H$. Therefore, $r^{-1}x^{f(H)}r \in K$. As $r \in G$ is arbitrary, we have $f(K) \mid f(H)$. To prove that f is surjective, for any $n \geq 0$, define H_n as the subgroup generated by $\{g^{-1}x^n g | g \in G\}$. Then, as x is of an infinite non-commutative order, then $f(H_n) = n$. Finally, to prove that f is a p-morphism, assume that $f(H) \mid m$, for some number $m \in \mathbb{N}$. Consider H_m. As $f(H_m) = m$, it is enough to prove that $H_m \leq H$. For any $g \in G$, as $g^{-1}x^{f(H)}g \in H$ and $f(H) \mid m$, we also have $g^{-1}x^m g \in H$, for any $g \in G$. Therefore, $H_m \subseteq H$ and hence $H_m \leq H$. □

Theorem 169 *Let G be a group that has an element of an infinite non-commutative order. Then, there exists a weakly-full* AB*-functor* $F : \mathbf{NM} \to \mathbf{Set}^G$. *Therefore,* $\mathbf{Set}^G$ *is existence-complete for* **NM**.

Proof By Lemma 157, it is enough to prove the existence of a weakly-full AB-functor from **NM** to $\text{Po}(\mathbf{Set}^G)$. By Theorem 165, we know that $\text{Po}(\mathbf{Set}^G)$ is $\mathbf{2}^{Sub(G)^{op}}$ and by Lemma 168, there is a surjective p-morphism from $Sub(G)^{op}$ to $(\mathbb{N}, |)$ which implies the existence of a weakly-full AB-functor from $\mathbf{2}^{(\mathbb{N},|)}$ to $\mathbf{2}^{Sub(G)^{op}}$. Therefore, it is just enough to find a weakly-full AB-functor from **NM** to $\mathbf{2}^{(\mathbb{N},|)}$ which is provided by Theorem 162. □

Remark 170 One may wonder if we can extend Theorem 169 to cover the whole **NJ**. The answer is negative. As we observed before, a poset in the form $\mathbf{2}^{(Q,\leq)}$ where $(Q, \leq)$ has a top element cannot be existence-complete for **NJ**. As $\mathrm{Po}(\mathbf{Set}^G)$ is $\mathbf{2}^{Sub(G)^{op}}$ and $Sub(G)^{op}$ has the trivial subgroup $\{e\}$ as its top element, we cannot expect $\mathbf{Set}^G$ to be existence-complete for **NJ**.

Corollary 171 (Läuchli realizability for MPC Läuchli (1968), Harnik and Makkai (1992)) *There exists a weakly-full* AB-*functor* $F : \mathbf{NM} \to \mathbf{Set}^{\mathbb{Z}}$. *Therefore,* $\mathbf{Set}^{\mathbb{Z}}$ *is existence-complete for* **NM**.

Remark 172 Läuchli realizability Läuchli (1968) was the first BHK style complete semantics for minimal logic. It is also possible to extend it to intuitionistic logic using a trick to interpret $\bot$ Harnik and Makkai (1992). The modern categorical version we present here is developed and elegantly presented in Harnik and Makkai (1992).

So far, we have computed the poset reflection of some categories of the form $\mathbf{Set}^C$. Let us conclude this subsection by a remark that such computation is not always an easy task. For instance, there are small categories C such that the poset reflection of $\mathbf{Set}^C$ is not even a set. To see such a situation, let X and Y be two sets and $f, g : X \to Y$ be two functions:

$$X \underset{g}{\overset{f}{\rightrightarrows}} Y$$

Then, define the category $C_{f,g:X\to Y}$ in the following manner. For the objects, use the elements in the disjoint union $X + Y = \{(0, x) \mid x \in X\} \cup \{(1, y) \mid y \in Y\}$ of X and Y and for the maps, define $\mathrm{Mor}(C_{f,g:X\to Y})$ as the union of the set of identity maps and the set

$$\{f_x : (0, x) \to (1, y) \mid f(x) = y\} \cup \{g_x : (0, x) \to (1, y) \mid g(x) = y\}$$

where for any pair (x, y) if $f(x) = y$ (resp. $g(x) = y$), there is a map $f_x : (0, x) \to (1, y)$ (resp. $g_x : (0, x) \to (1, y)$). As no non-trivial composition is possible with our maps, we define the composition in a trivial way. In Lawvere (1996), Lawvere claimed that there are sets X and Y and maps $f, g : X \to Y$ such that $\mathrm{Po}(\mathbf{Set}^{C_{f,g:X\to Y}})$ is not even a set, see Menni (2000).

7.2 *The Equivalence Problem*

For the equivalence problem, we only focus on one BCC rather than a family of BCC's. Here is the formal definition of the completeness of a BCC.

Definition 173 A BCC C is called *equivalence-complete* for **NJ**, when for any two maps $f, g : A \to B$ in **NJ**, if $F(f) = F(g)$, for any BC-functor $F : \mathbf{NJ} \to C$, then $f = g$. For **NM** and **NN**, a similar definition is in place, replacing the BC-structure by the AB- or CC-structures, respectively.

Remark 174 There is also a stronger version of the equivalence-completeness of C for **NJ** that simply demands the existence of a faithful BC-functor $F : \mathbf{NJ} \to C$. In fact, it is a uniform version of the equivalence-completeness, as we have just one BC-functor working for all formulas A and B.

7.2.1 Equivalence-Completeness for NN

Historically, the equivalence problem started with and still focused on the **NN** case and the CC-structure. The reason is the close connection this setting has with the simply typed lambda calculus as a very basic programming language. The first two important equivalence-completeness results are the following:

Theorem 175 (Čubrić (1998)) *Any* CC*-functor* $F : \mathbf{NJ} \to \mathbf{Set}$ *that maps the atoms of* $\mathcal{L}_p$ *to infinite sets is faithful. Therefore,* **Set** *is equivalence-complete for* **NN**.

Theorem 176 (Solovev (1981)) **FinSet** *is equivalence-complete for* **NN**.

The equivalence-completeness of **FinSet** is crucial as it helps to prove the decidability of the equivalence problem for **NN**.

Corollary 177 (Solovev (1981)) *The equivalence problem for* **NN** *is decidable.*

Proof It is enough to prove that the equivalence problem and its complement are both recursively enumerable. It is recursively enumerable as if two proofs are equal, *there exists* a sequence of basic $\beta\eta$-equalities transforming one to the other. For the complement, if two proofs are not equal, then by Theorem 176, *there is* an assignment of finite sets to the atoms in the proofs such that the resulting functions are not equal. □

It is worth mentioning that the two versions of the equivalence-completeness are not equivalent. For instance, although **FinSet** is equivalence-complete for **NN**, there is no faithful CC-functor $F : \mathbf{NN} \to \mathbf{FinSet}$. As explained in Simpson (1995a), the reason is the existence of infinitely many maps in $\mathrm{Hom}_{\mathbf{NN}}((p \to p) \wedge p, p)$ encoding the natural numbers. To explain, for any natural number $n \in \mathbb{N}$, define the derivation D_n of p from $(p \to p) \wedge p$, inductively by

$$\mathsf{D}_0\colon\ \dfrac{(p \to p) \wedge p}{p}\,\wedge_2 E \qquad \mathsf{D}_{n+1}\colon\ \dfrac{\dfrac{(p \to p) \wedge p}{p \to p}\,\wedge_1 E \qquad \begin{array}{c}(p \to p) \wedge p\\ \mathsf{D}_n\\ p\end{array}}{p}\to E$$

These proofs form an infinite family $\{\mathsf{D}_n\}_{n \in \mathbb{N}}$ of distinct derivations. To see why, it is enough to show that a CC-functor maps them to different maps. Using the freeness of **NN**, define $F : \mathbf{NN} \to \mathbf{Set}$ by mapping p to the set $\mathbb{N}$. Then, F maps D_n to the function $f_n : \mathbb{N}^{\mathbb{N}} \times \mathbb{N} \to \mathbb{N}$, defined by $f_n(\alpha, m) = \alpha^n(m)$, where α^n means iterating α for n many times. Substituting α as the successor function s and setting $m = 0$, it is clear that $f_n(s, 0) = n$, which implies that all f_n's are distinct.

In the rest of this subsubsection, we will present a necessary and sufficient condition for a CCC to be equivalence-complete for **NN** in any of the two mentioned senses. First, for the weaker version, it is clear that if C is a preordered set, then it cannot be complete for **NN**, as there are at least two non-equal projection morphisms in $\mathrm{Hom}_{\mathbf{NJ}}(p \wedge p, p)$, as discussed in Example 152. The good news is that this necessary condition is actually sufficient.

Theorem 178 (Simpson 1995a) *For any* CCC *C, it is equivalence-complete for* **NN** *iff it is not a preordered set.*

Philosophical Comment 179 By Theorem 178, if C is not equivalence-complete for **NN**, then it must be a preordered set. Reading any CCC as a proof system, it simply means that if two non-equivalent derivations in **NN** are considered as equal in C, then the whole notion of deduction in C collapses to the mere deductibility. In other words, it is impossible to add more equalities than what is governed by the $\beta\eta$-equivalences if we want to keep the notion of deductibility non-trivial. In this sense, the deduction equivalence of **NN** is optimal Došen and Petrić (2000).

For the other version of the equivalence-completeness for **NN**, let us first explain a necessary condition Simpson (1995a). A map $f : A \to A$ in a CCC C is called of an *infinite order* if $f^m \neq f^n$, for any $m \neq n$, where f^k means the result of k many compositions of f with itself. We show that the existence of a map $f : A \to A$ in C with an infinite order is a necessary condition, if we want a faithful functor $F : \mathbf{NN} \to C$. To see why, it is enough to find a map in **NN** with an infinite order. For that purpose, consider the following derivation D from $[(p \to p) \wedge p] \to p$ to itself:

$$\dfrac{\dfrac{\dfrac{\dfrac{[(p \to p) \wedge p]_1}{p \to p} \qquad \dfrac{\dfrac{[(p \to p) \wedge p]_1}{p} \qquad \dfrac{[(p \to p) \wedge p]_1}{p \to p}}{p}}{(p \to p) \wedge p} \qquad [(p \to p) \wedge p] \to p}{p}}{[(p \to p) \wedge p] \to p}\to I_1$$

This derivation is of an infinite order in **NN**. To see why, it is enough to find a CC-functor $F : \mathbf{NN} \to \mathbf{Set}$ such that $F(\mathsf{D}^m) \neq F(\mathsf{D}^n)$, for any $m \neq n$. For that purpose, use the freeness of **NN** and find a CC-functor $F : \mathbf{NN} \to \mathbf{Set}$ such that $F(p) = \mathbb{N}$. Then, one can see that $F(\mathsf{D})$ is the function $G : \mathbb{N}^{\mathbb{N}^{\mathbb{N}} \times \mathbb{N}} \to \mathbb{N}^{\mathbb{N}^{\mathbb{N}} \times \mathbb{N}}$ defined by $G(\alpha) = \lambda fk.\alpha(f, f(k))$. Computing G^n, we see that $G^n(\alpha) = \lambda fk.\alpha(f, f^n(k))$ and hence $G^n(p_1) = \lambda fk.f^n(k)$. Therefore, $G^n(p_1)$'s are distinct functions, for different n's which means that $G^n \neq G^m$, if $m \neq n$.

The following theorem shows that this necessary condition is actually sufficient:

Theorem 180 (Simpson 1995a) *For any* CCC C*, there exists a faithful* CC*-functor* $F : \mathbf{NN} \to C$ *iff* C *contains an endomorphism of an infinite order.*

7.2.2 Equivalence-Completeness for NJ

The structure of the proofs in **NJ** is usually more complex than what is present in **NN**. The reason simply is the rather complex behavior of the disjunction elimination rule. However, recently, the **NJ** counterpart of Theorem 176 was also proved:

Theorem 181 (Scherer 2017, Scherer (2016)) **FinSet** *is equivalence-complete for* **NJ**.

Corollary 182 *The equivalence problem for* **NJ** *is decidable.*

Proof The proof is similar to that of Corollary 177. □

Our final task in this subsection is to generalize Theorem 181 from **FinSet** to any non-preorder BCC. The basic idea behind the following line of argument is attributed to Alex Simpson (1995b). First, notice that using the argument we gave for **NN**, it is clear that if a BCC is equivalence-complete for **NJ**, then it cannot be a preordered set. For sufficiency, we need the following lemma:

Lemma 183 *Let* C *be a* BCC*. For any objects* A, B*, and* C*, if* $i_0 : A \to A + B$ *and* $i_1 : B \to A + B$ *are the injections, then:*

(i) i_0 *is monic, i.e., for any maps* $f, g : C \to A$ *if* $i_0 f = i_0 g : C \to A + B$*, then* $f = g$.
(ii) i_1 *is monic, i.e., for any maps* $f, g : C \to B$ *if* $i_1 f = i_1 g : C \to A + B$*, then* $f = g$.
(iii) *If* C *is not a preordered set, then for any maps* $f : 1 \to A$ *and* $g : 1 \to B$*, the maps* $i_0 f, i_1 g : 1 \to A + B$ *are distinct.*

Proof For (i), consider the map $i_0 \times id_A : A \times A \to (A + B) \times A$. First, we intend to define a map $j : (A + B) \times A \to A \times A$ such that $j \circ (i_0 \times id_A) = id_{A \times A}$:

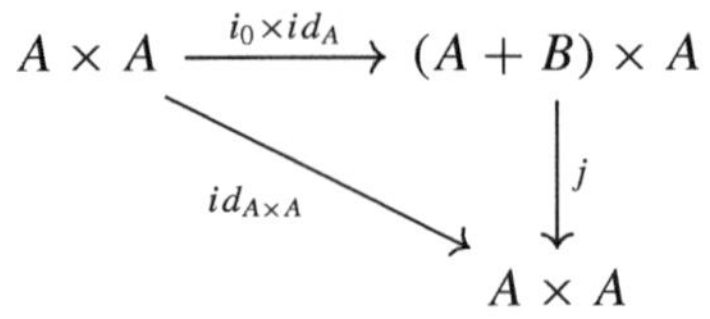

For that purpose, define j' as the following canonical map:

$$j' = (\lambda_A id_{A \times A}, \lambda_A \langle p_1, p_1 \rangle) : A + B \to [A, A \times A]$$

and set $j = ev\langle j' \times id_A \rangle : (A + B) \times A \to A \times A$. It is easy to see that $j \circ (i_0 \times id_A) = id_{A \times A}$. Now, let $f, g : C \to A$ satisfy $i_0 f = i_0 g$. Then,

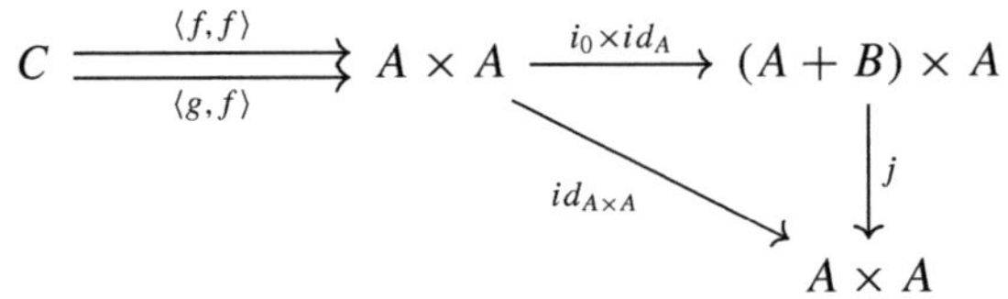

the two maps of the first row are equal. Hence, $\langle f, f \rangle = \langle g, f \rangle$ which implies $f = g$. The proof for (ii) is similar to (i).

For (iii), assume $i_0 f = i_1 g : 1 \to A + B$, for some $f : 1 \to A$ and $g : 1 \to B$. As C is not a preordered set, there are distinct morphisms $k, l : E \to F$ in C. Therefore, the two maps $\lambda_E k, \lambda_E l : 1 \to [E, F]$ are distinct. Now, consider the maps $\lambda_E k! : A \to [E, F]$ and $\lambda_E l! : B \to [E, F]$ and note that $\lambda_E k! f = \lambda_E k$ and $\lambda_E l! g = \lambda_E l$. Using the universal property for the coproduct $A + B$, we have a map $h : A + B \to [E, F]$ such that $h \circ i_0 = \lambda_E f!$ and $h \circ i_1 = \lambda_E l!$:

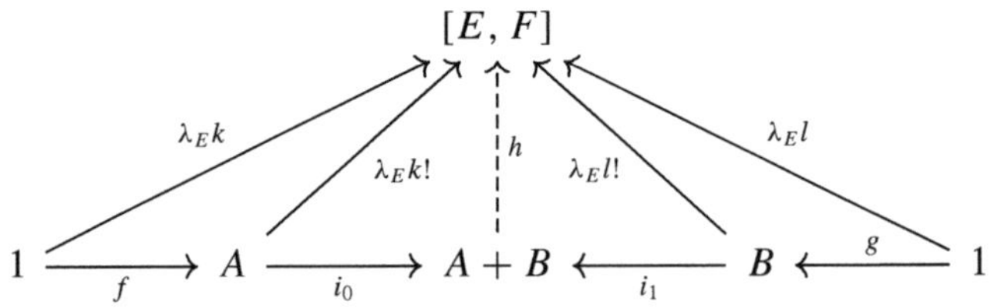

As $i_0 f = i_1 g$, we have $\lambda_E k = h i_0 f = h i_1 g = \lambda_E l$ which is a contradiction. □

Corollary 184 *Let C be a non-preorder* BCC. *Then, the maps $i_j : 1 \to \sum_{j=0}^{n} 1$ are distinct for different j's.*

Proof The proof is by an induction on n. For $n = 1$, by part (iii) of Lemma 183, we have $i_0 \neq i_1$. To prove the claim for $n + 1$, if $i_j = i_k$, for some $j, k \leq n$, we can use part (ii) of Lemma 183 to reduce the claim to the induction hypothesis. If $j \leq n$ and $k = n + 1$, we can use part (iii) of Lemma 183 to reach a contradiction. □

Lemma 185 *Let C be a non-preorder* BCC. *Then, there is a faithful* BC*-functor* $F : \mathbf{FinSet} \to C$.

Proof For any finite set A, define $F(A)$ as $\sum_{a\in A} 1$ and for any function $f : A \to B$ define $F(f) : \sum_{a\in A} 1 \to \sum_{b\in B} 1$ by the canonical map $(i_{f(a)})_{a\in A}$. It is easy to prove that F is a BC-functor. For faithfulness, assume that $f, g : A \to B$ are two maps such that $F(f) = F(g)$. Then, using the definition of $F(f)$ and $F(g)$, we have $i_{f(a)} = i_{g(a)} : 1 \to \sum_{b\in B} 1$, for any $a \in A$. As C is not a preordered set, by Corollary 184, we have $f(a) = g(a)$ and as $a \in A$ is arbitrary, $f = g$. □

Corollary 186 *A* BCC *C is equivalence-complete for* **NJ** *iff it is not a preordered set.*

Proof One direction is explained before. For the other, if $f \neq g : A \to B$ in **NJ**, by Theorem 181, there is a BC-functor $G : \mathbf{NJ} \to \mathbf{FinSet}$ such that $G(f) \neq G(g)$. Combined with the BC-functor $F : \mathbf{FinSet} \to C$ of Lemma 185, we reach the BC-functor $FG : \mathbf{NJ} \to C$ and as F is faithful, we have $FG(f) \neq FG(g)$. This completes the proof of the equivalence-completeness of C for **NJ**. □

Philosophical Comment 187 Similar to Philosophical Note 179, Corollary 186 implies that it is impossible to add more equalities than what is governed by the $\beta\eta$-equivalences to the **NJ**, if we want to keep the deductibility non-trivial. In this sense, the deduction equivalence of **NJ** is optimal.

7.3 The Identity Problem

In this subsection, we will focus on the last problem among the three we introduced in Sect. 1, i.e., the propositional identity problem. To identify all the pairs of isomorphic propositions in a given fragment, we have two approaches to follow. Either we ask syntactically for some basic isomorphisms from which all the other isomorphisms can be derived or we can go for a semantical approach of having a concrete complete category C such that two propositions are isomorphic iff all of their images in C are isomorphic. We will explain both of these approaches later. But, first, we need a diversion into the arithmetical world.

7.3.1 Tarski's High School Algebra Problem

Consider the first-order language $\{\times, +, (-)^{(-)}, 1\}$ for arithmetic and the equations in Table 3. All these equations clearly hold in the *standard model*, i.e., the set of natural numbers equipped with the usual addition, multiplication, and exponentiation functions and the number 1 as its designated constant. In Doner and Tarski (1969), Tarski asked if the equalities in Table 3 axiomatize all equalities in the language $\{\times, +, (-)^{(-)}, 1\}$ that hold over the set of positive natural numbers and conjectured that the answer is positive. For some fragments of the language, we know that this is really the case. For instance, for the fragment $\{1, \times, (-)^{(-)}\}$, the +-free equations in Table 3 axiomatize all the valid equations over the fragment. However, later Wilkie (2000) proved that for the full language, the answer is negative. Consider the following equation called Wilkie's equation:

$$(A^x + B^x)^y \times (C^y + D^y)^x = (A^y + B^y)^x \times (C^x + D^x)^y,$$

where $A = 1 + x$, $B = 1 + x + x^2$, $C = 1 + x^3$ and $D = 1 + x^2 + x^4$. Wilkie's equation is valid in the standard model. To see why, it is enough to note that we have the identity $AD = BC$. Therefore,

$$A^{xy}[(A^x + B^x)^y \times (C^y + D^y)^x] = (A^x + B^x)^y \times ((AC)^y + (AD)^y)^x =$$

$$(A^x + B^x)^y \times ((AC)^y + (BC)^y)^x = C^{xy}[(A^x + B^x)^y \times (A^y + B^y)^x].$$

Similarly, we have

$$A^{xy}[(A^y + B^y)^x \times (C^x + D^x)^y] = (A^y + B^y)^x \times ((AC)^x + (AD)^x)^y =$$

$$(A^y + B^y)^x \times ((AC)^x + (BC)^x)^y = C^{xy}[(A^y + B^y)^x \times (A^x + B^x)^y].$$

Therefore,

$$A^{xy}[(A^x + B^x)^y \times (C^y + D^y)^x] = A^{xy}[(A^y + B^y)^x \times (C^x + D^x)^y].$$

As A is always positive, we reach the equality we wanted to prove by canceling out the term A^{xy}. Notice that in all the steps of this proof except the canceling out

Table 3 High School identities

$1 \times x = x$	$x \times y = y \times x$	$x \times (y \times z) = (x \times y) \times z$
$x^1 = x$	$x^{(y \times z)} = (x^y)^z$	$1^x = 1$
$(x \times y)^z = x^z \times y^z$	$x + y = y + x$	$x + (y + z) = (x + y) + z$
$x \times (y + z) = x \times y + x \times z$	$x^{(y+z)} = x^y \times x^z$	

part, we only used the high school algebra identities in Table 3. Using a complex proof-theoretic argument, Wilkie (2000) showed that this equation is not provable by the axioms in Table 3. Later Gurevič (1985) provided a finite counter-model to reprove this unprovability. He also showed that the theory of all valid equalities is not finitely axiomatizable. More precisely, he considered the equation:

$$(A^x + B_n^x)^{2^x} \times (C_n^{2^x} + D_n^{2^x})^x = (A^{2^x} + B_n^{2^x})^x \times (C_n^x + D_n^x)^{2^x},$$

where $A = 1 + x$, $B_n = \sum_{i=0}^{n-1} x^i$, $C_n = 1 + x^n$ and $D_n = \sum_{i=0}^{n-1} x^{2i}$ and $n \geq 3$ is an odd number. Using the fact that $AD_n = B_n C_n$, one can see that the equation is valid in the standard model. Then, Gurevič showed that for any finite set of axioms, there exists an odd $n \geq 3$ such that the corresponding equation is not provable from the axioms. Although the theory of all valid equations is not finitely axiomatizable, it is known that it is decidable Macintyre (1979), Gurevič (1985).

7.3.2 The Propositional Identity

In this subsubsection, we will address the propositional identity problem for the systems **NN**, **NM**, and **NJ**. For **NN**, using the positive solution for Tarski's high school algebra problem for the $+$-free fragment, one can prove

Theorem 188 *The following are equivalent:*

(i) $A \cong B$ *is provable from the theory* $Th_{\top,\wedge,\to}$ *presented in Table 4.*
(ii) $A \cong B$ *as two objects in* **NN**.
(iii) $F(A) \cong F(B)$, *for any* CC*-functor* $F : \mathbf{NN} \to \mathbf{FinSet}$.

Proof Proving (ii) from (i) is a consequence of the fact that all the isomorphisms in Table 4 hold in any CCC including **NN**. Proving (iii) from (ii) is also easy, as functors respect isomorphisms. To prove (i) from (iii), first rewrite the formulas A and B as terms t_A and t_B in the language $\{\times, (-)^{(-)}, 1\}$ replacing $\top$, $\wedge$ and $\to$ with 1, $\times$ and $(-)^{(-)}$, respectively. Now, as the relation $\cong$ over **FinSet** is just the equality between the cardinals of the finite sets, (iii) implies the validity of $t_A = t_B$ in the standard model. Hence, $t_A = t_B$ is provable from the $+$-free high school algebra identities that are nothing but the axioms of $Th_{\top,\wedge,\to}$ in disguise. □

Table 4 $Th_{\top,\wedge,\to}$

$\top \wedge A \cong A$	$A \wedge B \cong B \wedge A$
$A \wedge (B \wedge C) \cong (A \wedge B) \wedge C$	$\top \to A \cong A$
$(A \wedge B) \to C \cong (A \to (B \to C))$	$A \to \top \cong \top$
$A \to (B \wedge C) \cong (A \to B) \wedge (A \to C)$	

Corollary 189 *The propositional identity problem in* **NN** *is decidable.*

Proof Using Theorem 188, $A \cong B$ holds in **NN** iff it is provable from a finitely axiomatizable theory. Hence, the relation $\cong$ is recursively enumerable. Moreover, the theorem also proves that $A \ncong B$ iff there exists a CC-functor $F : \mathbf{NN} \to \mathbf{FinSet}$ such that $F(A) \ncong F(B)$. As any CC-functor is essentially a numeral assignment to the atoms in A and B and $F(A) \ncong F(B)$ is just an inequality between natural numbers, we see that $\ncong$ is also recursively enumerable. Therefore, $\cong$ is decidable. □

For **NM**, the propositional identities is not finitely axiomatizable as shown in Fiore et al. (2006). Actually, all Gurevič's equations are also propositional identities, and hence, we can use the non-finite-axiomatizability of the valid equations in that language to prove the non-finite-axiomatizability of the propositional identities in **NM**. For **NJ**, in Cosmo et al. (2005), it is proved that the equational theory for the arithmetical language $\{\times, +, (-)^{(-)}, 0, 1\}$ is not finitely-axiomatizable, using again Gurevič's equations. Hence, the propositional identities in **NJ** are not finitely axiomatizable. Moreover, as observed in Fiore et al. (2006), for **NJ**, the equivalence between (ii) and (iii) in Theorem 188 breaks downs. To see why, it is enough to consider the propositions $\neg p \vee \neg\neg p$ and $\top$. Any of their interpretations in **FinSet** are isomorphic as both sides have one element. However, they are not even intuitionistically equivalent. Having that said, for negative propositions:

Theorem 190 (Fiore et al. (2006)) *The following are equivalent:*

(i) $\mathsf{IPC} \vdash \neg A \longleftrightarrow \neg B$.
(ii) $\neg A \cong \neg B$ *as two objects in* **NJ**.
(iii) $F(\neg A) \cong F(\neg B)$, *for any* BC-*functor* $F : \mathbf{NJ} \to \mathbf{FinSet}$.

Corollary 191 *The following are equivalent:*

(i) $A \cong 0$ *as two objects in* **NJ**.
(ii) $F(A) \cong 0$, *for any* BC-*functor* $F : \mathbf{NJ} \to \mathbf{FinSet}$.

To the best of our knowledge, the decidability of the propositional identity problem for **NM** and **NJ** is still open. For a survey on the results and the applications of the propositional identity problem, see (Di Cosmo 2005, 1995).

8 Higher-Order Theories

In the previous sections, we used categorical machinery to formalize a proof system over the *propositional language* as a structured category. While these investigations are intriguing, one must acknowledge that mathematics extends far beyond this restrictive language to form propositions about various *types* of entities, such as natural numbers, real numbers, points, lines, and more. It also addresses higher-order entities constructed from these *basic* entities. For example, we may consider functions on natural numbers, functions on those functions, and so forth. To formalize this richer language, we need to extend the propositional language with *types* over which quantifiers operate and to include *higher types* that encompass functions between these types. In this section, we will discuss these higher-order languages in general and introduce a specific language for discussing natural numbers.

Let $\{T_i\}_{i \in I}$ be a family of primitive symbols. We define the set of all *types* inductively as follows: each T_i is a type, for any $i \in I$, and for any two types σ and τ, the expressions $\sigma \times \tau$ and $\sigma \to \tau$ are also types. The basic types represent the fundamental sets of entities we work with, while the operations $\sigma \times \tau$ and $\sigma \to \tau$ formalize the cartesian product of σ and τ, and the set of all functions from σ to τ, respectively.

Next, we need to introduce *terms* as names for entities within the types. First, consider a family of function symbols $\{f_j^\tau(x_1^{\sigma_1}, x_2^{\sigma_2}, \ldots, x_{n_j}^{\sigma_{n_j}})\}_{j \in J}$, where $\sigma_1, \sigma_2, \ldots, \sigma_{n_j}$, and τ are types. The function symbol f_j^τ is intended to formalize a basic function of arity n_j, with inputs of types $\sigma_1, \sigma_2, \ldots, \sigma_{n_j}$ and output of type τ. Now consider:

- Infinitely many variables in the form x^σ, for any type σ.
- The function symbol $\mathbf{p}^{\sigma \times \tau}(x^\sigma, y^\tau)$, for any types σ and τ. The intended meaning of $\mathbf{p}^{\sigma \times \tau}(x^\sigma, y^\tau)$ is the pair of x^σ and y^τ.
- The function symbols $\mathbf{p_0}^\sigma(x^{\sigma \times \tau})$ and $\mathbf{p_1}^\tau(x^{\sigma \times \tau})$, for any types σ and τ. The intended meaning of $\mathbf{p_0}^\sigma(x^{\sigma \times \tau})$ and $\mathbf{p_1}^\tau(x^{\sigma \times \tau})$ are the projections on the first and second components, respectively.

Now, define *terms* inductively by

- any variable of type σ is a term of type σ,
- for any function symbol $f^\tau(x_1^{\sigma_1}, x_2^{\sigma_2}, \ldots, x_n^{\sigma_n})$, (including $\mathbf{p}^{\sigma \times \tau}(x^\sigma, y^\tau)$, $\mathbf{p_0}^\sigma(x^{\sigma \times \tau})$ and $\mathbf{p_1}^\tau(x^{\sigma \times \tau})$), if $\{t_k\}_{k=1}^n$ are terms of types $\{\sigma_k\}_{k=1}^n$, then $f^\tau(t_1, t_2, \ldots, t_n)$ is a term of type τ,
- $\lambda x^\sigma.t(x^\sigma)$ is a term of type $\sigma \to \tau$, for any types σ and τ and any term $t(x^\sigma)$ of type τ,
- ts is a term of type τ, for any terms t and s of types $\sigma \to \tau$ and σ, respectively.

The intended meaning of $\lambda x^\sigma.t(x^\sigma)$ is a function that maps x^σ to $t(x^\sigma)$ and sr means the application of the function s on the input r.

Finally, consider a family of predicate symbols $\{P_k(x_1^{\sigma_1}, x_2^{\sigma_2}, \ldots, x_{n_k}^{\sigma_{n_k}})\}_{k \in K}$. Then, define *formulas* inductively in the following way:

- $\top$ and $\bot$ are formulas,
- $t_1 =_\sigma t_2$ is a formula, if both t_1 and t_2 are terms of type σ,
- $P(t_1, t_2, \ldots, t_n)$ is a formula, if $\{t_r\}_{r=1}^n$ are terms of types $\{\sigma_r\}_{r=1}^n$, for any predicate symbol $P(x_1^{\sigma_1}, x_2^{\sigma_2}, \ldots, x_n^{\sigma_n})$,
- if A and B are formulas, $A \wedge B$, $A \vee B$, $A \to B$ are formulas,
- if $A(x^\sigma)$ is a formula, then $\forall x^\sigma A(x^\sigma)$ and $\exists x^\sigma A(x^\sigma)$ are formulas.

As it is clear from the definition, to specify a higher-order language $\mathcal{L}$, it is enough to specify its basic types and its families of function and predicate symbols. We denoted these three families by $B(\mathcal{L})$, $F(\mathcal{L})$ and $P(\mathcal{L})$, respectively. We also denote the set of all types of $\mathcal{L}$ by $T(\mathcal{L})$.

Define the logic IQC^{ω} over the language $\mathcal{L}$ as the theory consisting of the usual axioms of multi-sorted first-order logic with equality plus the following defining axioms:

- $\mathbf{p}^{\sigma\times\tau}(\mathbf{p_0}^{\sigma}(x^{\sigma\times\tau}), \mathbf{p_1}^{\tau}(x^{\sigma\times\tau})) =_{\sigma\times\tau} x^{\sigma\times\tau}$, $\mathbf{p_0}^{\sigma}(\mathbf{p}^{\sigma\times\tau}(x^{\sigma}, y^{\tau})) =_{\sigma} x^{\sigma}$, and $\mathbf{p_1}^{\tau}(\mathbf{p}^{\sigma\times\tau}(x^{\sigma}, y^{\tau})) =_{\tau} y^{\tau}$,
- $(\lambda x^{\sigma}.t(x^{\sigma}))y^{\sigma} =_{\tau} t(y^{\sigma})$, where $t(x^{\sigma})$ is of type τ and $t(y^{\sigma})$ is the substitution of y^{σ} in $t(x^{\sigma})$ for x^{σ}.

There are two non-logical axioms of interest in this chapter. The first is the axiom $\lambda y^{\sigma}.(x^{\sigma\to\tau}\cdot y^{\sigma}) = x^{\sigma\to\tau}$ or equivalently the *axiom of extensionality*

$$\mathsf{EXT}: \quad \forall f^{\sigma\to\tau} g^{\sigma\to\tau}[\forall x^{\sigma}(fx =_{\tau} gx) \to (f =_{\sigma\to\tau} g)],$$

stating that a function is uniquely determined by its action. The second is the *axiom of choice*

$$\mathsf{AC}: \quad \forall x^{\sigma}\exists y^{\tau} A(x, y) \to \exists f^{\sigma\to\tau}\forall x^{\sigma} A(x, fx),$$

stating that if for any x of type σ, there is a y of type τ satisfying $A(x, y)$, one can find a function f of type $\sigma \to \tau$ to choose such a y of type τ, for any x of type σ.

Remark 192 Here are some remarks. First, when it is clear from the context, we omit the types, including in x^{σ}, function and predicate symbols, and $=_{\sigma}$. Second, if we eliminate the type constructors $\times$ and $\to$, the function symbols $\mathbf{p}$, $\mathbf{p_0}$, $\mathbf{p_1}$, and the term constructors λ and application, along with their corresponding equalities, the result is multi-sorted first-order logic, denoted by IQC. Third, sometimes the zero type 0 and the sum type $+$ are added to represent the empty set and disjoint union, respectively. However, we will not include these in this chapter to keep the presentation simpler. Fourth, the language can be restricted to its fragments, as we did for the propositional language. One fragment of interest is the one without $\bot$ and $\vee$, as it simplifies the treatment of arithmetic as we will see later.

There is a canonical classical interpretation of higher-order languages. It is enough to assign a non-empty set to any basic type in $B(\mathcal{L})$ and interpret the type σ by the set M^{σ} recursively defined by $M^{\sigma\times\tau} = M^{\sigma} \times M^{\tau}$ and $M^{\sigma\to\tau} = [M^{\sigma}, M^{\tau}]$, where $\times$ and $[-, -]$ are the cartesian product and the exponentiation in **Set**. Then, we must assign functions to the function symbols in $F(\mathcal{L})$ respecting the types and extend the interpretation to all terms by composition. Note that the function symbols $\mathbf{p}$, $\mathbf{p_0}$ and $\mathbf{p_1}$ must be interpreted as the pairing functions and the projections, respectively, $\lambda x^{\sigma}.t(x)$ must be read as a function mapping $a \in M^{\sigma}$ to $t(a) \in M^{\tau}$, where τ is the type of $t(x)$ and ts must be read as the usual application function. Finally, we must assign relations to the predicate symbols in $P(\mathcal{L})$ respecting the types. It is easy to see that $\mathsf{IQC}^{\omega} + \mathsf{EXT} + \mathsf{AC}$ is valid under any such interpretation.

For a constructive interpretation of a higher-order theory, we offer an informal interpretation here, which will be formalized later in Sect. 9. We start with an informal concept of construction. To interpret types in a higher-order language, we change the category of sets in the classical interpretation to the category of *constructed sets*, i.e., sets where all elements are constructed with *constructed functions* serving as the morphisms. For interpreting formulas, we need to extend the BHK interpretation from Sect. 5 to include equality and quantifiers. This extension can take various forms, and we outline the one that we plan to formalize later:

- a proof of $t = s$ is a verification that t and s are equal,
- a proof of $\forall x^\sigma A(x)$ is a construction that transforms any construction of a to a proof of $A(a)$, for any a of type σ,
- a proof of $\exists x^\sigma A(x)$ is a pair of a construction of an element a of the type σ and a proof of $A(a)$.

Even at this informal level, we can verify the provability of some axioms. It is straightforward to see that every theorem of IQC^ω is provable through this BHK interpretation. The axiom EXT has a proof under the BHK interpretation because if we have a proof of $\forall x^\sigma(fx = gx)$, then we reach a proof of $fx = gx$ for any x of type σ, which implies $f = g$, thereby establishing its provability. For the axiom of choice, we need to make an additional assumption. Let us assume that each element in σ has a unique construction. Then, if we have a construction for $\forall x^\sigma \exists y^\tau A(x, y)$, it provides an element b of type τ along with its construction for any construction of any element a of type σ. Since constructions for the elements of type σ are unique, we can ignore the dependence on the construction of a. This effectively gives us a function $f : \sigma \to \tau$ by mapping a's to b's. Note that the function f is induced by a construction and hence must be interpreted as a constructed function.

8.1 Constructive Theories of Arithmetic

In this subsection, we will introduce some theories of constructive arithmetic. Let us start with the simpler first-order case. Fix the language $\mathcal{L}_{\mathsf{HA}} = \{0, S, +, \cdot\}$ and assume that the language does not have the connectives $\vee$ and $\bot$. This is no restriction on the theory as it is known that the propositions $S(0) = 0$ and $\exists z[(z = 0 \to A) \wedge (z \neq 0 \to B)]$ can play the role of $\bot$ and $A \vee B$, respectively Troelstra and Van Dalen (1988a). Define *Heyting (resp. Peano) arithmetic*, denoted by HA (resp. PA) as the intuitionistic (resp. classical) logic with the equality extended by the axioms:

- $\forall x(S(x) \neq 0)$,
- $\forall xy(S(x) = S(y) \to x = y)$,
- $\forall x(x + 0 = x)$,
- $\forall xy(x + S(y) = S(x + y))$,

- $\forall x(x \cdot 0 = 0)$,
- $\forall xy(x \cdot S(y) = x \cdot y + x)$,
- $A(0) \wedge \forall x(A(x) \to A(S(x))) \to \forall x A(x)$, for any formula $A(x)$.

It is clear that the set of natural numbers with the canonical interpretation of the language $\mathcal{L}_{\mathsf{HA}}$ is a model of PA. This model is called the *standard model* and will be denoted by $\mathfrak{N}$. Here are some basic facts about the theories HA and PA. First, defining the relation $x \leq y$ as $\exists z(x + z = y)$, it is possible to prove the usual properties of the ordering in HA such as its totality and the fact that the operations $+$ and $\cdot$ respect the ordering. Any quantifier in the form $\forall x(x \leq t \to A(x))$ and $\exists x(x \leq t \wedge A(x))$, for some term t is called a *bounded quantifier* and will be abbreviated by $\forall x \leq t\, A(x)$ and $\exists x \leq t\, A(x)$, respectively. A formula is called *bounded* if all of its quantifiers can be replaced by a bounded quantifier up to provability in HA. Any bounded formula $A(\bar{x})$ is decidable in HA, i.e., $\mathsf{HA} \vdash \forall \bar{x}(A(\bar{x}) \vee \neg A(\bar{x}))$. The theory PA is conservative over HA for formulas in the form $\forall \bar{x} \exists \bar{y} A(\bar{x}, \bar{y})$, where $A(\bar{x}, \bar{y})$ is bounded.[8] A formula is called $\exists$-*free* if it has no occurrence of an existential quantifier. It is easy to see that any bounded formula is equivalent to an $\exists$-free formula in HA. To prove, it is just enough to change any $\exists x \leq t\, A(x)$ to $\neg \forall x \leq t\, \neg A(x)$. These two formulas are equivalent as bounded formulas are decidable in HA. It is possible to formalize some basic recursion theory in HA. More precisely, there is a $\exists$-free formula $T(e, x, w, y)$ such that in the standard model it means that the Turing machine with the code e halts on the input x with the computation w and the output y.

For the higher-order version of Heyting arithmetic, denoted by HA^{ω}, use the higher-order language with the basic type N, the function symbols $\{0, S(x^N)\}$ of type N, the function symbols $\mathbf{R}^{\sigma}(x^{\sigma}, y^{N \times \sigma \to \sigma}, z^N)$, for any type σ and add no predicate symbol other than equality. Define HA^{ω} as IQC^{ω} plus the following axioms:

- $\mathbf{R}^{\sigma}(x, y, 0) = x$ and $\mathbf{R}^{\sigma}(x, y, S(z)) = y\mathbf{p}(z, \mathbf{R}^{\sigma}(x, y, z))$,
- $S(0) \neq 0$ and $S(x) = S(y) \to x = y$,
- $A(0) \wedge \forall x^N(A(x) \to A(S(x)) \to \forall x^N A(x)$, for any formula $A(x)$ in the extended language.

We also define the extension HA^{ω}_2 of HA^{ω} that we use in a restricted way later. First, add a new basic type 2, two constants 0_2 and 1_2 of type 2 and for any type σ, a function symbol $c^{\sigma}(x^{\sigma}, y^{\sigma}, z^2)$. The intended meaning of 2 is the set $\{0, 1\}$ and $c^{\sigma}(x^{\sigma}, y^{\sigma}, z^2)$ is the function that reads the elements x^{σ} and y^{σ} from σ and z^2 from $\{0, 1\}$ and returns x^{σ} if $z = 0$ and y^{σ} if $z = 1$. Extend the theory HA^{ω} by the equations $c^{\sigma}(x, y, 0_2) = x$, $c^{\sigma}(x, y, 1_2) = y$ and $c^{\sigma}(w0_2, w1_2, z) = wz$, for any w of type $2 \to \sigma$. The resulting theory is denoted by HA^{ω}_2. Notice that the induction axiom operates over the whole new language including the formulas with terms of type 2.

It is clear that HA can be interpreted inside HA^{ω}, reading any variable x as a variable of the type N, interpreting 0 and S as the corresponding constants and defining the addition and multiplication by the function symbol $\mathbf{R}$ using the usual

[8] This means that $\mathsf{PA} \vdash \forall \bar{x} \exists \bar{y} A(\bar{x}, \bar{y})$ implies $\mathsf{HA} \vdash \forall \bar{x} \exists \bar{y} A(\bar{x}, \bar{y})$, if $A(\bar{x}, \bar{y})$ is bounded.

recursive definitions. For more on the definition and the basic properties of the defined arithmetical theories, see Troelstra and Van Dalen (1988b).

Similar to HA, the theories HA^ω and HA_2^ω have a canonical classical interpretation that we also call the *standard model* and denote by $\mathfrak{N}$. To define the model, it is enough to map the basic type N to the set of natural numbers $\mathbb{N}$ and read terms 0 and $S(x^N)$ as the zero element and the successor function on $\mathbb{N}$. For $\mathbf{R}$, we must use the canonical function defined by the recursive definition the axioms for $\mathbf{R}$ suggest. Later in Sect. 10, we will provide a more constructive interpretation using the BHK interpretation.

8.2 Non-classical Axioms and Consistency

We began this chapter with a brief history of Hilbert's program and its consistency problems, which motivated the formal and precise investigation of the concept of proof. This investigation ultimately led to the development of categorical proof theory, providing concrete and distinct models for proof systems, as demonstrated in the previous sections. To close the circle, it would be interesting to apply categorical proof theory to prove the consistency of mathematical theories in general, and arithmetical theories in particular.

For the usual arithmetical theories such as HA and PA, one might argue that since mathematics is no longer in crisis, the urgency of addressing consistency problems has diminished. Today, we consider the standard model to be both secure and intuitively acceptable, making it reasonable to assert the consistency of arithmetic based on the existence of this model. However, the philosophical and mathematical discourse surrounding consistency problems is more nuanced than this straightforward objection suggests, although the objection does hold some validity. To address the objection, it is important to note that proving consistency in the intuitive manner described above is essentially verifying the soundness of the theory with respect to our classical image. Constructive mathematics, however, encompasses various alternative theories that describe alternative worlds inconsistent with our classical understanding. Thus, proving consistency is not merely a philosophical endeavor but involves establishing the plausibility of such hypothetical worlds, where classical intuition may not be as effective as we assume. To illustrate this, we provide two examples of alternative worlds within constructive mathematics.

The primary claim of constructive mathematics is that all mathematical entities are constructions. Within this framework, a challenge arises when dealing with infinitary objects, such as functions over natural numbers or infinite sequences of zeros and ones, which exceed the finite and bounded capacity of human construction. There are numerous approaches to handling these objects, with two extreme ends of the spectrum: the *law-like* and *lawless* approaches.

In the law-like approach, all infinitary objects are required to be law-like, meaning that despite their infinitary nature, they are generated by a finite set of rules or laws. The natural candidate for such a finite law is an algorithm. Consequently, this interpretation of constructive mathematics confines itself to the computable world, where everything, including functions on numbers, must be computable. This variant of constructive mathematics is known as *Russian constructivism*. A typical axiom in Russian constructivism is the *Church-Turing thesis*.

$$\mathsf{CT}: \quad \forall f^{N \to N} \exists e^N \forall x^N \exists w^N \exists y^N [T(e, x, w, y) \wedge (fx = y)],$$

stating that *all total functions on natural numbers are computable*. More precisely, it asserts that for any function $f : \mathbb{N} \to \mathbb{N}$, there exists a Turing machine with the code e that computes f, meaning that for any input x, there is a computation w of e on x producing the output $y = f(x)$. Now, regarding our consistency problem, we might ask whether $\mathsf{HA}^{\omega} + \mathsf{CT}$ is consistent. The answer is unclear because, in the usual classical framework, some functions, such as the characteristic function for the halting predicate, are uncomputable. Therefore, the theory $\mathsf{HA}^{\omega} + \mathsf{CT}$ is inconsistent with our classical understanding of the world.

At the opposite end of the spectrum is the assumption that all infinite objects are *lawless*. For example, when considering functions over natural numbers, we may work with any function whose values are chosen in a lawless manner, one after another, over time. At any given moment, we can only comprehend a finite number of values in our finite minds, yet we can still discuss potentially infinite sequences of numbers constructed in this lawless fashion. One might argue that even if we include these mathematical objects in our constructive mathematics, we cannot derive any conclusions about them due to their lawless and hence completely unpredictable nature. Brouwer's insightful idea was that if all functions are lawless, the only meaningful statements we can make about them must involve only finitely many values of the function. For instance, stating that the first element of a lawless sequence α is five is meaningful, whereas claiming that all its elements are zero is not. In this framework, we can accommodate lawless sequences if we restrict ourselves to discussing the *continuous* properties of the functions where continuity ensures that the property is actually about a finite part of the input. This type of continuity is a defining characteristic of what is commonly referred to as *Brouwerian constructivism*. A typical axiom in this form of constructivism is the *continuity principle*[9]:

$$\mathsf{WC\text{-}N}: \quad \forall f^{N^N \to N} \forall \alpha^{N \to N} \exists x^N \forall \beta^{N \to N} (\alpha =_x \beta \to f\alpha = f\beta),$$

where $\alpha =_x \beta$ abbreviates $\forall y^N (y < x \to \alpha y = \beta y)$, stating that the *value of a total numeral function at α depends only on finitely many values of α*. The x bounding the cardinality of these values is called the *continuity module*. Similar to what we encountered with CT, this axiom is also inconsistent with our usual classical image. It asserts that all total functions $f : \mathbb{N}^{\mathbb{N}} \to \mathbb{N}$ are continuous. However, using the axiom of the excluded middle, we can observe that the characteristic function of the

[9] This axiom is formally called the *weak continuity principle for numbers*. Hence, the abbreviation WC–N.

singleton $\{\alpha_0\}$, where α_0 is the constant zero function, is total but not continuous. In other words, the value of $f(\alpha)$ depends on all values of $\alpha(n)$, not just a finite number of them. Therefore, we face a consistency question again, this time about the theory $\mathsf{HA}^\omega + \mathsf{WC{-}N}$. For more on non-classical constructive theories, see (Troelstra and Van Dalen 1988a, b).

One natural strategy for proving such consistency claims is providing a model—an alternative world where the anti-classical axioms hold. To construct these models, we use categories of constructed sets, interpreting constructions in a way that aligns with the consistency problem at hand. We then apply the BHK interpretation to validate the theory and demonstrate its consistency. Consider, for instance, the case of $\mathsf{HA}^\omega + \mathsf{CT}$. Here, we naturally expect to interpret the informal notion of construction as an algorithm or a computable function. The category of constructed sets then provides a *computable world*, where everything, including all functions on natural numbers, is computable. This setup is precisely what we need to support the Church-Turing thesis, and by leveraging this observation, we aim to show that the BHK interpretation validates CT.

However, under the BHK interpretation, the axiom

$$\mathsf{CT}: \quad \forall f^{N\to N} \exists e^N \forall x^N \exists w^N \exists y^N [T(e, x, w, y) \wedge (fx = y)],$$

asserts more than just "all total functions over natural numbers are computable." It also requires that the process of finding the code e for a function f from its construction $\mathtt{f}$ be computable. Later, we will explain how to weaken CT to align with what we originally intended. For now, let us examine the implications of working with this current version.

As we saw above, we have two options for interpreting constructions: either as an algorithm or as a computable function, and the choice significantly impacts the validation of CT. If we interpret $\mathtt{f}$ as a computable function, i.e., as f itself, then the requirement to recursively find a code for it becomes impossible, as we will explain later. On the other hand, if we interpret $\mathtt{f}$ as an algorithm for f, then CT simply asserts that if we have an algorithm $\mathtt{f}$ for f, we can find a code e for f in a computable manner. This is straightforward, as e can simply be set as the algorithm $\mathtt{f}$ itself.

Let us now investigate the axiom

$$\mathsf{WC{-}N}: \quad \forall f^{N^N \to N} \forall \alpha^{N\to N} \exists x^N \forall \beta^{N\to N} (\alpha =_x \beta \to f\alpha = f\beta)$$

from the same lens. Here, we must interpret construction as a continuous entity to get a *continuous world*. Similar to CT, the axiom not only asserts that $f(\alpha)$ depends on x many initial values of α, but also demands the continuity of the process of determining x based on the constructions of f and α. When the constructions of f and α are read as the functions themselves, this requirement becomes impossible again. However, there exists a notion of an "algorithm" that operates on infinitary data. By interpreting the constructions of f and α as such algorithms, one can validate the axiom, as it is possible to determine x from the algorithms Troelstra (1998).

In general, there are essentially two ways to interpret a construction: *intensionally* or *extensionally*. The intensional approach treats constructions as *instructions* to be executed, while the extensional approach views them as *functions in the traditional sense*. For example, in the above examples, the intensional interpretation considers a construction as an *algorithm*, whereas the extensional interpretation views it as a computable or continuous *function*.

In this chapter, we adopt the *extensional* approach. A key point about this approach is that it typically tends to validate the axioms EXT and AC. As mentioned in the last paragraph prior to Sect. 8.1, the validity of EXT is straightforward. Regarding AC, the extensional interpretation usually treats a function as its own construction, implying that constructions are generally unique. This uniqueness leads to the validation of AC. In contrast, under the intensional approach, a function can correspond to many different algorithms, which influences the validation of the axiom of choice.

In the following theorem, we will show that CT and WC−N are inconsistent with EXT and AC. Consequently, we demonstrate what we previously promised: that the extensional BHK interpretation cannot validate the axioms CT and WC−N.

Theorem 193 *The theory* $\mathsf{HA}^\omega + \mathsf{EXT} + \mathsf{AC}$ *is inconsistent with either of the axioms* CT *or* WC−N.

Proof We argue informally inside the theory $\mathsf{HA}^\omega + \mathsf{EXT} + \mathsf{AC}$. For CT, use AC on CT to get a functional $F : N^N \to N$ that maps a total function $f : N \to N$ to one of its codes. We claim that $f = g$ iff $F(f) = F(g)$. One direction is clear. For the other, if $F(f) = F(g)$, as $F(f)$ is the code for f, we have $fx = F(f) \cdot x = F(g) \cdot x = gx$, where $e \cdot x$ is the output of applying the machine with the code e on the input x. Note that the predicate $e \cdot x = y$ is definable in arithmetic by the definition $\exists w^N T(e, x, w, y)$. Therefore, $fx = gx$, for any x of type N and thus $f = g$, by EXT. Now, using the equivalence between $f = g$ and $F(f) = F(g)$ and the decidability of the equality over the type N in HA^ω, we have the decidability of the equality over N^N, i.e.,

$$\forall f^{N \to N} g^{N \to N}[(f = g) \vee \neg(f = g)]. \qquad (*)$$

Set x as a parameter and define f as the map that reads w and outputs the truth value of the predicate $T(x, x, w, 1)$ and g as the constant zero function. This is possible as T is decidable and hence f is total. Note that $f = g$ iff $\forall w^N \neg T(x, x, w, 1)$. Therefore, by $(*)$, we have

$$\forall x^N (\forall w^N \neg T(x, x, w, 1) \vee \neg \forall w^N \neg T(x, x, w, 1)).$$

Using this decidability, we can define a total characteristic function $h : N \to N$ that reads x and outputs 1 if $\forall w^N \neg T(x, x, w, 1)$, and 0 if $\neg \forall w^N \neg T(x, x, w, 1)$. By CT, there is a code for the function h. Call it e. Then, either $e \cdot e = 1$ or $e \cdot e = 0$. In the first case, we have $e \cdot e = 1$ which means that $\exists w^N T(e, e, w, 1)$. However, we are in the first case. Hence, we must have $\forall w^N \neg T(e, e, w, 1)$ which is impossible. If $e \cdot e = 0$, then as the value of $e \cdot e$ is unique, it is not 1 and hence it is easy to prove that $\forall w^N \neg T(e, e, w, 1)$. However, we are in the second case. Hence, $\neg \forall w^N \neg T(e, e, w, 1)$ which is again a contradiction.

For WC–N, we will reproduce the proof provided in Hötzel Escardó and Xu (2015), only altering the type-theoretic language to our arithmetical language. Since functions on natural numbers and sequences of natural numbers are essentially the same entities, we will use them interchangeably to simplify our presentation. Let $0^n k^\omega$ denote the sequence of n many zeros followed by infinitely many k's, for any k and n of type N. Then, notice that $0^n k^\omega =_n 0^\omega$ and $0^n k^\omega =_{(n+1)} k$. By applying AC on WC–N, we have a functional F such that for any $f^{N^N \to N}$, $\alpha^{N \to N}$ and $\beta^{N \to N}$ if $\alpha =_{F(f,\alpha)} \beta$ then $f\alpha = f\beta$. By substituting $\alpha = 0^\omega$, we have

$$\forall f^{N^N \to N} \forall \beta^{N \to N} [0^\omega =_{M(f)} \beta \to f0^\omega = f\beta], \qquad (**)$$

where $M(f) = F(f, 0^\omega)$. Let $m = M(\lambda\gamma.0)$ and define $f^{N^N \to N}$ by $f\beta = M(\lambda\gamma.\beta(\gamma m))$. Note that $0^\omega(\gamma m) = 0$, for the variable $\gamma^{N \to N}$. Hence, by EXT, we have $\lambda\gamma.0^\omega(\gamma m) = \lambda\gamma.0$. Thus, by definition, we have $f(0^\omega) = M(\lambda\gamma.0^\omega(\gamma m)) = M(\lambda\gamma.0) = m$. Therefore, by $(**)$, we have

$$\forall \beta^{N \to N} [0^\omega =_{M(f)} \beta \to m = f\beta]. \qquad (\dagger)$$

On the other hand, for any $\beta^{N \to N}$, we can define $g_\beta = \lambda\gamma.\beta(\gamma m)$. Note that $M(g_\beta) = f(\beta)$, by the definition of f. Moreover, notice that $g_\beta 0^\omega = \beta(0)$ and $g_\beta(\gamma) = \beta(\gamma m)$. Applying $(**)$ on g_β, we have

$$\forall \gamma^{N \to N} [0^\omega =_{f\beta} \gamma \to \beta 0 = \beta(\gamma m)]. \qquad (\dagger\dagger)$$

Substituting $\beta = 0^{M(f)+1} 1^\omega$ in $(\dagger)$ and using the fact that $0^\omega =_{M(f)} \beta$, we have $f(\beta) = m$. Using the same $\beta = 0^{M(f)+1} 1^\omega$ in $(\dagger\dagger)$ and the equality $f(\beta) = m$, we have

$$\forall \gamma^{N \to N} [0^\omega =_m \gamma \to \beta 0 = \beta(\gamma m)].$$

Substituting $\gamma = 0^m (M(f) + 1)^\omega$, we have $0^\omega =_m \gamma$. Hence, $0 = \beta 0 = \beta(\gamma m) = \beta(M(f) + 1) = 1$ which is a contradiction. $\square$

As a concluding part of this subsection, we will explain how to weaken the axioms CT and WC–N to reflect our original intent. Our approach is to modify the existential quantifiers to require only the *mere* existence of the code and the continuity module, rather than their constructibility via a recursive or continuous process. In this revised formulation, CT aligns with our original understanding, stating that for any computable function f, there exists a code e for f, with no requirement for the process of finding this code to be computable. Similarly, WC–N would assert that for any function f and any α, there exists an x such that the value of $f(\alpha)$ depends only on the first x values of α, and the process of finding x need not be continuous.

To achieve this, we need a method to *collapse* proofs of a proposition into a single proof, preserving only the fact of provability while discarding additional details. This process is known as *propositional truncation* Awodey and Bauer (2004), The Univalent Foundations Program (2013). However, for simplicity, we will use a similar mechanism adapted to the specific BHK interpretation employed in this chapter.

First, note that if we formalize the BHK interpretation in a set-like manner without any temporal structure, it becomes apparent that for any sentence A, if there is no proof of A, then *any* construction can serve as a proof of $\neg A$, since there is no proof of A to transform into a proof of $\bot$. Consequently, if A does have a proof, then $\neg A$ cannot have a proof and hence *any* construction serves as a proof of $\neg\neg A$, and if A is not provable, then $\neg A$ is provable implying that $\neg\neg A$ has *no* proof. Thus, a proof of $\neg\neg A$ conveys no information beyond the mere provability of A, meaning that $\neg\neg$ can function as a collapsing mechanism. Using this collapsing machine, the weaker versions of the axioms CT and $\mathsf{WC{-}N}$ can be formulated as

$$\mathsf{CT}_{\neg\neg}: \quad \forall f^{N\to N} \neg\neg\exists e^N \forall x^N \exists w^N \exists y^N [T(e, x, w, y) \wedge (fx = y)],$$

$$\mathsf{WC{-}N}_{\neg\neg}: \quad \forall f^{N^N\to N} \forall \alpha^{N\to N} \neg\neg\exists x^N \forall \beta^{N\to N} (\alpha =_x \beta \to f\alpha = f\beta)$$

where the existence of the code or the continuity module x is asserted, but the requirement for finding them in a recursive or continuous manner is relaxed. Note that even these weakened versions remain in contradiction with our classical image, as double negation has no impact in classical mathematics. In Sect. 10, we will formalize the arguments provided here to construct models for the theories $\mathsf{HA}^\omega + \mathsf{EXT} + \mathsf{AC} + \mathsf{CT}\neg\neg$ and $\mathsf{HA}^\omega + \mathsf{EXT} + \mathsf{AC} + \mathsf{WC{-}N}\neg\neg$, thereby demonstrating their consistency.

8.3 *Information Extraction*

There is also another application of categorical proof theory that is relevant not only to the alternative theories discussed in the previous subsection but also to conventional theories compatible with our classical perspective. To illustrate this, consider a simple question: Does a proof of a proposition contain more information than just its mere truth? The answer appears to be affirmative. For example, in the context of HA, knowing a proof of a proposition of the form $\forall x \exists y A(x, y)$ tells us that for any $n \in \mathbb{N}$, there exists an $m \in \mathbb{N}$ such that $A(n, m)$ holds in the standard model. However, the proof might contain additional information. Specifically, it may be possible to *extract* an algorithm from the proof that, given $n \in \mathbb{N}$, computes an $m \in \mathbb{N}$ such that $A(n, m)$.

The traditional black-and-white approach to proofs often overlooks the possibility of intermediate solutions. It either retains all proofs with their syntactical complexities intact or shifts to a semantic view, discarding the entire structure of a proof in favor of greater flexibility. The trade-off is that this approach loses all information contained within the proof, leaving only the bare truth of its conclusion. While ignoring the proof structure can be useful in some contexts, it may be overly simplistic in others. Therefore, we may need to find a middle ground that balances flexibility with the retention of useful information. Specifically, developing *models of proofs* that preserve certain aspects of the proof structure while allowing for some flexibility could offer a more nuanced approach. This way, we can maintain essential information while adapting the proofs for practical use.

This is precisely where categorical proof theory, and specifically the BHK interpretation, becomes highly valuable. For instance, a BHK proof of the sentence $\forall x^{\sigma} \exists y^{\tau} A(x, y)$ is a construction that maps any construction of any element a of type σ to a pair consisting of a construction of an element b of type τ and a proof of $A(a, b)$. Thus, if the elements in types σ and τ are constructions for themselves, having a BHK proof for the sentence $\forall x^{\sigma} \exists y^{\tau} A(x, y)$ enables us to find a method to compute b from a. Note that the notion of construction is not predetermined; it can be tailored to suit specific needs. For example, if we choose to interpret constructions as computable functions, then for $\sigma = \tau = N$, a BHK proof of $\forall x^{N} \exists y^{N} A(x, y)$ yields a recursive function that takes a natural number $n \in \mathbb{N}$ and computes a natural number $m \in \mathbb{N}$ such that $A(n, m)$ holds. Similarly, if we choose to interpret constructions as continuous functions, then for $\sigma = N \to N$ and $\tau = N$, a BHK proof of $\forall x^{N \to N} \exists y^{N} A(x, y)$ provides a continuous function that, given a function $\alpha : \mathbb{N} \to \mathbb{N}$, computes a natural number $m \in \mathbb{N}$ such that $A(\alpha, m)$ holds. We will revisit these applications in Sect. 10.

9 Realizability

In the previous sections, we explained the categorical formalization of propositional deductive systems. The natural next step is to extend the formalization to the higher-order setting. However, this task is far more demanding than what we covered for the propositional case, and thus it goes beyond what we can include in this chapter.[10] To provide a taste of this, though, we will present a special yet powerful enough setting called *realizability*, based on the BHK interpretation we discussed earlier.

Choosing this special setting has its own benefits. First, it provides the categorical presentation of one of the most powerful techniques in proof theory of the same name. Second, our setting uses less structure, which translates to gaining more insights into the theories. For instance, we will show that any computable function provably total in HA and PA is representable in any CCC with enough structure. In this section, we establish the general setting for an arbitrary higher-order language, and in Sect. 10, we will apply the machinery to arithmetical theories.

[10] To see how such generalized settings work, see (Palmgren 2004; Hofmann 1995b; Jacobs 1999).

9.1 Propositional Realizability

To formalize the higher-order BHK interpretation, we follow the format explained in Sect. 5 and extended in Sect. 8. However, this time, we relax the temporal structure and change our notion of construction from set functions to maps in a BCC. First, let us start with the propositional language again.

Notation 194 In the realizability parts of this chapter, namely the present section and Sect. 10, we use Typewriter font (as in $\mathtt{f} : \mathtt{N} \to \mathtt{N}$) for objects and maps in our BCC to distinguish them from set-theoretical objects and functions.

Definition 195 Let C be a BCC. By a *C-interpretation* $\mathfrak{I}$ for the propositional language, we mean an assignment that maps each atom p in the language to a pair $(|p|, R_p)$, where $|p|$ is an object in C and $R_p \subseteq \mathrm{Hom}_C(\mathtt{1}, |p|)$ is a set.

By recursion, we extend $\mathfrak{I}$ to assign an object $|A|$ to any proposition A in the following way:

- $|\top| = \mathtt{1}$, $|\bot| = \mathtt{0}$,
- $|A \wedge B| = |A| \times |B|$, $|A \vee B| = |A| + |B|$, $|A \to B| = [|A|, |B|]$.

Then, inductively define the relation $\mathtt{r} \Vdash A$ between the map $\mathtt{r} \in \mathrm{Hom}_C(\mathtt{1}, |A|)$ and the proposition A in the following way:

- $\mathtt{r} \Vdash p$ if $\mathtt{r} \in R_p$, for any $\mathtt{r} : \mathtt{1} \to |p|$,
- $!_\mathtt{1} \Vdash \top$ for the unique map $!_\mathtt{1} : \mathtt{1} \to \mathtt{1}$,
- $\mathtt{r} \nVdash \bot$ for any map $\mathtt{r} : \mathtt{1} \to \mathtt{0}$,
- $\mathtt{r} \Vdash A \wedge B$ iff $\mathtt{p}_0\mathtt{r} \Vdash A$ and $\mathtt{p}_1\mathtt{r} \Vdash B$, for any $\mathtt{r} : \mathtt{1} \to |A| \times |B|$,
- $\mathtt{r} \Vdash A \vee B$ iff there exists $\mathtt{s} : \mathtt{1} \to |A|$ such that $\mathtt{r} = \mathtt{i}_0\mathtt{s}$ and $\mathtt{s} \Vdash A$ or there exists $\mathtt{t} : \mathtt{1} \to |B|$ such that $\mathtt{r} = \mathtt{i}_1\mathtt{t}$ and $\mathtt{t} \Vdash B$, for any $\mathtt{r} : \mathtt{1} \to |A| + |B|$,
- $\mathtt{r} \Vdash A \to B$ iff for any $\mathtt{s} : \mathtt{1} \to |A|$, if $\mathtt{s} \Vdash A$ then $\mathtt{r} \cdot \mathtt{s} \Vdash B$, for any $\mathtt{r} : \mathtt{1} \to [|A|, |B|]$.

For any *finite* set Γ of formulas, define $|\Gamma|$ as $\prod_{\gamma \in \Gamma} |\gamma|$. We use an arbitrary order on the formulas in Γ, and since the product is commutative up to an isomorphism, the chosen order is actually immaterial. Then, by $\Gamma \Vdash_\mathfrak{I} A$, we mean the existence of a map $\mathtt{r} : |\Gamma| \to |A|$ such that $\mathtt{r} \circ \mathtt{s} \Vdash A$, for any $\mathtt{s} : \mathtt{1} \to |\Gamma|$ satisfying $\mathtt{s} \Vdash \bigwedge \Gamma$. This $\mathtt{r}$ is called a *realizer* for $\Gamma \Vdash_\mathfrak{I} A$. When Γ is empty, $|\Gamma| = \mathtt{1}$, by definition and $\mathtt{r} : \mathtt{1} \to |A|$ is called a realizer for A. When the C-interpretation $\mathfrak{I}$ is clear from the context, we drop the subscript $\mathfrak{I}$ in $\Vdash_\mathfrak{I}$. Finally, by $\Gamma \Vdash_C A$, we mean $\Gamma \Vdash_\mathfrak{I} A$, for any C-interpretation $\mathfrak{I}$.

Remark 196 Notice that $\mathtt{r} : |\Gamma| \to |A|$ realizes $\Gamma \Vdash_\mathfrak{I} A$ iff for any sequence $\{\mathtt{s}_\gamma : \mathtt{1} \to |\gamma|\}_{\gamma \in \Gamma}$ of maps such that $\mathtt{s}_\gamma \Vdash \gamma$, for any $\gamma \in \Gamma$, we have $\mathtt{r} \circ \langle \mathtt{s}_\gamma \rangle_{\gamma \in \Gamma} \Vdash A$. This presentation is easier to work with in practice.

Example 197 For any $\mathcal{C}$-interpretation $\mathfrak{I}$, the identity map $\mathtt{id}_{|A|} : |A| \to |A|$ realizes $A \Vdash A$ because if $\mathtt{r} \Vdash A$ then $\mathtt{id}_{|A|} \circ \mathtt{r} = \mathtt{r} \Vdash A$. The unique map $!_{|A|} : 0 \to |A|$ realizes $\bot \Vdash A$. To prove that, we have to show that for any $\mathtt{r} : 1 \to 0$, if $\mathtt{r} \Vdash \bot$ then $!_{|A|} \circ \mathtt{r} \Vdash A$. However, $\mathtt{r} \Vdash \bot$ is impossible, by definition. The unique map $!_{|A|} : |A| \to 1$ realizes $A \Vdash \top$ because if $\mathtt{r} \Vdash A$ then $!_{|A|} \circ \mathtt{r} : 1 \to 1$. As 1 is terminal, we have $!_{|A|} \circ \mathtt{r} = !_1$ and as $!_1 \Vdash \top$, we reach $!_{|A|} \circ \mathtt{r} \Vdash \top$.

Example 198 The map $\mathtt{p}_0 : |A| \times |B| \to |A|$ realizes $\{A, B\} \Vdash A$. Let $\mathtt{r} : 1 \to |A \times B| = |A| \times |B|$. Then, $\mathtt{p}_0 \circ \mathtt{r} \Vdash A$, by definition. Similarly, one can show that $\mathtt{i}_1 : |B| \to |A| + |B|$ realizes $B \Vdash A \vee B$.

Example 199 If $C \Vdash A$ and $C \Vdash B$ then $C \Vdash A \wedge B$. Let $\mathtt{r} : |C| \to |A|$ and $\mathtt{s} : |C| \to |B|$ be the realizers, respectively. Then, $\langle \mathtt{r}, \mathtt{s} \rangle : |C| \to |A| \times |B|$ realizes $C \Vdash A \wedge B$. To prove that, let $\mathtt{u} : 1 \to |C|$ realize C. Then, by definition, $\mathtt{r} \circ \mathtt{u} \Vdash A$ and $\mathtt{s} \circ \mathtt{u} \Vdash B$. As $\langle \mathtt{r}, \mathtt{s} \rangle \circ \mathtt{u} = \langle \mathtt{r} \circ \mathtt{u}, \mathtt{s} \circ \mathtt{u} \rangle$, we reach $\langle \mathtt{r}, \mathtt{s} \rangle \circ \mathtt{u} \Vdash A \wedge B$, by definition.

Example 200 If $A \Vdash C$ and $B \Vdash C$ then $A \vee B \Vdash C$. Let $\mathtt{r} : |A| \to |C|$ and $\mathtt{s} : |B| \to |C|$ be the realizers, respectively. Then, $(\mathtt{r}, \mathtt{s}) : |A| + |B| \to |C|$ realizes $A \vee B \Vdash C$. To prove that, let $\mathtt{u} : 1 \to |A| + |B|$ realize $A \vee B$. Then, by definition, either there is $\mathtt{v} : 1 \to |A|$ such that $\mathtt{u} = \mathtt{i}_0\mathtt{v}$ and $\mathtt{v} \Vdash A$ or there is $\mathtt{w} : 1 \to |B|$ such that $\mathtt{u} = \mathtt{i}_1\mathtt{w}$ and $\mathtt{w} \Vdash B$. In the former case, we have $(\mathtt{r}, \mathtt{s}) \circ \mathtt{u} = (\mathtt{r}, \mathtt{s}) \circ \mathtt{i}_0\mathtt{v} = \mathtt{r} \circ \mathtt{v} \Vdash C$ and in the latter case $(\mathtt{r}, \mathtt{s}) \circ \mathtt{u} = (\mathtt{r}, \mathtt{s}) \circ \mathtt{i}_1\mathtt{w} = \mathtt{s} \circ \mathtt{w} \Vdash C$.

Example 201 If $C \Vdash A$ and $C \Vdash A \to B$ then $C \Vdash B$. Let $\mathtt{r} : |C| \to |A|$ and $\mathtt{s} : |C| \to [|A|, |B|]$ be the realizers, respectively. Then, the map $\mathrm{ev}\langle \mathtt{s}, \mathtt{r} \rangle : |C| \to |B|$ realizes $C \Vdash B$. To prove that, let $\mathtt{u} : 1 \to |C|$ be a map such that $\mathtt{u} \Vdash C$. Then, $\mathtt{s} \circ \mathtt{u} \Vdash A \to B$ and $\mathtt{r} \circ \mathtt{u} \Vdash A$. Hence, $(\mathtt{s} \circ \mathtt{u}) \cdot (\mathtt{r} \circ \mathtt{u}) \Vdash B$. As $\mathrm{ev}\langle \mathtt{s}, \mathtt{r} \rangle \circ \mathtt{u} = (\mathtt{s} \circ \mathtt{u}) \cdot (\mathtt{r} \circ \mathtt{u})$, we reach $\mathrm{ev}\langle \mathtt{s}, \mathtt{r} \rangle \circ \mathtt{u} \Vdash B$.

Example 202 If $\{C, A\} \Vdash B$ then $C \Vdash A \to B$. Let $\mathtt{r} : |C| \times |A| \to |B|$ be a realizer for $\{C, A\} \Vdash B$. We claim that the map $\lambda_{|A|}\mathtt{r} : |C| \to [|A|, |B|]$ realizes $C \Vdash A \to B$. To show that, let $\mathtt{s} : 1 \to |C|$ be a map such that $\mathtt{s} \Vdash C$. We have to show that $(\lambda_{|A|}\mathtt{r}) \circ \mathtt{s} \Vdash A \to B$. For that purpose, let $\mathtt{t} : 1 \to |A|$ be a map such that $\mathtt{t} \Vdash A$. Then, as $((\lambda_{|A|}\mathtt{r}) \circ \mathtt{s}) \cdot \mathtt{t} = \mathtt{r} \circ \langle \mathtt{s}, \mathtt{t} \rangle$ and we know that $\mathtt{r} \circ \langle \mathtt{s}, \mathtt{t} \rangle \Vdash B$, the proof is complete.

As might already be clear from the provided examples, our goal is to transform proofs in IPC into realizers, i.e., maps in $\mathcal{C}$. This transformation allows us to import the proofs into $\mathcal{C}$ and leverage the structure of $\mathcal{C}$ to gain insights about proofs.

Theorem 203 (Soundness-completeness) *For any set $\Gamma \cup \{A\}$ of formulas, $\Gamma \vdash_{\mathsf{IPC}} A$ iff $\Gamma \Vdash_{\mathcal{C}} A$ for any* BCC $\mathcal{C}$.

Proof For soundness, it is enough to use an induction on the length of the proof of $\Gamma \vdash_{\mathsf{IPC}} A$. Some cases are checked in the above examples. The rest is left to the reader. For completeness, set $C = \mathbf{NJ}$ and $\mathfrak{I}(p) = (p, \mathrm{Hom}_{\mathbf{NJ}}(\top, p))$. Then, as $\Gamma \Vdash_{\mathfrak{I}} A$, it has a realizer map $\mathtt{r} : |\Gamma| \to |A|$ in $\mathbf{NJ}$. Hence, there is a derivation of A from Γ in $\mathbf{NJ}$ which implies $\Gamma \vdash_{\mathsf{IPC}} A$. □

9.2 C-sets and Higher-Order Realizability

So far, we have covered our alternative formalization for the propositional BHK interpretation. To extend this formalization to higher-order languages, we need a notion of sets whose elements are constructed by maps in C. The following provides such a notion.

Definition 204 Let C be a BCC. By a *C-set*, we mean a tuple in the form $S = (|S|, \|S\|, \Vdash_S)$, where $|S|$ is an object in C, $\|S\|$ is a set and $\Vdash_S \subseteq \mathrm{Hom}_C(\mathtt{1}, |S|) \times \|S\|$ is a binary relation such that for any $x \in \|S\|$, there exists $\mathtt{x} : \mathtt{1} \to |S|$ such that $\mathtt{x} \Vdash_S x$. A *C-set map* $f : (|S|, \|S\|, \Vdash_S) \to (|T|, \|T\|, \Vdash_T)$ is a function $f : \|S\| \to \|T\|$ induced (or realized) by a map $\mathtt{f} : |S| \to |T|$ in C, i.e., if $\mathtt{x} \Vdash_S x$ then $\mathtt{fx} \Vdash_T f(x)$, for any $x \in \|S\|$ and any $\mathtt{x} : \mathtt{1} \to |S|$. It is easy to see that C-sets together with C-set maps form a category, denoted by $C\text{–}\mathbf{Set}$.[11]

Philosophical Comment 205 Interpreting C as the world of constructions, a C-set can be viewed as a *constructed set*. More precisely, the C-set $S = (|S|, \|S\|, \Vdash_S)$ consists of the set $\|S\|$ equipped with a type $|S|$ of constructions and a relation $\mathtt{x} \Vdash_S x$ indicating that $\mathtt{x} : \mathtt{1} \to |S|$ is a construction of x. The surjectivity condition ensures that every element has a construction. Similarly, a C-set map is a function induced by changing constructions via compositions with maps within C.

The category $C\text{–}\mathbf{Set}$ is a BCC. We only present the BC-structure without the proof that it really is the BC-structure. For more details, see (Birkedal 2002, 1999; Bauer 2000, 2005). The terminal object of $C\text{–}\mathbf{Set}$ is the C-set $1 = (\mathtt{1}, \{*\}, \Vdash_1)$, where $\mathtt{1}$ is the terminal object of C and $\Vdash_1$ is defined by $id_{\mathtt{1}} \Vdash_1 *$. The initial object of $C\text{–}\mathbf{Set}$ is the C-set $0 = (\emptyset, \varnothing, \Vdash_0)$, where $\emptyset$ is the initial object of C and $\Vdash_0$ is the empty relation.

For products, coproducts and exponentials, let $S = (|S|, \|S\|, \Vdash_S)$ and $T = (|T|, \|T\|, \Vdash_T)$ be two C-sets. Then, for the product of S and T, consider the tuple $S \times T = (|S| \times |T|, \|S\| \times \|T\|, \Vdash_{S\times T})$, where $\mathtt{x} \Vdash_{S\times T} x$ iff $\mathtt{p_0x} \Vdash_S p_0(x)$ and $\mathtt{p_1x} \Vdash_T p_1(x)$, for any $\mathtt{x} : \mathtt{1} \to |S| \times |T|$ and any $x \in \|S\| \times \|T\|$. It is easy to see that $S \times T$ is a C-set. Moreover, it is clear that the set projection functions are induced by the projection maps in C and hence they are C-set maps. The data of $S \times T$,

[11] Following the usual convention in the literature, C-sets must be called assemblies over C. We use the term C-set because it better aligns with the intuition that it is a set whose elements are constructed via the maps in C.

together with the projections is the binary product of S and T. For the coproduct of S and T, consider the tuple $S + T = (|S| + |T|, \|S\| + \|T\|, \Vdash_{S+T})$, where $\mathtt{x} \Vdash_{S+T} x$ iff there are $\mathtt{y} : 1 \to |S|$ and $y \in \|S\|$ such that $\mathtt{x} = \mathtt{i}_0\mathtt{y}$, $x = (0, y)$ and $\mathtt{y} \Vdash_S y$ or there are $\mathtt{z} : 1 \to |T|$ and $z \in \|T\|$ such that $\mathtt{x} = \mathtt{i}_1\mathtt{z}$, $x = (1, z)$ and $\mathtt{z} \Vdash_S z$. It is easy to see that $S + T$ is a C-set. Moreover, the set injection functions are induced by the injection maps in C. Therefore, we can consider them as C-set maps. The data of $S + T$ together with the injections is the binary coproduct of S and T. Finally, for the exponential $[S, T]$, consider the tuple

$$[S, T] = ([|S|, |T|], \mathrm{Hom}_{C\text{-}\mathbf{Set}}(S, T), \Vdash_{[S,T]}),$$

where $\mathtt{f} \Vdash_{[S,T]} f$ iff $\mathtt{x} \Vdash_S x$ implies $\mathtt{f} \cdot \mathtt{x} \Vdash_T f(x)$, for any $\mathtt{x} : 1 \to |S|$ and any $x \in \|S\|$. It is easy to see that $[S, T]$ is a C-set. Moreover, the usual evaluation map is induced by the evaluation map in C. Therefore, we can consider it as a C-set map. The data of $[S, T]$ together with the evaluation map is the exponentiation of T by S.

With a suitable notion of constructed sets in place, we can now extend the BHK interpretation to higher-order languages.

Definition 206 Let $\mathcal{L}$ be a higher-order language and C be a BCC. By a *C-interpretation* of $\mathcal{L}$, we mean a tuple

$$\mathfrak{I} = (\{M^\sigma\}_{\sigma \in T(\mathcal{L})}, \{f^\mathfrak{I}\}_{f \in F(\mathcal{L})}, \{P^\mathfrak{I}\}_{P \in P(\mathcal{L})}),$$

where M^σ is a C-set such that $\|M^\sigma\| \neq \varnothing$ and $M^{\sigma \times \tau} = M^\sigma \times M^\tau$ and $M^{\sigma \to \tau} = [M^\sigma, M^\tau]$, for any types σ and τ, $f^\mathfrak{I} : M^{\sigma_1} \times \cdots \times M^{\sigma_n} \to M^\tau$ is a C-set map, for any n-ary function symbol $f(x_1^{\sigma_1}, \cdots, x_n^{\sigma_n})$ of type τ and $P^\mathfrak{I} = (|P|, R_P)$, where $|P|$ is an object in C and $R_P \subseteq \mathrm{Hom}_C(1, |P|) \times \|M^{\sigma_1}\| \times \cdots \times \|M^{\sigma_n}\|$ is a relation, for any n-ary predicate symbol $P(x_1^{\sigma_1}, \cdots, x_n^{\sigma_n})$.

The C-interpretation $\mathfrak{I}$ assigns a C-set map to any function symbol of the language. We can extend the assignment to all terms of the language. However, let us first extend the language $\mathcal{L}$ by adding a constant of type σ, for any $a \in \|M^\sigma\|$. This is a usual technique to have a name for any element in the domain. For simplicity, we denote this constant by a itself and call the extended language $\mathcal{L}(\mathfrak{I})$. Now, we interpret any $\mathcal{L}(\mathfrak{I})$-term $t(\bar{x}) = t(x^{\sigma_1}, \ldots, x^{\sigma_n})$ of type τ as a C-set map $t(\bar{x})^\mathfrak{I} : M^{\sigma_1} \times \cdots \times M^{\sigma_n} \to M^\tau$ using the interpretation of the function symbols and composition. For the new constants, we interpret the constant $a \in \|M^\sigma\|$ as the map $1 \to M^\sigma$ that picks a from $\|M^\sigma\|$. The next step is extending the C-interpretation $\mathfrak{I}$ to all *sentences* in the language $\mathcal{L}(\mathfrak{I})$. First, assign an object in C to any $\mathcal{L}(\mathfrak{I})$*-formula* in the following way:

- $|\top| = 1$, $|\bot| = 0$, $|t = s| = 1$, $|P(t_1, \ldots, t_n)| = |P|$,
- $|A \wedge B| = |A| \times |B|$, $|A \vee B| = |A| + |B|$, $|A \to B| = [|A|, |B|]$,
- $|\forall x^\sigma A(x^\sigma)| = [|M^\sigma|, |A(x^\sigma)|]$, $|\exists x^\sigma A(x^\sigma)| = |M^\sigma| \times |A(x^\sigma)|$.

Remark 207 Note that for any formula $A(\bar{x})$ and any two families of terms $\bar{t}$ and $\bar{s}$ of the appropriate types, we have $|A(\bar{t})| = |A(\bar{s})|$. This demonstrates a restriction in our approach, highlighting that our object-assignment is essentially propositional and does not account for the dependency of the formulas on the variables. To emphasize this fact, we use the fixed name $|A|$ to denote $|A(\bar{t})|$, regardless of the family of terms $\bar{t}$.

Define the relation $\mathtt{r} \Vdash A$ between the map $\mathtt{r} \in \mathrm{Hom}_C(\mathtt{1}, |A|)$ and the $\mathcal{L}(\mathfrak{I})$-*sentence* A in the following way:

- $!_{\mathtt{1}} \Vdash \top$, for the unique map $!_{\mathtt{1}} : \mathtt{1} \to \mathtt{1}$,
- $\mathtt{r} \nVdash \bot$, for any map $\mathtt{r} : \mathtt{1} \to \mathtt{0}$,
- $!_{\mathtt{1}} \Vdash t = s$ iff $t^{\mathfrak{I}} = s^{\mathfrak{I}}$, for the unique map $!_{\mathtt{1}} : \mathtt{1} \to \mathtt{1}$,
- $\mathtt{r} \Vdash P(t_1, \cdots, t_n)$ iff $(\mathtt{r}, t_1^{\mathfrak{I}}, \ldots, t_n^{\mathfrak{I}}) \in R_P$, for any $\mathtt{r} : \mathtt{1} \to |P|$, where the C-set map $t_i^{\mathfrak{I}} : 1 \to M^{\sigma_i}$ is interpreted as an element in $\|M^{\sigma_i}\|$,
- $\mathtt{r} \Vdash A \wedge B$ iff $\mathtt{p}_0\mathtt{r} \Vdash A$ and $\mathtt{p}_1\mathtt{r} \Vdash B$, for any $\mathtt{r} : \mathtt{1} \to |A| \times |B|$,
- $\mathtt{r} \Vdash A \vee B$ iff there exists $\mathtt{s} : \mathtt{1} \to |A|$ such that $\mathtt{r} = \mathtt{i}_0\mathtt{s}$ and $\mathtt{s} \Vdash A$ or there exists $\mathtt{t} : \mathtt{1} \to |B|$ such that $\mathtt{r} = \mathtt{i}_1\mathtt{t}$ and $\mathtt{t} \Vdash B$, for any $\mathtt{r} : \mathtt{1} \to |A| + |B|$,
- $\mathtt{r} \Vdash A \to B$ iff for any $\mathtt{s} : \mathtt{1} \to |A|$, if $\mathtt{s} \Vdash A$ then $\mathtt{r} \cdot \mathtt{s} \Vdash B$, for any $\mathtt{r} : \mathtt{1} \to [|A|, |B|]$,
- $\mathtt{r} \Vdash \forall x^\sigma A(x^\sigma)$ iff for any $\mathtt{a} : \mathtt{1} \to |M^\sigma|$ and any $a \in \|M^\sigma\|$, if $\mathtt{a} \Vdash_{M^\sigma} a$ then $\mathtt{r} \cdot \mathtt{a} \Vdash A(a)$, for any $\mathtt{r} : \mathtt{1} \to [|M^\sigma|, |A|]$,
- $\mathtt{r} \Vdash \exists x^\sigma A(x^\sigma)$ iff there is $a \in \|M^\sigma\|$ such that $\mathtt{p}_0\mathtt{r} \Vdash_{M^\sigma} a$ and $\mathtt{p}_1\mathtt{r} \Vdash A(a)$, for any $\mathtt{r} : \mathtt{1} \to |M^\sigma| \times |A|$.

Let $\bar{x}$ be a *finite* set of variables of types $\sigma_1, \ldots, \sigma_n$. For any finite set of formulas $\Gamma(\bar{x})$, define $|\Gamma|$ as $\prod_{\gamma \in \Gamma} |\gamma|$. We use an arbitrary order on the formulas in Γ, and since the product is commutative up to an isomorphism, the chosen order is actually immaterial. Then, by $\Gamma(\bar{x}) \Vdash_{\mathfrak{I}} A(\bar{x})$, we mean the existence of a map $\mathtt{r} : |M^{\sigma_1}| \times \cdots \times |M^{\sigma_n}| \times |\Gamma| \to |A|$ such that $\mathtt{r} \circ \langle \langle \mathtt{a}_i \rangle_{i=1}^n, \mathtt{s} \rangle \Vdash A(\bar{a})$, for any $\{\mathtt{a}_i : \mathtt{1} \to |M^{\sigma_i}|\}_{i=1}^n$, any $\{a_i \in \|M^{\sigma_i}\|\}_{i=1}^n$ and any $\mathtt{s} : \mathtt{1} \to |\Gamma|$ such that $\mathtt{a}_i \Vdash_{M^{\sigma_i}} a_i$, for any $1 \le i \le n$ and $\mathtt{s} \Vdash \bigwedge \Gamma(\bar{a})$. This $\mathtt{r}$ is called a *realizer* for $\Gamma(\bar{x}) \Vdash_{\mathfrak{I}} A(\bar{x})$. When $\Gamma(\bar{x})$ is empty, $\mathtt{r}$ is called a realizer for $A(\bar{x})$. Again, we used an arbitrary order on the variables in $\bar{x}$ which is immaterial. For an infinite set $\Gamma(\bar{x})$, by $\Gamma(\bar{x}) \Vdash_{\mathfrak{I}} A(\bar{x})$, we mean the existence of a finite set $\Delta(\bar{x}) \subseteq \Gamma(\bar{x})$ such that $\Delta(\bar{x}) \Vdash_{\mathfrak{I}} A(\bar{x})$. When the C-interpretation $\mathfrak{I}$ is clear from the context, we drop the subscript $\mathfrak{I}$ in $\Vdash_{\mathfrak{I}}$. Finally, by $\Gamma(\bar{x}) \Vdash_C A(\bar{x})$, we mean $\Gamma(\bar{x}) \Vdash_{\mathfrak{I}} A(\bar{x})$, for any C-interpretation $\mathfrak{I}$.

Remark 208 The definition of $\Gamma(\bar{x}) \Vdash A(\bar{x})$ can be somewhat ambiguous since it is not immediately clear which variables are considered free in the formulas in $\Gamma(\bar{x}) \cup A(\bar{x})$. Naturally, we would choose $\bar{x}$ to be the set of all free variables in $\Gamma(\bar{x}) \cup A(\bar{x})$. However, there are cases where other variables need to be treated as free. For instance, in the formula $A(x) \wedge B(y)$, the set of free variables is $\{x, y\}$. However, the definition of realizability for this formula requires us to consider $A(x)$ with free variables $\{x, y\}$ rather than just $\{x\}$. Fortunately, extending the set of free variables does not impact realizability. This is because $\|M^\tau\| \neq \varnothing$ for any type τ, which implies that there is always a map $\mathtt{a} : 1 \to |M^\tau|$. Thus, using $\mathtt{a}$, the existence of a realizer $\mathtt{r} : |M^{\sigma_1}| \times \cdots \times |M^{\sigma_n}| \times |\Gamma| \to |A|$ is equivalent to the existence of a realizer $\mathtt{s} : |M^\tau| \times |M^{\sigma_1}| \times \cdots \times |M^{\sigma_n}| \times |\Gamma| \to |A|$, when the variable of type τ is a dummy in $A(\bar{x})$.

Example 209 It is straightforward to verify that the examples of propositional realizability we previously discussed extend to the case where variables are present. For instance, the following examples hold: $[A(\bar{x}) \Vdash A(\bar{x})]$, $[\bot \Vdash A(\bar{x})]$, $[A(\bar{x}) \Vdash \top]$, $[\{A(\bar{x}), B(\bar{x})\} \Vdash A(\bar{x})]$, and $[B(\bar{x}) \Vdash A(\bar{x}) \vee B(\bar{x})]$.

Example 210 Let $\mathtt{p} : |M^\sigma| \times |M^\sigma| \times 1 \times |A| \to |A|$ be the projection map on the last component. Then $\mathtt{p}$ realizes $(x^\sigma = y^\sigma), A(x^\sigma) \Vdash A(y^\sigma)$. The reason is that for any $\mathtt{a}, \mathtt{b} : 1 \to |M^\sigma|$, any $a, b \in \|M^\sigma\|$, any $\mathtt{r} : 1 \to 1$ and any $\mathtt{s} : 1 \to |A|$, if $\mathtt{a} \Vdash_{M^\sigma} a, \mathtt{b} \Vdash_{M^\sigma} b, \mathtt{r} \Vdash a = b$ and $\mathtt{s} \Vdash A(a)$, we must have $\mathtt{r} = !_1$ and hence $a = b$, by definition. Therefore, $\mathtt{p} \circ \langle \mathtt{a}, \mathtt{b}, \mathtt{r}, \mathtt{s} \rangle = \mathtt{s} \Vdash A(b)$.

Example 211 Let $t(\bar{x}) = t(x_1^{\sigma_1}, \ldots, x_n^{\sigma_n})$ be an $\mathcal{L}(\mathfrak{J})$-term of type τ. The interpretation of $t(\bar{x})$ is the C-set map $t^{\mathfrak{J}} : M^{\sigma_1} \times \cdots \times M^{\sigma_n} \to M^\tau$ and as it is a C-set map, it is induced by a map $\mathtt{t} : |M^{\sigma_1}| \times \cdots \times |M^{\sigma_n}| \to |M^\tau|$ in C. Then, for any $\{\mathtt{a}_i : 1 \to |M^{\sigma_i}|\}_{i=1}^n$ and $\{a_i \in \|M^{\sigma_i}\|\}_{i=1}^n$ such that $\mathtt{a}_i \Vdash_{M^{\sigma_i}} a_i$, for any $1 \leq i \leq n$, it is easy to see that $\mathtt{t}\langle \mathtt{a}_i \rangle_{i=1}^n \Vdash_{M^\tau} t^{\mathfrak{J}}(\bar{a})$. This claim holds for any such $\mathtt{t}$.

Example 212 Let $t(\bar{x}) = t(x_1^{\sigma_1}, \ldots, x_n^{\sigma_n})$ be an $\mathcal{L}(\mathfrak{J})$-term of type τ. We claim that $A(t(\bar{x}), \bar{x}) \Vdash \exists y^\tau A(y, \bar{x})$. Using the previous example, let $\mathtt{t} : |M^{\sigma_1}| \times \cdots \times |M^{\sigma_n}| \to |M^\tau|$ be a map inducing $t^{\mathfrak{J}}$. Now, define the map $\mathtt{r} = \langle \mathtt{tp}, \mathtt{q} \rangle : |M^{\sigma_1}| \times \cdots \times |M^{\sigma_n}| \times |A| \to |M^\tau| \times |A|$, where $\mathtt{p}$ is the projection on $|M^{\sigma_1}| \times \cdots \times |M^{\sigma_n}|$ and $\mathtt{q}$ is the projection on $|A|$. Then, $\mathtt{r}$ realizes $A(t(\bar{x}), \bar{x}) \Vdash \exists y^\tau A(y, \bar{x})$. To prove that, let $\{\mathtt{a}_i : 1 \to |M^{\sigma_i}|\}_{i=1}^n$, $\{a_i \in \|M^{\sigma_i}\|\}_{i=1}^n$, and $\mathtt{s} : 1 \to |A|$ such that $\mathtt{a}_i \Vdash_{M^{\sigma_i}} a_i$, for any $1 \leq i \leq n$ and $\mathtt{s} \Vdash A(t(\bar{a}), \bar{a})$. As $\mathtt{t}\langle \mathtt{a}_i \rangle_{i=1}^n \Vdash_{M^\tau} t^{\mathfrak{J}}(\bar{a})$, we reach $\langle \mathtt{t}\langle \mathtt{a}_i \rangle_{i=1}^n, \mathtt{s} \rangle \Vdash \exists y^\tau A(y, \bar{a})$. As $\mathtt{r} \circ \langle \langle \mathtt{a}_i \rangle_{i=1}^n, \mathtt{s} \rangle = \langle \mathtt{t}\langle \mathtt{a}_i \rangle_{i=1}^n, \mathtt{s} \rangle$, the proof is complete.

Example 213 If $A(\bar{x}) \Vdash B(\bar{x}, y)$ and $A(\bar{x})$ has no free variable y, then $A(\bar{x}) \Vdash \forall y^\tau B(\bar{x}, y)$. Let $\mathtt{r} : |M^{\sigma_1}| \times \cdots \times |M^{\sigma_n}| \times |M^\tau| \times |A| \to |B|$ be the realizer for $A(\bar{x}) \Vdash B(\bar{x}, y)$. Using the commutativity of the product, we can transform $\mathtt{r}$ to the map $\mathtt{s} : |M^{\sigma_1}| \times \cdots \times |M^{\sigma_n}| \times |A| \times |M^\tau| \to |B|$. Then, the map $\lambda_{|M^\tau|}\mathtt{s} : |M^{\sigma_1}| \times \cdots \times |M^{\sigma_n}| \times |A| \to [|M^\tau|, |B|]$ realizes $A(\bar{x}) \Vdash \forall y^\tau B(\bar{x}, y)$. The proof follows a similar method to the one used for implications, as detailed in Example 202.

Theorem 214 (Soundness) *For any set* $\Gamma(\bar{x}) \cup \{A(\bar{x})\}$ *of formulas, if* $\Gamma(\bar{x}) \vdash_{\mathsf{IQC}^\omega + \mathsf{EXT}} A(\bar{x})$ *then* $\Gamma(\bar{x}) \Vdash_C A(\bar{x})$, *for any* BCC C.

Proof The proof is by induction on the length of the proof of $\Gamma(\bar{x}) \vdash_{\mathsf{IQC}^\omega + \mathsf{EXT}} A(\bar{x})$. We have already realized the propositional rules as shown in Theorem 203. For the quantifier rules, as well as the axioms for equality and EXT, some specific instances have been addressed in the examples above. The remaining cases are left as an exercise for the reader. □

Since our notion of realizability is a *special case* of the semantics for higher-order languages, we do not anticipate completeness. The following example illustrates a concrete counterexample.

Example 215 (*Incompleteness of Realizability*) For any sentence A, any formula $B(x^\sigma)$ and any C-interpretation $\mathfrak{I}$, we show that $(A \to \exists x^\sigma B(x)) \Vdash_{\mathfrak{I}} \exists x^\sigma (A \to B(x))$. Notice that $(A \to \exists x^\sigma B(x)) \nvdash \exists x^\sigma (A \to B(x))$ in intuitionistic logic. This provides a counterexample to the completeness of our realizability notion. To prove the realizability part, note that $|A \to \exists x^\sigma B(x)| = [|A|, |M^\sigma| \times |B|]$ and $|\exists x^\sigma (A \to B(x))| = |M^\sigma| \times [|A|, |B|]$. There are two cases to consider. Either $\Vdash_{\mathfrak{I}} A$ or $\nVdash_{\mathfrak{I}} A$. In the first case, let $\mathtt{r} \Vdash_{\mathfrak{I}} A$. Then, define

$$\mathtt{F}_0 = \quad [|A|, |M^\sigma| \times |B|] \xrightarrow{\langle \mathtt{id}, \mathtt{r}\circ ! \rangle} [|A|, |M^\sigma| \times |B|] \times |A| \xrightarrow{\mathtt{p}_0 \circ \mathtt{ev}} |M^\sigma|$$

$$\mathtt{F}_1 = \quad [|A|, |M^\sigma| \times |B|] \times |A| \xrightarrow{\mathtt{id} \times (\mathtt{r}\circ !)} [|A|, |M^\sigma| \times |B|] \times |A| \xrightarrow{\mathtt{p}_1 \circ \mathtt{ev}} |B|$$

We claim that $\mathtt{F} = \langle \mathtt{F}_0, \lambda_{|A|} \mathtt{F}_1 \rangle$ realizes $(A \to \exists x^\sigma B(x)) \Vdash \exists x^\sigma (A \to B(x))$. Let $\mathtt{s} \Vdash A \to \exists x^\sigma B(x)$. Then, $\mathtt{s} \cdot \mathtt{r} \Vdash \exists x^\sigma B(x)$ which implies the existence of $a \in \|M^\sigma\|$ such that $\mathtt{p}_0(\mathtt{s} \cdot \mathtt{r}) \Vdash_{M^\sigma} a$ and $\mathtt{p}_1(\mathtt{s} \cdot \mathtt{r}) \Vdash B(a)$. Now, we have to show that $\mathtt{F} \circ \mathtt{s} \Vdash \exists x^\sigma (A \to B(x))$. First, notice that $\mathtt{p}_0(\mathtt{F} \circ \mathtt{s}) = \mathtt{F}_0 \circ \mathtt{s} = \mathtt{p}_0(\mathtt{s} \cdot \mathtt{r}) \Vdash_{M^\sigma} a$. Second, we want to show that $\mathtt{p}_1(\mathtt{F} \circ \mathtt{s}) \Vdash A \to B(a)$. Let $\mathtt{t} \Vdash A$. Then, $\mathtt{p}_1(\mathtt{F} \circ \mathtt{s}) \cdot \mathtt{t} = \mathtt{p}_1(\mathtt{s} \cdot \mathtt{r}) \Vdash B(a)$ which completes the proof.

In the second case, we assume $\nVdash_{\mathfrak{I}} A$. As $\|M^\sigma\| \neq \varnothing$, we can pick a map $\mathtt{a} : 1 \to |M^\sigma|$ and an element $a \in \|M^\sigma\|$ such that $\mathtt{a} \Vdash_{M^\sigma} a$. Define $\mathtt{F}_0 = \mathtt{a}\circ ! : [|A|, |M^\sigma| \times |B|] \to |M^\sigma|$ and $\mathtt{F}_1 = \mathtt{p}_1 \circ \mathtt{ev} : [|A|, |M^\sigma| \times |B|] \times |A| \to |B|$. We claim that $\mathtt{F} = \langle \mathtt{F}_0, \lambda_{|A|} \mathtt{F}_1 \rangle$ realizes $(A \to \exists x^\sigma B(x)) \Vdash \exists x^\sigma (A \to B(x))$. Let $\mathtt{s} \Vdash A \to \exists x^\sigma B(x)$. First, notice that $\mathtt{p}_0(\mathtt{F} \circ \mathtt{s}) = \mathtt{F}_0 \circ \mathtt{s} = \mathtt{a} \Vdash_{M^\sigma} a$. Second, we have $\mathtt{p}_1(\mathtt{F} \circ \mathtt{s}) \Vdash A \to B(a)$ as there is no $\mathtt{t} \Vdash A$.

9.3 Projective C-sets and the Axiom of Choice

In this subsection, we investigate the situations where the axiom of choice is realizable. To do this, we first need to introduce projective C-sets.

Definition 216 A C-set $S = (|S|, \|S\|, \Vdash_S)$ is called *projective* if $\mathtt{x}_0 \Vdash_S x$ and $\mathtt{x}_1 \Vdash_S x$ implies $\mathtt{x}_0 = \mathtt{x}_1$, for any $\mathtt{x}_0, \mathtt{x}_1 : 1 \to |S|$ and any $x \in \|S\|$.

Theorem 217 *Let $\mathfrak{I}$ be a C-interpretation that maps the type σ to a projective C-set. Then, the sentence $\forall x^\sigma \exists y^\tau A(x, y) \to \exists f^{\sigma \to \tau} \forall x^\sigma A(x, fx)$ is realizable.*

Proof Note that $|\forall x^\sigma \exists y^\tau A(x, y)| = [|M^\sigma|, |M^\tau| \times |A|]$ and $|\exists f^{\sigma\to\tau} \forall x^\sigma A(x, fx)| = |M^{\sigma\to\tau}| \times [|M^\sigma|, |A|]$. Consider the maps $\mathtt{F}_0$ and $\mathtt{F}_1$ defined as below

$$\mathtt{F}_0 = \ [|M^\sigma|, |M^\tau| \times |A|] \times |M^\sigma| \xrightarrow{\ \mathtt{ev}\ } |M^\tau| \times |A| \xrightarrow{\ \mathtt{p}_0\ } |M^\tau|$$

$$\mathtt{F}_1 = \ [|M^\sigma|, |M^\tau| \times |A|] \times |M^\sigma| \xrightarrow{\ \mathtt{ev}\ } |M^\tau| \times |A| \xrightarrow{\ \mathtt{p}_1\ } |A|$$

We claim that the map

$$\mathtt{F} = \langle \lambda_{|M^\sigma|}\mathtt{F}_0, \lambda_{|M^\sigma|}\mathtt{F}_1 \rangle : [|M^\sigma|, |M^\tau| \times |A|] \to |M^{\sigma\to\tau}| \times [|M^\sigma|, |A|]$$

realizes

$$\forall x^\sigma \exists y^\tau A(x, y) \Vdash \exists f^{\sigma\to\tau} \forall x^\sigma A(x, fx).$$

Let $\mathtt{r} \Vdash \forall x^\sigma \exists y^\tau A(x, y)$. As M^σ is projective, for any $a \in \|M^\sigma\|$, there is a unique $\mathtt{a}_0 : 1 \to |M^\sigma|$ such that $\mathtt{a}_0 \Vdash_{M^\sigma} a$. Hence, we have $\mathtt{r} \cdot \mathtt{a}_0 \Vdash \exists y^\tau A(a, y)$ which implies the existence of $b \in \|M^\tau\|$ such that $\mathtt{p}_0(\mathtt{r} \cdot \mathtt{a}_0) \Vdash_{M^\tau} b$ and $\mathtt{p}_1(\mathtt{r} \cdot \mathtt{a}_0) \Vdash A(a, b)$. For any $a \in \|M^\sigma\|$, choose such a $b \in \|M^\tau\|$ and define $g : \|M^\sigma\| \to \|M^\tau\|$ by setting $g(a) = b$. We claim that $\mathtt{p}_0(\mathtt{F} \circ \mathtt{r}) \Vdash_{[M^\sigma, M^\tau]} g$. For the proof, let $\mathtt{a} \Vdash_{M^\sigma} a$. By the projectivity of M^σ, we have $\mathtt{a} = \mathtt{a}_0$. As $\mathtt{p}_0(\mathtt{F} \circ \mathtt{r}) \cdot \mathtt{a} = ((\lambda_{|M^\sigma|}\mathtt{F}_0) \circ \mathtt{r}) \cdot \mathtt{a}_0 = \mathtt{F}_0 \circ \langle \mathtt{r}, \mathtt{a}_0 \rangle = \mathtt{p}_0(\mathtt{r} \cdot \mathtt{a}_0)$ and $\mathtt{p}_0(\mathtt{r} \cdot \mathtt{a}_0) \Vdash_{M^\tau} b = g(a)$, we reach $\mathtt{p}_0(\mathtt{F} \circ \mathtt{r}) \cdot \mathtt{a} \Vdash_{M^\tau} g(a)$. Hence, $\mathtt{p}_0(\mathtt{F} \circ \mathtt{r}) \Vdash_{[M^\sigma, M^\tau]} g$. Similarly, we can prove that $\mathtt{p}_1(\mathtt{F} \circ \mathtt{r}) \Vdash \forall x^\sigma A(x, gx)$. Therefore, $\mathtt{F} \circ \mathtt{r} \Vdash \exists f^{\sigma\to\tau} \forall x^\sigma A(x, fx)$. □

Unfortunately, the exponentiation of projective C-sets is not necessarily projective. Therefore, even if the C-interpretation maps all basic types to projective C-sets, this does not guarantee that all M^σ's are projective. However, in certain special cases, we can ensure that all M^σ are projective, which in turn implies that the axiom of choice is realizable. To explain this, we need two definitions.

Definition 218 A category C with a terminal object 1 is called *well-pointed* if, for any two maps $f, g : A \to B$, having $f \circ h = g \circ h$ for any $h : 1 \to A$ implies $f = g$.

Example 219 The categories **Set** and **Top** are well-pointed while $\mathbf{Set}^\mathbf{2}$ is not. For the latter claim, consider the inclusion function $i : \{0\} \to \{0, 1\}$ as an object in $\mathbf{Set}^\mathbf{2}$ and the map $\alpha : i \to i$:

$$\begin{array}{ccc} \{0, 1\} & \xrightarrow{\ \alpha_1\ } & \{0, 1\} \\ \uparrow{\scriptstyle i} & & \uparrow{\scriptstyle i} \\ \{0\} & \xrightarrow[\ \alpha_0\ ]{} & \{0\} \end{array}$$

defined by $\alpha_0 = id_{\{0\}}$ and α_1 as the constant zero function. It is easy to see that the only map from the terminal object $1 = \{*\} \to \{*\}$ to i is

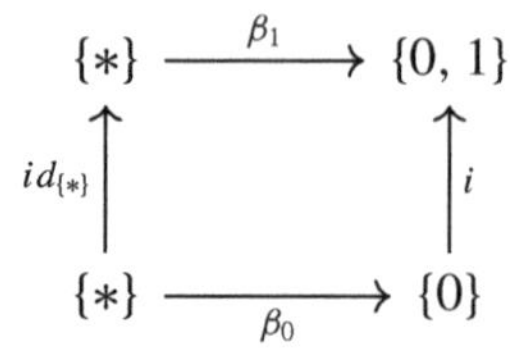

where $\beta_0(*) = \beta_1(*) = 0$. As $\alpha \circ \beta = id_i \circ \beta$ and $\alpha \neq id_i$, the category $\mathbf{Set^2}$ is not well-pointed.

Remark 220 Let C be a well-pointed CCC and $f, g : 1 \to [A, B]$ be two maps. Then, having $f \cdot h = g \cdot h$, for any $h : 1 \to A$, implies $f = g$. To prove this claim, as any map $k : 1 \to [A, B]$ is in the form $\lambda_A k'$, for some $k' : A \to B$, it is enough to pick $f', g' : A \to B$ such that $f = \lambda_A f'$ and $g = \lambda_A g'$ and show that $f' = g'$. Now, let $h : 1 \to A$. Then, $f \cdot h = (\lambda_A f') \cdot h = f' \circ h$. Similarly, $g \cdot h = g' \circ h$. As $f \cdot h = g \cdot h$, we have $f' \circ h = g' \circ h$, for any $h : 1 \to A$. As C is well-pointed, $f' = g'$ which implies $f = g$.

Definition 221 A C-set $S = (|S|, \|S\|, \Vdash_S)$ is called *total* if for any $\mathtt{x} : 1 \to |S|$ there is $x \in \|S\|$ such that $\mathtt{x} \Vdash_S x$.

Lemma 222 *Let S and T be C-sets. Then:*

(i) *If S and T are projective (resp. total), then $S \times T$ is also projective (resp. total).*
(ii) *If C is well-pointed, S is total and T is projective, $[S, T]$ is projective.*
(iii) *If S is projective and T is total, $[S, T]$ is total.*

Proof (i) is easy to prove. For (ii), let $\mathtt{f} \Vdash_{[S,T]} h$, $\mathtt{g} \Vdash_{[S,T]} h$, and $\mathtt{x} : 1 \to |S|$ be an arbitrary map. As S is total, there is an element $x \in \|S\|$ such that $\mathtt{x} \Vdash_S x$. Hence, we have $\mathtt{f} \cdot \mathtt{x} \Vdash_T h(x)$ and $\mathtt{g} \cdot \mathtt{x} \Vdash_T h(x)$. As T is projective, we have $\mathtt{f} \cdot \mathtt{x} = \mathtt{g} \cdot \mathtt{x}$. As $\mathtt{x} : 1 \to |S|$ is arbitrary and C is well-pointed, we have $\mathtt{f} = \mathtt{g}$, by Remark 220. For (iii), let $\mathtt{f} : 1 \to [|S|, |T|]$. We define a function $f : \|S\| \to \|T\|$ such that $\mathtt{f} \Vdash_{[S,T]} f$. As S is projective, for any $x \in \|S\|$, there is a unique $\mathtt{x} : 1 \to |S|$ such that $\mathtt{x} \Vdash_S x$. Hence, $\mathtt{f} \cdot \mathtt{x} : 1 \to |T|$. As T is total, there is an element $y \in \|T\|$ such that $\mathtt{f} \cdot \mathtt{x} \Vdash_T y$. Choose such a y and define $f(x) = y$. To prove $\mathtt{f} \Vdash_{[S,T]} f$, let $\mathtt{z} \Vdash_S x$. As S is projective, $\mathtt{z} = \mathtt{x}$. Then, by definition, $f(x)$ is chosen such that $\mathtt{f} \cdot \mathtt{z} = \mathtt{f} \cdot \mathtt{x} \Vdash_T f(x)$. □

Using Lemma 222, if C is well-pointed and the C-interpretation $\mathfrak{I}$ maps all basic types to total projective C-sets, then $\mathfrak{I}$ maps all types to total projective C-sets. Consequently, the axiom of choice is realizable for any such C-interpretation. We will use this fact later in Sect. 10.

10 Realizability for Arithmetical Theories

In this section, we intend to apply the realizability machinery to the arithmetical theories introduced in Sect. 8. To achieve this, we need two additional components. First, we must introduce two new BCC's to capture the worlds of computable and continuous constructions that we informally presented in Sect. 8. Second, we need a categorical way of representing the set of natural numbers using a categorical gadget called the *natural numbers object*. These two components will be covered in Sects. 10.1 and 10.2, respectively. Then, in Sect. 10.3, we will use the previously introduced BCC's to prove some arithmetical consistencies and to extract information from the arithmetical proofs.

10.1 A Recursive and a Continuous World

In this subsection, we will introduce two categories to formalize the previously mentioned computable and continuous worlds. For the latter, the canonical candidate is the category **Top**. However, **Top** does not have all the exponentials, making it somewhat cumbersome to work with. To achieve a BCC with a continuous nature similar to that of **Top**, we need to extend topological spaces by incorporating their own local equality relations.

Definition 223 An *equilogical space* is a pair $(X, \sim_X)$, where X is a topological space and $\sim_X$ is an equivalence relation on X. One can read the equilogical space $(X, \sim_X)$ as a proposition whose finitary proofs are stored in X and considered up to the equivalence $\sim_X$. An *equilogical map* $f : (X, \sim_X) \to (Y, \sim_Y)$ is a continuous function $f : X \to Y$ preserving the equivalence relation, i.e., if $x \sim_X y$ then $f(x) \sim_Y f(y)$, for any $x, y \in X$. Two equilogical maps $f, g : (X, \sim_X) \to (Y, \sim_Y)$ are considered equivalent, denoted by $f \sim g$, if $f(x) \sim_Y g(x)$, for any $x \in X$. The relation $\sim$ on the equilogical maps from $(X, \sim_X)$ to $(Y, \sim_Y)$ is an equivalence relation, respecting the composition of functions, i.e., if $f_1 \sim f_2$ and $g_1 \sim g_2$, then $f_1 \circ g_1 \sim f_2 \circ g_2$, for any $f_1, f_2 : (Y, \sim_Y) \to (Z, \sim_Z)$ and $g_1, g_2 : (X, \sim_X) \to (Y, \sim_Y)$. The collection of equilogical spaces and equilogical maps (up to the equivalence) with the evident composition and identity constitutes a category, denoted by **Equ**.

Philosophical Comment 224 An equilogical space should be understood as a constructive space, where both the set of elements and their equality are primitive notions. In constructive mathematics, every mathematical entity is a mental construction, and thus, there is no universal notion of equality applicable everywhere. One must define the corresponding equality locally for each constructed entity. Consequently, an equilogical map is simply a constructive continuous function that preserves the local identities as expected. The equivalence between equilogical maps serves as an *extensionality* criterion, ensuring that a map is uniquely determined by its values.

The category **Equ** is a BCC. The terminal and the initial objects are $(\{*\}, =_{\{*\}})$ and $(\varnothing, =_\varnothing)$, respectively. The product of $(X, \sim_X)$ and $(Y, \sim_Y)$ is the pair $(X \times Y, \sim_{X\times Y}$), equipped with the usual projections, where $X \times Y$ is the cartesian product with the product topology, and $(x_1, y_1) \sim_{X\times Y} (x_2, y_2)$ iff $x_1 \sim_X x_2$ and $y_1 \sim_Y y_2$. Note that the projection maps are continuous and respect the equivalence relation, thus they live in **Equ**. The coproduct of $(X, \sim_X)$ and $(Y, \sim_Y)$ is $(X + Y, \sim_{X+Y})$, equipped with the usual injections, where $X + Y$ is the disjoint union of X and Y with its canonical topology and $(0, x_1) \sim_{X+Y} (0, x_2)$ iff $x_1 \sim_X x_2$ and $(1, y_1) \sim_{X+Y} (1, y_2)$ iff $y_1 \sim_Y y_2$. Note that the injections are continuous and respect the equivalence relation, thus they live in **Equ**.

For the exponential objects in **Equ**, their detailed construction is unfortunately beyond the scope of this chapter and the interested reader can refer to Bauer et al. (2004), Bauer (2000) for more information. The only property of the construction we need here is the fact that the exponentials in **Top** (if they exist) remain intact by moving to **Equ**. More precisely, define the functor Eq : **Top** $\to$ **Equ** mapping the space X to the equilogical space $(X, =_X)$ and the function $f : X \to Y$ to the class of itself in **Equ**. Then, if the exponential $[X, Y]$ with the usual set-theoretical evaluation map exists in **Top**, then the exponential $[(X, =_X), (Y, =_Y)]$ is just $([X, Y], =_{[X,Y]})$ with the same evaluation map. For instance, the exponential $[(\mathbb{N}, =), (\mathbb{N}, =)]$ in **Equ** is just $(\mathbb{N}^\mathbb{N}, =)$ with the usual evaluation map.

The second category formalizes the computable world and is defined similarly to **Equ**. However, in this category, spaces and continuous functions are replaced with subsets of $\mathbb{N}$ and computable functions, respectively.

Example 225 Consider the collection of pairs $(A, \sim_A)$, where A is a subset (not necessarily decidable or computably enumerable) of $\mathbb{N}$ and $\sim_A$ is an equivalence relation on A. An *equivariant map* $f : (A, \sim_A) \to (B, \sim_B)$ is a *partial* computable function $f : \mathbb{N} \rightharpoonup \mathbb{N}$ that maps the elements of A to the elements of B while preserving the equivalence relation. Two equivariant maps are considered equivalent if they induce the same function on the equivalence classes. We can interpret the pair $(A, \sim_A)$ as representing the set of all *finitary proofs* of a proposition ϕ_A assigned to A, with $\sim_A$ capturing the equivalence between these proofs. Thus, an equivariant map is viewed as an *algorithmic* procedure that transforms proofs of ϕ_A into proofs of ϕ_B, respecting the equivalences among these proofs. The collection of all pairs $(A, \sim_A)$ as the objects and equivariant maps (up to the equivalence) as the morphisms together with the evident composition and identity constitutes a category, denoted by **Rec**.

The category **Rec** is a BCC. Similar to **Equ**, its terminal and initial objects are $(\{*\}, =_{\{*\}})$ and $(\varnothing, =_\varnothing)$, respectively. The prodcuts and coproducts are defined similar to **Equ**. However, as in **Rec**, we restrict ourselves to the subsets of $\mathbb{N}$, we need to encode the cartesian product and the disjoint union of two subsets of $\mathbb{N}$ by another subset. More precisely, in **Rec**, the product of $(A, \sim_A)$ and $(B, \sim_B)$ is the pair $(C, \sim_C)$, together with the maps $p_0 : C \to A$ and $p_1 : C \to B$, where $C = \{2^a(2b + 1) \mid a \in$

$A, b \in B\}$, $2^a(2b+1) \sim_C 2^c(2d+1)$ iff $a \sim_A c$ and $b \sim_B d$, and p_0 and p_1 are defined by $p_0(2^n(2m+1)) = n$ and $p_1(2^n(2m+1)) = m$. Notice that both p_0 and p_1 are computable functions and hence live in **Rec**. Similarly, for the coproduct of $(A, \sim_A)$ and $(B, \sim_B)$, we use $(C, \sim_C)$, together with the maps $i_0 : A \to C$ and $i_1 : B \to C$, where

$$C = \{2^a(2b+1) \mid (a = 0 \text{ and } b \in A) \text{ or } (a = 1 \text{ and } b \in B)\},$$

$2^a(2b+1) \sim_C 2^c(2d+1)$ if $(a = c = 0$ and $b \sim_A d)$ or $(a = c = 1$ and $b \sim_B d)$ and i_0 and i_1 are defined by $i_0(n) = 2^0(2n+1) = 2n+1$ and $i_1(n) = 2(2n+1)$. Notice that both i_0 and i_1 are computable functions and hence live in **Rec**.

For the exponential object $[(A, \sim_A), (B, \sim_B)]$, we use $(C, \sim_C)$, together with the map $ev : C \times A \to B$, where

$$C = \{e \in \mathbb{N} \mid \forall n \in A\ (e \cdot n \in B)\ \&\ \forall mn \in A\ (m \sim_A n \to e \cdot m \sim_B e \cdot n)\},$$

$e \sim_C f$ iff $e \cdot n \sim_B f \cdot n$, for any $n \in A$ and $ev(2^e(2a+1)) = e \cdot a$, where $e \cdot a$ is the output of a universal machine applying the code e of a Turing machine to the input a. Recall that for any partial computable map $f : \mathbb{N} \rightharpoonup \mathbb{N}$, by the S_{mn}-theorem, if e is one of the codes for the function $(m, n) \mapsto f((2^m(2n+1))$, we have $S(e, m) \cdot n = f((2^m(2n+1))$, for any $m, n \in \mathbb{N}$. Now, for any class $[f] : (D, \sim_D) \times (A, \sim_A) \to (B, \sim_B)$, it is enough to define $\lambda_A[f] : (D, \sim_D) \to (C, \sim_C)$ as the class of the computable function $d \mapsto S(e, d)$. One can verify that this map is well-defined, equivariant, and meets the required property in the definition of the exponential object. Moreover, it is unique with respect to this property.

10.2 Natural Numbers Object

In this subsection, we will introduce a categorical version of the set of natural numbers and represent numbers and numeral functions, i.e., the multi-variable functions over the natural numbers, by this categorical gadget.

Definition 226 (*Natural numbers object*) Let $\mathcal{C}$ be a CCC. A *natural numbers object* (an NNO, for short) in $\mathcal{C}$ is an object N together with maps $Z : 1 \to N$ and $s : N \to N$ such that for any object A and any maps $a : 1 \to A$ and $f : A \to A$, there exists a unique map $g : N \to A$ such that the following diagram commutes:

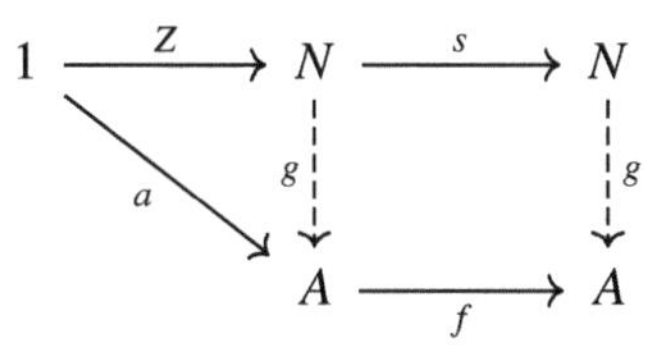

It is easy to see that any two natural numbers objects in a category are canonically isomorphic. For any maps $a : 1 \to A$ and $f : A \to A$, we denote the unique g by $R(f, a)$. Therefore, we have $R(f, a) \circ Z = a$ and $R(f, a) \circ s = f \circ R(f, a)$. Intuitively, the map $R(f, a) : N \to A$ is the map defined by iterating f over the initial values a and the equations are just the description of this recursive definition.

Remark 227 Despite what we had for products, coproducts, and expoentials, it is not clear how to rewrite the uniqueness condition for an NNO as an equality. This can be problematic in constructing the free CCC with an NNO. We will come back to this point later.

Remark 228 Notice that the universal definition of the natural numbers object roughly states that the diagram

$$1 \xrightarrow{Z} N \xrightarrow{s} N$$

is the "least" between all such diagrams with the fixed node 1.

Example 229 In **Set**, the set $\mathbb{N}$ of natural numbers together with the morphism $Z : \{*\} \to \mathbb{N}$ mapping $*$ to 0 and $s : \mathbb{N} \to \mathbb{N}$ mapping any number to its successor is an NNO. In a CCC poset $(P, \leq)$, the NNO is the terminal 1 with the canonical map $! : 1 \to 1$ as both Z and s. In both $\mathbf{Set}^{\mathbb{N}}$ and $\mathbf{Set}^{\mathbb{Z}}$, the pair $(\mathbb{N}, id_{\mathbb{N}})$ together with the same maps as before is an NNO. Note that both the maps Z and s preserve the identity and hence are equivariant. Moreover, note that for any $a : (\{*\}, id_{\{*\}}) \to (A, \sigma_A)$ and $f : (A, \sigma_A) \to (A, \sigma_A)$, it is easy to see that $\sigma_A(f^n(a(*))) = f^n(a(*))$. Therefore, the map $g : \mathbb{N} \to A$ defined by $g(n) = f^n(a(*))$ is actually a map from the dynamic system $(\mathbb{N}, id_{\mathbb{N}})$ to the dynamic system (A, σ_A). It is clear that this g is the unique $R(f, a)$ we are looking for. More generally, the constant functor $\Delta_{\mathbb{N}} : C \to \mathbf{Set}$ together with Δ_Z and Δ_s is an NNO in $\mathbf{Set}^C$. In **Equ**, the pair $(\mathbb{N}, =_{\mathbb{N}})$ along with the set functions Z and s is an NNO, where $\mathbb{N}$ is the discrete space of natural numbers. Note that both Z and s are continuous and equivariant as they preserve the equivalence relation. Moreover, for any $[x] : (\{*\}, =_{\{*\}}) \to (X, \sim_X)$ and $[f] : (X, \sim_X) \to (X, \sim_X)$, it is easy to see that the class of the map $g : \mathbb{N} \to X$ defined by $g(n) = f^n(a(*))$ is independent of the choice of f and a. The map g is clearly a continuous equivariant map from $(\mathbb{N}, =_{\mathbb{N}})$ to $(X, \sim_X)$. It is clear that the class of g is the unique $R(f, a)$ we are looking for. The same construction applies to **Rec**, with the important point being that the class of all computable functions is closed under primitive recursion, allowing the definition of g from a and f.

For clarity in presenting some arguments, it is useful to consider the category **FinOrd**, which consists of natural numbers as objects and functions between finite sets as morphisms. Specifically, an object in **FinOrd** is a natural number $m \in \mathbb{N}$, and a morphism from m to n is a function from the set $\{0, 1, \ldots, m-1\}$ to the set $\{0, 1, \ldots, n-1\}$. The identity and composition of morphisms are defined in the standard way. The category **FinOrd** is a BCC, with the terminal object being 1, the initial object being 0, and the product, coproduct, and exponential operations corresponding to the usual operations on natural numbers. Essentially, **FinOrd** provides a simpler version of the category **FinSet**, where we work with the cardinalities of finite sets rather than the sets themselves. In fact, by choosing an arbitrary bijection $\alpha_A : A \to \{0, 1, \dots, Card(A) - 1\}$ for any finite set A, we can define a functor $Card$: **FinSet** $\to$ **FinOrd**. This functor maps a finite set A to its cardinality $Card(A)$ and a function $f : A \to B$ to the function $\alpha_B \circ f \circ \alpha_A^{-1}$: $\{0, 1, \dots, Card(A) - 1\} \to \{0, 1, \dots, Card(B) - 1\}$. The functor $Card$ is a BC-functor.

Example 230 (*Non-existence of the* NNO) The category **FinOrd** has no NNO. For the sake of contradiction, let the number n together with the maps $Z : \{0\} \to \{0, \dots, n-1\}$ and $s : \{0, \dots, n-1\} \to \{0, \dots, n-1\}$ be an NNO. Then, as $\{0, \dots, n-1\}$ is finite, there are distinct natural numbers $p < q \in \mathbb{N}$ such that $s^p Z = s^q Z$. Now, consider the number $q + 1$ as an object in **FinOrd** and the maps $a : \{0\} \to \{0, \dots, q\}$ picking the element 0 in $\{0, \dots, q\}$ and $f : \{0, \dots, q\} \to \{0, \dots, q\}$ defined as the successor function, except that for q, we define $f(q) = 0$. Then, as (n, Z, s) is an NNO, there is a map $g : n \to q + 1$ such that $gs^k Z(0) = k$, for any natural number $k \leq q$. Therefore, as $gs^p Z = gs^q Z$, we have $p = q$ which is impossible. Note that we only used the existence of the map $R(f, a)$ to reach a contradiction and not its uniqueness.

Example 231 Let C be a BCC such that there is a BC-functor $F : C \to$ **FinOrd**. We claim that C has no NNO. First, note that F is surjective on objects, because F maps $\sum_{i=0}^{n-1} 1$ in C to n in **FinOrd**. Moreover, F is a full functor as for any map $f : \{0, \dots, m-1\} \to \{0, \dots, n-1\}$ in **FinOrd**, there is a map $g : \sum_{i=0}^{m-1} 1 \to \sum_{j=0}^{n-1} 1$ in C that uses the injections to mimic the behavior of f and hence $F(g) = f$. Using these two facts, it is easy to see that if (N, Z, s) is an NNO in C, then $(F(N), F(Z), F(s))$ in **FinOrd** has all the properties of an NNO except probably its uniqueness. However, the existence of such data is impossible as observed in Example 230. Therefore, C has no NNO. One can use this observation to prove that the categories **FinSet** and **NJ** have no NNO. For the former, use the BC-functor $Card$: **FinSet** $\to$ **FinOrd**. For the latter, by mapping the atoms p_i to some numbers, we can use Theorem 111 to find a BC-functor F : **NJ** $\to$ **FinOrd**.

In any CCC equipped with an NNO (N, Z, s), there is a standard way to represent natural numbers and *some* of numeral functions. For the former, the natural number n is represented by the map $\bar{n} = s^n \circ Z : 1 \to N$, iterating the successor map for n many times. For the latter, we have

Definition 232 Let C be a CCC with the NNO (N, Z, s). A numeral function $f : \mathbb{N}^k \to \mathbb{N}$ is called *representable* in C if there is a map $F : N^k \to N$ in C such that for any $(n_1, \dots, n_k) \in \mathbb{N}^k$, we have

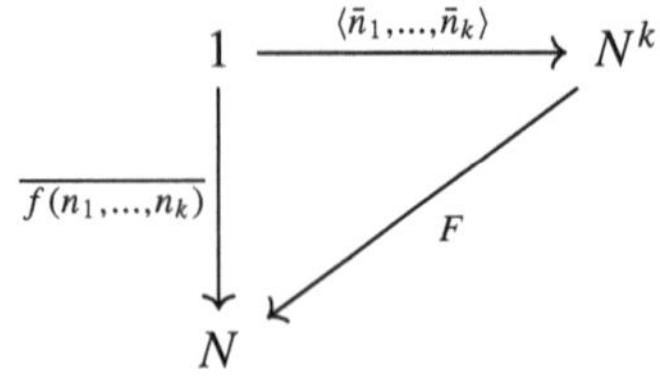

Example 233 Let C be a CCC with the NNO (N, Z, s). It is clear that the function $c_0 : \mathbb{N}^0 \to \mathbb{N}$ mapping its one and only input to zero is representable by the map $Z : 1 \to N$. Similarly, the successor function is representable by the map $s : N \to N$. The projection function $I_i^k : \mathbb{N}^k \to \mathbb{N}$ mapping a k-tuple to its i-th argument is representable by the projection map $p_i : \prod_{i=0}^{k-1} N \to N$. For a less trivial example, consider the map $f : N \to N$ as $f = R(s \circ s, Z)$:

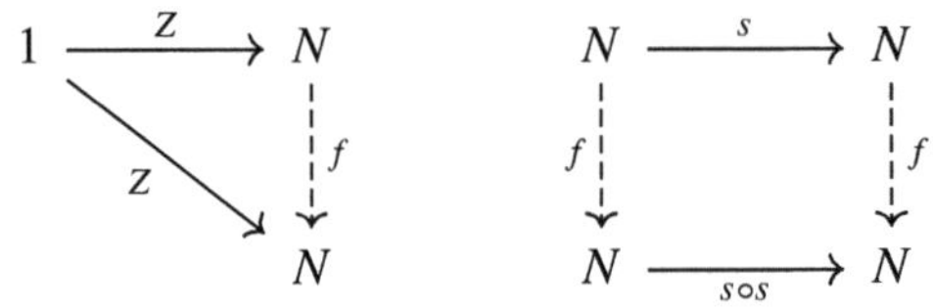

It is easy to see that for any $n \in \mathbb{N}$, the following diagram commutes:

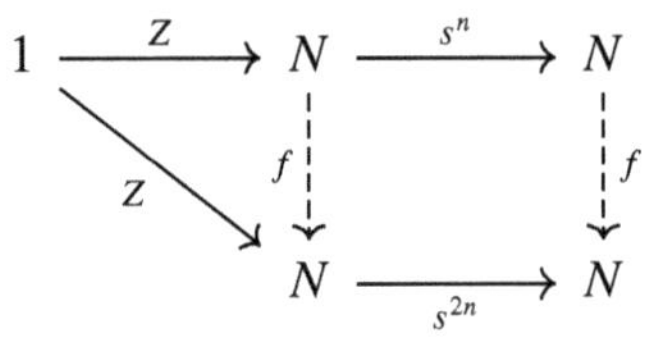

Therefore, the map $f : N \to N$ represents the numeral function $d : \mathbb{N} \to \mathbb{N}$ defined by $d(n) = 2n$.

Example 234 In **Set**, **Set**^{C}, and **Equ**, all numeral functions are representable. In **Rec**, however, the representable functions are the total computable functions.

In **Set**, we often use the inductive nature of $\mathbb{N}$ to define new functions from the existing ones by what is called *primitive recursion*. More precisely, given functions $g : X \to A$ and $h : \mathbb{N} \times X \times A \to A$, for some sets X and A, we can use recursion on n to define a function $f : \mathbb{N} \times X \to A$ satisfying $f(0, x) = g(x)$ and $f(n + 1, x) = h(n, x, f(n, x))$. A similar situation occurs for any NNO in any CCC. Here, the inductive nature of $\mathbb{N}$ is mimicked by the universality of the NNO.

Lemma 235 (Primitive recursion) *Let* C *be a* CCC *with the* NNO (N, Z, s). *Then, for any map* $g : X \to A$ *and any* $h : N \times X \times A \to A$, *there exists a unique* $f : N \times X \to A$ *such that*

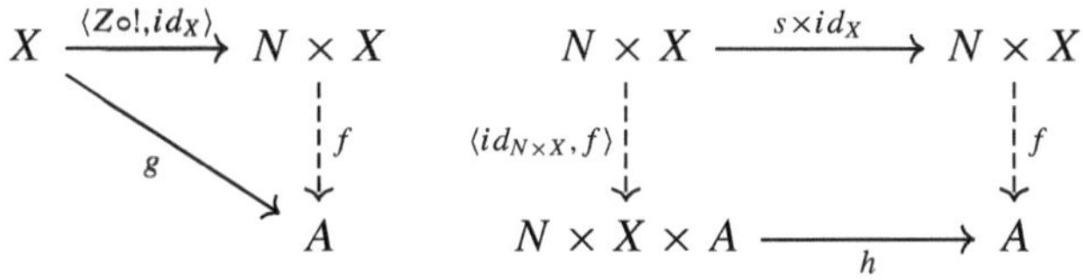

Proof Define $G : 1 \to (N \times A^X)$ as $\langle Z, \lambda_X g\rangle$ and $H' : N \times A^X \times X \to A$ as $h\langle p_0, p_2, ev\langle p_1, p_2\rangle\rangle$ and set $H : (N \times A^X) \to (N \times A^X)$ as $\langle s p_0, \lambda_X H'\rangle$. By the universality of the NNO, there exists a unique $F : N \to N \times A^X$ such that

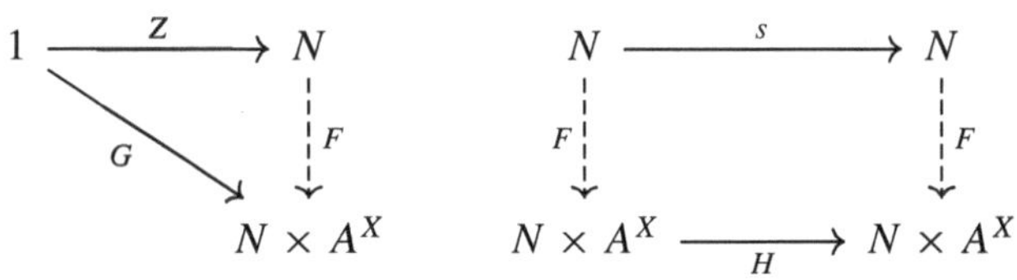

Define $f : N \times X \to A$ by $ev\langle p_1^{N,A^X} F p_0^{N,X}, p_1^{N,X}\rangle$. It is easy to see that this f satisfies the claimed property and it is unique in that property. □

Remark 236 Note that in the above proof, we used exponentiation to manage the parameter object X. If the category is cartesian but not cartesian closed, one must modify the definition of the NNO to a stronger form, known as a *parametrized* NNO, which directly incorporates the parameter X into the definition.

Theorem 237 *All primitive recursive functions are representable in any* CCC *with an* NNO.

Proof As observed in Example 233, all basic functions are clearly representable. It is also easy to see that the composition of representable functions is representable. For primitive recursion, it is sufficient to use Lemma 235. □

It is clear that in a poset CCC, all numbers are represented with the map $! : 1 \to 1$. This situation is annoying as we expect to represent numbers in an injective way. The next lemma shows that this poset case is actually the only problematic case.

Lemma 238 *Let* C *be a non-preorder* CCC *with an* NNO. *If* $\bar{m} = \bar{n}$, *then* $m = n$, *for any* $m, n \in \mathbb{N}$.

Proof Let $pd : N \to N$ be a representation for the predecessor function in C and assume $\bar{m} = \bar{n}$, for some $m > n$. By composing pd with $\bar{m}$ and $\bar{n}$ for n many times, we have $\overline{m - n} = \bar{0}$. Therefore, it is enough to prove that $\bar{k} \neq \bar{0}$, for any $k \neq 0$. Let $f, g : A \to B$ be two different maps between two objects in C. Define $h : N \to [A, B]$ such that

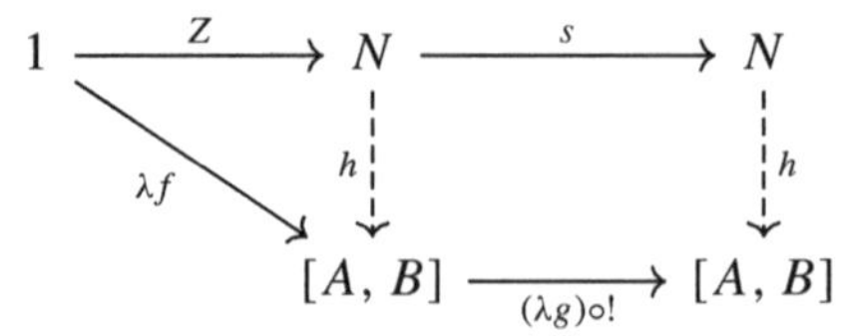

It is easy to see that $h \circ \bar{0} = \lambda f$ while $h \circ \bar{k} = \lambda g$, for any $k > 0$. Therefore, if $\bar{k} = \bar{0}$, by composition with h, we reach $\lambda f = \lambda g$ which implies $f = g$. This is a contradiction. □

In a similar way to what we had for **NJ**, it is also possible to come up with the "least" CCC with an NNO, where every object is constructed from N and 1 by using product and exponentiation, and any morphism is constructed from the very basic morphisms such as $!$, p_0, p_1, ev, Z, s, and the operations $\langle -, - \rangle$, λ, and R. To state a formal version of this claim, we need the following definition:

Definition 239 Let C and $\mathcal{D}$ be two CCC's and $F : C \to \mathcal{D}$ be a CC-functor. We say that F *preserves the* NNO, if for any NNO

$$1 \xrightarrow{Z} N \xrightarrow{s} N$$

in C, the diagram

$$F(1) \xrightarrow{F(Z)} F(N) \xrightarrow{F(s)} F(N)$$

is an NNO in $\mathcal{D}$.

Theorem 240 (Free CCC with an NNO) *There exists a free* CCC *with an* NNO*, i.e., a* CCC C *with an* NNO *such that for any* CCC $\mathcal{D}$ *with an* NNO*, there is a unique (up to isomorphism)* CC*-functor* $F : C \to \mathcal{D}$*, preserving the* NNO*. We denote this* CCC *by* Λ_N.

Proof As the conditions of being an NNO go beyond simple equalities, we cannot apply the usual method of generating all required objects and maps and then using a quotient to form the equivalence relation generated by the equalities. To address this, we should weaken the notion of an NNO to something representable by equalities, construct the free CCC with that weak NNO, and then demonstrate that the weak NNO is actually an NNO in that category. For more details on this strategy, see (Lambek 2000). □

Remark 241 The category Λ_N is not a preorder. To prove this, we assume otherwise. Thus, the two maps $Z : 1 \to N$ and $sZ : 1 \to N$ must be equal. Use the freeness of Λ_N to get a CC-functor $F : \Lambda_N \to$ **Set** preserving the NNO. Hence, $F(N) \cong \mathbb{N}$. As $Z = sZ$ in Λ_N, we have $F(Z) = F(s)F(Z) : \{*\} \to F(N)$, which implies $0 = 1$ in $\mathbb{N}$, leading to a contradiction. Hence, Λ_N is not a preorder.

Definition 242 A numerical function is called a *primitive recursive functional* if it is representable in Λ_N.

It is reasonable to claim that the numerical primitive recursive functionals are exactly the functions that are representable in *all* CCC's with an NNO, simply because Λ_N is the "smallest" of such categories. This claim can be proved using the following lemma:

Lemma 243 *Let $\mathcal{C}$ and $\mathcal{D}$ be two* CCC*'s with an* NNO *and $G : \mathcal{C} \to \mathcal{D}$ is a* CC-*functor preserving the* NNO*. Then, if the map $F : N_{\mathcal{C}}^k \to N_{\mathcal{C}}$ in $\mathcal{C}$ represents the function $f : \mathbb{N}^k \to \mathbb{N}$, then $G(F) : N_{\mathcal{D}}^k \to N_{\mathcal{D}}$ in $\mathcal{D}$ also represents f.*

Proof As F represents f, for any natural number $n \in \mathbb{N}$, we have $F \circ \bar{n}_{\mathcal{C}} = \overline{f(n)}_{\mathcal{C}}$. As G is a CC-functor respecting the NNO, we have $G(\bar{n}_{\mathcal{C}}) = \bar{n}_{\mathcal{D}}$. Therefore, by applying G on $F \circ \bar{n}_{\mathcal{C}} = \overline{f(n)}_{\mathcal{C}}$, we get $G(F) \circ \bar{n}_{\mathcal{D}} = \overline{f(n)}_{\mathcal{D}}$. □

Corollary 244 *A numerical function is a primitive recursive functional iff it is representable in all* CCC*'s with an* NNO*.*

Proof One direction is clear as Λ_N is itself a CCC with an NNO. For the other direction, let $\mathcal{C}$ be a CCC with an NNO. Define G as the unique CC-functor from Λ_N to $\mathcal{C}$ that preserves the NNO. By Lemma 243, it follows that if a function is representable in Λ_N, it is also representable in $\mathcal{C}$. □

In the remainder of this subsection, we aim to evaluate the computational power of numerical primitive recursive functionals. To begin, we demonstrate that all such functionals are computable.

Theorem 245 *All numerical primitive recursive functionals are recursive.*

Proof Let $f : \mathbb{N}^k \to \mathbb{N}$ be a primitive recursive functional. Then, by Corollary 244, f is representable in **Rec**. Hence, it is recursive, by Example 234. □

Remark 246 Having established Theorem 245, one might assume that all total recursive functions are representable in Λ_N. However, this is not the case. The reason is tied to the well-known informal argument that no formal system can capture all total recursive functions. For a formal proof, see (Lambek and Scott 1988).

We have shown that all primitive recursive functions are representable in any CCC with an NNO. Consequently, these functions are all primitive recursive functionals. However, the power of Λ_N extends well beyond that of primitive recursion. This is because Λ_N supports primitive recursion not only on numerical functions but also on functionals. For example, consider the Ackermann function $A(m, n)$, defined by $A(0, n) = n + 1$ and $A(m + 1, n) = A_m^{n+1}(1)$, where A_m is just the function $A(m, -) : \mathbb{N} \to \mathbb{N}$. This function is *not* primitive recursive. However, it is a primitive recursive functional as one can use a higher-order recursion on m to define A_m using the equations $A_0 = s$ and $A_{m+1} = \lambda n.A_m^{n+1}(1)$.

To formally prove the representability of the Ackermann function in Λ_N, we first consider the function $iter : \mathbb{N} \times \mathbb{N}^{\mathbb{N}} \to \mathbb{N}$ that maps (n, f) to $f^{n+1}(1)$. Using the recursive definition of $iter$, which is given by the equations $iter(0, f) = f(1)$ and $iter(n + 1, f) = f(iter(n, f))$, and applying Lemma 235, we can construct a map $i : N \times N^N \to N$ in Λ_N that simulates the behavior of $iter$. More precisely, set $X = N^N$ and $A = N$ in Lemma 235 and define $g : N^N \to N$ by $ev\langle id_{N^N}, \bar{1}\circ!\rangle$ and $h : N \times N^N \times N \to N$ by $ev\langle p_1, p_2\rangle$. By Lemma 235, there is a map $i : N \times N^N \to N$ such that

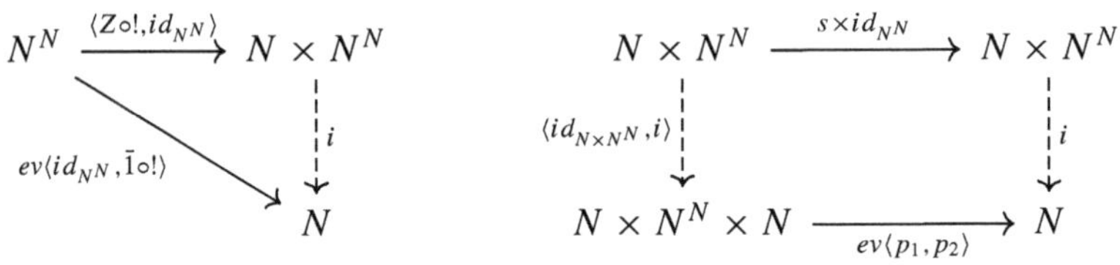

Now, define $F : N \to N^N$ as a map induced by the universality of the NNO:

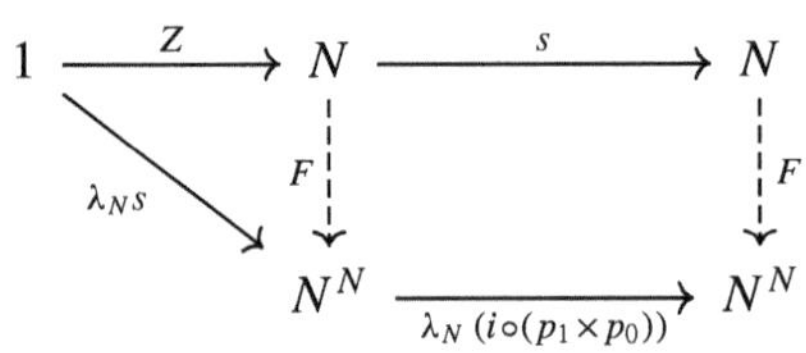

It is easy to see that $F : N \to N^N$ mimics the definition of A_{-}. Therefore, the map $ev\langle F, id_N\rangle : N \times N \to N$ represents the Ackermann function in Λ_N.

10.3 Realizability

From the introduction of Sect. 8.1, recall that the language of arithmetic does not include the connectives $\vee$ and $\bot$. Therefore, it suffices to work with CCC's rather than BCC's for its interpretation. Furthermore, since we aim to realize arithmetical proofs, it is natural to assume that the CCC has an NNO. Thus, in this subsection, we assume that C is a CCC equipped with the NNO $(\mathtt{N}, \mathtt{Z}, \mathtt{s})$. For any such C, there is a canonical C-interpretation called the *standard C-interpretation* that captures the intended meaning of the arithmetical proofs. Formally, by the *standard C-interpretation* $\mathfrak{S}$ of the language of HA^{ω}, we mean a C-interpretation that maps:

- the basic type N to the C-set $N = (\mathtt{N}, \mathbb{N}, \Vdash_N)$, where $\mathtt{n} \Vdash_N n$, for any natural number $n \in \mathbb{N}$ in which $\mathtt{n} : \mathtt{1} \to \mathtt{N}$ is the representation of the natural number n,
- the function symbol 0 to the C-set map $0^{\mathfrak{S}} : 1 \to N$ picking the element $0 \in \mathbb{N}$. Note that this map is realized by $\mathtt{Z} : \mathtt{1} \to \mathtt{N}$,

- the function symbol $S(x^N)$ of type N to the C-set map $S^{\mathfrak{S}} : N \to N$ defined by $S^{\mathfrak{S}}(n) = n + 1$. Note that this map is realized by $\mathtt{s} : \mathtt{N} \to \mathtt{N}$,
- the function symbol $\mathbf{R}^\sigma(x^\sigma, y^{N\times\sigma\to\sigma}, z^N)$ of type σ to the C-set map $(\mathbf{R}^\sigma)^{\mathfrak{S}} : M^\sigma \times [N \times M^\sigma, M^\sigma] \times N \to M^\sigma$ defined recursively by

$$(\mathbf{R}^\sigma)^{\mathfrak{S}}(a, f, 0) = a \quad \text{and} \quad (\mathbf{R}^\sigma)^{\mathfrak{S}}(a, f, n + 1) = f(n, (\mathbf{R}^\sigma)^{\mathfrak{S}}(a, f, n)).$$

  Using Lemma 235, it is easy to see that this map is realized by a map in C constructed via a primitive recursion.

If the basic type 2 is also present in the language and the coproduct $\mathtt{2} = \mathtt{1} + \mathtt{1}$ exists in C, the standard C-interpretation must map:

- the type 2 to the C-set $(\mathtt{2}, \{0, 1\}, \Vdash_2)$, where $\mathtt{i}_0 \Vdash_2 0$ and $\mathtt{i}_1 \Vdash_2 1$,
- the constants 0_2 and 1_2 of type 2 to the C-set maps $0_2^{\mathfrak{S}} : 1 \to 2$ and $1_2^{\mathfrak{S}} : 1 \to 2$ picking the elements 0 and 1 in $\{0, 1\}$, respectively. Note that these functions are realized by the maps $\mathtt{i}_0 : \mathtt{1} \to \mathtt{2}$ and $\mathtt{i}_1 : \mathtt{1} \to \mathtt{2}$, respectively,
- the function symbol $c^\sigma(x^\sigma, y^\sigma, z^2)$ to the C-set map $(c^\sigma)^{\mathfrak{S}} : M^\sigma \times M^\sigma \times 2 \to M^\sigma$ defined by $(c^\sigma)^{\mathfrak{S}}(a, b, 0) = a$ and $(c^\sigma)^{\mathfrak{S}}(a, b, 1) = b$. It is easy to see that the function $(c^\sigma)^{\mathfrak{S}}$ is realized by a map in C.

In the rest of this subsection, we will use only the standard C-interpretation for realizability. Thus, we will omit the subscript $\mathfrak{S}$ in $\Vdash_{\mathfrak{S}}$ everywhere. Additionally, by a sentence, we always mean an $\mathcal{L}(\mathfrak{S})$-sentence.

Remark 247 Here are some remarks. First, as the language does not have the connective $\bot$, one can use induction to prove that for any sentence A, the set $\mathrm{Hom}_C(\mathtt{1}, |A|)$ is non-empty. Second, for any $\exists$-free sentence, i.e., a sentence A with no existential quantifiers, the object $|A|$ is always a terminal object. Therefore, there is only one map in $\mathrm{Hom}_C(\mathtt{1}, |A|)$. Third, as we interpreted $\bot$ as $S(0) =_N 0$ and $|S(0) =_N 0| = \mathtt{1}$, we have $|\neg B| = |B \to \bot| = [|B|, \mathtt{1}] \cong \mathtt{1}$. Therefore, for any negative sentence A, we also have $|A| \cong \mathtt{1}$. One can read this property as an incarnation of the usual intuition that these sentences have no computational content because such sentences can have at most one realizer, encoding whether they are provable or not.

Example 248 (*Realizability for Negative Formulas*) For any formula $A(x^\sigma)$, we have $\Vdash \neg A(x^\sigma)$ iff $\nVdash A(a)$, for any $a \in \|M^\sigma\|$. For the forward direction, let $\mathtt{r} \Vdash \neg A(x^\sigma)$ and $a \in \|M^\sigma\|$. There is $\mathtt{a} : \mathtt{1} \to |M^\sigma|$ such that $\mathtt{a} \Vdash_{M^\sigma} a$. Hence, $\mathtt{r} \circ \mathtt{a} \Vdash \neg A(a)$. If $\mathtt{s} \Vdash A(a)$, we reach $(\mathtt{r} \circ \mathtt{a}) \cdot \mathtt{s} \Vdash S(0) = 0$ which implies $S(0) = 0$. Hence, $\nVdash A(a)$. For the backward direction, assume $\nVdash A(a)$, for any $a \in \|M^\sigma\|$. Consider $! : |M^\sigma| \to |\neg A| \cong \mathtt{1}$. We claim that $! \Vdash \neg A(x^\sigma)$. To prove that, we have to show that for any $\mathtt{a} : \mathtt{1} \to |M^\sigma|$, any $a \in \|M^\sigma\|$ and any $\mathtt{s} : \mathtt{1} \to |A|$, if $\mathtt{s} \Vdash A(a)$ then $(! \circ \mathtt{a}) \cdot \mathtt{s} \Vdash S(0) =_N 0$. As there is no $\mathtt{s} : \mathtt{1} \to |A|$ satisfying $\mathtt{s} \Vdash A(a)$, we get $! \Vdash \neg A(x^\sigma)$ which complete the proof. A similar thing is true for $\Vdash A(\bar{x})$ for a set $\bar{x}$ of variables. Especially, if A is a *sentence*, then $\Vdash \neg A$ iff $\nVdash A$.

Remark 249 (*Double Negation as a Collapsing Operator*) Using Example 248, one can see that $\Vdash \neg\neg A(x^\sigma)$ iff $\Vdash A(a)$, for any $a \in \|M^\sigma\|$. A similar thing is true for $\Vdash \neg\neg A(\bar{x})$ for a set $\bar{x}$ of variables. Notice the difference between $\Vdash \neg\neg A(x^\sigma)$ and $\Vdash A(x^\sigma)$. In the latter, one must come up with a map $\mathtt{r} : |M^\sigma| \to |A|$ such that for any $\mathtt{a} : 1 \to |M^\sigma|$ and any $a \in \|M^\sigma\|$ satisfying $\mathtt{a} \Vdash_{M^\sigma} a$, we have $\mathtt{r} \circ \mathtt{a} \Vdash A(a)$. This simply means the existence of a *uniform* way to find a realizer for $A(a)$ for $a \in \|M^\sigma\|$. In this sense, one can say that $\Vdash \neg\neg A(x^\sigma)$ is actually the non-uniform version of $\Vdash A(x^\sigma)$, where we only care about the existence of the realizer and not its uniform construction. This non-uniformity is exactly what we get from the double negation operator. We discussed this role of double negation before when we talked about the axioms CT and $\mathsf{WC-N}$ in Sect. 8.

Example 250 (*Realizability for Disjunctive Formulas*) $\Vdash A(x^\sigma) \vee B(x^\sigma)$ iff there are maps $\mathtt{r} : |M^\sigma| \to \mathtt{N}$, $\mathtt{s} : |M^\sigma| \to |A|$ and $\mathtt{t} : |M^\sigma| \to |B|$ such that for any $\mathtt{a} : 1 \to |M^\sigma|$ and any $a \in \|M^\sigma\|$, if $\mathtt{a} \Vdash_{M^\sigma} a$, either $\mathtt{r} \circ \mathtt{a} = 0$ and $\mathtt{s} \circ \mathtt{a} \Vdash A(a)$ or $\mathtt{r} \circ \mathtt{a} = \mathtt{n}$, for some $n \geq 1$ and $\mathtt{t} \circ \mathtt{a} \Vdash B(a)$. We only prove the forward direction and leave the rest to the reader. To prove it, substitute the formula $A(x^\sigma) \vee B(x^\sigma)$ with its definition $\exists y^N[(y = 0 \to A(x^\sigma)) \wedge (y \neq 0 \to B(x^\sigma))]$ and let $\mathtt{u}$ be its realizer. Therefore, for any $\mathtt{a} : 1 \to |M^\sigma|$ and any $a \in \|M^\sigma\|$, if $\mathtt{a} \Vdash_{M^\sigma} a$, there exists $n \in \mathbb{N}$ such that $\mathtt{p}_0(\mathtt{u} \circ \mathtt{a}) \Vdash_N n$ and $\mathtt{p}_1(\mathtt{u} \circ \mathtt{a}) \Vdash [(n = 0 \to A(a)) \wedge (n \neq 0 \to B(a))]$. There are two cases. Either $n = 0$ or $n \neq 0$. In the first case, as $\mathtt{p}_0(\mathtt{u} \circ \mathtt{a}) \Vdash_N n$, we have $\mathtt{p}_0(\mathtt{u} \circ \mathtt{a}) = 0$. Moreover, $\mathtt{p}_0\mathtt{p}_1(\mathtt{u} \circ \mathtt{a}) \Vdash (n = 0 \to A(a))$. As $n = 0$, we have $! \Vdash n = 0$. Hence, $\mathtt{p}_0\mathtt{p}_1(\mathtt{u} \circ \mathtt{a})\cdot! \Vdash A(a)$. In the second case, as $\mathtt{p}_0(\mathtt{u} \circ \mathtt{a}) \Vdash_N n$ and $n \neq 0$, we have $\mathtt{p}_0(\mathtt{u} \circ \mathtt{a}) = \mathtt{n}$, for some $n \geq 1$. Moreover, $\mathtt{p}_1\mathtt{p}_1(\mathtt{u} \circ \mathtt{a}) \Vdash (n \neq 0 \to B(a))$. As $n \neq 0$, we have $\nVdash n = 0$ which implies $! \Vdash n \neq 0$, by Example 248. Hence, $\mathtt{p}_1\mathtt{p}_1(\mathtt{u} \circ \mathtt{a})\cdot! \Vdash B(a)$. Now, define $\mathtt{r} = \mathtt{p}_0\mathtt{u}$, $\mathtt{s} = \mathtt{ev}\langle \mathtt{p}_0\mathtt{p}_1\mathtt{u}, !\rangle$ and $\mathtt{t} = \mathtt{ev}\langle \mathtt{p}_1\mathtt{p}_1\mathtt{u}, !\rangle$. Then, as $\mathtt{r} \circ \mathtt{a} = \mathtt{p}_0(\mathtt{u} \circ \mathtt{a})$, $\mathtt{s} \circ \mathtt{a} = (\mathtt{p}_0\mathtt{p}_1(\mathtt{u} \circ \mathtt{a}))\cdot!$, and $\mathtt{t} \circ \mathtt{a} = (\mathtt{p}_1\mathtt{p}_1(\mathtt{u} \circ \mathtt{a}))\cdot!$, it is clear that either $\mathtt{r} \circ \mathtt{a} = 0$ and $\mathtt{s} \circ \mathtt{a} \Vdash A(a)$ or $\mathtt{r} \circ \mathtt{a} = \mathtt{n}$ for some $n \geq 1$ and $\mathtt{t} \circ \mathtt{a} \Vdash B(a)$. A similar thing is true for $\Vdash A(\bar{x}) \vee B(\bar{x})$ for a set $\bar{x}$ of variables.

Example 251 (*Incompleteness of Realizability*) For any *sentence* A, the intuitionistically invalid sentence $A \vee \neg A$ is realizable. To prove, note that there are two cases to consider. Either $\Vdash A$ or $\nVdash A$. In the first case, define $\mathtt{r} = 0$ and set $\mathtt{s} : 1 \to |A|$ as one of the realizers of A. In the second case, set $\mathtt{r} = 1$ and $\mathtt{t} =! : 1 \to |\neg A| \cong 1$. By Example 248, as $\nVdash A$, we have $\mathtt{t} \Vdash \neg A$. Finally, by Example 250, we have $\Vdash A \vee \neg A$.

So far, we have investigated the behavior of realizability for negative and disjunctive formulas. In the next lemma, we connect the realizability to the validity in the standard model $\mathfrak{N}$. First, we need a definition. In the language of arithmetic, a type is called a 0-*type* if it is only constructed from the basic type N using $\times$ but not $\to$. It is called a 1-*type*, if it is either a 0-type or it is in the form $\sigma \to \tau$, for two 0-types σ and τ. A formula is called a 0-*formula* (resp. 1-*formula*) if all terms in it (including variables, constants and function symbols) are of 0-types (resp. 1-types). A formula is called *standard* if $\|M^\sigma\| = \mathfrak{N}^\sigma$, for any type σ occurring in A as the type of a term in it. For instance, any 0-formula is a standard formula. To have another example, as the exponentiation $[K, L]$ exists in **Top**, for any finite powers K and L of $\mathbb{N}$, the exponentiation of $[(K, =), (L, =)]$ in **Equ** is $([K, L], =)$. Additionally, as K is discrete, all functions from K to L are continuous. Hence, if $C = \mathbf{Equ}$, for any 1-type σ, we have $\|M^\sigma\| = \mathfrak{N}^\sigma$. Therefore, all 1-formulas are standard, when $C = \mathbf{Equ}$.

Remark 252 Note that for $C = \mathbf{Equ}$ and any 1-types σ and τ, any $f \in \|M^{\sigma\to\tau}\|$ is actually a continuous function $f : \mathfrak{N}^\sigma \to \mathfrak{N}^\tau$ realized by itself as a map in **Equ**. Therefore, we can write $f \Vdash_{M^{\sigma\to\tau}} f$.

Lemma 253 (Realizability Versus Truth) *Let A be an $\exists$-free standard sentence. Then, $\Vdash A$ iff $\mathfrak{N} \vDash A$. Especially, for any $\exists$-free 0-sentence A, we have $\Vdash A$ iff $\mathfrak{N} \vDash A$.*

Proof First, recall from Remark 247 that for any $\exists$-free sentence A, there is only one map in $\mathrm{Hom}_C(\mathtt{1}, |A|)$ that we denote by θ_A. Therefore, we have to show that $\theta_A \Vdash A$ iff $\mathfrak{N} \vDash A$. The proof is by induction on A. For the atomic formulas, the claim holds by definition and the fact that for any type in A we have $\|M^\sigma\| = \mathfrak{N}^\sigma$. The conjunction case is easy. For the implication $A = B \to C$, if $\theta_{B\to C} \Vdash B \to C$ and $\mathfrak{N} \vDash B$, then by the induction hypothesis, $\theta_B \Vdash B$. Therefore, by definition, $\theta_{B\to C} \cdot \theta_B \Vdash C$. As there is just one map in $\mathrm{Hom}(\mathtt{1}, |C|)$, we have $\theta_{B\to C} \cdot \theta_B = \theta_C$. Hence, $\theta_C \Vdash C$. Again, by the induction hypothesis, $\mathfrak{N} \vDash C$. Therefore, $\mathfrak{N} \vDash B \to C$. For the other direction, assume $\mathfrak{N} \vDash B \to C$. To prove $\theta_{B\to C} \Vdash B \to C$, we must show that $\theta_{B\to C} \cdot \mathtt{r} \Vdash C$, for any $\mathtt{r} \Vdash B$. Since, there is only one map in $\mathrm{Hom}(\mathtt{1}, |B|)$, we have $\mathtt{r} = \theta_B$. Since $\theta_B \Vdash B$, by the induction hypothesis $\mathfrak{N} \vDash B$. As $\mathfrak{N} \vDash B \to C$, we have $\mathfrak{N} \vDash C$. By the induction hypothesis, $\theta_C \Vdash C$. As there is only one map in $\mathrm{Hom}(\mathtt{1}, |C|)$, we have $\theta_{B\to C} \cdot \mathtt{r} = \theta_C$. Hence, $\theta_{B\to C} \cdot \mathtt{r} = \theta_C \Vdash C$. Therefore, $\theta_{B\to C} \Vdash B \to C$. For the universal quantifier $A = \forall x^\sigma B(x)$, if $\theta_{\forall x^\sigma B(x)} \Vdash \forall x^\sigma B(x)$, then for any $a \in \|M^\sigma\| = \mathfrak{N}^\sigma$, pick $\mathtt{a} : 1 \to |M^\sigma|$ such that $\mathtt{a} \Vdash_{M^\sigma} a$. Then, $\theta_{\forall x^\sigma B(x)} \cdot \mathtt{a} \Vdash B(a)$. By the induction hypothesis, $\mathfrak{N} \vDash B(a)$, for any $a \in \mathfrak{N}^\sigma$. Hence, $\mathfrak{N} \vDash \forall x^\sigma B(x)$. Conversely, if $\mathfrak{N} \vDash \forall x^\sigma B(x)$ then $\mathfrak{N} \vDash B(a)$, for any $a \in \mathfrak{N}^\sigma = \|M^\sigma\|$. By the induction hypothesis $\theta_{B(a)} \Vdash B(a)$. Now, similar to the argument for the implication, we have $\theta_{\forall x^\sigma B(x)} \Vdash \forall x^\sigma B(x)$. For the second part, note that if A is a 0-sentence, all types σ occurring in A are in the form $N \times N \times \ldots \times N$. Hence, $\|M^\sigma\| = \mathbb{N}^k$. □

Example 254 (*Non-classicality of Realizability*) We saw that $A \vee \neg A$ is always realizable, for any sentence A. In this example, we show that $A(x) \vee \neg A(x)$ is *not* necessarily realizable if $A(x)$ is a formula with free variables. For instance, let $A(x) = \forall w^N y^N \neg T(x^N, x^N, w^N, y^N)$ stating that if the code x of a Turing machine applies to x itself, the computation does not halt. Set $C = \mathbf{Rec}$. We claim that $\nVdash A(x) \vee \neg A(x)$. To see why, assume $\Vdash A(x) \vee \neg A(x)$. By Example 250, there is a map $\mathtt{r} : \mathtt{N} \to \mathtt{N}$ in **Rec** such that for any natural number $n \in \mathbb{N}$, either $\mathtt{r} \circ \mathtt{n} = \mathtt{0}$ and $\Vdash A(n)$ or $\mathtt{r} \circ \mathtt{n} = \mathtt{m}$, for some $m \geq 1$ and $\Vdash \neg A(n)$. As both $A(n)$ and $\neg A(n)$ are $\exists$-free 0-sentences, by Lemma 253, realizability coincides with the truth. Therefore, as $\mathtt{m} \neq \mathtt{0}$, for any $m \geq 1$, deciding $\mathtt{r} \circ \mathtt{n} = \mathtt{0}$ also decides the truth of $A(n)$ which is the halting problem for the input n. As $\mathtt{r} : \mathtt{N} \to \mathtt{N}$ is a map in **Rec**, it is a total computable function. Hence, $\mathtt{r} \circ \mathtt{n} = \mathtt{0}$ is a decidable predicate for the input n. This implies that the halting problem is also decidable which is a contradiction. Hence, $\nVdash A(x) \vee \neg A(x)$.

Example 255 (*Undecidability of Equality*) Let x and y be two variables of type $N \to N$. We show that $(x = y) \vee (x \neq y)$ is not necessarily realizable. Set $C =$ **Equ** and assume $\Vdash (x = y) \vee (x \neq y)$. Then, by Example 250, there is a map $\mathtt{r} : [\mathtt{N}, \mathtt{N}] \times [\mathtt{N}, \mathtt{N}] \to \mathtt{N}$ in **Equ** such that for any functions $\alpha, \beta : \mathbb{N} \to \mathbb{N}$, if we also read them as maps in **Equ** realizing themselves, either $\mathtt{r} \circ \langle \alpha, \beta \rangle = \mathtt{0}$ and $\Vdash \alpha = \beta$ or $\mathtt{r} \circ \langle \alpha, \beta \rangle = \mathtt{m}$, for some $m \geq 1$ and $\Vdash \alpha \neq \beta$. As both $\alpha = \beta$ and $\alpha \neq \beta$ are $\exists$-free 1-sentences, by Lemma 253, realizability coincides with the truth. Therefore, as $\mathtt{m} \neq \mathtt{0}$, for any $m \geq 1$, deciding $\mathtt{r} \circ \langle \alpha, \beta \rangle = \mathtt{0}$ also decides the equality $\alpha = \beta$. However, $\mathtt{r} : [\mathtt{N}, \mathtt{N}] \times [\mathtt{N}, \mathtt{N}] \to \mathtt{N}$ is a map in **Equ**. Hence, it is a continuous function and as $\{0\}$ is open in the discrete topology of $\mathbb{N}$, the subset $\Delta = \{(\alpha, \alpha) \mid \alpha : \mathbb{N} \to \mathbb{N}\}$ must be open in $\mathbb{N}^{\mathbb{N}} \times \mathbb{N}^{\mathbb{N}}$. Let 0^ω be the constant zero function on $\mathbb{N}$. As Δ is open, there is a basic open neighborhood $U = \prod_{i=0}^{\omega} U_i \subseteq \Delta$ around $(0^\omega, 0^\omega)$. As $U_i = \mathbb{N}$ except for finitely many i's, U always intersects with $\mathbb{N}^{\mathbb{N}} \times \mathbb{N}^{\mathbb{N}} - \Delta$ which is a contradiction. Therefore, Δ is not open and hence $\nVdash (x = y) \vee (x \neq y)$.

So far, we have seen some examples of (un)realizability of arithmetical formulas. Next theorem shows that any provable sentence in $\mathsf{HA}^\omega + \mathsf{EXT}$ is realizable. To extend the same claim to cover the axiom of choice, we need the following definition:

Definition 256 The NNO $(\mathtt{N}, \mathtt{Z}, \mathtt{s})$ in a CCC is called *total* if for any map $\mathtt{a} : \mathtt{1} \to \mathtt{N}$, there is a natural number $n \in \mathbb{N}$ such that $\mathtt{a} = \mathtt{n}$.

For instance, the NNO in the categories **Set**, **Rec**, and **Equ** is total.

Theorem 257 *Let C be a* CCC *with an* NNO. *If* $\mathsf{HA}^\omega + \mathsf{EXT} \vdash A$, *then* $\Vdash A$. *If* $\mathtt{2}$ *exists in C, the same also holds for* $\mathsf{HA}_2^\omega + \mathsf{EXT}$. *If C is also well-pointed non-preorder category with the total* NNO, *we have a similar claim for* $\mathsf{HA}^\omega + \mathsf{EXT} + \mathsf{AC}$. *If M^2 is also total, we have the same for* $\mathsf{HA}_2^\omega + \mathsf{EXT} + \mathsf{AC}$.

Proof Using Theorem 214, the logical part is realizable. The only thing to prove is the realizability of the non-logical axioms of the theories. The realizability of the defining axioms for the function symbol $\mathbf{R}^\sigma$ is easy to check. For $S(0) \neq 0$ and $\forall x^N (S(x) = S(y) \to (x = y))$, as they are 0-sentences, their truth implies their realizability, by Lemma 253. For the induction

$$A(0) \wedge \forall x^N (A(x) \to A(Sx)) \to \forall x^N A(x),$$

set $X = |A| \times [N, [|A|, |A|]]$. It is enough to find a map $\mathtt{f} : N \times X \to |A|$ such that for any $\mathtt{r} : 1 \to X$ and any $n \in \mathbb{N}$, if $\mathtt{r} \Vdash A(0) \wedge \forall x^N (A(x) \to A(Sx))$, then $\mathtt{f} \circ \langle \mathtt{n}, \mathtt{r} \rangle \Vdash A(n)$. Consider $\mathtt{p}_0 : X \to |A|$ and $\mathtt{p}_1 : X \to [N, [|A|, |A|]]$ as the projection maps and $\mathtt{e} : N \times [N, [|A|, |A|]] \times |A| \to |A|$ as the canonical evaluation map. Then, define $\mathtt{h} : N \times X \times |A| \to |A|$ as

$$N \times X \times |A| \xrightarrow[\mathtt{id}_N \times \mathtt{p}_1 \times \mathtt{id}_{|A|}]{} N \times [N, [|A|, |A|]] \times |A| \xrightarrow[\mathtt{e}]{} |A|$$

Using Lemma 235, there is a map $\mathtt{f} : N \times X \to |A|$ such that

$$\begin{array}{ccc} X & \xrightarrow{\langle \mathtt{Z} \circ !, \mathtt{id}_X \rangle} & N \times X \\ & \searrow^{\mathtt{p}_0} & \downarrow \mathtt{f} \\ & & |A| \end{array} \qquad \begin{array}{ccc} N \times X & \xrightarrow{\mathtt{s} \times \mathtt{id}_X} & N \times X \\ \downarrow \langle \mathtt{id}_{N \times X}, \mathtt{f} \rangle & & \downarrow \mathtt{f} \\ N \times X \times |A| & \xrightarrow[\mathtt{h}]{} & |A| \end{array}$$

We claim that this $\mathtt{f}$ works. First, for any map $\mathtt{r} : 1 \to X$ and any number $n \in \mathbb{N}$, note that $\mathtt{f} \circ \langle 0, \mathtt{r} \rangle = \mathtt{p}_0 \mathtt{r}$ and $\mathtt{f} \circ \langle \mathtt{n} + 1, \mathtt{r} \rangle = (\mathtt{p}_1 \mathtt{r} \cdot \mathtt{n}) \cdot \mathtt{f}(\mathtt{n}, \mathtt{r})$. Now, if $\mathtt{r} \Vdash A(0) \wedge \forall x (A(x) \to A(S(x)))$, then by induction on natural numbers, it is easy to see that $\mathtt{f} \circ \langle \mathtt{n}, \mathtt{r} \rangle \Vdash A(n)$, for any natural number $n \in \mathbb{N}$.

For the second part, as the NNO is total, the C-set N is total. As C is not a preordered set, the C-set N is projective, by Lemma 238. Now, as C is well-pointed, the standard C-interpretation maps all types to projective total C-sets, by Lemma 222. Hence, by Theorem 217, the axiom of choice is realized. The case involving the type 2 is easy. In the second case, it is enough to use an argument similar to the proof of part (iii) of Lemma 183 to show that M^2 is projective. The rest is similar to the previous case. □

Using Theorem 257, one can easily establish some consistency and unprovability statements. First, we show that the negation of some classical tautologies are consistent with $\mathsf{HA}^\omega + \mathsf{EXT} + \mathsf{AC}$.

Corollary 258 *There are* 0*-formulas* $A(x)$ *and* $B(x)$ *such that* $\neg \forall x^N (A(x) \vee \neg A(x))$ *and* $\neg [\forall x^N \neg\neg B(x) \to \neg\neg \forall x^N B(x)]$ *are consistent with* $\mathsf{HA}^\omega + \mathsf{EXT} + \mathsf{AC}$.

Proof Set $A(x) = \forall w^N y^N \neg T(x, x, w, y)$ and $B(x) = (A(x) \vee \neg A(x))$. As **Rec** is a well-pointed non-preorder category with the total NNO, we have the soundness of $\mathsf{HA}^\omega + \mathsf{EXT} + \mathsf{AC}$. Moreover, in Example 254, we showed that $\forall x^N (A(x) \vee \neg A(x))$ is not realizable for the standard **Rec**-interpretation. Hence, by Example 248, we reach $\Vdash \neg \forall x^N (A(x) \vee \neg A(x))$. For the other part, we claim that $\nVdash \forall x^N \neg\neg B(x) \to \neg\neg \forall x^N B(x)$. Assume otherwise. Then, as $\forall x^N \neg\neg B(x)$ is an intuitionistic tautology, we have $\Vdash \forall x^N \neg\neg B(x)$. Therefore, we must have $\Vdash \neg\neg \forall x^N B(x)$ which is equivalent to $\Vdash \forall x^N B(x)$. However, we showed that $\nVdash \forall x^N (A(x) \vee \neg A(x))$. □

The next consistency statement is the consistency of Church-Turing thesis:

Corollary 259 *The theory* $\mathsf{HA}^{\omega} + \mathsf{EXT} + \mathsf{AC} + \mathsf{CT}_{\neg\neg}$ *is consistent.*

Proof Set $C = \mathbf{Rec}$. As **Rec** is a well-pointed non-preorder category with the total NNO, using Theorem 257, it is enough to show that

$$\mathsf{CT}_{\neg\neg}: \quad \forall f^{N \to N} \neg\neg \exists e^N \forall x^N \exists w^N \exists y^N [T(e, x, w, y) \wedge (fx = y)],$$

is realizable. For that purpose, first notice that the elements in $\|[N, N]\|$ are just total recursive functions. Now, using Remark 249, it is enough to show that for any total recursive function $g : \mathbb{N} \to \mathbb{N}$, the statement

$$\exists e^N \forall x^N \exists w^N \exists y^N [T(e, x, w, y) \wedge (gx = y)],$$

is realizable. As g is computable, it has a code $m \in \mathbb{N}$. It is easy to show that $\Vdash \forall x^N \exists w^N \exists y^N [T(m, x, w, y) \wedge (gx = y)]$. □

Remark 260 As explained before, to prove the consistency of the anti-classical theory $\mathsf{HA}^{\omega} + \mathsf{EXT} + \mathsf{AC} + \mathsf{CT}_{\neg\neg}$, one cannot use the usual classical image. However, as observed in Corollary 259, one can take an alternative world, i.e., the computable world **Rec** and interpret arithmetical proofs in that world. In fact, the possibility of realizing Church-Turing thesis was quite expected in such a world as every "function" in that world is computable. One can go even further to say that the theory $\mathsf{HA}^{\omega} + \mathsf{EXT} + \mathsf{AC} + \mathsf{CT}_{\neg\neg}$ with its Church-Turing thesis is actually a device to describe such computable worlds.

To have a similar consistency result for $\mathsf{HA}^{\omega} + \mathsf{EXT} + \mathsf{AC} + \mathsf{WC{-}N}_{\neg\neg}$, we need the following lemma:

Lemma 261 (i) *There exists a function* $M : [\mathbb{N}^{\mathbb{N}}, \mathbb{N}] \times \mathbb{N}^{\mathbb{N}} \to \mathbb{N}$ *that reads continuous functions* $f : \mathbb{N}^{\mathbb{N}} \to \mathbb{N}$ *and* $\alpha : \mathbb{N} \to \mathbb{N}$ *and returns a number* $M(f, \alpha)$ *such that for any* $\beta : \mathbb{N} \to \mathbb{N}$, *if* $\alpha =_{M(f,\alpha)} \beta$, *then* $f(\alpha) = f(\beta)$.

(ii) *There exists a* continuous *function* $M : [2^{\mathbb{N}}, \mathbb{N}] \to \mathbb{N}$ *that reads a continuous function* $f : 2^{\mathbb{N}} \to \mathbb{N}$ *and returns a number* $M(f)$ *such that for any* $\alpha, \beta : \mathbb{N} \to 2$, *if* $\alpha =_{M(f)} \beta$, *then* $f(\alpha) = f(\beta)$.

Proof First, recall from Example 89 that the function space $\mathbb{N}^{\mathbb{N}}$ as a topological space is the same as the product space $\mathbb{N}^{\omega}$. Therefore, we work with the latter as it is easier to work with. For (i), for any f and α, as f is continuous, $f^{-1}(\{f(\alpha)\})$ must be open. Hence, $\alpha \in f^{-1}(\{f(\alpha)\})$ has a neighborhood in the form $\{\alpha_0\} \times \cdots \times \{\alpha_{n-1}\} \times \mathbb{N}^{\omega}$ around α in $f^{-1}(f(\alpha))$, where $\alpha_i \in \mathbb{N}$ is the i-th argument of α. Therefore, for any $\beta \in \mathbb{N}^{\mathbb{N}}$, if $\alpha =_n \beta$, we have $f(\alpha) = f(\beta)$. Define $M(f, \alpha)$ as the least such number n. By definition, it is clear that $M(f, \alpha)$ has the required property.

For (ii), for any finite string $w = w_0 w_1 \ldots w_n$ of zeros and ones, define $K_w = \{w_0\} \times \cdots \times \{w_n\} \times \{0, 1\}^{\omega}$ as a subset of $2^{\mathbb{N}}$ and notice that it is both open and compact. Now, as the first step, we prove that for any $f : 2^{\mathbb{N}} \to \mathbb{N}$, there is at least one $m \in \mathbb{N}$ such that for any $\alpha, \beta \in 2^{\mathbb{N}}$, if $\alpha =_m \beta$, then $f(\alpha) = f(\beta)$. Note that $2^{\mathbb{N}}$ is a compact space and as f is continuous, $I = f[2^{\mathbb{N}}]$ is a compact subspace of the discrete space $\mathbb{N}$. Hence, it is finite. Moreover, as $\{k\}$ is both open and closed, then so is $f^{-1}(k)$, for any $k \in I$. Therefore, $f^{-1}(k)$ is compact. As it is open, it is a union of opens in the form K_w, for some finite strings w. By compactness, we can assume that the number of w's is finite. Hence, $f^{-1}(k) = \bigcup_{i=1}^{n_k} K_{w_{k,i}}$. Pick m as the maximum of the lengths of all $w_{k,i}$'s, for all $k \in I$ and $i \leq n_k$. Therefore, it is clear that if $\alpha =_m \beta$, then $\alpha \in K_{w_{k,i}}$ iff $\beta \in K_{w_{k,i}}$ which implies $\alpha \in f^{-1}(k)$ iff $\beta \in f^{-1}(k)$ and hence $f(\alpha) = f(\beta)$. This completes the proof of claim. Now, define $M(f)$ as the least such number m. By definition, it is clear that $M(f)$ has the claimed property. The only thing to prove is the continuity of M. As $M(f)$ is the least possible number with the property

$$\forall \alpha\beta \in 2^{\mathbb{N}}\, [\alpha =_m \beta \to f(\alpha) = f(\beta)],$$

it is easy to see that $M^{-1}(m)$ is the set of $f : 2^{\mathbb{N}} \to \mathbb{N}$ such that f is constant over K_w, for all $w \in \{0,1\}^m$ but there are distinct $\alpha, \beta \in 2^{\mathbb{N}}$ such that $\alpha =_{m-1} \beta$, while $f(\alpha) \neq f(\beta)$. Therefore, one can represent $M^{-1}(m)$ as the intersection of

$$\bigcup_{n\in\mathbb{N}} \bigcap_{w\in\{0,1\}^m} \{f : 2^{\mathbb{N}} \to \mathbb{N} \mid f[K_w] = \{n\}\}$$

and

$$\bigcup_{w\in\{0,1\}^{m-1}} \bigcup_{\alpha\neq\beta\in K_w} \bigcup_{n\neq n'\in\mathbb{N}} [\{f : 2^{\mathbb{N}} \to \mathbb{N} \mid f(\alpha) = n\} \cap \{f : 2^{\mathbb{N}} \to \mathbb{N} \mid f(\beta) = n'\}].$$

Note that the topology of $[2^{\mathbb{N}}, \mathbb{N}]$ is the usual compact-open topology, as observed in Example 89. Therefore, using the continuity of the evaluation function and the compactness of K_w, it is easy to prove that $M^{-1}(m)$ is open. Hence, M is continuous. □

Corollary 262 *The theory* $\mathsf{HA}_2^\omega + \mathsf{EXT} + \mathsf{AC}$ *is consistent with the axioms* $\mathsf{WC{-}N}_{\neg\neg}$ *and* $\mathsf{UWC{-}N}$, *where*

$$\mathsf{UWC{-}N}: \quad \forall f^{2^N \to N} \exists x^N \forall \alpha^{N\to 2} \beta^{N\to 2} (\alpha =_x \beta \to f\alpha = f\beta),$$

is the uniform continuity principle, *stating that for any map $f : 2^N \to N$, the value of f only depends on finitely many of the values of its input.*

Proof Set $C = \mathbf{Equ}$. Note that C is a non-preorder well-pointed category and its NNO is total. Moreover, M^2 is total. Therefore, it is enough to show that both $\mathsf{WC{-}N}_{\neg\neg}$ and $\mathsf{UWC{-}N}$ are realizable. For $\mathsf{WC{-}N}_{\neg\neg}$, using Remark 249, it is enough to show that for any continuous functions $f : \mathbb{N}^{\mathbb{N}} \to \mathbb{N}$ and any function $\alpha : \mathbb{N} \to \mathbb{N}$, the statement $\exists x^N \forall \beta^{N\to N} [\alpha =_x \beta \to f\alpha = f\beta]$ is realizable. Using Lemma 261, we know that $M(f,\alpha)$ provides the witness for x. The sentence $\forall \beta^{N\to N}[\alpha =_{M(f,\alpha)} \beta \to f\alpha = f\beta]$ is a true $\exists$-free 1-sentence. Hence, its truth implies its realizability. For (ii), it is enough to use the function M introduced in Lemma 261 for $2^{\mathbb{N}}$. Note that M is continuous. □

Remark 263 Similar to the case of Church-Turing thesis, to prove the consistency of the anti-classical theory $\mathsf{HA}^\omega + \mathsf{EXT} + \mathsf{AC} + \mathsf{WC} - \mathsf{N}_{\neg\neg}$, we can interpret arithmetical proofs in an alternative world, i.e., the continuous world **Equ**. Again, the interpretation was quite expected and one can say that the theory with its continuity principle is actually a device to describe such continuous worlds.

In the rest of this subsection, we use the realizability machinery to extract some computational information from the arithmetical proofs. First, recall the notion of representable functions defined in Definition 232. Then, we have

Theorem 264 (Function Extraction) *Let $A(x, y)$ be an $\exists$-free 0-formula and $\mathsf{HA}^\omega + \mathsf{EXT} \vdash \forall x^N \exists y^N A(x, y)$. Then, for any non-preorder* CCC *C with an* NNO, *there exists a function $f : \mathbb{N} \to \mathbb{N}$ representable in* C *such that $\mathfrak{N} \models \forall x^N A(x, f(x))$. The same also holds for* HA *and if $A(x, y)$ is bounded, for* PA.

Proof For the first part, notice that $|\forall x^N \exists y^N A(x, y)| = [\mathtt{N}, \mathtt{N} \times |A|]$. Therefore, by Theorem 257, there is a map $\mathtt{r} : \mathtt{1} \to [\mathtt{N}, \mathtt{N} \times |A|]$ such that $\mathtt{r} \Vdash \forall x^N \exists y^N A(x, y)$. Therefore, for any natural number $n \in \mathbb{N}$, as $\mathtt{n} \Vdash n$, we have $\mathtt{r} \cdot \mathtt{n} \Vdash \exists y^N A(n, y)$ which means that there is a natural number $m \in \mathbb{N}$ such that $\mathtt{p}_0(\mathtt{r} \cdot \mathtt{n}) \Vdash_N m$ and $\mathtt{p}_1(\mathtt{r} \cdot \mathtt{n}) \Vdash A(n, m)$. As $A(n, m)$ is an $\exists$-free 0-sentence, by Lemma 253, we have $\mathfrak{N} \models A(n, m)$. As C is non-preorder, the number $m \in \mathbb{N}$ is uniquely determined by $\mathtt{p}_0(\mathtt{r} \cdot \mathtt{n})$. Define the function $f : \mathbb{N} \to \mathbb{N}$ by $f(n) = m$. It is clear that $A(n, f(n))$ holds in the standard model, for any $n \in \mathbb{N}$. The only thing to note is that the function $f : \mathbb{N} \to \mathbb{N}$ is representable by the map $\mathtt{p}_0 \circ (\mathtt{r} \cdot \mathtt{id}_{\mathbb{N}})$. For the second part, it is enough to embed HA into HA^ω and note that any formula in the language of HA is translated into a 0-formula as its terms are all of type N. For PA, use the conservativity of PA over HA for the formulas in the form $\forall x \exists y A(x, y)$, where A is bounded and the fact that any bounded formula is equivalent to an $\exists$-free formula in HA. □

Corollary 265 *If* $\mathsf{HA}^\omega + \mathsf{EXT} \vdash \forall x^N \exists y^N A(x, y)$, *for an $\exists$-free 0-formula $A(x, y)$, then there exists a recursive function $f : \mathbb{N} \to \mathbb{N}$ such that $\mathfrak{N} \models \forall x^N A(x, f(x))$. The same also holds for* HA.

Proof Set $C =$ **Rec** in Theorem 264 and recall that only recursive functions are representable in **Rec**. □

To find the strongest possible application of Theorem 264, we must work with the "least" possible CCC with an NNO, i.e., the free category Λ_N.

Corollary 266 *If* $\mathsf{HA}^\omega + \mathsf{EXT} \vdash \forall x^N \exists y^N A(x, y)$, *for an $\exists$-free 0-formula, then there exists a primitive recursive functional $f : \mathbb{N} \to \mathbb{N}$ such that $\mathfrak{N} \models \forall x^N A(x, f(x))$. The same also holds for* HA *and if $A(x, y)$ is bounded, for* PA.

Proof Set $C = \Lambda_N$ in Theorem 264. By Remark 241, Λ_N is not a preorder. Therefore, it is enough to recall that the numeral primitive recursive functionals are by definition the functions representable in the category Λ_N. □

Finally, we can apply the extraction technique to the higher functionals:

Corollary 267 *If* $\mathsf{HA}^{\omega} + \mathsf{EXT} + \mathsf{AC} \vdash \forall x^{N \to N} \exists y^N A(x, y)$, *for an* $\exists$*-free* 1*-formula* $A(x, y)$, *then there exists a continuous function* $f : \mathbb{N}^{\mathbb{N}} \to \mathbb{N}$ *such that* $\mathfrak{N} \vDash A(\alpha, f(\alpha))$, *for any function* $\alpha : \mathbb{N} \to \mathbb{N}$.

Proof As **Equ** is a well-pointed non-preorder with the total NNO, in Theorem 257, set $C =$ **Equ**. Notice that $|\forall x^{N \to N} \exists y^N A(x, y)| = [\mathrm{N}^{\mathrm{N}}, \mathrm{N} \times |A|]$. Therefore, there is a map $\mathtt{r} : \mathtt{1} \to [\mathrm{N}^{\mathrm{N}}, \mathrm{N} \times |A|]$ such that $\mathtt{r} \Vdash \forall x^{N \to N} \exists y^N A(x, y)$. By definition, for any function $\alpha : \mathbb{N} \to \mathbb{N}$, we have $\mathtt{r} \cdot \alpha \Vdash \exists y^N A(\alpha, y)$ which means that there is a natural number $m \in \mathbb{N}$ such that $\mathtt{p}_0(\mathtt{r} \cdot \alpha) \Vdash_N m$ and $\mathtt{p}_1(\mathtt{r} \cdot \alpha) \Vdash A(\alpha, m)$. As **Equ** is not a preordered set, the value of m is uniquely determined by $\mathtt{p}_0(\mathtt{r} \cdot \alpha)$. Define $f : \mathbb{N}^{\mathbb{N}} \to \mathbb{N}$ by $f(\alpha) = m$. Note that f is a continuous map, as it is realized by $\mathtt{p}_0(\mathtt{r} \cdot \alpha)$ living in **Equ**. As $A(\alpha, f(\alpha))$ is an $\exists$-free 1-formula and hence standard, by Lemma 253, we have $\mathfrak{N} \vDash A(\alpha, f(\alpha))$, for any function $\alpha : \mathbb{N} \to \mathbb{N}$. □

Acknowledgements We wish to thank the organizers of the TACL 2022 summer school and Rosalie Iemhoff for their invitation, the participants of the summer school and the fundamental computing team at the Bernoulli institute who organized and actively participated in the lectures on the present text and the editors of the present volume, especially Maria Manuel Clementino, for their support and patience without which this chapter would never come to the existence. We are also grateful to Steve Awodey, Andrej Bauer, David Pym, Pino Rosolini, Alex Simpson, Thomas Streicher, and the anonymous referees for their very helpful comments, suggestions, and corrections on the first draft. Support by the FWF project P 33548, the Czech Academy of Sciences (RVO 67985840), and the GAČR grant 23-04825S is also gratefully acknowledged.

References

Abramsky S, Coecke B (2004) A categorical semantics of quantum protocols. In: Proceedings of the 19th annual IEEE symposium on logic in computer science. IEEE, Turku, Finland, pp 415–425. https://doi.org/10.1109/LICS.2004.1319636

Abramsky S, Coecke B (2009) Categorical quantum mechanics. In: Engesser K, Gabbay DM, Lehmann D (eds) Handbook of quantum logic and quantum structures–quantum logic. Elsevier/North-Holland, Amsterdam, pp 261–323. https://doi.org/10.1016/B978-0-444-52869-8.50010-4

Awodey S, Bauer A (2004) Propositions as [types]. J Logic Comput 14(4):447–471. https://doi.org/10.1093/logcom/14.4.447

Barr M (1979) $*$-autonomous categories, vol 752. Lecture Notes in Mathematics, Springer, Berlin

Barr M (1991) $*$-autonomous categories and linear logic. Math Struct Comput Sci 1(2):159–178. https://doi.org/10.1017/S0960129500001274

Bauer A (2000) The realizability approach to computable analysis and topology. PhD thesis, Carnegie Mellon University

Bauer A (2005) Realizability as the connection between computable and constructive mathematics. In: Proceedings of CCA

Bauer A, Birkedal L, Scott DS (2004) Equilogical spaces. Theoret Comput Sci 315(1):35–59. https://doi.org/10.1016/j.tcs.2003.11.012

Bellin G, Hyland M, Robinson E, Urban C (2006) Categorical proof theory of classical propositional calculus. Theoret Comput Sci 364(2):146–165. https://doi.org/10.1016/j.tcs.2006.08.002

Bénabou J (1974) Logique catégorique. A series of lectures presented at the Séminaire de Mathématiques Supérieur, Montreal

Benton N, Bierman G, de Paiva V, Hyland M (1993) Linear λ-calculus and categorical models revisited. In: Bürger E, Jäger G, Büning HK, Martini S, Richter MM (eds) Computer science logic (San Miniato, 1992), Lecture Notes in Computer Science, vol 702. Springer, Berlin, pp 61–84. https://doi.org/10.1007/3-540-56992-8_6

Benton N, Bierman G, de Paiva V, Hyland M (1993) A term calculus for intuitionistic linear logic. In: Bezem M, Groote JF (eds) Typed lambda calculi and applications (Utrecht, 1993), Lecture Notes in Computer Science, vol 664. Springer, Berlin, pp 75–90. https://doi.org/10.1007/BFb0037099

Bezhanishvili G, Bezhanishvili N, Lucero-Bryan J, van Mill J (2018) A new proof of the McKinsey-Tarski theorem. Studia Logica 106(6):1291–1311. https://doi.org/10.1007/s11225-018-9789-5

Bierman GM (1994) On intuitionistic linear logic. PhD thesis, University of Cambridge, Computer Laboratory

Bierman GM (1995) What is a categorical model of intuitionistic linear logic? In: Dezani-Ciancaglini M, Plotkin G (eds) (1995) Typed Lambda Calculi and applications—second international conference on Typed Lambda Calculi and Applications, TLCA'95 Edinburgh, United Kingdom, 10–12 Apr 1995 Proceedings 2, Lecture Notes in Computer Science, vol 902. Springer, Berlin, pp 78–93. https://doi.org/10.1007/BFb0014046

Birkedal L (1999) Developing theories of types and computability via realizability. PhD thesis, Carnegie Mellon University Pittsburgh, PA

Birkedal L (2002) A general notion of realizability. Bull Symbol Logic 8(2):266–282. https://doi.org/10.2307/2693966

Borceux F (1994) Handbook of Categorical Algebra 2–Categories and Structures, Encyclopedia of Mathematics and its Applications, vol 51. Cambridge University Press, Cambridge

Brouwer LEJ (1981) Brouwer's Cambridge lectures on intuitionism. Cambridge University Press, Cambridge, New York

Chagrov A, Zakharyaschev M (1997) Modal logic, Oxford Logic Guides, vol 35. The Clarendon Press, Oxford University Press, New York

Cockett JRB (1997) Seely RAG (1997) Weakly distributive categories. J Pure Appl Algebra 114(2):133–173. https://doi.org/10.1016/0022-4049(95)00160-3

Čubrić D (1998) Embedding of a free Cartesian-closed category into the category of sets. J Pure Appl Algebra 126(1–3):121–147. https://doi.org/10.1016/S0022-4049(96)00137-5

Crole RL (1993) Cambridge mathematical textbooks. Cambridge University Press, Cambridge

De Paiva VCV (1989) The Dialectica categories. In: Categories in computer science and logic (Boulder, CO, 1987), Contemporary mathematics, vol 92. American Mathematical Society, Providence, RI, pp 47–62. https://doi.org/10.1090/conm/092/1003194

Di Cosmo R (1995) Isomorphisms of types: from λ-calculus to information retrieval and language design. Progress in theoretical computer science. Birkhäuser Boston Inc, Boston, MA. https://doi.org/10.1007/978-1-4612-2572-0

Di Cosmo, R.: A short survey of isomorphisms of types. Math. Structures Comput. Sci. 15(5), 825–838 (2005). https://doi.org/10.1017/S0960129505004871

Di Cosmo R, Dufour T (2005) The equational theory of $\langle \mathcal{N}, 0, 1, +, \times, \uparrow \rangle$ is decidable, but not finitely axiomatisable. In: Baader F, Voronkov A (eds), Logic for programming, artificial intelligence, and reasoning. Lecture Notes in Computer Science, vol 3452. Springer, Berlin, pp 240–256. https://doi.org/10.1007/978-3-540-32275-7_17

Doner J, Tarski A (1969) An extended arithmetic of ordinal numbers. Fund Math 65:95–127. https://doi.org/10.4064/fm-65-1-95-127

Došen K (1999) Cut elimination in categories, Trends in Logic-Studia Logica Library, vol 6. Springer Science and Business Media. https://doi.org/10.1007/978-94-017-1207-1

Došen K (2003) Identity of proofs based on normalization and generality. Bull Symbol Logic 9(4):477–503. https://doi.org/10.2178/bsl/1067620091

Došen K (2011) A prologue to the theory of deduction. The logica yearbook 2010. Coll Publications, London, pp 65–80

Došen K, Petrić Z (2000) The maximality of the typed lambda calculus and of Cartesian closed categories. Publ Inst Math (Beograd) (NS) 68(82):1–19

Došen K, Petrić Z (2004) Proof-theoretical coherence. Student Log (London), vol 1. King's College Publications, London

Fiore M, Di Cosmo R, Balat V (2006) Remarks on isomorphisms in typed lambda calculi with empty and sum types. Ann Pure Appl Logic 141(1–2):35–50. https://doi.org/10.1016/j.apal.2005.09.001

Friedman H (1975) Equality between functionals. Logic Colloquium (Boston, MA, 1972–1973), vol 453. Lecture Notes in Mathematics. Springer, Berlin-New York, pp 22–37

Fuhrmann C, Pym D (2004) On the geometry of interaction for classical logic. In: Proceedings of the 19th annual IEEE symposium on logic in computer science, 2004. IEEE, pp 211–220. https://doi.org/10.1109/LICS.2004.1319615

Führmann C, Pym D (2006) Order-enriched categorical models of the classical sequent calculus. J Pure Appl Algebra 204(1):21–78. https://doi.org/10.1016/j.jpaa.2005.03.016

Führmann C, Pym D (2007) On categorical models of classical logic and the geometry of interaction. Math Struct Comput Sci 17(5):957–1027. https://doi.org/10.1017/S0960129507006287

Girard JY, Taylor P, Lafont Y (1989) Proofs and types, Cambridge tracts in theoretical computer science, vol 7. Cambridge University Press, Cambridge

Goodman ND (1978) Relativized realizability in intuitionistic arithmetic of all finite types. J Symbol Logic 43(1):23–44. https://doi.org/10.2307/2271946

Gurevič R (1985) Equational theory of positive numbers with exponentiation. Proc Am Math Soc 94(1):135–141. https://doi.org/10.2307/2044966

Harnik V, Makkai M (1992) Lambek's categorical proof theory and Läuchli's abstract realizability. J Symbolic Logic 57(1):200–230. https://doi.org/10.2307/2275186

Heunen C, Vicary J (2019) Categories for quantum theory—an introduction, Oxford Graduate Texts in Mathematics, vol 28. Oxford University Press, Oxford. https://doi.org/10.1093/oso/9780198739623.001.0001

Hofmann M (1995) Sound and complete axiomatisations of call-by-value control operators. Math Struct Comput Sci 5(4):461–482. https://doi.org/10.1017/S0960129500001195

Hofmann M (1997) Syntax and semantics of dependent types. In: Semantics and logics of computation (Cambridge, 1995), Publications of the Newton Institute, vol 14. Cambridge University Press, Cambridge, pp 79–130. https://doi.org/10.1017/CBO9780511526619.004

Hofmann M, Streicher T (2002) Completeness of continuation models for λ_μ-calculus. Inform Comput 179(2):332–355. LICS'97 (Warsaw). https://doi.org/10.1006/inco.2001.2947

Hötzel Escardó M, Xu C (2015) The inconsistency of a Brouwerian continuity principle with the Curry-Howard interpretation. In: Altenkirch T (ed) 13th international conference on Typed Lambda Calculi and applications, LIPIcs. Leibniz international proceedings in informatics, vol 38, pp 153–164. Schloss Dagstuhl. Leibniz-Zentrum Information, Wadern (2015)

Howard WA (1980) The formulae-as-types notion of construction. In: Seldin JP, Hindley JR (eds) To H B Curry–essays on combinatory logic, lambda calculus and formalism. Academic Press, London-New York, pp 479–490

Hyland M (1982) The effective topos. In: Troelstra AS, van Dalen D (eds) The LEJ Brouwer Centenary Symposium (Noordwijkerhout, 1981), vol 110. Studies in Logic and the Foundations of Mathematics. North-Holland, Amsterdam-New York, pp 165–216

Hyland M (2002) Proof theory in the abstract. Ann Pure Appl Logic 114(1–3):43–78 (2002). doi.org/10.1016/S0168-0072(01)00075-6

Hyland M (2004) Abstract interpretation of proofs: classical propositional calculus. In: Marcinkowski J, Tarlecki A (eds) Computer science logic: 18th international workshop, CSL

2004, 13th annual conference of the EACSL, Karpacz, Poland, 20–24 Sept 2004. Proceedings 18, Lecture Notes in Computer Science, vol 3210. Springer, Berlin, pp 6–21. https://doi.org/10.1007/978-3-540-30124-0_2

Jacobs B (1999) Categorical logic and type theory, Studies in Logic and the Foundations of Mathematics, vol 141. North-Holland Publishing Co., Amsterdam

Joyal A, Moerdijk I (1990) Toposes as homotopy groupoids. Adv Math 80(1):22–38. https://doi.org/10.1016/0001-8708(90)90013-D

Kelly GM, Laplaza ML (1980) Coherence for compact closed categories. J Pure Appl Algebra 19:193–213. https://doi.org/10.1016/0022-4049(80)90101-2

Kleene SC (1945) On the interpretation of intuitionistic number theory. J Symbolic Logic 10(4):109–124. https://doi.org/10.2307/2269016

Kock A (2006) Synthetic differential geometry, London Mathematical Society Lecture Note Series, vol 333, 2nd edn. Cambridge University Press, Cambridge. https://doi.org/10.1017/CBO9780511550812

Lambek J (1968) Deductive systems and categories. I. Syntactic calculus and residuated categories. Math Syst Theory 2:287–318. https://doi.org/10.1007/BF01703261

Lambek J (1969) Deductive systems and categories. II. Standard constructions and closed categories. In: Hilton PJ (ed) Category theory, homology theory and their applications, I (Battelle Institute Conference, Seattle, Wash, 1968, vol One), Lecture Notes in Mathematics, vol 86. Springer, Berlin, New York, pp 76–122. https://doi.org/10.1007/BFb0079385

Lambek J (1972) Deductive systems and categories. III. Cartesian closed categories, intuitionist propositional calculus, and combinatory logic. In: Lawvere FW (ed) Toposes, algebraic geometry and logic (Conf, Dalhousie Univ, Halifax, NS, 1971), Lecture Notes in Mathematics, vol 274. Springer, Berlin, New York, pp 57–82. https://doi.org/10.1007/BFb0073965

Lambek J (2000) An equational variant of Lawvere's natural numbers object. J Pure Appl Algebra 154(1–3):265–272. https://doi.org/10.1016/S0022-4049(99)00178-4

Lambek J, Scott PJ (1988) Introduction to higher-order categorical logic, vol 7. Cambridge University Press, Cambridge

Läuchli H (1970) An abstract notion of realizability for which intuitionistic predicate calculus is complete. In: Kino A, Myhill J, Vesley RE (eds) Intuitionism and Proof Theory (Proc Conf, Buffalo, NY, 1968), Studies in Logic and the Foundations of Mathematics. North-Holland, Amsterdam-London, pp 227–234 (1970)

Lawvere FW (1970) Equality in hyperdoctrines and comprehension schema as an adjoint functor. In: Applications of categorical algebra (Proc Sympos Pure Math, Vol XVII, New York, 1968), Proceedings of a Symposium in Pure Mathematics, vol XVII. American Mathematical Society, Providence, RI, pp 1–14

Lawvere FW (1996) Adjoints in and among bicategories. In: Ursini A, Agliano P (eds) Logic and algebra (Pontignano, 1994), vol 180. Lecture notes in pure and applied mathematics. Dekker, New York, pp 181–189

Lawvere FW (2004) Functorial semantics of algebraic theories and some algebraic problems in the context of functorial semantics of algebraic theories. Repr Theory Appl Categ 5:1–121

Lawvere FW (2006) Adjointness in foundations. Repr Theory Appl Categ 16:1–16

Mac Lane S (1963) Natural associativity and commutativity. Rice Univ Stud 49(4):28–46

Mac Lane S (1982) Why commutative diagrams coincide with equivalent proofs. In: Amitsur SA, Saltman DJ, Seligman GB (eds) Algebraists' homage: papers in ring theory and related topics (New Haven, Conn, 1981), Contemporary Mathematics, vol 13. American Mathematical Society. Providence, RI, pp 387–401

Mac Lane S (1997) Despite physicists, proof is essential in mathematics. Synthese 111(2):147–154. https://doi.org/10.1023/A:1004918402670

Mac Lane S (1998) Categories for the working mathematician. Graduate texts in mathematics, vol 5, 2nd edn. Springer, New York

Mac Lane S, Moerdijk I (1994) Sheaves in geometry and logic. Universitext. Springer, New York (1994). A first introduction to topos theory, Corrected reprint of the 1992 edition

Macintyre A (1981) The laws of exponentiation. In: Model theory and arithmetic (Paris, 1979–1980), Lecture Notes in Mathematics, vol 890. Springer, Berlin, New York, pp 185–197 (1981)

Makkai M (1993a) The fibrational formulation of intuitionistic predicate logic I: completeness according to Gödel, Kripke, and Läuchli. I. Notre Dame J Formal Logic 34(3):334–377. https://doi.org/10.1305/ndjfl/1093634727

Makkai M (1993b) The fibrational formulation of intuitionistic predicate logic I: completeness according to Gödel, Kripke, and Läuchli. II. Notre Dame J Formal Logic **34**(4):471–498. https://doi.org/10.1305/ndjfl/1093633902

Martin-Löf P (1984) Intuitionistic type theory, Studies in Proof Theory. Lecture Notes, vol 1. Bibliopolis, Naples. Notes by Giovanni Sambin

McKinley R (2006) Categorical models of first-order classical proofs. PhD thesis, University of Bath, UK

McKinsey JCC, Tarski A (1944) The algebra of topology. Ann Math (2)45:141–191. https://doi.org/10.2307/1969080

Menni M (2000) Exact completions and toposes. PhD thesis, University of Edinburgh

Ong CHL (1996) A semantic view of classical proofs: type-theoretic, categorical, and denotational characterizations. In: 11th annual IEEE symposium on logic in computer science (New Brunswick, NJ, 1996). IEEE Computer Society Press, Los Alamitos, CA, pp 230–241. https://doi.org/10.1109/LICS.1996.561323

van Oosten J (2008) Realizability–an introduction to its categorical side, Studies in Logic and the Foundations of Mathematics, vol 152. Elsevier B V, Amsterdam

Palmgren E (2004) A categorical version of the Brouwer-Heyting-Kolmogorov interpretation. Math Struct Comput Sci 14(1):57–72. https://doi.org/10.1017/S0960129503003955

Parigot M (1992) $\lambda\mu$-calculus: an algorithmic interpretation of classical natural deduction. In: Voronkov A (ed) Logic programming and automated reasoning: international conference LPAR'92 (St Petersburg, 1992). Proceedings 3, Lecture Notes in Computer Science, vol 624, pp 190–201. Springer, Berlin. https://doi.org/10.1007/BFb0013061

Plotkin GD (1973) Lambda-definability and logical relations. Technical Report SAIRM-4, School of Artificial Intelligence, University of Edinburgh

Plotkin GD (1980) Lambda-definability in the full type hierarchy. To HB Curry: essays on combinatory logic, lambda calculus and formalism. Academic Press, London-New York, pp 363–373

Prawitz D (1971) Ideas and results in proof theory. In: Fenstad JE (ed) Proceedings of the second scandinavian logic symposium (Univ Oslo, Oslo, 1970), vol 63. Studies in Logic and the Foundations of Mathematics. North-Holland, Amsterdam-London, pp 235–307

Prawitz D (1974) On the idea of a general proof theory. Synthese 27:63–77. https://doi.org/10.1007/BF00660889

Pym D (2001) Ritter E (2001) On the semantics of classical disjunction. J Pure Appl Algebra 159(2–3):315–338. https://doi.org/10.1016/S0022-4049(00)00161-4

Pym D, Ritter E, Robinson E (2014) A proof-theoretic analysis of the classical propositional matrix method. J Logic Comput 24(1):283–301. https://doi.org/10.1093/logcom/exs045

Rasiowa H, Sikorski R (1963) The mathematics of metamathematics. Monografie Matematyczne, Tom 41. Państwowe Wydawnictwo Naukowe, Warsaw

Robinson E (2003) Proof nets for classical logic. J Logic Comput 13(5):777–797. https://doi.org/10.1093/logcom/13.5.777

Scherer G (2016) Which types have a unique inhabitant? PhD thesis, Université Paris-Diderot

Scherer G (2017) Deciding equivalence with sums and the empty type. In: Proceedings of the 44th annual ACM SIGPLAN symposium on principles of programming languages, POPL '17 (Paris, France 2017), ACM SIGPLAN Notices, vol 52. Association for Computing Machinery (ACM), New York, NY, USA, pp 374–386. https://doi.org/10.1145/3009837.3009901

Seely RAG (1977) Hyperdoctrines and natural deduction. PhD thesis, University of Cambridge

Seely RAG (1983) Hyperdoctrines, natural deduction and the Beck condition. Z Math Logik Grundlag Math 29(6):505–542. https://doi.org/10.1002/malq.19830291005

Seely RAG (1989) Linear logic, $*$-autonomous categories and cofree coalgebras. In: Gray JW, Scedrov A (eds) Categories in computer science and logic (Boulder, CO, 1987), Contemporary Mathematics, vol 92. Am Math Soc, Providence, RI, pp 371–382. https://doi.org/10.1090/conm/092/1003210

Selinger P (2001) Control categories and duality: on the categorical semantics of the lambda-mu calculus. Math Struct Comput Sci 11(2):207–260. https://doi.org/10.1017/S096012950000311X

Simpson A (1995) Categorical completeness results for the simply-typed lambda-calculus. In: Dezani-Ciancaglini M, Plotkin G (eds) Typed lambda calculi and applications (Edinburgh, 1995), Lecture Notes in Computer Science, vol 902. Springer, Berlin, pp 414–427. https://doi.org/10.1007/BFb0014068

Simpson A (1995) TLCA list of problems. Problem 15. https://tlca.di.unito.it/opltlca

Simpson A (2012) Measure, randomness and sublocales. Ann Pure Appl Logic 163(11):1642–1659. https://doi.org/10.1016/j.apal.2011.12.014

Solovev SV (1981) The category of finite sets and Cartesian closed categories. Zap Nauchn Sem Leningrad Otdel Mat Inst Steklov (LOMI) 105(174–194):200

Sørensen MH, Urzyczyn P (2006) Lectures on the Curry-Howard isomorphism, Studies in Logic and the Foundations of Mathematics, vol 149. Elsevier

van Stigt WP (1990) Brouwer's intuitionism, Studies in the History and Philosophy of Mathematics, vol 2. North-Holland Publishing Co, Amsterdam

Straßburger L (2011) Towards a theory of proofs of classical logic. PhD thesis, Université Paris-Diderot-Paris VII

Szabo ME (1978) Algebra of proofs, Studies in Logic and the Foundations of Mathematics, vol 88. North-Holland Publishing Co, Amsterdam-New York

Taylor P (1999) Practical foundations of mathematics, Cambridge Studies in Advanced Mathematics, vol 59. Cambridge University Press, Cambridge

Program TUF (2013) Homotopy type theory-univalent foundations of mathematics. The Univalent Foundations Program, Princeton, NJ, Institute for Advanced Study (IAS), Princeton, NJ

Troelstra AS (1998) Realizability. In: Buss SR (ed) Handbook of proof theory, vol 137. Studies in Logic and the Foundations of Mathematics, North-Holland, Amsterdam, pp 407–473. https://doi.org/10.1016/S0049-237X(98)80021-9

Troelstra AS, Schwichtenberg H (2000) Basic proof theory, Cambridge Tracts in Theoretical Computer Science, vol 43, 2nd edn. Cambridge University Press, Cambridge. https://doi.org/10.1017/CBO9781139168717

Troelstra AS, Van Dalen D (1988) Constructivism in mathematics, vol 1. Studies in Logic and the Foundations of Mathematics, vol 121. North-Holland Publishing Co, Amsterdam

Troelstra AS, Van Dalen D (1988) Constructivism in mathematics, vol 2. Studies in Logic and the Foundations of Mathematics, vol 123. North-Holland Publishing Co, Amsterdam

Wilkie AJ (2000) On exponentiation-a solution to Tarski's high school algebra problem. Connections between model theory and algebraic and analytic geometry, Quad Mat, vol 6. Seconda University Napoli, Caserta, Department Mathematics, pp 107–129

From Metric Spaces to Quantale-Enriched Categories

Walter Tholen

Abstract The goal of this article is to show how the study of metric spaces and of quantale-enriched categories interact, through generalization and specialization. We replace the recipient interval $[0, \infty]$ of a (potentially ∞-valued) metric on a set by a commutative quantale, that is, by a complete lattice $\mathcal{V}$ that comes with a commutative monoid operation interacting with the lattice structure. As the quantale $\mathcal{V}$ may be seen as a thin cocomplete and symmetric monoidal-closed category, with arrows given by order, sets equipped with a generalized $\mathcal{V}$-valued metric will then become small $\mathcal{V}$-enriched categories. In this way, we explore how the theories of (not necessarily symmetric or separated) metric spaces and of enriched categories enhance each other. Our focus is on the enriched completion theory of these structures and on the important role that distributors between them play for them, and we offer a few glances at the many applications beyond the original metric theory, owing to the wide range of interesting quantales.

2010 Mathematics Subject Classification 18A40 · 18C20 · 18D30 · 18F60

1 Introduction

Our digital devices remind many of us every day that the distance from A to B, measured in travel time, may be far from symmetric. Also, some locations B may not be reachable from A by the chosen mode of travel, in which case it seems reasonable to declare the distance from A to B to be infinite. Finally, our distance measure may not be fine enough to distinguish between locations A and B when

The author gratefully acknowledges the financial assistance received from the Centre of Mathematics of the University of Coimbra and from the Natural Sciences and Engineering Council of Canada under its Discovery Grants Program (no. 501260).

W. Tholen (✉)
Department of Mathematics and Statistics, York University, Toronto, ON, Canada
e-mail: tholen@yorku.ca

M. M. Clementino et al. (eds.), *Topology, Algebra and Categories in Logic*,
Coimbra Mathematical Texts 5, https://doi.org/10.1007/978-3-032-13760-9_5

they are very close to each other, so that they be declared to have distance 0 in either direction, even though the two locations do not coincide.

It was Lawvere (1973) who emphasized the need to pay attention to such generalized metric spaces which, other than admitting infinite distances, are deprived of the symmetry and separation conditions appearing in Fréchet's original notion (Fréchet (1906)). Lawvere's great insight was to interpret the quantity $d(A, B)$ as $\hom(A, B)$ in the enriched categorical sense—at a time when enriched category theory was still in its infancy, despite Bénabou's early communications Bénabou (1963), Bénabou (1965) and the well-disseminated work of Eilenberg and Kelly (1965)—the book-style presentations, such as Kelly (1982) or Borceux (1994), were still a decade or two away. Lawvere (1973) also embraced Bénabou's *distributors*, referring to Bénabou (1973) and renaming them *bimodules*. He showed their great utility not only for metric spaces, for instance, by representing equivalence classes of Cauchy sequences in a space as adjoint distributors, but also to advance enriched category theory in general.

There is an intermediate level of generality between the study of Lawvere's metric spaces and general enriched category theory, whereby the cocomplete symmetric monoidal-closed category $\mathcal{V}$, in which the "homs" of a $\mathcal{V}$-enriched category must reside as objects, is just a small, skeletal, and thin category; that is, there is only a set of objects in $\mathcal{V}$, and its morphisms are given by order, making $\mathcal{V}$ a poset. Cocompleteness of $\mathcal{V}$ then amounts to the existence of all suprema which, by monoidal closedness, must distribute over the given commutative and associative binary operation $\otimes$ of $\mathcal{V}$, and there is a special element $\mathrm{k} \in \mathcal{V}$ which is neutral w.r.t. the $\otimes$-operation. In other words, $\mathcal{V}$ is a commutative and unital *quantale* $(\mathcal{V}, \leq, \otimes, \mathrm{k})$, a term coined by Mulvey (1986), as a contraction of "quantum locale." In this chapter, the Boolean quantale $\mathsf{2} = (\{\bot, \top\}, \Rightarrow, \&, \top)$ and the Lawvere quantale $([0, \infty], \geq, +, 0)$ serve as role models for this structure which, since its inception, has been studied and utilized by mathematicians, logicians, and some scientists, also after the appearance of the first book Rosenthal (1990) devoted to them.

This is the level of generality which we adopt in this survey. It greatly simplifies the machinery of enriched category theory, as all coherence conditions disappear: in a category in which morphisms are given by a (pre)order, all (legitimately drawn) diagrams commute! Of course, this simplification remains intact also for non-commutative quantales (as largely adopted in Hofmann and Seal (2014)), and even when $\mathcal{V}$ is a *quantaloid* (the thin bicategorical generalization[1] of a quantale, as presented in Rosenthal (1996), but introduced for bicategorical enrichment already in Walters (1981), and pursued further in Stubbe (2005), Stubbe (2006), Stubbe (2007), and Tholen (2019)). But such generalization comes at the cost of having to

[1] Another way of differentiating between quantales and quantaloids is to regard the latter as small categories enriched in the symmetric monoidal-closed category Sup of complete lattices and their suprema-preserving maps; the former are then just the one-object categories, i.e., the monoid objects in Sup.

distinguish between two internal homs from the beginning. We find that our range of examples does not justify the effort of carrying this additional feature throughout this chapter.

Our aim then is to give an introduction to quantale-enriched category theory, with a particular focus on completion theory. Frequent stops at the role model of generalized metric spaces serve as an illustration of the theory. We assume the reader's familiarity with basic ordinary category theory (limits, adjunctions, monads), and the notion of 2-category. To give some context to the more advanced reader, we sometimes use 2-categorical or enriched categorical terms, without drawing on the respective theory. Hence, we hope that the understandability of the mathematics of our text by readers not familiar with such terms has not been compromised. (The use of lax idempotency of the presheaf monad in Theorem 68 may be an exception, but an easy reference has been given there to complete the argumentation.)

Readers in search of results believed to be new are directed particularly to Theorem 57, where we discuss a two-dimensional universal property of the hom objects of the category of $\mathcal{V}$-distributors, as well as to Sect. 7. There we show how the important Joyal–Tierney presentation Joyal and Tierney (1984) of the category Sup of complete lattices and sup-preserving maps as *$*$-autonomous* (Barr 1979, 1999) extends to the quantale-enriched setting, and thereby provide an alternative proof of Theorem 148, i.e., of a result first established in Eklund et al. (2018), which is in the spirit of first constructing a universal bi-morphism as generally suggested by Banaschewski and Nelson (1976). As a particular consequence, there is a metric version of the $*$-autonomous category Sup, whereby order is replaced by numerical distance, and it behaves just as well as Sup, also as a concrete monadic category.

2 A Brief Record of Fréchet's Axioms for a Metric and Its Variants

2.1 Basic Definitions and Notation

In search of well-behaved categories that contain all classical metric spaces as introduced by Fréchet (1906), we start by writing the axioms of a metric $d : X \times X \to [0, \infty]$ less compactly than usual, as follows ($x, y, z \in X$):

- *Point inequality*: $0 \geq d(x, x)$ (which, given the range of d, means $0 = d(x, x)$).
- *Triangle inequality*: $d(x, y) + d(y, z) \geq d(x, z)$ (with $+$ extended to $[0, \infty]$).
- *Symmetry*: $d(x, y) \geq d(y, x)$ (which, by x-y-exchange, means $d(x, y) = d(y, x)$).
- *Separation*: If $0 \geq d(x, y)$ and $0 \geq d(y, x)$, then $x = y$.
- *Finiteness*: $\infty > d(x, y)$.

While, as we will see, this sequence of conditions becomes increasingly cumbersome from a categorical point of view, there can be little dispute about the choice of the "right" morphisms $f : (X, d) \to (Y, d')$ for the objects under consideration, even if we dispose of some of the above axioms: they should be *non-expansive* (also called *short* or 1-*Lipschitz*) maps $f : X \to Y$ and, hence, satisfy the condition of

- *Non-expansion*: $d(x, y) \geq d'(fx, fy)$.

With these morphisms we obtain the category Met of *generalized metric spaces* (X, d), *required to satisfy only the point and triangle inequalities*. We allow ourselves to often refer to generalized metrics just as metrics, and then to use the term Fréchet metric for the classical notion. But it is important to remember that the absence of symmetry and separation makes d a "one-way metric" which, in addition, allows distinct points to have distance 0. Important special morphisms in Met are of course the *isometries, i.e.*, the maps f for which "=" holds in the non-expansion condition, rather than just "≥."

Met contains the full subcategories $\mathsf{Met}_{\mathrm{sym}}$, $\mathsf{Met}_{\mathrm{sep}}$, and $\mathsf{Met}_{\mathrm{sym,sep}}$ of objects satisfying, respectively, the symmetry, the separation, and both of these conditions; and $\mathsf{Met}_{\mathrm{Fréchet}}$ denotes the subcategory of all classically defined metric spaces. We will also consider two prominent full subcategories of Met, namely, the category BMet of *bounded (generalized) metric spaces*, and UMet of *(generalized) ultrametric spaces*, respectively, satisfying the additional conditions:

- *Boundedness*: $1 \geq d(x, y)$.
- *Triangle ultra-inequality*: $\max\{d(x, y), d(y, z)\} \geq d(x, z)$.

They, in turn, contain the self-explanatory subcategories $\mathsf{BMet}_{\mathrm{sym}}$, $\mathsf{UMet}_{\mathrm{sym}}$, *etc.*

In order to prepare the reader early on for Lawvere's (1973) categorical interpretation of distances as internal homs, and to avoid the introduction of new names when forming new metrics from old, we often adopt his very economical notation.

NOTATIONAL CONVENTION: For a generalized metric space (X, d) and $x, y \in X$, we often write $X(x, y)$ instead of $d(x, y)$ and refer to the space just as X. Hence, its basic axioms read as

$$0 \geq X(x, x) \qquad \text{and} \qquad X(x, y) + X(y, z) \geq X(x, z),$$

and a morphism $f : X \to Y$ must satisfy

$$X(x, y) \geq Y(fx, fy).$$

As done in this inequality, we often write fx instead of $f(x)$, and the composition of maps just by juxtaposition from right to left. This is consistent with the categorical interpretation of elements $x \in X$ as maps $\mathrm{E} \to X$, with a fixed one-element set E.

2.2 *Some Categorical Properties and Shortcomings of* $\mathsf{Met}_{\text{Fréchet}}$

Since the forgetful functor $\mathsf{Met}_{\text{Fréchet}} \to \mathsf{Set}$ is representable (by the trivial singleton space E, which also serves as a terminal object), any existing limits in $\mathsf{Met}_{\text{Fréchet}}$ must be formed by providing the Set-limit with an appropriate metric. Hence, equalizers are trivially obtained as subspaces, and one gets the product of objects X, Y by endowing their Set-product with the max-metric:

$$(X \times Y)((x, y), (x', y')) = \max\{X(x, x'), Y(y, y')\}.$$

However, infinite products will generally fail to exist; not even a product P of the countably many two-point subspaces $2_n = \{0, n\}$ $(n \in \mathbb{N})$ of the real line $\mathbb{R}$ exists, as every projection p_n would have to be non-expansive, so that $P((0)_{m\in\mathbb{N}}, (m)_{m\in\mathbb{N}}) \geq 2_n(0, n) = n$, for all $n \in \mathbb{N}$, in contradiction to the finiteness condition.

Even more strikingly, no pair of non-empty spaces X, Y has a coproduct in $\mathsf{Met}_{\text{Fréchet}}$: if we had such coproduct $X \xrightarrow{\ i\ } C \xleftarrow{\ j\ } Y$, considering the constant maps $f : X \to \mathbb{N}$ and $g_n : Y \to \mathbb{N}$ into the Euclidean space of non-negative integers with, respectively, value 0 and n $(n \in \mathbb{N})$, for the induced morphism $h_n : C \to \mathbb{N}$ and any $x \in X, y \in Y$ we would have $C(ix, jy) \geq \mathbb{N}(h_n(ix), h_n(jy)) = \mathbb{N}(0, n) = n$, for all $n \in \mathbb{N}$—again a violation of the finiteness condition. Nevertheless, $\mathsf{Met}_{\text{Fréchet}}$ does have coequalizers, as follows more generally from Example 25.

The Cartesian structure of $\mathsf{Met}_{\text{Fréchet}}$ must be carefully distinguished from its symmetric monoidal structure that requires the tensor product of X and Y to provide the *set* $X \times Y$ with the +-metric:

$$(X \otimes Y)((x, y), (x', y')) = X(x, x') + Y(y, y').$$

But $\mathsf{Met}_{\text{Fréchet}}$ fails to be closed (Mac Lane (1971)) with respect to this monoidal structure (or to its Cartesian structure), again because infinite distances are forbidden. Indeed, first note that, if we have a function space F with universal morphism $e : F \otimes X \to Y$ for the given objects X, Y, we may assume the underlying set of F to be the hom set $\mathsf{Met}_{\text{Fréchet}}(X, Y)$ (since the morphisms $\mathrm{E} \to F$ must correspond naturally and bijectively to the morphisms $\mathrm{E} \otimes X \cong X \to Y$, for the terminal and $\otimes$-neutral space E). Now, for $X = Y = \mathbb{N}$ and the morphisms $\mathrm{id}_{\mathbb{N}}$ and $0 : \mathbb{N} \to \mathbb{N}$ with constant value 0, we would have to have

$$F(0, \mathrm{id}_{\mathbb{N}}) = (F \otimes \mathbb{N})((0, n), (\mathrm{id}_{\mathbb{N}}, n)) \geq \mathbb{N}(e(0, n), e(\mathrm{id}_{\mathbb{N}}, n)) = \mathbb{N}(0, n) = n,$$

for all $n \in \mathbb{N}$, which is disallowed.

2.3 *Some Properties of* Met, BMet, *and* UMet

Not surprisingly, avoiding the finiteness conditions overcomes many of the shortcomings of $\mathsf{Met}_{\text{Fréchet}}$. Indeed, let's first look at the largest category introduced, Met, which distinguishes itself by many convenient properties, also since it is then easy to see to which extent they are inherited by the subcategories we have mentioned above.

(1) The forgetful functor Met $\to$ Set is *topological* (see Adámek et al. (2006), Hofmann and Seal (2014)), making Met behave in many ways very much like the category Top of topological spaces, that is, given a family Y_i $(i \in I)$ of objects in Met of any size (so that I may be empty or even a proper class) that comes with a set X and maps $f_i : X \to Y_i$ $(i \in I)$, then there is a so-called *initial* Met-structure on X, characterized by the property that for any $Z \in$ Met, a Set-map $h : Z \to X$ is a morphism in Met if, and only if, every composite map $f_i h : Z \to Y$ is a morphism; explicitly:

$$X(x, y) = \sup_{i \in I} Y_i(f_i x, f_i y).$$

(2) As topologicity of a functor is a self-dual property, one also has that, given any-size family $X_i \in$ Met $(i \in I)$ that comes with a set Y and maps $f_i : X_i \to Y$, then there is a so-called *final* Met-structure on Y, characterized by the property that for any $Z \in$ Met, a Set-map $h : Y \to Z$ is a morphism in Met if, and only if, every map hf_i is a morphism. (But unlike in (1), this time we do not have a generally valid convenient formula describing the structure on Y.)

(3) Considering the case $I = \emptyset$ in both (1) and (2), one obtains, respectively, a right adjoint and a left adjoint for Met $\to$ Set, putting, respectively, the *indiscrete* and *discrete* structure on a set X, described by $X(x, y) = 0$ and $X(x, y) = \infty$ whenever $x \neq y$ in X.

(4) As a further consequence of (1) and (2), Met is *complete* and *cocomplete*, with the limit (colimit) of a diagram in Met to be formed by putting the initial (final) structure on the limit (colimit, respectively) of its shadow in Set. We note that coproduct injections are isometries, while distances between points of distinct summands are infinite.

(5) Putting the +-metric on the cross-product of objects makes Met a *symmetric monoidal* category which, unlike $\mathsf{Met}_{\text{Fréchet}}$, is *closed*, so that for every $X \in$ Met, the endofunctor $- \otimes X$ has a right adjoint $[X, -]$ which, for every $Y \in$ Met, provides the set $[X, Y]$ of all Met-morphisms $X \to Y$ with the sup-metric

$$[X, Y](f, g) = \sup_{x \in X} Y(fx, gx).$$

(6) Following Hausdorff (1914), for $X \in \mathsf{Met}$ one may make the power set of X an object of Met again, here denoted in his honor by $\mathcal{H}X$, by providing it with the *directed Hausdorff metric*

$$\mathcal{H}X(A, B) = \sup_{x \in A} \inf_{y \in B} X(x, y),$$

for all $A, B \subseteq X$. It turns out that $\mathcal{H}$ carries the structure of a monad on Met, which is a lifting of the power-set monad from Set, the significance of which we explore in Sect. 6.7. Here we note only that, even if we start with a Fréchet metric space, in general only the point and triangle inequalites will survive the passage from X to $\mathcal{H}X$ (as actually already discussed by Hausdorff (1914)).

The category BMet is reflective in Met (with the identical reflection morphisms $X \to X_{\restriction 1}$ modifying the metric of X to $X_{\restriction 1}(x, y) = \min\{X(x, y), 1\}$), but not coreflective (since the formation of coproducts differs). UMet is both reflective and coreflective in Met (since UMet is closed under products and coproducts in Met, as well as under subspaces and quotient spaces—as follows from Sect. 4). More importantly, just like Met, also BMet and UMet are topological over Set, with initial structures formed as in Met. The symmetric monoidal-closed structure of Met restricts to BMet, but not to UMet, since the tensor product of ultrametric spaces may fail to be ultrametric. This failure, however, is not to be seen as such since UMet is in fact *Cartesian closed*; in detail: by replacing the $+$-metric by the max-metric one shows that the Cartesian product gives a closed structure in UMet, with the function spaces $[X, Y]$ of ultrametric spaces to be formed like in Met via the sup-metric, so that they serve as the internal hom objects also in UMet. The Hausdorff distance of a generalized metric space that is bounded or ultrametric maintains these additional properties.

In Examples 18, we will see the general reason for the common properties of the three categories. It is not to be seen as a matter of BMet and UMet inheriting certain properties from Met; rather, all three categories turn out to be instances of a common generalization.

2.4 The Impact of Symmetry and Separation

Let us now clarify to which extent the full subcategories $\mathsf{Met}_{\mathrm{sym}}$ and $\mathsf{Met}_{\mathrm{sep}}$ inherit categorically relevant properties from their parent category Met.

(1) $\mathsf{Met}_{\mathrm{sym}}$ is both reflective and coreflective in Met. With the (co)universal morphisms carried by identity maps, the coreflective symmetrization X_{csym} of $X \in \mathsf{Met}$ is given by $X_{\mathrm{csym}}(x, y) = \max\{X(x, y), X(y, x)\}$, while the reflective symmetrization X_{rsym} is described by (see Grandis (2009))

$$X_{\mathrm{rsym}}(x, x') = \inf_{x = x_0, \ldots, x_n = x'} \Sigma_{j=1}^{n} \min\{X(x_{j-1}, x_j), X(x_j, x_{j-1})\}.$$

(2) As a generally valid consequence of (1) (see, for example, Theorem 2.5.10.3 of Hofmann and Seal (2014)), $\mathsf{Met}_{\mathrm{sym}}$ is topological over Set, with both initial and final structures formed as in Met. The complete and cocomplete category $\mathsf{Met}_{\mathrm{sym}}$ is also symmetric monoidal closed, with all constructions to be carried out like in Met.

(3) $\mathsf{Met}_{\mathrm{sep}}$ is reflective in Met, but not coreflective, since $\mathsf{Met}_{\mathrm{sep}}$ is not closed under colimits. Indeed, considering the colimit D_∞ of the sequence D_n of the Fréchet metric spaces of just two points, with distance $\frac{1}{n}$ $(n \geq 1)$, we see that their distance becomes 0 in D_∞ while staying distinct. The reflection morphism p of $X \in \mathsf{Met}$ projects X isometrically onto the quotient space $X/\simeq$ defined by

$$x \simeq y \iff X(x, y) = 0 = X(y, x),$$

so that $X/\simeq(px, py) = X(x, y)$. Note that $X/\simeq$ is symmetric if X is so, whence $\mathsf{Met}_{\mathrm{sym,sep}}$ is reflective in $\mathsf{Met}_{\mathrm{sym}}$. Also, the closed symmetric monoidal structure of Met restricts to $\mathsf{Met}_{\mathrm{sep}}$ (and $\mathsf{Met}_{\mathrm{sym,sep}}$). But just like for any of the standard separation axioms in Top, forming initial structures does not preserve the separation condition and, in fact, $\mathsf{Met}_{\mathrm{sep}}$ fails to be topological over Set.

(4) We already noted in 2.3(6) that neither symmetry nor separation gets transferred from X to $\mathcal{H}X$; rather, one has to enforce these properties through coreflection or reflection. In fact, the coreflectively symmetrized formula

$$(\mathcal{H}X)_{\mathrm{csym}}(A, B) = \max\{\sup_{x\in A}\inf_{y\in B} X(x, y),\ \sup_{y\in B}\inf_{x\in A} X(y, x)\}$$

gives the traditionally used *Hausdorff metric* for Fréchet metric spaces X, under the restriction that A, B must be non-empty compact sets in X, in order to ensure the separation and finiteness conditions. A variation of this formula, whereby the maximum is replaced by the sum, originates in a special context with Pompeiu's thesis (Pompeiu (1905) as referred to by Hausdorff (1914)).

(5) The non-symmetric *Gromov distance* Gromov (1981) between arbitrary metric spaces X and Y is obtained by taking the infimum of all their directed Hausdorff distances in any metric space Z into which X and Y may be isometrically embedded as disjoint subspaces:

$$\mathcal{GH}Z(X, Y) = \inf_Z \mathcal{H}Z(X, Y).$$

In full generality, this distance will often render only the trivial values 0 or ∞, but it turns out to be useful when X and Y are symmetric and the range of the infimum is restricted to symmetric spaces Z. Alternatively, one may use $\mathcal{H}_{\mathrm{csym}}$ instead of $\mathcal{H}$ on both sides of the above formula and obtain a generally meaningful distance even for non-symmetric metric spaces; see Akhvlediani et al. (2010).

3 Quantales and Quantale-Valued Relations

3.1 Basic Definitions and Examples

Which are the structural features and properties of the extended interval $[0, \infty]$ that are needed to express the axioms of a metric $X \times X \to [0, \infty]$ on a set X and establish a satisfactory basic theory for the resulting spaces? Replacing $[0, \infty]$ by a set $\mathcal{V}$, we would at least need that $\mathcal{V}$ carries the structure of a complete lattice, as well as a binary operation $\otimes$ (replacing the addition in $[0, \infty]$). It then seems reasonable to ask $\otimes$ to make $\mathcal{V}$ a commutative[2] monoid (i.e., a commutative semigroup with neutral element). But in which way should the tensor interact with the order? The specification of this interaction is the most important part of the notion of quantale.

Definition 1 A *quantale* $\mathcal{V} = (\mathcal{V}, \leq, \otimes, \mathrm{k})$ is simultaneously a complete lattice $(\mathcal{V}, \leq)$ and a commutative monoid $(\mathcal{V}, \otimes, \mathrm{k})$ (with k denoting the $\otimes$-neutral element), such that the binary operation $\otimes$ distributes over arbitrary suprema (= joins):

$$(\bigvee_{i \in I} u_i) \otimes v = \bigvee_{i \in I} (u_i \otimes v),$$

for all v, u_i $(i \in I)$ in $\mathcal{V}$.

Note that the index set I is permitted to be empty, in which case the distributivity law reads as $\bot \otimes v = \bot$, with $\bot$ denoting the least element in $\mathcal{V}$. We generally impose no condition on the top element $\top$ in $\mathcal{V}$ vis-à-vis the neutral element k, but if $\mathrm{k} = \top$ holds, we say that the quantale $\mathcal{V}$ is *integral*. Clearly, $\mathcal{V}$ is *trivial* (i.e., $|\mathcal{V}| = 1$) if, and only if, $\mathrm{k} = \bot$.

Before exhibiting some important examples, we state the first key property of quantales. For that, recall that a join-preserving map $f : X \to Y$ of a complete (pre)ordered set X into a (pre)ordered set Y has a right adjoint $g : Y \to X$; up to order-equivalence, it is determined by the equivalence $(x \leq gy \iff fx \leq y)$ for all $x \in X, y \in Y$ or, equivalently, by $gy \simeq \bigvee\{x \in X \mid fx \leq y\}$ for all $y \in Y$. An application of this fact to the join-preserving map $- \otimes v : \mathcal{V} \to \mathcal{V}$ for every element v in a quantale $\mathcal{V}$ gives us an important tool:

Proposition 2 *For every pair of elements v, w in a quantale $\mathcal{V}$, there is an element $[v, w] \in \mathcal{V}$ that is uniquely determined by satisfying the equivalence*

$$u \leq [v, w] \iff u \otimes v \leq w$$

for all $u \in \mathcal{V}$; explicitly, $[v, w] = \bigvee\{u \in \mathcal{V} \mid u \otimes v \leq w\}$.

Of course, this proposition may also be seen as a consequence of the following remark, which justifies calling the element $[v, w]$ the *internal hom of* $v, w \in \mathcal{V}$, also

[2] Imposing commutativity is, at least initially, more a matter of convenience than of necessity.

written as hom(v, w), often with "hom" annotated by $\mathcal{V}$, or by its tensor product when we consider different quantalic structures on the same lattice.

Remark 3 Considered as a small *thin* category with its partial order (so that there is precisely one arrow $u \to v$ when $u \leq v$, and none otherwise), its monoid structure makes a quantale $\mathcal{V}$ a cocomplete, symmetric monoidal-closed category. Conversely, a small and thin, cocomplete, symmetric monoidal-closed category whose underlying preorder is separated (i.e., a partial order) is a quantale.

As a consequence of Proposition 2 and Remark 3 we record some easily established rules for internal homs in a quantale $(\mathcal{V}, \leq, \otimes, \mathrm{k})$:

Corollary 4 $[\mathrm{k}, v] = v, \quad [v_1 \otimes v_2, w] = [v_1, [v_2, w]] = [v_2, [v_1, w]],$

$$[\bigvee_{i \in I} v_i, w] = \bigwedge_{i \in I}[v_i, w], \quad [v, \bigwedge_{i \in I} w_i] = \bigwedge_{i \in I}[v, w_i] \, .$$

Examples 5 Here are the quantales of principal interest in this article:

1. Every *frame* (i.e., every complete lattice in which the binary meet $\wedge$ distributes over arbitrary joins) is an integral quantale. We call *Cartesian* any quantale $(\mathcal{V}, \leq, \otimes, \mathrm{k})$ with $\otimes = \wedge$ (and then necessarily $\mathrm{k} = \top$). In particular, the power set of any set S gives the Cartesian quantale $(\mathcal{P}S, \subseteq, \cap, S)$; for S empty, one has the trivial quantale $\mathsf{1} = \{\bot = \top\}$, and for S a singleton set the *Boolean quantale* $\mathsf{2} = \{\bot < \top\}$. In a Cartesian quantale the internal hom $[v, w]$ is given by the Heyting operation $v \Rightarrow w = \bigvee\{u \mid u \wedge v \leq w\}$; in particular, in $\mathcal{P}S$ one has $\mathrm{hom}_{\cap}(A, B) = \{x \in S \mid (x \in A) \Rightarrow (x \in B)\} = (S \setminus A) \cup B$, for all $A, B \subseteq S$.
2. In the *Lawvere quantale* $[0, \infty]_+ = ([0, \infty], \geq, +, 0)$ the order is given by the inversion of the natural order $\leq$ of the extended real line. We continue to use "inf" and "sup" to denote infima and suprema with respect to the natural order, keeping in mind that, in the Lawvere quantale, they need to be understood as "$\bigvee$" and "$\bigwedge$," respectively. Likewise, when using "max" or "min," we refer to the natural order. Hence, the internal hom of $\alpha, \beta \in [0, \infty]_+$ is computed as the truncated difference
$$\mathrm{hom}_+(\alpha, \beta) = \inf\{\gamma \in [0, \infty] \mid \gamma + \alpha \geq \beta\} = \max\{\beta - \alpha, 0\}$$
with the real operation "$-$" (necessarily) extended by $\beta - \infty = 0$ for all $\beta \in [0, \infty]$, and $\infty - \alpha = \infty$ for $\alpha < \infty$.
3. As a complete linearly ordered set, the underlying lattice $([0, \infty], \geq)$ of the Lawvere quantale gives the Cartesian quantale $[0, \infty]_{\max} = ([0, \infty], \geq, \max, 0)$, so that its tensor product is now given by $\max\{\alpha, \beta\}$. Consequently, the internal hom must now be computed as $\mathrm{hom}_{\max}(\alpha, \beta) = \beta$ if $\alpha < \beta$, and 0 otherwise.
4. The unit interval with its natural order and the multiplication of real numbers becomes the quantale $[0, 1]_{\bullet} = ([0, 1], \leq, \cdot, 1)$. The internal hom is computed as $\mathrm{hom}_{\bullet}(v, w) = \frac{w}{v}$ for $w \leq v \neq 0$, and as 1 otherwise.

The operation $\min\{u, v\}$ gives the Cartesian quantale $[0, 1]_{\min} = ([0, 1], \leq, \min, 1)$. Instead of the ordinary multiplication or the minimum operation, one may also consider the *Łukasiewicz operation* $u \odot v = \max\{u + v - 1, 0\}$ for all $u, v \in [0, 1]$ and obtain the quantale $[0, 1]_{\odot} = ([0, 1], \leq, \odot, 1)$. The internal hom of $[0, 1]_{\odot}$ is computed as $\hom_{\odot}(v, w) = \min\{1, 1 - v + w\} = 1 - \max\{0, v - w\}$.
Instead of these three operations, often one may consider *any* binary operation & on $[0, 1]$ that makes $[0, 1]_{\&} = ([0, 1], \leq, \&, 1)$ an integral quantale; we call such an operation on $[0, 1]$ *quantalic* (also known as a *t-norm*).

5. We now combine the Lawvere quantale $[0, \infty]_+$ and the quantale $[0, 1]_{\&}$, for any quantalic operation & on $[0, 1]$, and form the complete lattice

$$\Delta = \{\varphi\colon [0, \infty] \to [0, 1] \mid \forall \alpha \in [0, \infty] : \varphi(\alpha) = \sup_{\beta < \alpha} \varphi(\beta)\}$$

of all left-continuous $[0, 1]$-valued functions on $[0, \infty]$, provided with the point-wise natural order inherited from $[0, 1]$. We think of these functions as *distance distribution functions* and compose them with the help of the *convolution* operation:

$$(\varphi \,\&\, \psi)(\gamma) = \sup_{\alpha+\beta<\gamma} \varphi(\alpha) \,\&\, \psi(\beta)\,.$$

The distance distribution function κ with $\kappa(0) = 0$ and $\kappa(\alpha) = 1$ for $\alpha > 0$ is &-neutral, and we obtain the integral quantale $\Delta_{\&} = (\Delta, \leq, \&, \kappa)$. Inside it, we find "isomorphic copies" (in the sense of Definition 6) of $[0, \infty]_+$ and $[0, 1]_{\&}$, thanks to the embeddings

$$[0, \infty]_+ \xrightarrow{\ \sigma_+\ } \Delta_{\&} \xleftarrow{\ \tau_{\&}\ } [0, 1]_{\&}\ ,$$

whose values $\sigma_+(\alpha)$ and $\tau_{\&}(u)$ are step functions, defined by $\sigma_+(\alpha)(\gamma) = 0$ if $\gamma \leq \alpha$, and 1 otherwise, and by $\tau_{\&}(u)(\gamma) = u$ if $\gamma > 0$, and 0 otherwise. The significance of these step functions lies in the fact that every distance distribution function φ may be built with them via the formula:

$$\varphi = \sup_{0\leq\alpha\leq\infty} \sigma_+(\alpha) \,\&\, \tau_{\&}(\varphi(\alpha)) = \sup_{0<\alpha<\infty} \sigma_+(\alpha) \,\&\, \tau_{\&}(\varphi(\alpha)).$$

6. For any commutative monoid $(M, *, \eta)$, the power set $\mathcal{P}M$, again ordered by inclusion, becomes a quantale when one lets

$$A * B = \{\alpha * \beta \mid \alpha \in A,\ \beta \in B\}$$

play the role of the tensor product, for all $A, B \subseteq M$. As $\{\eta\}$ is then tensor-neutral, the quantale $(\mathcal{P}M, \subseteq, *, \{\eta\})$ is integral (or Cartesian) only if the monoid M is trivial. For $A, B \subseteq M$, the set $\hom_*(A, B)$ collects the elements $\gamma \in M$ for which the translation $\gamma * -\colon M \to M$ maps A into B.

3.2 Lax and Strict Homomorphisms

We have a brief look at the maps of quantales which respect the quantalic structure.

Definition 6 A *lax homomorphism* $\varphi\colon \mathcal{V} \to \mathcal{W}$ of quantales $\mathcal{V} = (\mathcal{V}, \leq, \otimes, \mathrm{k})$ and $(\mathcal{W}, \leq, \boxtimes, \mathrm{j})$ is a mapping that laxly preserves all quantalic structure, as follows:

$$\bigvee_{i\in I} \varphi(u_i) \leq \varphi\big(\bigvee_{i\in I} u_i\big), \qquad \varphi(u) \boxtimes \varphi(v) \leq \varphi(u \otimes v), \qquad \mathrm{j} \leq \varphi(\mathrm{k}).$$

Of course, the lax preservation of joins by φ just means that φ must preserve the order, i.e., be *monotone*. If all three inequalities may be replaced by equalities, then φ is a *homomorphism* or, for emphasis, a *strict homomorphism*, of quantales.

We denote by Qnt the category of quantales and their strict homomorphisms. It comes with forgetful functors $\mathsf{Qnt} \to \mathsf{Sup}$ and $\mathsf{Qnt} \to \mathsf{CMon}$ which leave the underlying sets fixed; here Sup denotes the category of complete lattices and their suprema-preserving maps, and CMon is the category of commutative monoids and their homomorphisms.

Examples 7 Here are some lax homomorphisms of the quantales of Examples 5.

1. For every quantale $(\mathcal{V}, \leq, \otimes, \mathrm{k})$, the only homomorphism $\iota : \mathsf{2} = \{\bot, \top\} \to \mathcal{V}$ must map $\bot$ to $\bot_{\mathcal{V}}$ and $\top$ to k. The Boolean quantale $\mathsf{2}$ therefore serves as an initial object in Qnt while the trivial quantale $\mathsf{1}$ is terminal. As a monotone map, ι has a right adjoint, $\mathrm{p}\colon \mathcal{V} \to \mathsf{2}$, and when $\mathcal{V}$ is integral, also a left adjoint, $\mathrm{o}\colon \mathcal{V} \to \mathsf{2}$, defined by

$$(\ \mathrm{p}(v) = \top \iff \mathrm{k} \leq v\ ) \quad \text{and} \quad (\ \mathrm{o}(v) = \bot \iff v = \bot_{\mathcal{V}}\ )$$

   for all $v \in \mathcal{V}$. The monotone map p (dubbed the "pessimist's map"—only values v beyond k are declared true) is actually a lax homomorphism, but generally not strict, whereas, for the monotone map o (the "optimist's map," which declares every value v to be true unless it is certainly false) to be a lax homomorphism, the quantale $\mathcal{V}$ must be non-trivial and allow $u \otimes v = \bot$ for any $u, v \in \mathcal{V}$ only if $u = \bot$ or $v = \bot$, in which case o is actually strict.
2. The real function $e^{-x} : [0, \infty] \to [0, 1]$ (extended by $e^{-\infty} = 0$) gives an isomorphism $[0, \infty]_+ \cong [0, 1]_\cdot$ of quantales. The same function makes $[0, \infty]_{\max}$ isomorphic to the quantale $[0, 1]_{\min} = ([0, 1], \leq, \min, 1)$. Similarly, the function $1 - x : [0, 1] \to [0, 1]$ makes the Łukasiewicz quantale $[0, 1]_{\odot}$ isomorphic to the quantale $[0, 1]_{\oplus} = ([0, 1], \geq, \oplus, 0)$, with the quantalic operation $u \oplus v = \min\{1, u + v\}$.
3. The function $\min\{1, x\}$ gives a strict homomorphism $[0, \infty]_+ \to [0, 1]_{\oplus}$. As a monotone map it is left adjoint to the inclusion map, which is a lax homomorphism $[0, 1]_{\oplus} \to [0, \infty]_+$.

4. The injective maps σ_+ and $\tau_\&$ of Example 5.5 embed their domains isomorphically as subquantales into $\Delta_\&$, for every quantalic operation & on [0, 1].
5. By taking direct images, every homomorphism $\varphi : M \to N$ of commutative monoids induces a strict homomorphism $\mathcal{P}M \to \mathcal{P}N$ of quantales.

Remarks 8 1. The quantale $\mathcal{P}M$ is freely generated by the monoid M, i.e., $\mathcal{P}$ becomes left adjoint to the forgetful functor $\mathsf{Qnt} \to \mathsf{CMon}$.
2. With the formula given in Example 5.5, it is easy to show that the quantale $\Delta_\&$ is a coproduct of $[0, \infty]_+$ and $[0, 1]_\&$ in Qnt, for every quantalic operation & on [0, 1].

3.3 Quantale-Valued Relations

We introduced quantales expecting that they will capture the essential structure of $[0, \infty]$ for studying metrics $d : X \times X \to [0, \infty]$ on a set X. Having replaced the interval by a general quantale $\mathcal{V}$, we should now have another look at the map d itself. In the case of the Boolean quantale $\mathsf{2}$, such a map describes precisely a relation on the set X. In what follows, we adapt this "relational" point of view also in the general case. For instance, the distance $d(x, y)$ may be viewed as a numerical truth value for the statement "x is d-related to y," with the value 0 confirming the statement as true, while the value ∞ means that it is certainly false; in other words, we may think of $d(x, y)$ as a measure of the extent to which the statement fails.

Definition 9 Let $\mathcal{V} = (\mathcal{V}, \leq, \otimes, \mathrm{k})$ be a quantale. For any sets X, Y, a *$\mathcal{V}$-relation* r from X to Y is a map $r : X \times Y \to \mathcal{V}$, also written as an *$(X \times Y)$-matrix* of elements $r(x, y)$ $(x \in X, y \in Y)$ in $\mathcal{V}$. We usually use the special arrow notation

$$r : X \looparrowright Y$$

for a $\mathcal{V}$-relation. In fact, we may compose r with a $\mathcal{V}$-relation $s : Y \looparrowright Z$ via matrix multiplication which, in case $\mathcal{V} = \mathsf{2}$, amounts to the standard composition of relations:

$$(s \cdot r)(x, z) = \bigvee_{y \in Y} r(x, y) \otimes s(y, z) .$$

This defines the category $\mathcal{V}$-Rel of sets and their $\mathcal{V}$-relations. Its identity morphism on X is the $\mathcal{V}$-graph of the identity map 1_X of X; here, more generally, one defines the *$\mathcal{V}$-graph* $f_\circ : X \looparrowright Y$ of a mapping $f : X \to Y$ by

$$f_\circ(x, y) = \begin{cases} \mathrm{k} & \text{if } f(x) = y, \\ \bot & \text{otherwise;} \end{cases}$$

this defines the identity-on-objects functor $-_\circ : \mathsf{Set} \to \mathcal{V}\text{-}\mathsf{Rel}$. The category $\mathcal{V}\text{-}\mathsf{Rel}$ comes with the *involution* $-^\circ : (\mathcal{V}\text{-}\mathsf{Rel})^{\mathrm{op}} \to \mathcal{V}\text{-}\mathsf{Rel}$, defined by

$$r^\circ(y, x) = r(x, y)$$

for all $x \in X$, $y \in Y$; it makes the category $\mathcal{V}\text{-}\mathsf{Rel}$ self-dual. Composing the involution with the dual of the functor $-_\circ$ gives the functor $-^\circ : \mathsf{Set}^{\mathrm{op}} \to \mathcal{V}\text{-}\mathsf{Rel}$ that sends every map $f : X \to Y$ to its $\mathcal{V}$-*cograph*

$$f^\circ = (f_\circ)^\circ : Y \nrightarrow X\,.$$

For $(1_X)^\circ = (1_X)_\circ$ we write just 1_X°, and for $\mathcal{V}$-relations $s : Y \nrightarrow Z$ and $t : Z \nrightarrow Y$, their composites with the $\mathcal{V}$-relations $f_\circ$ and f° of a map f are simply computed by

$$(s \cdot f_\circ)(x, z) = s(fx, z) \quad \text{and} \quad (f^\circ \cdot t)(z, x) = t(z, fx)\,.$$

The hom-sets $\mathcal{V}\text{-}\mathsf{Rel}(X, Y)$ come equipped with the pointwise order

$$r \le r' \iff \forall x \in X,\ y \in Y : r(x, y) \le r'(x, y),$$

inherited from $\mathcal{V}$, which makes $\mathcal{V}\text{-}\mathsf{Rel}$ a 2-category. The involution $-^\circ$ preserves the order of the hom-sets, i.e., it is a 2-functor (contravariant on 1-cells, covariant on 2-cells). For every map $f : X \to Y$ one has

$$1_X^\circ \le f^\circ \cdot f_\circ \quad \text{and} \quad f_\circ \cdot f^\circ \le 1_Y^\circ\,,$$

so that one has the adjunction $f_\circ \dashv f^\circ$ in the 2-category $\mathcal{V}\text{-}\mathsf{Rel}$; i.e., $f_\circ$ is a *map* in Lawvere's 2-categorical sense. As $\mathcal{V}$ is complete, so are the hom-sets $\mathcal{V}\text{-}\mathsf{Rel}(X, Y)$, and the matrix multiplication preserves arbitrary suprema in each variable:

$$\Big(\bigvee_{i\in I} s_i\Big) \cdot r = \bigvee_{i\in I}(s_i \cdot r) \quad \text{and} \quad s \cdot \bigvee_{i\in I} r_i = \bigvee_{i\in I} s \cdot r_i\,.$$

These rules make $\mathcal{V}\text{-}\mathsf{Rel}$ a Sup-enriched category.

Remarks 10 (1) Sup-enriched categories are known as *quantaloids* (see Rosenthal (1996), Stubbe (2005)). With its tensor product as the category composition, every quantale becomes a quantaloid. In fact, our quantales are precisely the small and commutative one-object quantaloids. By considering every element of a quantale $\mathcal{V}$ as a (1×1)-matrix (with a singleton set 1) one obtains an embedding $\mathcal{V} \to \mathcal{V}\text{-}\mathsf{Rel}$, which preserves all quantaloidal structure, i.e., it is a *homomorphism of quantaloids*.

(2) Although, in our context, we consider the category $\mathcal{V}\text{-}\mathsf{Rel}$ only as a temporary syntactical tool, without much interest in its properties, let us mention that it has coproducts formed as in Set via disjoint union. As a self-dual category, it therefore

has also products, again given by disjoint union. More importantly then, while the Cartesian product of sets does not serve as the categorical product, it does provide a symmetric monoidal structure for $\mathcal{V}$-Rel which, moreover, is closed, with the internal hom objects again resting on the Cartesian products of sets since, trivially, there are natural isomorphisms

$$\mathcal{V}\text{-}\mathsf{Rel}(X \times Y, Z) \cong \mathcal{V}^{X\times Y\times Z} \cong \mathcal{V}\text{-}\mathsf{Rel}(X, Y \times Z)\,.$$

In fact, with any singleton set serving as a dualizing object, in this way $\mathcal{V}$-Rel becomes even a *compact ∗-autonomous category* in the sense of Barr (1979, 1999); see also Remark 56.

3.4 Extensions and Liftings of $\mathcal{V}$-Relations

We have seen in Definition 9 and Remarks 10(1) that the matrix multiplication of $\mathcal{V}$-relations extends the monoid operation $\otimes$ of the quantale $(\mathcal{V}, \leq, \otimes, \mathrm{k})$. Just like $\otimes$, it preserves arbitrary suprema in each variable separately, but unlike $\otimes$, it generally fails to be commutative. Hence, while the sup-preservation led us to *the* internal hom in $\mathcal{V}$, in $\mathcal{V}$-Rel we must distinguish between two distinct types of "internal homs," arising as right adjoints to the monotone maps given by pre- and post-composition with a fixed $\mathcal{V}$-relation, as follows.

Definition 11 For $\mathcal{V}$-relations $r : X \to\!\!\!\!\!+\; Y$ and $s : Y \to\!\!\!\!\!+\; Z$, let us denote the right adjoints of the monotone maps

$$-\cdot r\colon \mathcal{V}\text{-}\mathsf{Rel}(Y, Z) \longrightarrow \mathcal{V}\text{-}\mathsf{Rel}(X, Z) \quad\text{and}\quad s\cdot -\colon \mathcal{V}\text{-}\mathsf{Rel}(X, Y) \longrightarrow \mathcal{V}\text{-}\mathsf{Rel}(X, Z),$$

respectively, by $[r, -]$ and $]s, -[$. For $t : X \to\!\!\!\!\!+\; Z$, one calls

$[r, t] : Y \to\!\!\!\!\!+\; Z$ the *extension of* t *along* r, and $]s, t[: X \to\!\!\!\!\!+\; Y$ the *lifting of* t *along* s :

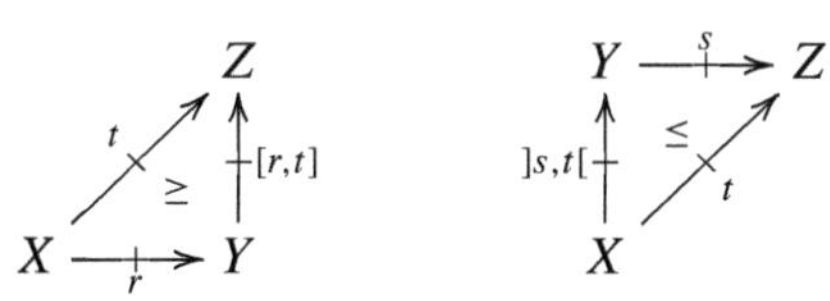

By definition, these $\mathcal{V}$-relations are, respectively, determined by the equivalences

$$s \leq [r, t] \iff s\cdot r \leq t \quad\text{and}\quad s\cdot r \leq t \iff r \leq\,]s, t[,$$

in the case of $[r, t]$ to be universally quantified over s, and for $]s, t[$ over r. Their pointwise computation in terms of the internal hom of $\mathcal{V}$ proceeds as follows:

Proposition 12

$$[r, t](y, z) = \bigwedge_{x\in X}[r(x, y), t(x, z)], \qquad]s, t[(x, y) = \bigwedge_{z\in Z}[s(y, z), t(x, z)].$$

Proof The first identity follows dually from the second since $[r, t] =]r^\circ, t^\circ[^\circ$; indeed, this duality formula holds since, for all s, one has

$$s \le]r^\circ, t^\circ[^\circ \iff s^\circ \le]r^\circ, t^\circ[\iff r^\circ \cdot s^\circ \le t^\circ \iff s \cdot r \le t.$$

To establish the second identity, we verify that the $\mathcal{V}$-relation $\tilde{r}$ defined by the meet on the right satisfies the defining condition of $]s, t[$: for all $r\colon X \nrightarrow Y$ one has

$$\begin{aligned} r \le \tilde{r} &\iff \forall x \in X, y \in Y : \ r(x, y) \le \tilde{r}(x, y) \\ &\iff \forall x \in X, y \in Y, z \in Z : \ r(x, y) \le [s(y, z), t(x, z)] \\ &\iff \forall x \in X, y \in Y, z \in Z : \ r(x, y) \otimes s(y, z) \le t(x, z) \\ &\iff \forall x \in X, z \in Z : \ (s \cdot r)(x, z) \le t(x, z) \\ &\iff s \cdot r \le t\,. \end{aligned}$$

□

Remark 13 If $X = Y = Z = \mathrm{E}$ is a singleton set, $\mathcal{V}$-relations are just elements of $\mathcal{V}$, and the Proposition (once again) confirms that extensions and liftings both coincide with $\hom_{\mathcal{V}}$. Not surprisingly then, we may extend the rules of Corollary 4 to $\mathcal{V}$-relations, but must keep in mind that extra care must be given when the non-commutative composition of $\mathcal{V}$-relations is involved. Briefly, the defining adjunctions for extensions and liftings yield the following rules whenever the terms make sense:

$$[1_X^\circ, t] = t =]1_Z^\circ, t[, \quad [r' \cdot r, t] = [r', [r, t]], \quad]s \cdot s', t[=]s',]s, t[[,$$

$$[\bigvee_{i\in I} r_i, t] = \bigwedge_{i\in I}[r_i, t], \quad [r, \bigwedge_{i\in I} t_i] = \bigwedge_{i\in I}[r, t_i], \quad]\bigvee_{i\in I} s_i, t[= \bigwedge_{i\in I}]s_i, t[,$$

$$]s, \bigwedge_{i\in I} t_i[= \bigwedge_{i\in I}]s, t_i[\,.$$

Example 14 For a metric d on a set X, one has $d(x, z) = \inf_y(d(x, y) + d(y, z))$ for all $x, z \in X$ and, hence, $d \cdot d = d$ for $d : X \nrightarrow X$ in $[0, \infty]_+$-Rel. With the formulae of Proposition 12 one easily shows that $[d, d]$ and $]d, d[$ are again generalized metrics since, in fact, $[d, d] = d =]d, d[$. A considerable generalization of this fact is shown in Corollary 52.

4 Quantale-Enriched Categories, Functors, and Distributors

4.1 $\mathcal{V}$-Valued Metrics as the Structures of $\mathcal{V}$-Categories

Replacing the Lawvere quantale $[0, \infty]_+ = ([0, \infty], \geq, +, 0)$ by an arbitrary quantale $\mathcal{V} = (\mathcal{V}, \leq, \otimes, \mathrm{k})$ one easily arrives at the objects of interest in this article, as follows.

Definition 15 A $\mathcal{V}$*-valued (generalized) metric* on a set X is a map $a : X \times X \to \mathcal{V}$ satisfying the point and triangle inequalities

$$\mathrm{k} \leq a(x, x) \quad \text{and} \quad a(x, y) \otimes a(y, z) \leq a(x, z)$$

for all $x, y, z \in X$. We may consider the structure a as the *internal hom* of a small $\mathcal{V}$*-category* with object set X, by writing these axioms in the form:

$$\mathrm{k} \leq X(x, x) \quad \text{and} \quad X(x, y) \otimes X(y, z) \leq X(x, z),$$

where "$\leq$" is now to be seen as an arrow in the thin monoidal-closed category $\mathcal{V}$ (as in Remark 3); occasionally we write $a = \hom_X$. Looking at the structure of X as a $\mathcal{V}$-relation $a : X \nrightarrow X$, these two conditions describe a as *reflexive* and *transitive* on X:

$$1_X^\circ \leq a \quad \text{and} \quad a \cdot a \leq a\ .$$

We note that the first of these inequalities actually forces equality in the second, since $a = a \cdot 1_X^\circ \leq a \cdot a$. The *small* $\mathcal{V}$-categories[3] are the objects of the (large) category

$$\mathcal{V}\text{-}\mathsf{Cat}$$

whose morphisms are the $\mathcal{V}$*-functors* $f : (X, a) \to (Y, b)$, i.e., the maps $f : X \to Y$ satisfying any of the following equivalent conditions ($x, x' \in X$):

$$a(x, x') \leq b(fx, fx'); \quad X(x, x') \leq Y(fx, fx'); \quad a \leq f^\circ \cdot b \cdot f_\circ; \quad f_\circ \cdot a \leq b \cdot f_\circ\ .$$

[3] In what follows, "$\mathcal{V}$-category" always means "small $\mathcal{V}$-category."

The $\mathcal{V}$-functor f is *fully faithful* if $a = f^\circ \cdot b \cdot f_\circ$ or, equivalently, $X(x, x') = Y(fx, fx')$, for all $x, x' \in X$.

Remark 16 We have chosen to denote the structure of a general $\mathcal{V}$-category X by a, rather than d, taking the view that it is a unary *algebraic* operation in $\mathcal{V}$-Rel on the set X; a $\mathcal{V}$-functor $f : X \to Y$ is then a map that makes its $\mathcal{V}$-graph a *lax homomorphism*:

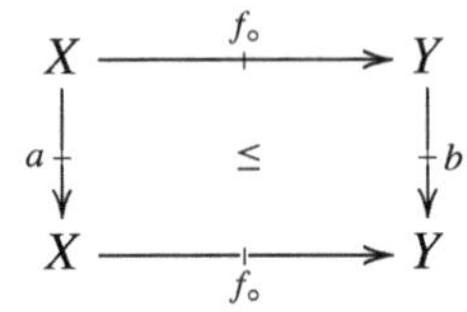

Since the $\mathcal{V}$-relation a of a $\mathcal{V}$-category satisfies $a = a \cdot 1_X^\circ \le a \cdot a \le a$, it is an idempotent element of the monoid $(\mathcal{V}\text{-}\mathsf{Rel}(X, X), \cdot, 1_X^\circ)$ that lies above the neutral element 1_X°. Note that, with its order, this monoid satisfies all of the conditions of a quantale, except for our commutativity requirement. When viewing $\mathcal{V}$-$\mathsf{Rel}(X, X)$ as a thin (non-symmetric) monoidal category (with the tensor given by composition), we may consider a $\mathcal{V}$-category structure on X simply as a monoid object in that category.

Before describing the category $\mathcal{V}$-Cat for specific quantales $\mathcal{V}$, let us mention some fundamental $\mathcal{V}$-categories and $\mathcal{V}$-functors in the general context:

Proposition 17 *With its internal hom, every quantale $\mathcal{V}$ becomes itself a $\mathcal{V}$-category $(\mathcal{V}, [\text{-}, \text{-}] : \mathcal{V} \times \mathcal{V} \to \mathcal{V})$. For every $\mathcal{V}$-category $X = (X, a)$, also $X^{\mathrm{op}} = (X, a^\circ)$ is a $\mathcal{V}$-category, so that $X^{\mathrm{op}}(y, x) = X(x, y)$ for all $x, y \in X$, and one obtains the $\mathcal{V}$-functors*

$$X(x, -) : X \to \mathcal{V} \quad \textit{and} \quad X(-, y) : X^{\mathrm{op}} \to \mathcal{V}\,.$$

Proof For all $u, v, w \in \mathcal{V}$, the trivial inequalities $\mathrm{k} \otimes v \le v$ and $[u, v] \le [u, v]$ translate to $\mathrm{k} \le [v, v]$ and $u \otimes [u, v] \le v$, so that first $u \otimes [u, v] \otimes [v, w] \le v \otimes [v, w] \le w$ and then $[u, v] \otimes [v, w] \le [u, w]$ follows, as is required for $\mathcal{V}$ to be a $\mathcal{V}$-category. The commutativity of $\otimes$ guarantees that with X also X^{op} is a $\mathcal{V}$-category. Finally, $X(x, y) \otimes X(y, y') \le X(x, y')$ translates to $X(y, y') \le [X(x, y), X(x, y')]$, so that $X(x, -) : X \to \mathcal{V}$ must be a $\mathcal{V}$ functor, as well as $X(-, y)$ since, by dualization, $X(-, y) = X^{\mathrm{op}}(y, -)$. □

Examples 18 In what follows, we refer to the quantales of Examples 5 and 7.

1. The category $\mathsf{1}$-Cat is isomorphic to Set, and $\mathsf{2}$-Cat is isomorphic to the category Ord of *preordered sets* and their monotone maps. In this paper, for simplicity we will refer to the objects of Ord as *ordered sets*; the more common *partially ordered sets* will appear as *separated* objects in Ord in Definition 26. More generally, for any set S, the objects of $\mathcal{P}S$-Cat may be described as sets X equipped with a family of orders $\le_s$ $(s \in S)$ on X, and morphisms are monotone with respect to every $\le_s$. As an object in $\mathcal{P}S$-Cat, $\mathcal{P}S$ itself carries the family of

orders $\leq_s$ ($s \in S$), whereby $A \leq_s B$ holds for $A, B \subseteq S$ precisely when the implication ($s \in A \Longrightarrow s \in B$) holds. For a singleton set S, this reproduces the order of the quantale $\mathsf{2} = \{\bot < \top\}$.

2. The categories $[0,\infty]_+$-**Cat**, $[0,\infty]_{\max}$-**Cat** and $[0,1]_\oplus$-**Cat** coincide, respectively, with the previously introduced categories **Met**, **UMet**, and **BMet**. The fully faithful $\mathcal{V}$-functors in these categories for the respective choices of $\mathcal{V}$ are the isometries.
   If, by Proposition 17, we consider $[0,\infty]$ as a $[0,\infty]_+$-category, it carries the truncated difference $\hom_+(\alpha,\beta) = \max\{\beta - \alpha, 0\}$ as a generalized metric (see Example 5(2)). Its coreflective symmetrization gives the ordinary Euclidean distance when restricted to $[0,\infty)$. Observe that, of course, a different metric gets imposed on $[0,\infty]$ when considered as a $[0,\infty]_{\max}$-category.
3. Objects in $[0,1]_\cdot$-**Cat** are sets X with a *probabilistic order* $p : X \times X \to [0,1]$, satisfying $p(x,x) = 1$ and $p(x,y)\cdot p(y,z) \leq p(x,z)$, with the value $p(x,y)$ to be interpreted as the probability that, in a random order $\leq$ on X, one would have $x \leq y$; morphisms $f : (X,p) \to (Y,q)$ must satisfy $p(x,x') \leq q(fx,fx')$. With the isomorphism $[0,1]_\cdot \cong [0,\infty]_+$ of quantales (see Examples 7) one easily obtains an isomorphism $[0,1]_\cdot$-**Cat** $\cong [0,\infty]_+$-**Cat** $=$ **Met** of categories (see Corollary 20). Likewise for the other quantalic structures on $[0,\infty]$ and $[0,1]$ introduced in Examples 5 and 7, as summarized by

$\mathcal{V}$	$[0,\infty]_+ \cong [0,1]_\cdot$	$[0,\infty]_{\max} \cong [0,1]_{\min}$	$[0,1]_\oplus \cong [0,1]_\odot$
$\mathcal{V}$-**Cat**	**Met**	**UMet**	**BMet**

4. For any quantalic operation & on $[0,1]$ (the standard choice being the ordinary multiplication), a $\Delta_\&$-category structure p on a set X gives a distance distribution function $p(x,y)\colon [0,\infty] \to [0,1]$ for all $x, y \in X$, subject to the rules:

$$p(x,x) = \kappa \quad \text{and} \quad p(x,y)\ \&\ p(y,z) \leq p(x,z).$$

   If one interprets the value $p(x,y)$ at $\alpha \in [0,\infty]$ as the probability that a random generalized metric d on X satisfies $d(x,y) < \alpha$ and therefore writes suggestively $p_\alpha(x,y)$ for $p(x,y)(\alpha)$, then the structure p is described by a family of functions $p_\alpha\colon X \times X \to [0,1]$ satisfying the conditions:

$$p_\alpha(x,x) = \begin{Bmatrix} 0 \text{ if } \alpha = 0 \\ 1 \text{ if } \alpha > 0 \end{Bmatrix} \text{ and } (\alpha + \beta < \gamma \Rightarrow p_\alpha(x,y)\ \&\ p_\beta(y,z) \leq p_\gamma(x,z)),$$

   for $\alpha, \beta, \gamma \in [0,\infty]$ and $x, y, z \in X$. A morphism $f\colon (X,p) \to (Y,q)$ in the category $\Delta_\&$-**Cat** of *(&-)probabilistic metric spaces* (originating with Menger (1942); see also Hofmann and Reis (2013)) is now a map satisfying

$$p_\alpha(x,x') \leq q_\alpha(fx,fx')$$

   for $\alpha \in [0,\infty],\ x, x' \in X$.

5. For a commutative monoid $(M, *, \eta)$, similar to the case of a $\mathcal{P}S$-category with a mere set S, a $(\mathcal{P}M, \subseteq, *, \{\eta\})$-structure on a set X may be described equivalently by a family of relations $\leq_\alpha$ $(\alpha \in M)$ that must satisfy the conditions

$$x \leq_\eta x \quad \text{and} \quad (x \leq_\alpha y,\ y \leq_\beta z \Longrightarrow x \leq_{\alpha*\beta} z)$$

for all $\alpha, \beta \in M$, but note that $\leq_\alpha$ is no longer guaranteed to be an order on X, unless $\alpha = \eta$. As in 1., $\mathcal{P}M$-functors are maps which preserve each of the relations.
Writing $x \xrightarrow{\alpha} y$ instead of $x \leq_\alpha y$, we see that a $\mathcal{P}M$-category may be considered as an ordinary small category $\mathbb{X}$ with object set X and composition of arrows given by $*$. Thinking of α as the "norm" or "length" of $x \xrightarrow{\alpha} y$, we see that $\mathbb{X}$ comes equipped with a faithful functor $\mathbb{X} \to M$, where M is considered as a one-object category. Conversely, the object set of any small category equipped with a faithful functor to M gives us a $\mathcal{P}M$-category, and $\mathcal{P}M$-functors are equivalently described as ordinary norm-preserving functors. In other words, $\mathcal{P}M$-Cat is equivalent to the comma category $\mathsf{Cat}_{\mathrm{ff}}/M$ of faithful functors to M.

Lax homomorphisms $\varphi : \mathcal{V} \to \mathcal{W}$ of quantales facilitate *change-of-base functors*:

Proposition 19 *For quantales $\mathcal{V}$ and $\mathcal{W}$, a mapping $\varphi : \mathcal{V} \to \mathcal{W}$ induces a functor*

$$\mathcal{B}_\varphi : \mathcal{V}\text{-}\mathsf{Cat} \longrightarrow \mathcal{W}\text{-}\mathsf{Cat}, \qquad (X, a) \longmapsto (X, \varphi a),\ f \longmapsto f,$$

if, and only if, φ is a lax homomorphism of quantales.

Proof With φa denoting the composite of φ preceded by the map $a : X \times X \to \mathcal{V}$, and with a lax homomorphism and $(X, a) \in \mathcal{V}$-Cat, confirming $(X, \varphi a) \in \mathcal{W}$-$\mathsf{Cat}$ is straightforward; likewise for the transformation of $\mathcal{V}$-functors into $\mathcal{W}$-functors. Conversely, assuming $\mathcal{B}_\varphi$ to be a well-defined functor, for the $\mathcal{V}$-category $(\mathcal{V}, [\text{-},\text{-}])$ one has in particular that $\mathcal{B}_\varphi\mathcal{V}$ is $\mathcal{W}$-category, and one then routinely shows that this means precisely that φ is a lax homomorphism. □

Corollary 20 *If for lax homomorphisms $\varphi : \mathcal{V} \to \mathcal{W}$ and $\psi : \mathcal{W} \to \mathcal{V}$ one has the adjunction $\varphi \dashv \psi$, then $\mathcal{B}_\varphi \dashv \mathcal{B}_\psi : \mathcal{W}$-$\mathsf{Cat} \to \mathcal{V}$-$\mathsf{Cat}$ are adjoint functors. If $\mathcal{V}$ and $\mathcal{W}$ are isomorphic, so are $\mathcal{V}$-Cat and $\mathcal{W}$-Cat.*

Examples 21 For notation, we refer to Examples 7 and 18.

1. The adjunction $\iota \dashv \mathrm{p} : \mathcal{V} \to \mathsf{2}$ induces the adjunction $\mathcal{B}_\iota \dashv \mathcal{B}_{\mathrm{p}} : \mathcal{V}$-$\mathsf{Cat} \to \mathsf{Ord}$. Here $\mathcal{B}_{\mathrm{p}}$ equips every $\mathcal{V}$-category X with its *underlying order*, defined by

$$x \leq y \iff \mathrm{k} \leq X(x, y),$$

for all $x, y \in X$. In particular, every $\mathcal{V}$-functor $f : X \to Y$ becomes a monotone map w.r.t. the induced orders of X and Y.

The left adjoint $\mathcal{B}_\iota$ makes every ordered set $(X, \leq)$ a $\mathcal{V}$-category, via $X(x, y) = \mathrm{k}$ if $x \leq y$, and $X(x, y) = \bot_{\mathcal{V}}$ otherwise. Hence, for $\mathcal{V}$ not trivial, $\mathcal{B}_\iota$ embeds Ord into $\mathcal{V}$-Cat as a full reflective subcategory. In particular, Ord gets reflectively embedded into Met, and the reflector orders a metric space X via $(x \leq y \iff X(x, y) = 0)$.

2. The homomorphism $\tau_{\bullet} : [0, 1]_{\bullet} \to \Delta_{\bullet}$ of quantales with $\tau_{\bullet}(u)(\gamma) = u$ for $\gamma > 0$ has, as a monotone map, a left adjoint, sup, which is again a homomorphism of quantales. Hence, $\mathcal{B}_{\tau_{\bullet}}$ embeds the category $\mathsf{Met} \cong \mathsf{ProbOrd} := [0, 1]_{\bullet}$-$\mathsf{Cat}$ of metric spaces, interpreted as probabilistic orders, into the category $\mathsf{ProbMet} := \Delta_{\bullet}$-$\mathsf{Cat}$ of probabilistic metric spaces as a full reflective subcategory.

4.2 *The Concrete Category* $\mathcal{V}$-Cat *and Its Subcategories* $\mathcal{V}$-$\mathsf{Cat}_{\mathrm{sym}}$ *and* $\mathcal{V}$-$\mathsf{Cat}_{\mathrm{sep}}$

We continue to work with a quantale $\mathcal{V} = (\mathcal{V}, \leq, \otimes, \mathrm{k})$ and list some properties of the category $\mathcal{V}$-Cat and its forgetful functor $\mathcal{O} : \mathcal{V}$-$\mathsf{Cat} \to \mathsf{Set}$. Not surprisingly, these are generalizations of what has been observed for $\mathsf{Met} = [0, \infty]_+$-$\mathsf{Cat}$ in Sect. 2.3. We omit the straightforward proofs.

Theorem 22 (1) *The functor* $\mathcal{O} : \mathcal{V}$-$\mathsf{Cat} \to \mathsf{Set}$ *is topological, with the initial structure a on a set X w.r.t a family of maps $f_i : X \to Y_i$ $(i \in I)$ into $\mathcal{V}$-categories $Y_i = (Y_i, b_i)$ given by*

$$a = \bigwedge_{i \in I} f_i^\circ \cdot b_i \cdot (f_i)_\circ \quad \textit{or, equivalently,} \quad X(x, y) = \bigwedge_{i \in I} Y_i(f_i x, f_i y) \quad (x, y \in X).$$

(2) *For every $\mathcal{V}$-relation $r : X \nrightarrow X$ on a set X, there is a least $\mathcal{V}$-category structure $\overline{r} \geq r$ on X, namely:*

$$\overline{r} = \bigvee_{n \geq 0} r^n, \quad \textit{with } r^0 = 1_X^\circ,\ r^{n+1} = r^n \cdot r.$$

(3) *The final structure b on a set Y w.r.t. a family of maps $g_i : X_i \to Y$ $(i \in I)$ of $\mathcal{V}$-categories $X_i = (X_i, a_i)$ is given by*

$$b = \overline{\bigvee_{i \in I} (g_i)_\circ \cdot a_i \cdot g_i^\circ}.$$

(4) $\mathcal{V}$-Cat *is complete and cocomplete, with the limit and colimit of a diagram in* $\mathcal{V}$-Cat *to be formed as in* Set *and then provided with, respectively, the initial and final structures w.r.t. the limit and colimit maps.*

(5) *The functor* $\mathcal{O}$ *has both, a left and right adjoint, providing every set* X *with, respectively, its* discrete *structure* 1_X° *and its* indiscrete *structure, defined by* $X(x,y) = \top_{\mathcal{V}}\ (x, y \in X)$.

Corollary 23 *For* $X_i \in \mathcal{V}\text{-}\mathsf{Cat}$, $i \in I$, *one has*

$$(\prod_{i\in I} X_i)((x_i)_I, (y_i)_I) = \bigwedge_{i\in I} X_i(x_i, y_i),\ (\coprod_{i\in I} X_i)(x,y) = \begin{Bmatrix} X_i(x,y) & \text{if } x, y \in X_i, i \in I, \\ \bot_{\mathcal{V}} & \text{otherwise.} \end{Bmatrix}$$

To form the equalizer of $f, g : X \to Y$ in $\mathcal{V}\text{-}\mathsf{Cat}$, one just restricts the structure of X to the equalizer in Set. To obtain the coequalizer of f, g, one forms the coequalizer $p : Y \to Z$ in Set and puts the final structure $\overline{r}$ on Z with respect to p, where the $\mathcal{V}$-relation r on Z is given by

$$r(z,w) = \bigvee_{x\in p^{-1}z,\ y\in p^{-1}w} Y(x,y)\ ,$$

for all $z, w \in Z$. For $x_0, y_0, x_1, y_1, \ldots, x_n, y_n \in Y\ (n \geq 0)$, let us use the abbreviation

$$(x_0, \ldots, y_n) : z \rightsquigarrow w \iff z = p(x_0),\ p(y_0) = p(x_1),\ p(y_1) = p(x_2), \ldots,\ p(y_n) = w,$$

in order to describe the $\mathcal{V}$-category structure $\overline{r}$ on Z more elaborately, as follows:

Corollary 24 *For a surjective map* $p\colon Y \to Z$ *of a* $\mathcal{V}$*-category* Y*, the final* $\mathcal{V}$*-category structure on* Z *is described, for all* $z, w \in Z$*, by*

$$Z(z,w) = \bigvee_{(x_0,\ldots,y_n):z\rightsquigarrow w} Y(x_0,y_0) \otimes \ldots \otimes Y(x_n,y_n)\ .$$

Example 25 A surjective map $p : Y \to Z$ of metric spaces is a regular epimorphism in Met if, and only if, for all $z, w \in Z$,

$$Z(z,w) = \inf_{(x_0,\ldots,y_n):z\rightsquigarrow w} Y(x_0,y_0) + \ldots + Y(x_n,y_n)\ .$$

Definition 26 Let $X = (X, a)$ be a $\mathcal{V}$-category.
(1) X is *symmetric* if $X = X^{\mathrm{op}}$; that is, if $a = a^\circ$ or, equivalently,

$$X(x,y) = X(y,x)\ \ (x, y \in X).$$

(2) X is *separated* if the induced order on X is antisymmetric; that is, if

$$\mathsf{k} \leq X(x,y) \text{ and } \mathsf{k} \leq X(y,x) \text{ imply } x = y\ \ (x, y \in X).$$

The resulting full subcategories of $\mathcal{V}$-Cat are, respectively, denoted by $\mathcal{V}$-$\mathsf{Cat}_{\mathrm{sym}}$ and $\mathcal{V}$-$\mathsf{Cat}_{\mathrm{sep}}$, and their intersection by $\mathcal{V}$-$\mathsf{Cat}_{\mathrm{sym,sep}}$.

One easily establishes that these subcategories enjoy the same properties in $\mathcal{V}$-Cat as their role models $\mathsf{Met}_{\mathrm{sym}}$, $\mathsf{Met}_{\mathrm{sep}}$, and $\mathsf{Met}_{\mathrm{sym,sep}}$ do in Met. For a $\mathcal{V}$-category X, we mention here explicitly only its

(1) *coreflective symmetrization* $X_{\mathrm{csym}} \to X$ with $X_{\mathrm{csym}}(x, y) = X(x, y) \wedge X(y, x)$;
(2) *reflective separation* $p : X \to X/\simeq$ with $X/\simeq(px, py) = X(x, y)$, where

$$x \simeq y \iff x \leq y \text{ and } y \leq x \quad (x, y \in X).$$

Remark 27 (1) While separation depends only on the underlying order of a $\mathcal{V}$-category X, we note that symmetry of its underlying order will generally not guarantee the symmetry of X as a $\mathcal{V}$-category.

(2) The formula for the structure of $X/\simeq$ greatly simplifies the general formula for a quotient structure in $\mathcal{V}$-Cat as given in Corollary 24, because, whenever one has $px = px'$ and $py = py'$, then

$$\begin{aligned} X(x', y') = \mathrm{k} \otimes X(x', y') \otimes \mathrm{k} &\leq X(x, x') \otimes X(x', y') \otimes X(y', y) \\ &\leq X(x, y) \leq \ldots \leq X(x', y'). \end{aligned}$$

4.3 *The Monoidal-Closed 2-Category* $\mathcal{V}$-Cat

Ordering $\mathcal{V}$-functors $f, f' : X \to Y$ pointwise by

$$f \leq f' \iff \forall x \in X : fx \leq f'x \iff \forall x \in X : \mathrm{k} \leq Y(fx, f'x)$$

one makes $\mathcal{V}$-$\mathsf{Cat}(X, Y)$ an ordered set and, hence, a small category, with arrows given by order, such that every composition map

$$\mathcal{V}\text{-}\mathsf{Cat}(X, Y) \times \mathcal{V}\text{-}\mathsf{Cat}(Y, Z) \longrightarrow \mathcal{V}\text{-}\mathsf{Cat}(X, Z)$$

becomes monotone. Hence, as an easy observation we have

Proposition 28 $\mathcal{V}$-Cat *is a 2-category, with 2-cells given by order.*

Using "co" to indicate the dualization with respect to 2-cells in a 2-category, we obtain

Corollary 29 *The functor*

$$(-)^{\mathrm{op}} : \mathcal{V}\text{-}\mathsf{Cat} \longrightarrow \mathcal{V}\text{-}\mathsf{Cat}^{\mathrm{co}},\ (f : X \to Y) \mapsto (f^{\mathrm{op}} : X^{\mathrm{op}} \to Y^{\mathrm{op}}),$$

is an isomorphism of 2-categories.

Like in every 2-category we have the notion of adjunction which, in the case of $\mathcal{V}$-Cat, since 2-cells are given by order, simplifies, as follows: a $\mathcal{V}$-functor $f : X \to Y$ is *left adjoint* to a $\mathcal{V}$-functor $g : Y \to X$, and g is *right adjoint* to f, written as $f \dashv g$ or $g \vdash f$, if

$$1_X \leq gf \ \text{ and } \ fg \leq 1_Y.$$

Not surprisingly, adjointness of f and g is determined at the level Ord, as long as f and g are $\mathcal{V}$-functors. Noteworthy, however, is that the usual hom-formula for adjunctions implies their $\mathcal{V}$-functoriality. More precisely:

Proposition 30 *Let $X = (X, a)$ and $Y = (Y, b)$ be $\mathcal{V}$-categories. Then the following are equivalent for maps $f : X \to Y,\ g : Y \to X$ in* Set*:*

(i) $b \cdot f_\circ = a \cdot g^\circ$, *that is,* $Y(fx, y) = X(x, gy)\ \ (x \in X, y \in Y)$*;*
(ii) f, g *are $\mathcal{V}$-functors, and $f \dashv g$ in $\mathcal{V}$-*Cat*;*
(iii) f, g *are $\mathcal{V}$-functors, and $f \dashv g$ in* Ord*.*

Proof (i)$\Longrightarrow$(ii): From $k \leq Y(fx, fx) = X(x, gfx)$ one obtains $x \leq gfx$ for all $x \in X$. With the monotonicity of the $\mathcal{V}$-functor $X(x, -) : X \to \mathcal{V}$ one obtains $X(x, x') \leq X(x, gfx') = Y(fx, fx')$ for all $x' \in X$ and, hence, the $\mathcal{V}$-functoriality of f. Likewise, (i) implies also $fgy \leq y$ for all $y \in Y$ and the $\mathcal{V}$-functoriality of g. (ii)$\Longrightarrow$(iii) is trivial. (iii)$\Longrightarrow$(i): From $x \leq gfx$, the monotonicity of the $\mathcal{V}$-functor $X(-, gy) : X^{\mathrm{op}} \to \mathcal{V}$, and from the $\mathcal{V}$-functoriality of g, one obtains $Y(fx, y) \leq X(gfx, gy) \leq X(x, gy)$; likewise, $X(x, gy) \leq Y(fx, y)$, for all $x \in X, y \in Y$. $\square$

Example 31 A $\mathcal{V}$-functor $f : X \to Y$, when regarded just as a monotone map, may have a right adjoint g in Ord, but this monotone map $g : Y \to X$ may fail to be a $\mathcal{V}$-functor. For instance, when $\mathcal{V}$ is not integral (so that $\mathrm{k} < \top_{\mathcal{V}}$), we may consider any non-empty set X made into a $\mathcal{V}$-category by $X(x, x') = \mathrm{k}$ for all $x, x' \in X$. Then the $\mathcal{V}$-functor $f : X \to 1 = (\{*\}, \top)$ to the terminal $\mathcal{V}$-category certainly has a right adjoint as a monotone map; in fact, any map $1 \to X$ serves in that capacity, but none of them is a $\mathcal{V}$-functor.

Following the role model Met and the $+$-metric on $X \times Y$, one establishes the monoidal structure of $\mathcal{V}$-Cat: for all $\mathcal{V}$-categories X, Y, making essential use of the commutativity of $\otimes$ in $\mathcal{V}$, one confirms easily that the set $X \times Y$ becomes a $\mathcal{V}$-category $X \otimes Y$ with

$$(X \otimes Y)((x, y), (x', y')) = X(x, x') \otimes Y(y, y'),$$

for all $x, x' \in X,\ y, y' \in Y$. This structure must be carefully distinguished from the product $X \times Y$ in $\mathcal{V}$-Cat. Likewise, unless $\mathcal{V}$ is integral, the $\otimes$-neutral object E with underlying set $\{*\}$ and

$$\mathrm{E}(*, *) = \mathrm{k}$$

must be distinguished from the terminal object 1 in $\mathcal{V}$-Cat which, with the same underlying set, satisfies $1(*, *) = \top$.

Like for Met, one sees more generally that $\mathcal{V}$-Cat is monoidal closed: the hom set $\mathcal{V}\text{-Cat}(X, Y)$ becomes the $\mathcal{V}$-category $[X, Y]$ with

$$[X, Y](f, g) = \bigwedge_{x \in X} Y(fx, gx)$$

for all $\mathcal{V}$-functors $f, g : X \to Y$. Indeed, in terms of the topological functor $\mathcal{O} : \mathcal{V}\text{-Cat} \to \mathsf{Set}$, this is the initial structure w.r.t. the family of evaluation maps

$$\mathrm{ev}_x : [X, Y] \longrightarrow Y, \quad f \longmapsto fx \quad (x \in X).$$

Theorem 32 *With its tensor product and its internal homs, $\mathcal{V}$-Cat is a symmetric monoidal-closed 2-category. For every $\mathcal{V}$-category X, the adjunction $- \otimes X \dashv [X, -] : \mathcal{V}\text{-Cat} \longrightarrow \mathcal{V}\text{-Cat}$ is $\mathcal{V}$-Cat-enriched, so that, for all $Y, Z \in \mathcal{V}$-Cat, there is a natural $\mathcal{V}$-Cat-isomorphism*

$$[Z \otimes X, Y] \cong [Z, [X, Y]]$$

which, by symmetry of $\otimes$, gives also the natural isomorphism $[X, [Y, Z]] \cong [Y, [X, Z]]$.

Proof We omit the routine verifications that the map

$$\Phi\colon [X \otimes Y, Z] \longrightarrow [X, [Y, Z]], \quad h \longmapsto (\Phi h\colon x \mapsto (h(x, \text{-})\colon Y \to Z)),$$

is well defined, bijective, and natural in Y, Z (as well as in X). That both, Φ and Φ^{-1}, are $\mathcal{V}$-functors may also be seen easily since, for all $\mathcal{V}$-functors $h, h'\colon X \otimes Y \to Z$,

$$\begin{aligned}[X \otimes Y, Z](h, h') &= \bigwedge_{x,y} Z(h(x, y), h'(x, y)) = \bigwedge_{x} [Y, Z](h(x, \text{-}), h'(x, \text{-})) \\ &= [X, [Y, Z]](h, h').\end{aligned}$$

□

Corollary 33 *If $\mathcal{V}$ is a Cartesian quantale, then $\mathcal{V}$-Cat is a Cartesian-closed category.*

In particular, $\mathsf{UMet} = [0, \infty]_{\max}$-$\mathsf{Cat}$ is Cartesian closed; however, the monoidal-closed category $\mathsf{Met} = [0, \infty]_{+}$-$\mathsf{Cat}$ is not, as the following example shows, and neither is $\mathsf{BMet} = [0, \infty]_{\oplus}$-$\mathsf{Cat}$.

Example 34 (M.M. Clementino) In order to show that Met fails to be Cartesian closed, it suffices to find a (generalized) metric space (Z, d), such that the product functor $- \times Z : \mathsf{Met} \to \mathsf{Met}$ fails to preserve regular epimorphisms, since such failure prevents the functor from being left adjoint.

To this end, we consider the subspaces $X = \{0\} \cup [\frac{1}{2}, \frac{3}{2}] \cup \{2\}$ and $Z = \{0, 1\}$ of the Euclidean line $\mathbb{R}$ and let $p : X \to Y$ be the quotient map that contracts the interval $[\frac{1}{2}, \frac{3}{2}]$ to a point, so that $Y = \{y_0, y_1, y_2\} = \{p(0), p(1), p(2)\}$ is a 3-point metric space with distances $d(y_0, y_1) = d(y_1, y_2) = \frac{1}{2}$ and $d(y_0, y_2) = 1$. Considering now the map $p \times 1_Z : X \times Z \longrightarrow Y \times Z$, we note that the distance between $(y_0, 0) = (p \times 1_Z)(0, 0)$ and $(y_2, 1) = (p \times 1_Z)(2, 1)$ in the product space $Y \times Z$ (with the max-metric) is 1, whereas the quotient metric with respect to the map $p \times 1_Z$ (as described by Corollary 24) assigns to them the distance $\frac{1}{2} + 1$.

The following fact indicates the prominent role of $(\mathcal{V}, [\text{-}, \text{-}])$ as an object of $\mathcal{V}$-Cat:

Proposition 35 *The quantale $(\mathcal{V}, \otimes, \mathrm{k})$ is an internal commutative monoid in the monoidal category $\mathcal{V}$-Cat.*

Proof We must show that the maps $\mathrm{E} \to \mathcal{V},\ * \mapsto \mathrm{k}$, and $\mathcal{V} \otimes \mathcal{V} \longrightarrow \mathcal{V},\ (u, v) \longmapsto u \otimes v$, are $\mathcal{V}$-functors. That is obvious for the first map, and for the second we observe that, for all $u, u', v, v' \in \mathcal{V}$, from $[u, u'] \otimes u \otimes [v, v'] \otimes v \leq u' \otimes v'$ we obtain

$$(\mathcal{V} \otimes \mathcal{V})((u, v), (u', v')) = [u, u'] \otimes [v, v'] \leq [u \otimes v, u' \otimes v'],$$

as required. □

Symmetry of $\mathcal{V}$-categories is obviously stable under the formation of both, tensor products and internal homs. The behavior of separation vis-à-vis tensor and internal hom in $\mathcal{V}$-Cat requires a closer look. Reassuringly though, since for all $\mathcal{V}$-functors $f, g : X \to Y$ one has

$$\mathrm{k} \leq [X, Y](f, g) = \bigwedge_{x \in X} Y(fx, gx) \iff \forall x \in X : \mathrm{k} \leq Y(fx, gx),$$

the induced order of $[X, Y]$ is the pointwise order inherited from Y, so that $[X, Y]$ is separated if Y is separated. However, the induced order of $X \otimes Y$ may be finer than the componentwise order of $X \times Y$, but they do coincide when $\mathcal{V}$ is integral (because, in that case, an inequality $\mathrm{k} \leq u \otimes v$ with any $u, v \in \mathcal{V}$ automatically implies $\mathrm{k} \leq u \otimes \mathrm{k} = u$, since $v \leq \mathrm{k}$ holds by integrality). Consequently, for $\mathcal{V}$ integral, $X \otimes Y$ is separated if X and Y are separated.

With these observations one concludes:

Corollary 36 *With the tensor product and the internal hom formed as in* $\mathcal{V}$-Cat, *the category* $\mathcal{V}$-$\mathsf{Cat}_{\mathrm{sym}}$ *is symmetric monoidal-closed and, if* $\mathcal{V}$ *is integral, so are the categories* $\mathcal{V}$-$\mathsf{Cat}_{\mathrm{sep}}$ *and* $\mathcal{V}$-$\mathsf{Cat}_{\mathrm{sym,sep}}$.

Here is an example of a non-integral quantale $\mathcal{V}$ with separated and symmetric $\mathcal{V}$-categories X, Y, whose tensor product $X \otimes Y$ fails to be separated.

Example 37 Let $\mathcal{V}$ be the quantale $\mathcal{P}M$ of Example 18(5), where we now choose for the monoid M the two-element group $\mathbb{Z}_2 = (\{0,1\}, +)$. We consider the two-element set $X = \{x, x'\}$ and the singleton set $Y = \{y\}$ as $\mathcal{V}$-categories with

$$X(x,x) = X(x',x') = \{0\},\ \ X(x,x') = X(x'x) = \{1\}\ \text{ and }\ Y(y,y) = \{0,1\}.$$

Then X and Y are separated and symmetric, but in $X \otimes Y$ one has the order-equivalent, but distinct, elements $(x,y) \simeq (x',y)$, since

$$X \otimes Y((x,y),(x',y)) = X(x,x') + Y(y,y) = \{1\} + \{0,1\} = \{0,1\} \supseteq \{0\}.$$

4.4 *$\mathcal{V}$-Distributors and the $\mathcal{V}$-Cat-Enriched 2-Category $\mathcal{V}$-Dist*

The $\mathcal{V}$-distributors of $\mathcal{V}$-categories generalize $\mathcal{V}$-functors in the same way as relations generalize functions. They turn out to be fundamental devices for the efficient treatment of presheaves on a $\mathcal{V}$-category.

Definition 38 Let $X = (X,a)$ and (Y,b) be $\mathcal{V}$-categories. A *$\mathcal{V}$-distributor* (also *$\mathcal{V}$-(bi)module* or *$\mathcal{V}$-profunctor*) from X to Y is a $\mathcal{V}$-relation $\rho : X \nrightarrow Y$ with $b \cdot \rho \cdot a \le \rho$; equivalently: if for all $x, x' \in X,\ y, y' \in Y$ one has

$$X(x',x) \otimes \rho(x,y) \otimes Y(y,y') \le \rho(x',y').$$

Since this inequality transcribes to $X^{\mathrm{op}}(x,x') \otimes Y(y,y') \le [\rho(x,y), \rho(x',y')]$, for the mapping $\rho : X \times Y \to \mathcal{V}$ to give a $\mathcal{V}$-module from X to Y means precisely that

$$\rho : X^{\mathrm{op}} \otimes Y \longrightarrow \mathcal{V}$$

is a $\mathcal{V}$-functor. We use the notation $\rho : X \nrightarrow\!\!\!\circ\ Y$ to tacitly indicate that the $\mathcal{V}$-relation $\rho : X \nrightarrow Y$is actually a $\mathcal{V}$-distributor.

Note that one actually has the equality $b \cdot \rho \cdot a = \rho$ for $\rho : X \nrightarrow\!\!\!\circ\ Y$.

Example 39 In the case $\mathcal{V} = [0, \infty]$, for metric spaces X, Y, a $\mathcal{V}$-distributor $\rho : X \twoheadrightarrow Y$ extends the metrics of X and Y to a *one-way* or *directed metric* on the disjoint union $X \sqcup Y$, in the sense that it gives new distances only from points of X to points of Y. Of course, we may trivially extend it to a metric on $X \sqcup Y$, by declaring all distances in the reverse direction to be ∞.

We collect some easily verified observations:

1. The structure a of a $\mathcal{V}$-category $X = (X, a)$ is a $\mathcal{V}$-distributor $a : X \twoheadrightarrow X$; it is neutral w.r.t. left or right composition with a $\mathcal{V}$-distributor; we write $1_X^* = a$.
2. $\mathcal{V}$-distributors are closed under composition in $\mathcal{V}$-Rel.
3. Every $\mathcal{V}$-distributor $\rho : X \twoheadrightarrow Y$ gives the $\mathcal{V}$-distributor $\rho^{\mathrm{op}} := \rho^\circ : Y^{\mathrm{op}} \twoheadrightarrow X^{\mathrm{op}}$.
4. $\mathcal{V}$-distributors are closed under arbitrary infima and suprema formed in $\mathcal{V}$-Rel.
5. Every element $u \in \mathcal{V}$ may be regarded as a $\mathcal{V}$-distributor $u : \mathrm{E} \twoheadrightarrow \mathrm{E}$, with $u(*, *) = u$. In this way one obtains an embedding $\mathcal{V} \to \mathcal{V}$-$\mathsf{Dist}$ which lets us identify the monoid $(\mathcal{V}, \otimes, \mathrm{k})$ with the monoid $(\mathcal{V}\text{-}\mathsf{Dist}(\mathrm{E}, \mathrm{E}), \cdot, 1_{\mathrm{E}}^*)$.

These observations imply the first two statements of the following theorem and facilitate the proof of the third. For $\rho, \rho' : X \twoheadrightarrow Y$, we sometimes use the abbreviation

$$\|\rho, \rho'\| := \mathcal{V}\text{-}\mathsf{Dist}(X, Y)(\rho, \rho') = \bigwedge_{x \in X,\, y \in Y} [\rho(x, y), \rho'(x, y)].$$

Theorem 40 *With the $\mathcal{V}$-relational composition and order, the $\mathcal{V}$-categories and their $\mathcal{V}$-distributors form a 2-subcategory of $\mathcal{V}$-Rel, denoted by*

$$\mathcal{V}\text{-}\mathsf{Dist}\,,$$

which, like $\mathcal{V}$-Rel, is self-dual[4]*, qua the isomorphism $(-)^{\mathrm{op}} : \mathcal{V}\text{-}\mathsf{Dist}^{op} \to \mathcal{V}\text{-}\mathsf{Dist}$. Likewise, also $\mathcal{V}$-Dist is Sup-enriched (i.e., a quantaloid). Moreover, with*

$$\mathcal{V}\text{-}\mathsf{Dist}(X, Y) = [X^{\mathrm{op}} \otimes Y, \mathcal{V}],$$

for all $\mathcal{V}$-categories X, Y, the category $\mathcal{V}$-Dist becomes also $\mathcal{V}$-Cat-enriched.

Proof For the last statement we show that the maps $\mathrm{E} \to \mathcal{V}\text{-}\mathsf{Dist}(X, X)$, $* \mapsto 1_X^*$, and

$$\mathcal{V}\text{-}\mathsf{Dist}(X, Y) \otimes \mathcal{V}\text{-}\mathsf{Dist}(Y, Z) \longrightarrow \mathcal{V}\text{-}\mathsf{Dist}(X, Z), \ (\rho, \sigma) \longmapsto \sigma \cdot \rho,$$

[4] But note that the isomorphism $(-)^{\mathrm{op}}$ is covariant on 2-cells, i.e., it preserves the order of $\mathcal{V}$-distributors.

are $\mathcal{V}$-functors, for all $\mathcal{V}$-categories X, Y, Z. This is trivial for the first map, and for the second map we use Corollary 4 and Proposition 35 to see that, for all $\rho, \rho' : X \to\!\!\!\!\!\!\circ\; Y,\ \sigma, \sigma' : Y \to\!\!\!\!\!\!\circ\; Z$, and all $x \in X,\ y \in Y,\ z \in Z$, we have

$$\begin{aligned}\|\rho, \rho'\| \otimes \|\sigma, \sigma'\| &\leq [\rho(x, y), \rho'(x, y)] \otimes [\sigma(y, z), \sigma'(y, z)]\\ &\leq [\rho(x, y) \otimes \sigma(y, z), \rho'(x, y) \otimes \sigma'(y, z)]\\ &\leq [\rho(x, y) \otimes \sigma(y, z), \textstyle\bigvee_{y' \in Y} \rho'(x, y') \otimes \sigma'(y', z)].\end{aligned}$$

Consequently,

$$\begin{aligned}\|\rho, \rho'\| \otimes \|\sigma, \sigma'\| &\leq \textstyle\bigwedge_{x,y,z}[\rho(x, y) \otimes \sigma(x, y), (\sigma' \cdot \rho')(x, z)]\\ &\leq \textstyle\bigwedge_{x,z}[\bigvee_y \rho(x, y) \otimes \sigma(x, y), (\sigma' \cdot \rho')(x, z)]\\ &= \textstyle\bigwedge_{x,z}[(\sigma \cdot \rho)(x, z), (\sigma' \cdot \rho')(x, z)] = \|\sigma \cdot \rho, \sigma' \cdot \rho'\|.\end{aligned}$$

This shows

$$(\mathcal{V}\text{-}\mathsf{Dist}(X, Y)\otimes\mathcal{V}\text{-}\mathsf{Dist}(Y, Z))((\rho, \sigma), (\rho', \sigma')) \leq \mathcal{V}\text{-}\mathsf{Dist}(X, Z)(\sigma \cdot \rho, \sigma' \cdot \rho').$$

□

Example 41 Considering $[0, \infty]_+$-distributors as directed metrics like in Example 39, we have the composite directed metric $\sigma \cdot \rho : X \to\!\!\!\!\!\!\circ\; Z$ of $\rho : X \to\!\!\!\!\!\!\circ\; Y$ followed by $\sigma : Y \to\!\!\!\!\!\!\circ\; Z$ which, for all $x \in X, z \in Z$, is given by

$$(\sigma \cdot \rho)(x, z) = \inf_y(\rho(x, y) + \sigma(y, z)).$$

The theorem tells us that the resulting category $[0, \infty]_+ - \mathsf{Dist}$ is enriched in Met, with the metric $\|-, -\|$ on its hom set $[0, \infty]_+\text{-}\mathsf{Dist}(X, Y)$ provided by the formula:

$$\|\rho, \rho'\| = \sup_{x,y}(\max\{\rho'(x, y) - \rho(x, y), 0\}).$$

4.5 $\mathcal{V}$-Functors as $\mathcal{V}$-Distributors, Adjunction of $\mathcal{V}$-Functors

Arguably the most important $\mathcal{V}$-distributors are given by $\mathcal{V}$-functors, as follows.

Proposition 42 *Every $\mathcal{V}$-functor $f : (X, a) \to (Y, b)$ induces the $\mathcal{V}$-distributors*

$$f_* = b \cdot f_\circ : X \to\!\!\!\!\!\!\circ\; Y \quad \textit{and} \quad f^* := f^\circ \cdot b : Y \to\!\!\!\!\!\!\circ\; X,$$

so that $f_*(x, y) = Y(fx, y)$ and $f^*(y, x) = Y(y, fx)$, for all $x \in X, y \in Y$. When composed with a $\mathcal{V}$-functor $g : Y \to Z$ one obtains

$$(1_X)_* = (1_X)^* = 1_X^* = a \quad \textit{and} \quad (gf)_* = g_* \cdot f_*,\ \ (gf)^* = f^* \cdot g^*,$$

and for every $\mathcal{V}$-functor $f' : X \to Y$ one has

$$f \le f' \iff f^* \le (f')^* \iff (f')_* \le f_* \,.$$

Proof Since $b \cdot f_* \cdot a = b \cdot b \cdot f_\circ \cdot a \le b \cdot f_\circ \cdot f^\circ \cdot b \cdot f_\circ \le b \cdot 1_Y^\circ \cdot b \cdot f_\circ \le b \cdot f_\circ = f_*$, we have the $\mathcal{V}$-distributor f_*. Similarly one checks for f^*, as well as the stated functorial properties of $(-)_*$ and $(-)^*$. Furthermore, the stated equivalences describing the order for $\mathcal{V}$-functors are a consequence of the contra-covariance of $Y(\text{-},\text{-}) : Y^{\mathrm{op}} \otimes Y \to \mathcal{V}$. □

With the 2-functor $(-)^{\mathrm{op}}$ of Corollary 29 we conclude:

Corollary 43 *The assignments $f \mapsto f_*$ and $f \mapsto f^*$ describe 2-functors*

$$(-)_* : \mathcal{V}\text{-}\mathsf{Cat} \longrightarrow \mathcal{V}\text{-}\mathsf{Dist}^{\mathrm{co}} \quad \textit{and} \quad (-)^* : \mathcal{V}\text{-}\mathsf{Cat} \longrightarrow \mathcal{V}\text{-}\mathsf{Dist}^{\mathrm{op}}$$

that map objects identically. They are dual to each other, in the sense of the following commutative diagram:

$$\begin{array}{ccc}
\mathcal{V}\text{-}\mathsf{Cat} & \xrightarrow[\cong]{(-)^{\mathrm{op}}} & \mathcal{V}\text{-}\mathsf{Cat}^{\mathrm{co}} \\
{\scriptstyle (-)^*}\big\downarrow & & \big\downarrow{\scriptstyle (-)_*^{\mathrm{co}}} \\
\mathcal{V}\text{-}\mathsf{Dist}^{\mathrm{op}} & \xrightarrow[\cong]{(-)^{\mathrm{op}}} & \mathcal{V}\text{-}\mathsf{Dist}
\end{array}$$

Remark 44 For $\mathcal{V}$-functors $f : X \to Y$ and $h : Z \to Y$, the $\mathcal{V}$-distributor $h^* \cdot f_* : X \nrightarrow Z$ may be conveniently evaluated at $x \in X, z \in Z$, as $(h^* \cdot f_*)(x, z) = Y(fx, hz)$. In particular, if we treat points x, y in a $\mathcal{V}$-category X as $\mathcal{V}$-functors $x, y : \mathrm{E} \to X$ and identify $\mathcal{V}$ with $\mathcal{V}\text{-}\mathsf{Dist}(\mathrm{E}, \mathrm{E})$ (as noted before Theorem 40), then

$$y^* \cdot x_* = X(x, y) \,.$$

Once transported into the 2-category $\mathcal{V}$-**Dist**, every $\mathcal{V}$-functor has an adjoint, and adjunctions in $\mathcal{V}$-**Cat** may conveniently be characterized in $\mathcal{V}$-**Dist**, as follows.

Proposition 45 *For every $\mathcal{V}$-functor $f : X \to Y$ one has $f_* \dashv f^*$, and f is fully faithful if, and only if, $f^* \cdot f_* = 1_X^*$. Furthermore, for every $\mathcal{V}$-functor $g : Y \to X$ one has the equivalences*

$$f \dashv g \iff f_* = g^* \iff g^* \dashv f^* \iff g_* \dashv f_* \,.$$

Proof With $X = (X, a)$ and $Y = (Y, b)$, from $(b \cdot f_\circ) \cdot (f^\circ \cdot b) \le b \cdot 1_Y \cdot b \le b$ and $(f^\circ \cdot b) \cdot (b \cdot f_\circ) \ge f^\circ \cdot b \cdot f_\circ \ge a$ one gets, respectively, $f_* \cdot f^* \le 1_Y^*$ and $f^* \cdot f_* \ge 1_X^*$, which means $f_* \dashv f^*$. By definition, f is fully faithful if, and only if, $1_X^* = f^\circ \cdot 1_Y^* \cdot f_\circ$, but the right-hand side equals $f^* \cdot f_*$. The equivalence $(f \dashv g \iff f_* = g^*)$ is an immediate consequence of Proposition 30. Since left adjoints to a $\mathcal{V}$-distributor are unique in $\mathcal{V}$-Dist, and since one always has $f_* \dashv f^*$, the equivalence $(f_* = g^* \iff g^* \dashv f^*)$ follows. Likewise for the equivalence $(f_* = g^* \iff g_* \dashv f_*)$. □

Corollary 46 *Let $f \dashv g : Y \to X$ be adjoint $\mathcal{V}$-functors. Then one has the equivalences*

$$g \text{ fully faithful} \iff f_* \cdot g_* = 1_Y^* \iff \forall y \in Y : fgy \simeq y \iff g^* \cdot f^* = 1_Y^*.$$

Proof Since, by Proposition 45, $f \dashv g$ means $f_* = g^*$, while g fully faithful reads as $g^* \cdot g_* = 1_Y^*$, the first equivalence is evident. The other two equivalences follow with Proposition 30. □

Having characterized fully faithful $\mathcal{V}$-functors $f : X \to Y$ as those $\mathcal{V}$-functors for which the unit of the adjunction $f_* \dashv f^*$ is an identity, let us briefly look at those $\mathcal{V}$-functors f for which the counit is an identity.

Definition 47 We call a $\mathcal{V}$-functor $f : X \to Y$ *dense* if $f_* \cdot f^* = 1_Y^*$; that is (since $f_* \cdot f^* \le 1_Y^*$ comes for free): if

$$Y(y, y') \le \bigvee_{x \in X} Y(y, fx) \otimes Y(fx, y'),$$

for all $y, y' \in Y$. The $\mathcal{V}$-functor f is *essentially surjective* if, for all $y \in Y$ there is some $x \in X$, with $fx \simeq y$.

Remark 48 Since every $\mathcal{V}$-functor $f : X \to Y$ satisfies $f_* = f_* \cdot f^* \cdot f_*$, one easily sees that f is dense if, and only if, f_* is an epimorphism in the category $\mathcal{V}$-Dist. Dually one has that f is fully faithful if, and only if, f^* is a monomorphism in $\mathcal{V}$-Dist.

Proposition 49 *Every essentially surjective $\mathcal{V}$-functor is dense.*

Proof Given $y, y' \in Y$, by hypothesis we have $y \simeq fx$ with some $x \in X$ and obtain

$$Y(y, y') = Y(fx, y') \le Y(fx, fx) \otimes Y(fx, y') \le \bigvee_{x' \in X} Y(y, fx') \otimes Y(fx', y'),$$

as desired. □

The converse statement generally fails:

Examples 50 (1) If $\mathcal{V} = \mathsf{2}$, a dense $\mathcal{V}$-functor is always essentially surjective. Indeed, a monotone map $f : X \to Y$ is dense if, and only if, f is *interpolating*, in the sense that $y \le y'$ in Y always implies $y \le fx \le y'$, for some $x \in X$; one just has to consider the case $y = y'$.

(2) For $\mathcal{V} = [0, \infty]_+$, the (topologically dense) embedding $\mathbb{Q} \hookrightarrow \mathbb{R}$ of the rationals into the Euclidean reals certainly fails to be essentially surjective, but it is dense (in the sense of Definition 47) since, for all $y, y' \in \mathbb{R}$,

$$|y - y'| = \inf_{x \in \mathbb{Q}} |y - x| + |x - y'|.$$

4.6 Extensions and Liftings of $\mathcal{V}$-Distributors

Since the $\mathcal{V}$-distributors are closed under $\mathcal{V}$-relational composition and the formation of suprema in $\mathcal{V}$-Rel, it is clear that, for $\mathcal{V}$-distributors $\rho : X \nrightarrow Y$ and $\sigma : Y \nrightarrow Z$, the monotone maps

$$- \cdot \rho \colon \mathcal{V}\text{-Dist}(Y, Z) \longrightarrow \mathcal{V}\text{-Dist}(X, Z) \quad \text{and} \quad \sigma \cdot - \colon \mathcal{V}\text{-Dist}(X, Y) \longrightarrow \mathcal{V}\text{-Dist}(X, Z)$$

are well defined and have right adjoints (i.e., the forgetful functor $\mathcal{V}$-Dist $\to$ $\mathcal{V}$-Rel is a homomorphism of quantaloids). But are these right adjoints formed as in $\mathcal{V}$-Rel? The following proposition gives an affirmative answer.

Proposition 51 *For $\mathcal{V}$-distributors $\rho : X \nrightarrow Y,\ \sigma : Y \nrightarrow Z,\ \tau : X \nrightarrow Z$, the extension $[\rho, \tau] : Y \nrightarrow Z$ and the lifting $]\sigma, \tau[: X \nrightarrow Y$ formed in $\mathcal{V}$-Rel are again $\mathcal{V}$-distributors.*

Proof By Proposition 12 one has $]\sigma, \tau[(x, y) = \bigwedge_z [\sigma(y, z), \tau(x, z)]$, for all $x \in X,\ y \in Y$. Using the $\mathcal{V}$-distributivity of τ, for all $x' \in X,\ z \in Z$ one consecutively obtains

$$X(x', x) \otimes [\sigma(y, z), \tau(x, z)] \otimes \sigma(y, z) \le X(x', x) \otimes \tau(x, z) \le \tau(x', z),$$

$$X(x', x) \otimes\,]\sigma, \tau[(x, y) \le X(x', x) \otimes [\sigma(y, z), \tau(x, z)] \le [\sigma(y, z), \tau(x', z)],$$

which implies $X(x, x') \otimes\,]\sigma, \tau[(x, y) \le\,]\sigma, \tau[(x', y)$ and finally $]\sigma, \tau[\cdot 1_X^* \le\,]\sigma, \tau[$. Similarly, now using the $\mathcal{V}$-distributivity of σ, for all $y' \in Y,\ z \in Z$, one has

$$[\sigma(y, z), \tau(x, z)] \otimes Y(y, y') \otimes \sigma(y', z) \le [\sigma(y, z), \tau(x, z)] \otimes \sigma(y, z) \le \tau(x, z),$$

$$]\sigma, \tau[(x, y) \otimes Y(y, y') \le [\sigma(y, z), \tau(x, z)] \otimes Y(y, y') \le [\sigma(y', z), \tau(x, z)],$$

which then gives $]\sigma, \tau[(x, y) \otimes Y(y, y') \le\,]\sigma, \tau[(x, y')$, or $1_Y^* \cdot]\sigma, \tau[\, \le\,]\sigma, \tau[$.

This completes the proof that the lifting $]\sigma, \tau[$ is a $\mathcal{V}$-distributor. Therefore, also the extension $[\rho, \tau]$ is a $\mathcal{V}$-distributor, by duality (see the proof of Proposition 12). □

Corollary 52 *For $\mathcal{V}$-distributors ρ, σ, τ as in* Proposition *51, the $\mathcal{V}$-relational extension $[\rho, \tau]$ and lifting $]\sigma, \tau[$ are, respectively, characterized by the equivalences*

$$\sigma \leq [\rho, \tau] \iff \sigma \cdot \rho \leq \tau \quad \text{and} \quad \sigma \cdot \rho \leq t \iff \rho \leq]\sigma, \tau[\,,$$

in the case of $[\rho, \tau]$ to be universally quantified over σ, and for $]\sigma, \tau[$ over ρ.

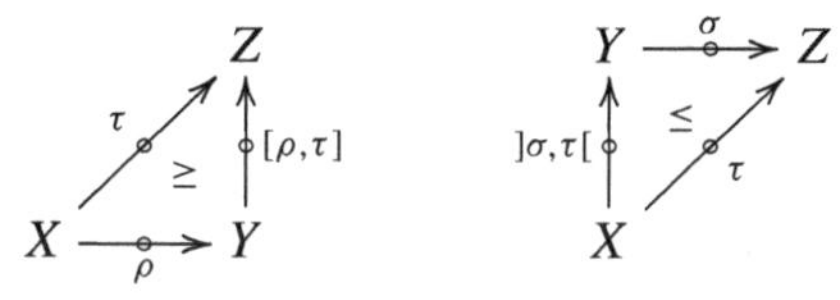

4.7 *The Tensor Product of $\mathcal{V}$-Distributors*

It is hardly surprising that the tensor product of $\mathcal{V}$-Cat may be extended functorially to $\mathcal{V}$-Dist, as follows.

Definition 53 For $\mathcal{V}$-distributors $\rho : X \nrightarrow Y$ and $\phi : S \nrightarrow T$ one defines their *tensor product* $\rho \otimes \phi : X \otimes S \nrightarrow Y \otimes T$ by

$$(\rho \otimes \phi)((x, s), (y, t)) = \rho(x, y) \otimes \phi(s, t),$$

for all $x \in X,\ y \in Y,\ s \in S,\ t \in T$.

Since this formula generalizes the formula for the structure of the tensor product $X \otimes S$, given by $X \otimes S((x, s), (x', s')) = X(x, x') \otimes S(s, s')$, one easily confirms that the tensor product preserves the $\mathcal{V}$-distributor property:

Proposition 54 *The $\mathcal{V}$-relation $\rho \otimes \phi : X \otimes S \nrightarrow Y \otimes T$ is a $\mathcal{V}$-distributor. Whenever the composites are defined, the middle-interchange rule*

$$(\sigma \otimes \psi) \cdot (\rho \otimes \phi) = (\sigma \cdot \rho) \otimes (\psi \cdot \phi)$$

holds, and for $\mathcal{V}$-functors $f : X \to Y,\ g : S \to T$ one has

$$1^*_{X \otimes S} = 1^*_X \otimes 1^*_S, \quad (f \otimes g)_* = f_* \otimes g_*, \quad (f \otimes g)^* = f^* \otimes g^* \,.$$

Proof The following computation shows that $\rho \otimes \phi$ is a $\mathcal{V}$-distributor:

$$\begin{aligned} &(X \otimes S)((x', s'), (x, s)) \otimes (\rho \otimes \phi)((x, s), (y, t)) \otimes (Y \otimes T)((y, t), (y', t')) \\ &= X(x', x) \otimes \rho(x, y) \otimes Y(y, y') \otimes S(s', s) \otimes \phi(s, t) \otimes T(t, t') \\ &\leq \rho(x', y') \otimes \phi(s', t') \;=\; (\rho \otimes \phi)((x', s'), (y', t')) \,. \end{aligned}$$

All of the other needed verifications proceed similarly. □

Corollary 55 *The tensor product makes* $\mathcal{V}$-Dist *a symmetric monoidal category, such that*

$$(-)_* : \mathcal{V}\text{-}\mathsf{Cat} \longrightarrow \mathcal{V}\text{-}\mathsf{Dist}^{\mathrm{co}} \quad \textit{and} \quad (-)^* : \mathcal{V}\text{-}\mathsf{Cat} \longrightarrow \mathcal{V}\text{-}\mathsf{Dist}^{\mathrm{op}}$$

are homomorphisms of monoidal categories.

Remark 56 It is fairly easy to see that (in generalization of Remark 10(2), representing the discrete case) the symmetric monoidal category $\mathcal{V}$-Dist is in fact *compact* $*$*-autonomous* in the sense of Barr (1979, 1999), but the internal hom may seem rather uninteresting. Indeed, since with Theorem 40 we have the natural isomorphisms

$$\mathcal{V}\text{-}\mathsf{Dist}(Z, X^{\mathrm{op}} \otimes Y) = [Z^{\mathrm{op}} \otimes (X^{\mathrm{op}} \otimes Y), \mathcal{V}] \cong [(Z \otimes X)^{\mathrm{op}} \otimes Y, \mathcal{V}] = \mathcal{V}\text{-}\mathsf{Dist}(Z \otimes X, Y)$$

for all $\mathcal{V}$-categories X, Y, Z, the functor $X^{\mathrm{op}} \otimes (-) : \mathcal{V}\text{-}\mathsf{Dist} \to \mathcal{V}\text{-}\mathsf{Dist}$ is right adjoint to $(-) \otimes X$, so that the $\mathcal{V}$-category $X^{\mathrm{op}} \otimes Y$ serves as the internal hom of X and Y in $\mathcal{V}$-Dist. Furthermore, since $X^{\mathrm{op}} \otimes \mathrm{E} \cong X^{\mathrm{op}}$, the tensor-neutral $\mathcal{V}$-category E serves as a dualizing object, making $\mathcal{V}$-Dist $*$-autonomous, with compactness following from the identification of the internal hom as $X^{\mathrm{op}} \otimes Y$, by the very definition of this property.

The question remains whether the $\mathcal{V}$-category $\mathcal{V}\text{-}\mathsf{Dist}(X, Y)$ has a role to play in $\mathcal{V}$-Dist vis-à-vis the tensor product $X \otimes Y$, similar to the role of $[X, Y]$ in $\mathcal{V}$-Cat. We can give here an affirmative answer to this (vague) question which, however, will become quite natural by Theorem 63; there we will prove that, for every $\mathcal{V}$-category X, not only the first and the second, but also the third of the consecutive 2-functors

$$\mathcal{V}\text{-}\mathsf{Cat}^{\mathrm{co}} \xrightarrow[\cong]{(-)^{\mathrm{op}}} \mathcal{V}\text{-}\mathsf{Cat} \xrightarrow{-\otimes X} \mathcal{V}\text{-}\mathsf{Cat} \xrightarrow{(-)_*} \mathcal{V}\text{-}\mathsf{Dist}^{\mathrm{co}}$$

has a right adjoint. Hence, after dualization with respect to 2-cells, the composite functor $\mathcal{V}\text{-}\mathsf{Cat} \longrightarrow \mathcal{V}\text{-}\mathsf{Dist}$, $Z \longmapsto Z^{\mathrm{op}} \otimes X$, has a right adjoint, which turns out to be given by the assignment $Y \longmapsto \mathcal{V}\text{-}\mathsf{Dist}(X, Y)$. In fact, the required natural correspondence between the $\mathcal{V}$-distributors $Z^{\mathrm{op}} \otimes X \to\!\!\!\!\circ\; Y$ and the $\mathcal{V}$-functors

$Z \longrightarrow \mathcal{V}\text{-}\mathsf{Dist}(X, Y)$ for all $\mathcal{V}$-categories Y, Z, producing this composite adjunction, follows quite easily from Theorems 32 and 40, and they, moreover, show that the adjunction is $\mathcal{V}$-Cat-enriched.

These observations provide the backbone of the proof of the following theorem.

Theorem 57 *Given $\mathcal{V}$-categories X, Y, Z, every $\mathcal{V}$-distributor $\rho : Z^{\mathrm{op}} \otimes X \nrightarrow Y$ factors as $\varepsilon \cdot (\sigma \otimes 1_X^*) = \rho$, for some $\mathcal{V}$-distributor $\sigma : Z^{\mathrm{op}} \nrightarrow (\mathcal{V}\text{-}\mathsf{Dist}(X, Y))^{\mathrm{op}}$, as in*

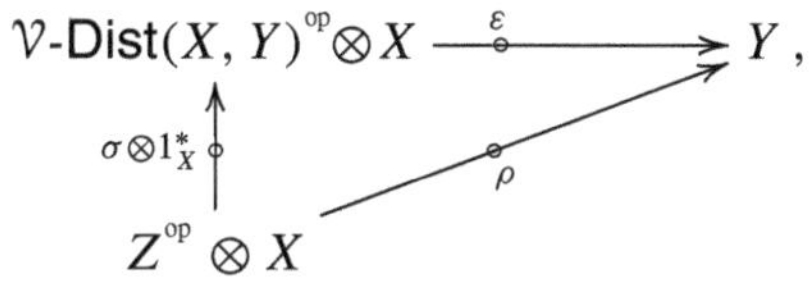

where ε is given by evaluation: $\varepsilon((\phi, x), y) = \phi(x, y)$, for all $\phi : X \nrightarrow Y, x \in X$, and $y \in Y$. Moreover, there is a largest such factorizing $\mathcal{V}$-distributor σ, given for all $z \in Z$ and ϕ by

$$\sigma(z, \phi) = \|\phi, \rho((z, \text{-}), \text{-})\| = \bigwedge_{x,y} [\phi(x, y), \rho((z, x), y)].$$

Proof Given X, the adjunction described above has counit ε at Y. The mate of the $\mathcal{V}$-distributor $\rho : Z^{\mathrm{op}} \otimes X \nrightarrow Y$ is the $\mathcal{V}$-functor $\widetilde{\rho} : Z \longrightarrow \mathcal{V}\text{-}\mathsf{Dist}(X, Y)$, defined by $\widetilde{\rho}(z)(x, y) = \rho((z, x), y)$; by adjunction, it satisfies $\varepsilon \cdot (\widetilde{\rho}^{\mathrm{op}} \otimes 1_X)_* = \varepsilon \cdot ((\widetilde{\rho}^{\mathrm{op}})_* \otimes 1_X^*) = \rho$.

To complete the proof, it suffices to show that any $\sigma : Z^{\mathrm{op}} \nrightarrow (\mathcal{V}\text{-}\mathsf{Dist}(X, Y))^{\mathrm{op}}$ with $\varepsilon \cdot (\sigma \otimes 1_X^*) \leq \rho$ satisfies $\sigma \leq (\widetilde{\rho}^{\mathrm{op}})_*$. Indeed, for all $z \in Z, x \in X, y \in Y$, we have

$$\begin{aligned}
\widetilde{\rho}(z)(x, y) = \rho((z, x), y) &\geq (\varepsilon \cdot (\sigma \otimes 1_X^*))((z, x), y) \\
&= \textstyle\bigvee_{\phi,x'} (\sigma \otimes 1_X^*)((z, x), (\phi, x')) \otimes \varepsilon((\phi, x'), y) \\
&= \textstyle\bigvee_{\phi,x'} (\sigma(z, \phi) \otimes X(x, x') \otimes \phi(x', y)) \\
&= \textstyle\bigvee_{\phi} \sigma(z, \phi) \otimes (\bigvee_{x'} X(x, x') \otimes \phi(x', y)) \\
&= \textstyle\bigvee_{\phi} \sigma(z, \phi) \otimes \phi(x, y) .
\end{aligned}$$

Consequently, for all $\phi : X \nrightarrow Y$, from $\sigma(z, \phi) \otimes \phi(x, y) \leq \widetilde{\rho}(z)(x, y)$ we obtain $\sigma(z, \phi) \leq [\phi(x, y), \widetilde{\rho}(z)(x, y)]$ and, thus,

$$\sigma(z, \phi) \leq \|\phi, \widetilde{\rho}(z)\| = (\mathcal{V}\text{-}\mathsf{Dist}(X, Y))^{\mathrm{op}}(\widetilde{\rho}(z), \phi) = (\widetilde{\rho}^{\mathrm{op}})_*(z, \phi),$$

as desired. □

Example 58 Here is a (somewhat vague) applied illustration of the theorem, with $\mathcal{V} = [0, \infty]_+$. With points x and y taken to be locations within the cities X and Y, we may think of $\phi : X \nrightarrow Y$ as the logistics/distribution company ϕ offering transportation services from X to Y, and of $\phi(x, y)$ as the surcharge for sending a truck from the actual location x in X to location y in Y. For a range of goods $z \in Z$, sorted by volume or weight, say, which company ϕ is able to transport, the value $\sigma(z, \phi)$ of $\sigma : Z \nrightarrow \mathcal{V}\text{-}\mathsf{Dist}(X, Y)$ gives their base price for sending the good z from city X to city Y, independently of the actual locations x and y within X and Y. Having scanned the offers of all companies, the customer arrives at the minimal market price

$$\rho(z, x, y) = \inf_{\phi}(\sigma(z, \phi) + \phi(x, y))$$

for shipping good z from location x to location y. The theorem says that, for company ϕ to stay competitive in the market of shipping the good z from city X to Y, they should offer

$$\sigma(z, \phi) = \sup_{x,y}(\rho(z, x, y) - \phi(x, y))$$

as their base price for transporting z from X to Y.

5 The Presheaf Monad and Its Category of Pseudo-Algebras

In this section, we introduce (co)presheaves as particular distributors and show that the formation of the $\mathcal{V}$-category of presheaves on a $\mathcal{V}$-category gives us a monad $\mathcal{P}$ on the 2-category $\mathcal{V}\text{-}\mathsf{Cat}$. The lax idempotency of this monad makes a pseudo-algebra structure w.r.t. $\mathcal{P}$ on a $\mathcal{V}$-category X unique up to order-equivalence, so that its existence becomes a property of X, here called $\mathcal{P}$-cocompleteness. This sets the stage for us in the following sections to explore the prominent role which the $\mathcal{V}$-category of presheaves on X plays when we investigate weighted colimits in X. Dually, copresheaves assume such role when we form weighted limits in $\mathcal{V}$-categories.

5.1 $\mathcal{V}$-Presheaves Represented by $\mathcal{V}$-Distributors

Having presented the $\mathcal{V}\text{-}\mathsf{Cat}$-enriched 2-category $\mathcal{V}\text{-}\mathsf{Dist}$ as a symmetric monoidal category (see Theorem 40 and Corollary 55), we should now pay attention to the functors

$$\mathcal{V}\text{-}\mathsf{Dist}^{\mathrm{op}} \xrightarrow{\mathcal{V}\text{-}\mathsf{Dist}(\text{-},\mathrm{E})} \mathcal{V}\text{-}\mathsf{Cat} \qquad \text{and} \qquad \mathcal{V}\text{-}\mathsf{Dist}^{\mathrm{co}} \xrightarrow{\mathcal{V}\text{-}\mathsf{Dist}^{\mathrm{co}}(\mathrm{E},\text{-})} \mathcal{V}\text{-}\mathsf{Cat}$$

represented by the tensor-neutral $\mathcal{V}$-category $\mathrm{E} = \{*\}$ with $\mathrm{E}(*, *) = \mathrm{k}$. Note that $\mathrm{E}^{\mathrm{op}} = \mathrm{E}$; furthermore, we will allow ourselves to *identify* X *with* $\mathrm{E} \otimes X$ *and* $X \otimes \mathrm{E}$, for every $\mathcal{V}$-category X.

Definition 59 For $X \in \mathcal{V}\text{-}\mathsf{Cat}$, one forms the $\mathcal{V}$-categories

$$\mathcal{P}_{\mathcal{V}}X := \mathcal{V}\text{-}\mathsf{Dist}(X, \mathrm{E}) = [X^{\mathrm{op}}, \mathcal{V}] \quad \text{and} \quad \mathcal{P}^{\sharp}_{\mathcal{V}}X := \mathcal{V}\text{-}\mathsf{Dist}^{\mathrm{co}}(\mathrm{E}, X) = [X, \mathcal{V}]^{\mathrm{op}},$$

and calls their objects $\mathcal{V}$-*presheaves* and $\mathcal{V}$-*copresheaves* on X, respectively. For every $x \in X$ one has the *representable* $\mathcal{V}$-(co)presheaves, given by $\hom_X$:

$$\mathbf{y}_X x := X(-, x) = X^{\mathrm{op}} \to \mathcal{V} \quad \text{and} \quad \mathbf{y}^{\sharp}_X x := X(x, -) : X \to \mathcal{V}.$$

For a $\mathcal{V}$-(co)presheaf σ on X and $x \in X$, we write σx, rather than $\sigma(x, *)$ (or $\sigma(*, x)$).

Remarks 60 (1) Considering $x \in X \in \mathcal{V}\text{-}\mathsf{Cat}$ as a $\mathcal{V}$-functor $x : \mathrm{E} \to X$ one has the $\mathcal{V}$-distributors $x^* : X \to\!\!\!\!\circ\;\, \mathrm{E}$ and $x_* : \mathrm{E} \to\!\!\!\!\circ\;\, X$; in fact,

$$x^* = \mathbf{y}_X x \quad \text{and} \quad x_* = \mathbf{y}^{\sharp}_X x.$$

(2) The two $\mathcal{V}$-categories just defined are dual to each other, in the sense that

$$\mathcal{P}_{\mathcal{V}}(X^{\mathrm{op}}) = (\mathcal{P}^{\sharp}_{\mathcal{V}}X)^{\mathrm{op}}.$$

Examples 61 (1) For $\mathcal{V} = \mathsf{2}$ and $X \in \mathsf{Ord}$, we can identify $\mathcal{P}_{\mathcal{V}}(X)$ with the set $\mathcal{D}X$ of down-closed subsets of X, ordered by inclusion $\subseteq$, and $\mathcal{P}^{\sharp}_{\mathcal{V}}X$ with the set $\mathcal{U}X$ of up-closed subsets of X, ordered by inverted inclusion $\supseteq$.

(2) For $\mathcal{V} = [0, \infty]_+$ and $X \in \mathsf{Met}$, the $\mathcal{V}$-(co)presheaves on X are the functions $\sigma : X \to [0, \infty]$ with $\sigma x - \sigma y \leq X(x, y)$ (or $\sigma x - \sigma y \leq X(y, x)$, respectively), for all $x, y \in X$. Distances between $\mathcal{V}$-(co)presheaves σ, σ' are given by the sup-metric: $\mathcal{P}_{\mathcal{V}}X(\sigma, \sigma') = \sup_x(\sigma' x - \sigma x)$ (or $\mathcal{P}^{\sharp}_{\mathcal{V}}X(\sigma, \sigma') = \sup_x(\sigma x - \sigma' x)$, respectively).
In the interpretation of Example 39, we can think of a $[0, \infty]_+$-(co)presheaf on X as an extension of the metric of X, providing distances *to* (*from*, respectively, but not simultaneously!) a new point $*$, in such a way that the triangle inequality remains intact:

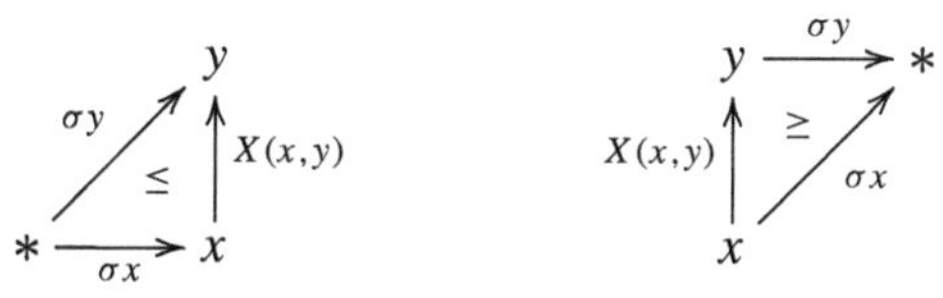

The following easy, but fundamental, lemma says that certain homs in the (co)presheaf category of X involving representable (co)presheaves are easily computable, by evaluation:

Lemma 62 (Yoneda Lemma) *For all $x \in X \in \mathcal{V}\text{-}\mathsf{Cat}$ and $\sigma \in \mathcal{P}_{\mathcal{V}}X,\ \tau \in \mathcal{P}^{\sharp}_{\mathcal{V}}X$,*

$$\mathcal{P}_{\mathcal{V}}X(\mathbf{y}_X x, \sigma) = \sigma x \qquad \text{and} \qquad \mathcal{P}^{\sharp}_{\mathcal{V}}X(\tau, \mathbf{y}^{\sharp}_X x) = \tau x.$$

Proof By Corollary 52, the extension $[1^*_X, \sigma]$ coincides trivially with $\sigma : X \nrightarrow \mathrm{E}$. Therefore, with Proposition 12, we compute

$$\sigma x = [1^*_X, \sigma]x = \bigwedge_{z \in X} [X(z, x), \sigma z] = \mathcal{P}_{\mathcal{V}}X(\mathbf{y}_X x, \sigma).$$

This proves the first formula; when applied to X^{op} instead of X, it gives the second. $\square$

Theorem 63 (Presheaf Adjunction) *For every $\mathcal{V}$-category X one has the fully faithful $\mathcal{V}$-functors*

$$\mathbf{y}_X : X \longrightarrow \mathcal{P}_{\mathcal{V}}X \qquad \text{and} \qquad \mathbf{y}^{\sharp}_X : X \longrightarrow \mathcal{P}^{\sharp}_{\mathcal{V}}X\,.$$

These are the respective units at X of the $\mathcal{V}\text{-}\mathsf{Cat}$-enriched (co)presheaf adjunctions

$$\mathcal{V}\text{-}\mathsf{Dist}^{\mathrm{op}} \underset{\mathcal{P}_{\mathcal{V}} = \mathcal{V}\text{-}\mathsf{Dist}(-,\mathrm{E})}{\overset{(-)^*}{\underset{\longrightarrow}{\overset{\longleftarrow}{\bot}}}} \mathcal{V}\text{-}\mathsf{Cat} \quad \text{and} \quad \mathcal{V}\text{-}\mathsf{Dist}^{\mathrm{co}} \underset{\mathcal{P}^{\sharp}_{\mathcal{V}} = \mathcal{V}\text{-}\mathsf{Dist}^{\mathrm{co}}(\mathrm{E},-)}{\overset{(-)_*}{\underset{\longrightarrow}{\overset{\longleftarrow}{\bot}}}} \mathcal{V}\text{-}\mathsf{Cat}\,.$$

Proof By the Yoneda Lemma we have $\mathcal{P}_{\mathcal{V}}X(\mathbf{y}_X x, \mathbf{y}_X y) = (\mathbf{y}_X y)x = X(x, y)$, which confirms that $\mathbf{y}_X$ is a fully faithful $\mathcal{V}$-functor. In order to establish the first adjunction we invoke Theorem 32 to obtain, for all $\mathcal{V}$-categories X, Y, the natural isomorphism

$$\mathcal{V}\text{-}\mathsf{Dist}(X, Y) = [X^{\mathrm{op}} \otimes Y, \mathcal{V}] \longrightarrow [Y, [X^{\mathrm{op}}, \mathcal{V}]] = \mathcal{V}\text{-}\mathsf{Cat}(Y, \mathcal{P}_{\mathcal{V}}X),$$

which assigns to every $\mathcal{V}$-distributor $\rho : X \nrightarrow Y$ the $\mathcal{V}$-functor $\rho^{\flat} : Y \to \mathcal{P}_{\mathcal{V}}X$ with $y \mapsto \rho(-, y)$. In particular, for $Y = X$ and $\rho = 1^*_X$ we obtain the adjunction unit $\rho^{\flat} = \mathbf{y}_X$ at X. (Likewise, considering $Y = \mathcal{P}_{\mathcal{V}}X$, we see that the inverse of the adjunction isomorphism produces the counit $(\mathbf{y}_X)_* : X \nrightarrow \mathcal{P}_{\mathcal{V}}X$ in $\mathcal{V}\text{-}\mathsf{Dist}$.)

Consequently, to confirm that the left adjoint of $\mathcal{P}_{\mathcal{V}}$ is correctly described by $(-)^*$, not only at the object level, but also at the morphism level, it suffices to note that, for every $\mathcal{V}$-functor $f : X \to Y$, the diagram

$$\begin{array}{ccc} X & \xrightarrow{\mathbf{y}_X} & \mathcal{P}_{\mathcal{V}}X \\ {\scriptstyle f}\big\downarrow & & \big\downarrow{\scriptstyle \mathcal{P}_{\mathcal{V}}f^*} \\ Y & \xrightarrow[\mathbf{y}_Y]{} & \mathcal{P}_{\mathcal{V}}Y \end{array}$$

commutes. Indeed, $(\mathcal{P}_{\mathcal{V}}f^*)(\mathbf{y}_X x) = \mathbf{y}_X x \cdot f^* = Y(\text{-}, fx) = \mathbf{y}_Y(fx)$ holds for all $x \in X$.

The second adjunction follows from the first by dualization. □

Remark 64 The $\mathcal{V}$-functors $\mathbf{y}_X$ and $\mathbf{y}_X^\sharp$ are commonly referred to as the *Yoneda embeddings* of X. But as one easily shows, these maps are injective only if the $\mathcal{V}$-category X is separated. In that case, however, X becomes isomorphic to a full $\mathcal{V}$-subcategory of its $\mathcal{V}$-(co)presheaf category.

Examples 65 (1) Every partially ordered set X is order-isomorphically embedded into the complete lattice $\mathcal{D}X$ of its down-closed subsets (and into the complete lattice $\mathcal{U}X$ of its up-closed subsets, with the reversed inclusion order).

(2) Every bounded Fréchet metric space X is isometrically embedded into the (Cauchy-)complete bounded Fréchet metric space of its $[0, 1]$-valued non-expanding functions. This is a specific consequence of the theorem applied to the quantale $[0, 1]_\oplus$ (as in Example 7, 2.).
Various generalizations and variations of this fact are obtained for generalized metric spaces when we apply the above theorem to other "metric quantales," like $[0, \infty]_+$, $[0, \infty]_{\max}$, *etc.*

5.2 The Isbell Duality

Beyond the formal duality noted in Remark 60(2), there is a more deeply rooted adjunction of $\mathcal{V}$-functors linking the presheaf $\mathcal{V}$-category and the copresheaf $\mathcal{V}$-category of any $\mathcal{V}$-category X, as follows:
For every presheaf $\sigma : X \mathrel{-\!\!\!\circ\!\!\!\rightarrow} \mathrm{E}$, one forms the copresheaf

$$L\sigma = [\sigma, 1_X^*] : \mathrm{E} \mathrel{-\!\!\!\circ\!\!\!\rightarrow} X, \quad (L\sigma)x = \bigwedge_{z \in X} [\sigma z, X(z, x)] \ (x \in X),$$

and for every copresheaf $\tau : \mathrm{E} \mathrel{-\!\!\!\circ\!\!\!\rightarrow} X$ one (dually) forms the presheaf

$$R\tau =]\tau, 1_X^*[\, : X \mathrel{-\!\!\!\circ\!\!\!\rightarrow} \mathrm{E}, \quad (R\tau)x = \bigwedge_{z \in X} [\tau z, X(x, z)] \ (x \in X).$$

Theorem 66 (Isbell Duality) *L and R are $\mathcal{V}$-functors, establishing the adjunction*

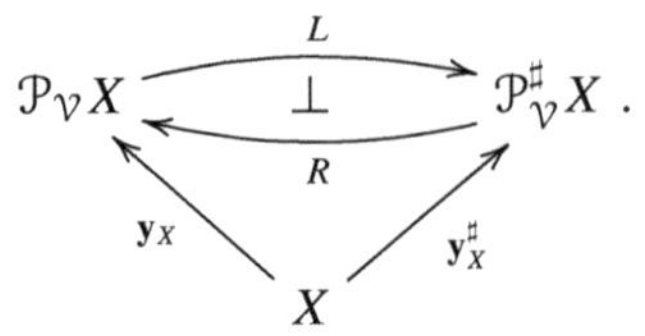

Moreover, L and R commute with the Yoneda maps as indicated by the above diagram.

Proof The repeated application of the rules of Corollary 4 gives, for all $\sigma \in \mathcal{P}_{\mathcal{V}}X$ and $\tau \in \mathcal{P}^{\sharp}_{\mathcal{V}}X$,

$$\begin{aligned}\mathcal{P}_{\mathcal{V}}X(\sigma, R\tau) &= \bigwedge_x [\,\sigma x, \bigwedge_y [\tau y, X(x,y)]\,] \\ &= \bigwedge_x \bigwedge_y [\,\sigma x, [\tau y, X(x,y)]\,] \\ &= \bigwedge_y \bigwedge_x [\,\tau y, [\sigma x, X(x,y)]\,] \\ &= \bigwedge_y [\,\tau y, \bigwedge_x [\sigma x, X(x,y)]\,] \\ &= \mathcal{P}^{\sharp}_{\mathcal{V}}X(L\sigma, \tau)\,.\end{aligned}$$

By Proposition 30, this confirms the $\mathcal{V}$-functoriality of L and R and their adjunction. Furthermore, with the Yoneda Lemma 62 we obtain for all $x, y \in X$

$$\begin{aligned}L(\mathbf{y}_X x)y &= [\mathbf{y}_X x, \mathbf{y}_X y] = X(x,y) = (\mathbf{y}^{\sharp}_X x)y, \\ R(\mathbf{y}^{\sharp}_X y)x &= [\mathbf{y}^{\sharp}_X y, \mathbf{y}^{\sharp}_X x] = X(x,y) = (\mathbf{y}_X y)x,\end{aligned}$$

so that $L\mathbf{y}_X = \mathbf{y}^{\sharp}_X$, $R\mathbf{y}^{\sharp}_X = \mathbf{y}_X$ follows. □

Example 67 Expanding on the notation used in Examples 65(1), for $\mathcal{V} = \mathsf{2}$ and an ordered set X, the Yoneda maps take the form:

$$\downarrow\colon X \longrightarrow \mathcal{D}X \quad \text{and} \quad \uparrow\colon X \longrightarrow \mathcal{U}X\,,$$

and the Isbell Duality gives the adjoint monotone maps $L \dashv R : \mathcal{U}X \to \mathcal{D}X$; here R assigns to an up-closed set U the set $\bigcap_{y \in U} \downarrow y$ of its lower bounds in X, and L sends a down-closed set D to the set $\bigcap_{x \in D} \uparrow x$ of its upper bounds in X.

5.3 *The Presheaf Monad and Its Restrictions to* $\mathcal{V}$-$\mathsf{Cat}_{\mathrm{sep}}$ *and* Set

Recall that every adjunction $F \dashv G : \mathcal{D} \to \mathcal{C}$ with unit η and counit ε induces the monad $(T = GF,\ \mu = G\varepsilon F,\ \eta)$ on $\mathcal{C}$. Consequently, the presheaf and copresheaf adjunctions of Theorem 63 give monads on $\mathcal{V}$-Cat. Induced by 2-functors, these two monads are actually 2-monads. For the record, we describe them in detail in the following two corollaries.

Corollary 68 *The adjunction* $(-)^* \dashv \mathcal{P}_{\mathcal{V}} : \mathcal{V}\text{-}\mathsf{Dist}^{\mathrm{op}} \to \mathcal{V}\text{-}\mathsf{Cat}$ *induces the* $\mathcal{V}$-*presheaf monad* $\mathcal{P} = (\mathcal{P}, \mathbf{s}, \mathbf{y})$ *on* $\mathcal{V}$-Cat. *Its endofunctor maps objects as* $\mathcal{P}_{\mathcal{V}}$ *does, and one has*

$$\mathcal{P}(X \xrightarrow{f} Y)(\sigma) = \sigma \cdot f^* \quad \textit{and} \quad \mathbf{s}_X(\Sigma) = \Sigma \cdot (\mathbf{y}_X)_* \,,$$

for all $\sigma \in \mathcal{P}_{\mathcal{V}}X$ *and* $\Sigma \in \mathcal{P}_{\mathcal{V}}\mathcal{P}_{\mathcal{V}}X$; *this means that, for all* $y \in Y$ *and* $x \in X$, *one has*

$$(\mathcal{P}f)(\sigma)(y) = \bigvee_{x \in X} Y(y, fx) \otimes \sigma x \quad \textit{and} \quad \mathbf{s}_X(\Sigma)(x) = \bigvee_{\sigma \in \mathcal{P}_{\mathcal{V}}X} \Sigma\sigma \otimes \sigma x.$$

Remark 69 Let us exhibit the action of the 2-functor $\mathcal{P}$ on 2-cells (given by order) explicitly: for $f, g : X \to Y$ in $\mathcal{V}$-Cat one has

$$f \le g \Longrightarrow f^* \le g^* \Longrightarrow \forall \sigma \in \mathcal{P}_{\mathcal{V}}X\ (\sigma \cdot f^* \le \sigma \cdot g^*) \Longrightarrow \mathcal{P}f \le \mathcal{P}g.$$

Moreover, by considering $\sigma = \mathbf{y}_X x$ $(x \in X)$ one sees that in this chain also the reverse implications hold, so that the 2-functor $\mathcal{P}$ is *locally fully faithful*. We will exploit the 2-categorical behavior of $\mathcal{P}$ further in Theorem 71.

The dualization of the above corollary reads as follows:

Corollary 70 *The adjunction* $(-)_* \dashv \mathcal{P}^{\sharp}_{\mathcal{V}} : \mathcal{V}\text{-}\mathsf{Dist}^{\mathrm{co}} \to \mathcal{V}\text{-}\mathsf{Cat}$ *induces the* $\mathcal{V}$-*copresheaf monad* $\mathcal{P}^{\sharp} = (\mathcal{P}^{\sharp}, \mathbf{s}^{\sharp}, \mathbf{y}^{\sharp})$ *on* $\mathcal{V}$-Cat. *Its locally fully faithful endofunctor maps objects as* $\mathcal{P}^{\sharp}_{\mathcal{V}}$ *does, and one has*

$$\mathcal{P}^{\sharp}(X \xrightarrow{f} Y)(\sigma) = f_* \cdot \sigma \quad \textit{and} \quad \mathbf{s}^{\sharp}_X(\Sigma) = (\mathbf{y}^{\sharp}_X)^* \cdot \Sigma \,,$$

for all $\sigma \in \mathcal{P}^{\sharp}_{\mathcal{V}}X$ *and* $\Sigma \in \mathcal{P}^{\sharp}_{\mathcal{V}}\mathcal{P}^{\sharp}_{\mathcal{V}}X$; *this means that, for all* $y \in Y$ *and* $x \in X$,

$$(\mathcal{P}^{\sharp}f)(\sigma)(y) = \bigvee_{x \in X} \sigma x \otimes Y(fx, y) \quad \textit{and} \quad \mathbf{s}^{\sharp}_X(\Sigma)(x) = \bigvee_{\sigma \in \mathcal{P}^{\sharp}_{\mathcal{V}}X} \Sigma\sigma \otimes \sigma x.$$

Recall that a 2-monad (T, μ, η) on a "locally thin" 2-category $\mathcal{C}$ (so that 2-cells are given by order) is *lax idempotent* (or *of Kock-Zöberlein type*; see, for example,

Hofmann and Seal (2014) 2.4.9) if, for all objects X in $\mathcal{C}$, one has $T\eta_X \le \eta_{TX}$. This is a profound property that we will exploit to its fullest in the rest of this chapter.

Theorem 71 *The presheaf 2-monad $\mathcal{P}$ on $\mathcal{V}$-*Cat *is lax idempotent.*

Proof For every $\mathcal{V}$-category X, we must verify the inequality

$$\mathcal{P}_{\mathcal{V}}(\mathbf{y}_X^*) \le \mathbf{y}_{\mathcal{P}_{\mathcal{V}}X} : \mathcal{P}_{\mathcal{V}}X \longrightarrow \mathcal{P}_{\mathcal{V}}\mathcal{P}_{\mathcal{V}}X,$$

which amounts to showing that, for all $\sigma, \tau \in \mathcal{P}_{\mathcal{V}}X$,

$$\mathcal{P}_{\mathcal{V}}(\mathbf{y}_X^*)(\sigma)(\tau) = \bigvee_{z\in X}(\mathcal{P}_{\mathcal{V}}X)(\tau, \mathbf{y}_X z) \otimes \sigma z \le (\mathcal{P}_{\mathcal{V}}X)(\tau, \sigma)$$

holds. Indeed, since $\sigma : X^{\mathrm{op}} \to \mathcal{V}$ is a $\mathcal{V}$-functor, for all $x, z \in X$ one first has

$$[\tau x, X(x, z)] \otimes \tau x \otimes \sigma z \le X(x, z) \otimes \sigma z \le \sigma x,$$

so that $[\tau x, X(x, z)] \otimes \sigma z \le [\tau x, \sigma x]$, and then

$$(\mathcal{P}_{\mathcal{V}}X)(\tau, \mathbf{y}_X z) \otimes \sigma z = \bigwedge_{x\in X}[\tau x, X(x, z)] \otimes \sigma z \le \bigwedge_{x\in X}[\tau x, \sigma x]$$

for all $z \in X$, which implies the desired inequality. □

Corollary 72 *The copresheaf 2-monad $\mathcal{P}^\sharp$ on $\mathcal{V}$-*Cat *is lax idempotent.*

The lax idempotence of a 2-monad (T, μ, η) on a locally thin 2-category $\mathcal{C}$ simplifies the detection of its *T-pseudo-algebras*: any morphism $a : TX \to X$ in $\mathcal{C}$ with $a \cdot \eta_X \simeq 1_X$ satisfies $a \dashv \eta_X$ and is therefore uniquely determined up to order-equivalence, as well as $a \cdot \mu_X \simeq a \cdot Ta$, which makes (X, a) a T-pseudo-algebra (see, for example, Proposition 2.4.9.1 of Hofmann and Seal (2014)). A *T-pseudo-homomorphism* $f : (X, a) \to (Y, b)$ of T-pseudo-algebras must satisfy $fa \simeq b(Tf)$.

Before exploiting these facts in the case of the (co)presheaf monad on $\mathcal{V}$-Cat, let us observe that the endofunctor $\mathcal{P}$ restricts to $\mathcal{V}$-$\mathsf{Cat}_{\mathrm{sep}}$ since presheaf categories are always separated (since $\mathcal{V}$ is). More elaborately, we may compose the adjunction $(-)^* \dashv \mathcal{P}_{\mathcal{V}}$ with the adjunction arising from the reflective embedding $\mathcal{V}$-$\mathsf{Cat}_{\mathrm{sep}} \hookrightarrow$ $\mathcal{V}$-Cat and obtain the restricted monad $\mathcal{P}_{\mathrm{sep}}$ as the monad induced by the composite adjunction. Dually one obtains the restriction $\mathcal{P}^\sharp_{\mathrm{sep}}$ of $\mathcal{P}^\sharp$ to $\mathcal{V}$-$\mathsf{Cat}_{\mathrm{sep}}$. Like $\mathcal{P}$ and $\mathcal{P}^\sharp$, also $\mathcal{P}_{\mathrm{sep}}$ and $\mathcal{P}^\sharp_{\mathrm{sep}}$ are lax idempotent.

Likewise, the *discretization of* $\mathcal{P}$ to Set may be obtained by composition of $(-)^* \dashv \mathcal{P}_{\mathcal{V}}$ with the adjunction $(-)_{\mathrm{d}} \dashv \mathcal{O} : \mathcal{V}$-$\mathsf{Cat} \to \mathsf{Set}$, where the left adjoint of the forgetful $\mathcal{O}$ provides a set X with its discrete $\mathcal{V}$-structure: $X_{\mathrm{d}} = (X, 1_X^\circ)$. Since the unit of the latter adjunction is an identity transformation, we may easily compute the monad on Set induced by the composite adjunction, by evaluating the $\mathcal{V}$-presheaf monad on discrete $\mathcal{V}$-categories, as follows:

Corollary 73 *The discrete $\mathcal{V}$-presheaf monad $\mathcal{P}_{\mathrm{d}} = (\mathcal{P}_{\mathrm{d}}, \mathbf{s}_{\mathrm{d}}, \mathbf{y}_{\mathrm{d}})$ on* Set *with $\mathcal{P}_{\mathrm{d}} = \mathcal{P}(-)_{\mathrm{d}}$ assigns to a set X the set $\mathcal{V}^X \cong \mathcal{V}\text{-}\mathsf{Rel}(X, \mathrm{E}) = \mathcal{V}\text{-}\mathsf{Dist}(X_{\mathrm{d}}, \mathrm{E})$ and, for all $x \in X$, $\sigma \in \mathcal{P}_{\mathrm{d}}X$ and $\Sigma \in \mathcal{P}_{\mathrm{d}}\mathcal{P}_{\mathrm{d}}X$, is described by*

$$\mathcal{P}_{\mathrm{d}}(X \xrightarrow{f} Y)(\sigma) = \sigma \cdot f^{\circ}, \quad (\mathbf{y}_{\mathrm{d}})_X x = \mathbf{y}_{X_{\mathrm{d}}} x = 1_X^{\circ}(\text{-}, x), \quad (\mathbf{s}_{\mathrm{d}})_X(\Sigma) = \mathbf{s}_{X_{\mathrm{d}}}(\Sigma) = \Sigma \cdot (\mathbf{y}_{X_d})_{\circ}.$$

Therefore, for all $y \in Y$ and $x \in X$,

$$(\mathcal{P}_{\mathrm{d}} f)(\sigma)(y) = \bigvee_{x \in f^{-1}y} \sigma x \quad \text{and} \quad (\mathbf{s}_{\mathrm{d}})_X(\Sigma)(x) = \bigvee_{\sigma \in \mathcal{V}^X} \Sigma\sigma \otimes \sigma x.$$

It is easy to see that the analogous discretization of the $\mathcal{V}$-copresheaf monad leads to the same Set-monad $\mathcal{P}$.

Example 74 For $\mathcal{V} = \mathsf{2}$ (the Boolean quantale), $\mathcal{P}_{\mathrm{d}}$ is the power-set monad on Set. Its Kleisli category is the category Rel of sets and relations, and its Eilenberg–Moore category of (strict) $\mathcal{P}_{\mathrm{d}}$-algebras is (isomorphic to) the category Sup of complete lattices with suprema-preserving maps. The presheaf monad $\mathcal{P}$ is the down-set monad $\mathcal{D}$ on $\mathsf{Ord} = \mathcal{V}\text{-}\mathsf{Cat}$ (see Example 65(1)), which restricts to its full reflective subcategory $\mathsf{Pos} = \mathcal{V}\text{-}\mathsf{Cat}_{\mathrm{sep}}$ of partially ordered sets. As the Eilenberg–Moore category over Pos, once again we obtain Sup; for additional explanation, see Example 80. Other than being monadic over Set and Pos, let us also note that Sup is a self-dual and symmetric monoidal-closed category; see Joyal and Tierney (1984) and Sect. 6.10.

One of our goals is to generalize these well-known facts, trading $\mathsf{2}$ for an arbitrary commutative quantale $\mathcal{V}$. To this end, we should first give a $\mathcal{V}$-categorical generalization of the category Sup, replacing the existence requirement of arbitrary suprema (and infima) for its objects by that of still-to-be-defined $\mathcal{V}$-categorial colimits (and limits). It turns out that Theorem 71 goes already a long way toward this goal, as we will show in the following sections.

5.4 The Kleisli and Eilenberg–Moore Categories of the Presheaf Monad

We recall that the Kleisli category of a monad (T, μ, η) on a category $\mathcal{C}$ has the same objects as $\mathcal{C}$, but a Kleisli morphism $X \to Y$ is given by a morphism $f : X \to TY$ in $\mathcal{C}$; it gets Kleisli-composed with $g : Y \to TZ$ via $g \circ f := \mu_Z(Tg)f$. In the case of the presheaf monad $\mathcal{P}$ on $\mathcal{V}$-Cat, a Kleisli morphism $X \to Y$ is a $\mathcal{V}$-functor $\varphi : X \to$

$\mathcal{P}Y$ which, by Theorem 32, corresponds bijectively and naturally to a $\mathcal{V}$-distributor $\varphi^\flat : Y \nrightarrow X$ (in the notation of the proof of Theorem 63). Once we have confirmed that the Kleisli composition translates to the composition of $\mathcal{V}$-distributors under this correspondence, we will have shown the important part of the following proposition.

Proposition 75 *The self-dual 2-category* $\mathcal{V}$-Dist *is isomorphic to the (dual of the) Kleisli category of the* $\mathcal{V}$*-presheaf monad on* $\mathcal{V}$-Cat, *and* $\mathcal{V}$-Rel *is isomorphic to the Kleisli category of the discrete presheaf monad on* Set.

Proof We confirm that the Kleisli composite of the $\mathcal{V}$-functors $\varphi : X \to \mathcal{P}Y$ and $\psi : Y \to \mathcal{P}Z$ coincides with the composite $Z \nrightarrow Y \nrightarrow X$ of their corresponding $\mathcal{V}$-distributors. Indeed, for all $x \in X$ and $z \in Z$, with Remark 44 and the Yoneda Lemma 62 one has

$$\begin{aligned}
\mathbf{s}_Z(\mathcal{P}\psi(\varphi x))(z) &= (\mathcal{P}\psi(\varphi x) \cdot (\mathbf{y}_Z)_*)(z) \\
&= (\varphi x \cdot \psi^* \cdot (\mathbf{y}_Z)_*)(z) \\
&= \bigvee_{y \in Y} (\varphi x)(y) \otimes (\psi^* \cdot (\mathbf{y}_Z)_*)(z, y) \\
&= \bigvee_{y \in Y} (\varphi x)(y) \otimes \mathcal{P}_{\mathcal{V}} Z(\mathbf{y}_Z z, \psi y) \\
&= \bigvee_{y \in Y} (\varphi x)(y) \otimes (\psi y)(z) \\
&= \bigvee_{y \in Y} \psi^\flat(z, y) \otimes \varphi^\flat(y, x).
\end{aligned}$$

For the self-duality of $\mathcal{V}$-Dist, see Corollary 43. The remaining (easier) verifications are straightforward. □

Next we describe the Eilenberg–Moore category of the (co)presheaf monad on $\mathcal{V}$-Cat. As mentioned after Corollary 72, the lax idempotence of $\mathcal{P}$ makes it easy to describe at least the $\mathcal{P}$-pseudo-algebras. As any $\mathcal{P}$-pseudo-algebra structure on $X \in \mathcal{V}$-Cat is unique up to order-equivalence, its existence becomes a property of X which deserves a special name and notation, the choice of which will be justified only later on.

Definition 76 Let X and Y be $\mathcal{V}$-categories and $f : X \to Y$ a $\mathcal{V}$-functor.

(1) We call X *$\mathcal{P}$-cocomplete* if X carries a $\mathcal{P}$-pseudo-algebra structure; that is, if there is a $\mathcal{V}$-functor $c : \mathcal{P}_{\mathcal{V}} X \longrightarrow X$ with $c\,\mathbf{y}_X \simeq 1_X$; for every $\omega \in \mathcal{P}_{\mathcal{V}} X$, we then write

$$\mathrm{colim}^{\omega} 1_X \simeq c(\omega).$$

(2) For X, Y $\mathcal{P}$-cocomplete, f is $\mathcal{P}$-*cocontinuous* if f is a $\mathcal{P}$-pseudo-homomorphism; that is, if, for all $\omega \in \mathcal{P}_{\mathcal{V}}X$,

$$f(\mathrm{colim}^{\omega} 1_X) \simeq \mathrm{colim}^{\omega \cdot f^*} 1_Y.$$

(3) The $\mathcal{V}$-category X is $\mathcal{P}^{\sharp}$-*complete* if X carries a $\mathcal{P}^{\sharp}$-pseudo-algebra structure; in that case, with $d : \mathcal{P}^{\sharp}_{\mathcal{V}}X \longrightarrow X$ satisfying $d\,\mathbf{y}^{\sharp}_X \simeq 1_X$ and $\upsilon \in \mathcal{P}^{\sharp}_{\mathcal{V}}X$, we write $\lim^{\upsilon} 1_X \simeq d(\upsilon)$.

(4) For X, Y $\mathcal{P}^{\sharp}$-complete, f is $\mathcal{P}^{\sharp}$-*continuous* if f is a $\mathcal{P}^{\sharp}$-pseudo-homomorphism; that is, $f(\lim^{\upsilon} 1_X) \simeq \lim^{f_* \cdot \upsilon} 1_Y$, for all $\upsilon \in \mathcal{P}^{\sharp}_{\mathcal{V}}X$.

Remark 77 A $\mathcal{V}$-category X is $\mathcal{P}^{\sharp}$-complete if, and only if, X^{op} is $\mathcal{P}$-cocomplete. Likewise, a $\mathcal{V}$-functor $f : X \to Y$ is $\mathcal{P}^{\sharp}$-continuous if, and only if, $f^{\mathrm{op}} : X^{\mathrm{op}} \to Y^{\mathrm{op}}$ is $\mathcal{P}$-cocontinuous.

The preceding observations (starting with those made after Corollary 72), combined with the fact that order-equivalence in separated $\mathcal{V}$-categories amounts to strict equality, give immediately:

Theorem 78 *The category of $\mathcal{P}$-pseudo-algebras and $\mathcal{P}$-pseudo-homomorphisms is equivalent to the category of $\mathcal{P}$-cocomplete $\mathcal{V}$-categories and their $\mathcal{P}$-cocontinuous $\mathcal{V}$-functors. Its full subcategory of separated objects is equivalent to the category of (strict) $\mathcal{P}_{\mathrm{sep}}$-algebras, for the restriction $\mathcal{P}_{\mathrm{sep}}$ of $\mathcal{P}$ to $\mathcal{V}$-$\mathsf{Cat}_{\mathrm{sep}}$.*

Analogously for the categories of $\mathcal{P}^{\sharp}$-(pseudo-)algebras and of $\mathcal{P}^{\sharp}$-complete $\mathcal{V}$-categories with their respective morphisms.

We note that, without restriction to separated $\mathcal{V}$-categories, the strict $\mathcal{P}$-algebras of $\mathcal{V}$-Cat and their homomorphisms are to be thought of as those $\mathcal{V}$-categories that come with a "choice of $\mathcal{P}$-colimits," with homomorphisms preserving that choice.

Here is a first criterion for $\mathcal{P}$-cocompleteness and its dual property. Recall that we may treat elements x of a $\mathcal{V}$-category X as $\mathcal{V}$-functors $x : \mathrm{E} = \mathrm{E}^{\mathrm{op}} \to X$.

Proposition 79 *Let X be a $\mathcal{V}$-category.*

(1) *X is $\mathcal{P}$-cocomplete if, and only if, for every $\omega \in \mathcal{P}_{\mathcal{V}}X$ there is some chosen element $c(\omega) \in X$ with $c(\omega)_* \simeq [\omega, 1^*_X]$; that is, for all $x \in X$,*

$$X(c(\omega), x) = \bigwedge_{z \in X} [\omega z, X(z, x)].$$

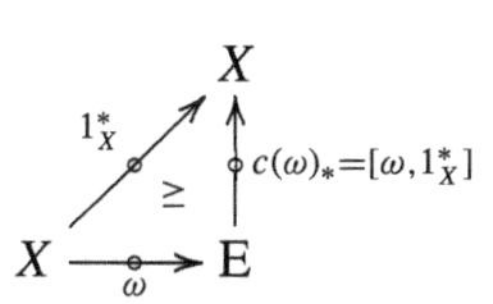

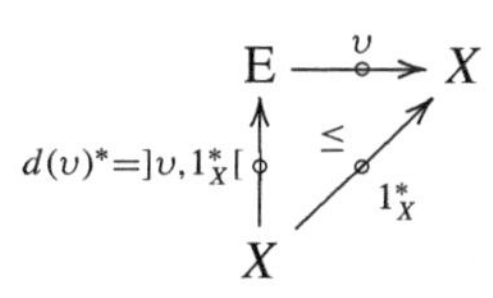

(2) *X is $\mathcal{P}^\sharp$-complete if, and only if, for every $\upsilon \in \mathcal{P}^\sharp_\mathcal{V} X$ there is some chosen element $d(\upsilon) \in X$ with $d(\upsilon)^* \simeq]\upsilon, 1^*_X[$; that is, for all $x \in X$,*

$$X(x, d(\upsilon)) = \bigwedge_{z \in X} [\upsilon z, X(x, z)].$$

Proof We show (1); then (2) follows dually. First assuming $\mathcal{P}$-cocompleteness of X, we have a $\mathcal{V}$-functor $c : \mathcal{P}_\mathcal{V} X \to X$ with $c\,\mathbf{y}_X \simeq 1_X$ and must show that, for every $\omega \in \mathcal{P}_\mathcal{V} X$, $c(\omega)_* : \mathrm{E} \to\!\!\!\!\!\circ\;\, X$ is an extension of 1^*_X along ω. Indeed, by the lax idempotence of the presheaf monad we have $c \dashv \mathbf{y}_X$ which, by Proposition 45, gives $c_* = \mathbf{y}^*_X$. Hence, $c_*(\omega, x) = \mathbf{y}^*_X(\omega, x)$ for all $\omega \in \mathcal{P}_\mathcal{V} X$ and $x \in X$, which transcribes to $X(c(\omega), x) = \mathcal{P}_\mathcal{V} X(\omega, \mathbf{y}_X x)$ and then, by Proposition 12, to

$$c(\omega)_*(x) = \bigwedge_{z \in X} [\omega z, X(z, x)] = [\omega, 1^*_X](x).$$

Conversely, assuming that for every $\omega \in \mathcal{P}_\mathcal{V} X$ we have a chosen element $c(\omega) \in X$ with $c(\omega)_* = [\omega, 1^*_X]$, one may reverse the above calculation and, noting that c must automatically be a $\mathcal{V}$-functor by Proposition 30, obtain $c_* = \mathbf{y}^*_X$. For the fully faithful Yoneda $\mathcal{V}$-functor, this gives $(c\,\mathbf{y}_X)_* = c_* \cdot (\mathbf{y}_X)_* = \mathbf{y}^*_X \cdot (\mathbf{y}_X)_* = 1^*_X$, or $c\,\mathbf{y}_X \simeq 1_X$. □

Example 80 A presheaf ω on $X \in \mathsf{Ord} = \mathsf{2}\text{-}\mathsf{Cat}$ may be considered as a down-closed subset of X. Having an element $c(\omega) \in X$ with $c(\omega) \simeq \mathrm{colim}^\omega 1_X$ translates to $X(c(\omega), x) = \bigwedge_{z \in X} [\omega(z), X(z, x)]$, or $(c(\omega) \le x \iff \omega \subseteq \downarrow x)$, for all $x \in X$. But this means precisely $c(\omega) \simeq \bigvee \omega$ in the ordered set X. Since a supremum of any subset of X may be presented as a supremum of the down-closure of that subset, $\mathcal{P}$-cocompleteness of X amounts precisely to the existence of all suprema in X, i.e., to the order-theoretic completeness of X.

Dually, $\mathcal{P}^\sharp$-completeness of X amounts to the existence of all infima in X, which again means order-theoretic completeness of X.

5.5 *$\mathcal{P}$-Cocompleteness of Presheaf Categories and the Role of $\mathcal{V}$ in $\mathcal{V}$-Cat*

It is high time for us to point to the most important general examples of $\mathcal{P}$-cocomplete $\mathcal{V}$-categories:

Proposition 81 *For every $\mathcal{V}$-category X, the $\mathcal{V}$-category $\mathcal{P}_\mathcal{V} X$ is $\mathcal{P}$-cocomplete, and $\mathcal{P}^\sharp_\mathcal{V} X$ is $\mathcal{P}^\sharp$-complete. Moreover, when X is $\mathcal{P}$-cocomplete, then* $\mathrm{colim}^{(\text{-})} 1_X : \mathcal{P}_\mathcal{V} X \to X$ *is $\mathcal{P}$-cocontinuous, and when X is $\mathcal{P}^\sharp$-complete, then* $\lim^{(\text{-})} 1_X : X \to \mathcal{P}^\sharp_\mathcal{V} X$ *is $\mathcal{P}^\sharp$-continuous.*

Proof For any monad (T, μ, η) on a category $\mathcal{C}$ and any object $X \in \mathcal{C}$, we have the (free) T-algebra (TX, μ_X), and when c is a T-algebra structure on X, then $c : (TX, \mu_X) \to (X, c)$ is a T-homomorphism. So, in view of Theorem 78, the claims of this proposition just reflect the cases $T = \mathcal{P}$ and $T = \mathcal{P}^\sharp$, with $\mathcal{C} = \mathcal{V}\text{-}\mathsf{Cat}$. □

The mounting evidence for the importance of (co)presheaf categories is rooted in the importance of the quantale $\mathcal{V}$ as a special object of $\mathcal{V}$-Cat, structured by its internal hom, here denoted by [-, -]. Recall that, as such, $\mathcal{V}$ is separated, but generally not symmetric. In Proposition 35, we showed that its tensor product makes $\mathcal{V}$ a commutative monoid in the symmetric monoidal category $\mathcal{V}$-Cat. Proposition 81 lets us add another important property:

Corollary 82 *The $\mathcal{V}$-category $\mathcal{V}$ is $\mathcal{P}$-cocomplete, and $\mathcal{V}^{\mathrm{op}}$ is $\mathcal{P}^\sharp$-complete. For all $\omega \in \mathcal{P}_\mathcal{V}\mathcal{V}$ and $\upsilon \in \mathcal{P}^\sharp_\mathcal{V}\mathcal{V}$ one has the formulae (to be read in $\mathcal{V}$)*

$$\mathrm{colim}^{\omega} 1_\mathcal{V} = \bigvee_{u \in \mathcal{V}} (\omega u) \otimes u \quad \textit{and} \quad \lim{}^{\upsilon} 1_{\mathcal{V}^{\mathrm{op}}} = \bigwedge_{u \in \mathcal{V}} (\upsilon u) \otimes u.$$

Proof With the identifications $\mathcal{V} = \mathcal{P}_\mathcal{V}\mathrm{E}$ and $\mathcal{V}^{\mathrm{op}} = \mathcal{P}^\sharp_\mathcal{V}\mathrm{E}$, one just applies Theorem 78 to $X = \mathrm{E}$. (Note that $\mathcal{V}^{\mathrm{op}}$ denotes the $\mathcal{V}$-category $(\mathcal{V}, [-,-]^\circ)$.) The given formulae follow from the fact that the $\mathcal{P}$- and $\mathcal{P}^\sharp$-algebra structures on $\mathcal{P}_\mathcal{V}\mathrm{E}$ and $\mathcal{P}^\sharp_\mathcal{V}\mathrm{E}$ are, respectively, given by the monad multiplications $\mathbf{s}_\mathrm{E}$ and $\mathbf{s}^\sharp_\mathrm{E}$, as described in Corollaries 68 and 70. (They are also easily obtained by direct calculation: for all $z \in \mathcal{V}$ one has

$$[\bigvee_{u \in \mathcal{V}} \omega u \otimes u,\ z] = \bigwedge_{u \in \mathcal{V}} [\omega u \otimes u, z] = \bigwedge_{u \in \mathcal{V}} [\omega u, [u, z]],$$

which implies the colimit formula, and similarly for the limit formula.) □

Next we show another important property of the object $\mathcal{V} \in \mathcal{V}$-Cat: it is *injective* with respect to fully faithful $\mathcal{V}$-functors. In fact, we prove:

Proposition 83 *Let $f : X \to Y$ and $\phi : X \to \mathcal{V}$ be $\mathcal{V}$-functors. If f is fully faithful, then there are a least and a largest $\mathcal{V}$-functor $\psi : Y \to \mathcal{V}$ with $\psi f = \phi$.*

Proof For any $\mathcal{V}$-functors f, ϕ, ψ, the following diagram on the left commutes if, and only if, the corresponding diagram of $\mathcal{V}$-distributors on the right commutes:

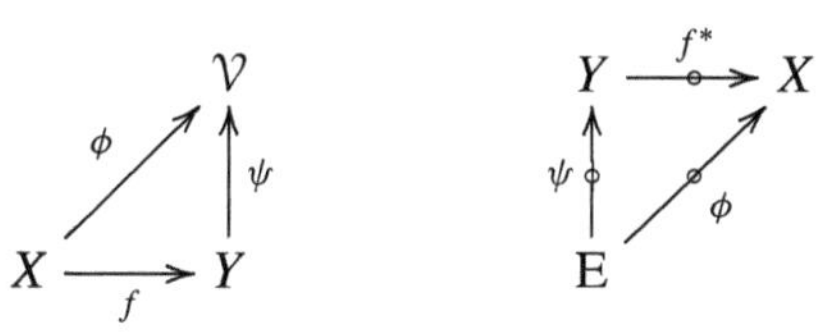

By adjunction, any f, ϕ, ψ with $f^* \cdot \psi = \phi$ give $\underline{\phi} := f_* \cdot \phi \leq \psi$, and $\psi \leq\,]f^*, \phi[=: \overline{\phi}$, and when f is fully faithful, one obtains $\underline{\phi}\, f = \phi$ and $\overline{\phi}\, f = \phi$. Indeed, the first of these two identities follows immediately from $f^* \cdot f_* \cdot \phi = 1_X^* \cdot \phi = \phi$, and for the second identity we invoke the first to obtain

$$]f^*, \phi[= \bigvee \{\psi \in \mathcal{P}^\sharp_{\mathcal{V}} Y \mid f^* \cdot \psi \leq \phi\} \geq f_* \cdot \phi,$$

which implies $f^* \cdot\,]f^*, \phi[\geq \phi$, with "$\leq$" holding trivially. □

Examples 84 (1) For $\mathcal{V} = 2$, the $\mathcal{V}$-copresheaves ϕ and ψ may, respectively, be treated as upward-closed subsets of the ordered sets X and Y, and the commutation with f then means $f^{-1}(\psi) = \phi$. If f is order-preserving and—reflecting (i.e., fully faithful), for every up-closed set ϕ in X we have a least and a largest up-closed set ψ in Y with $f^{-1}(\psi) = \phi$, namely, $\underline{\phi} = \uparrow f(\phi)$ and $\overline{\phi} = \bigcup \{U \in \mathcal{U}Y \mid f^{-1}(U) \subseteq \phi\}$ (in the notation of Example 65(1)).

(2) Let $f : X \to Y$ be an isometric embedding of generalized metric spaces, and $\phi : X \to [0, \infty]$ a non-expansive function. Then, as a morphism in $\mathsf{Met} = [0, \infty]_+\text{-}\mathsf{Cat}$, ϕ has (with respect to the natural order of the reals) a largest extension $\underline{\phi}$ to Y, and a smallest one, $\overline{\phi}$, given by the formulae

$$\underline{\phi}(y) = \inf_{x \in X} (\phi(x) + Y(f(x), y)) \quad \text{and} \quad \overline{\phi}(y) = \sup_{x \in X} (\phi(x) \ominus Y(y, f(x)))$$

for all $y \in Y$; here the notation $\beta \ominus \alpha = \max\{\beta - \alpha, 0\}$ is used for $\hom_{[0,\infty]_+}(\alpha, \beta)$.

(3) As noted in Lawvere (1973), statement (2) may actually be formulated more generally, as follows: any *Lipschitz function* $\phi : X \to [0, \infty]$ with Lipschitz constant λ $(0 \leq \lambda < \infty)$ has a largest and a smallest extension to Y, these being Lipschitz functions with the same Lipschitz constant λ. Indeed, multiplication by λ defines a (lax) homomorphism $\lambda : [0, \infty]_+ \to [0, \infty]_+$ of quantales. By Proposition 19, this gives us the change-of-base functor $\mathcal{B}_\lambda : \mathsf{Met} \to \mathsf{Met}$, providing a generalized metric space X with a new metric, whereby all old distances get multiplied by λ. Lipschitz functions $\phi : X \to [0, \infty]$ are then, by definition, precisely morphisms $\phi : \mathcal{B}_\lambda X \to [0, \infty]_+$ in Met, for which one finds the extensions $\underline{\phi}$, $\overline{\phi} : \mathcal{B}_\lambda Y \to [0, \infty]$ along the (still fully faithful) embedding $\mathcal{B}_\lambda f : \mathcal{B}_\lambda X \to \mathcal{B}_\lambda Y$, which maps points just like f does. The above formulae for $\underline{\phi}$ and $\overline{\phi}$ remain valid, except that the factor λ has to be inserted in front of Y.

Recall that an object A is a *regular cogenerator* in a category $\mathcal{C}$ with products if, for every object $X \in \mathcal{C}$, the canonical morphism $X \to A^{\mathcal{C}(X,A)}$ is a regular monomorphism. If every extremal monomorphism in $\mathcal{C}$ is regular, for A to be a regular cogenerator of $\mathcal{C}$, it then suffices that there be an extremal monomorphism $X \to A^I$ into *some* power of A. In $\mathcal{V}$-Cat, like in every topological category over Set, extremal monomorphisms are regular, and these are precisely the fully faithful injective $\mathcal{V}$-functors, i.e., the full embeddings. Therefore the following statement becomes evident:

Proposition 85 *The $\mathcal{V}$-category $\mathcal{V}$ is a regular cogenerator in $\mathcal{V}$-$\mathsf{Cat}_{\mathrm{sep}}$. As a consequence, every separated $\mathcal{V}$-category embeds fully into an injective object of $\mathcal{V}$-$\mathsf{Cat}_{\mathrm{sep}}$.*

Proof For every $\mathcal{V}$-category X, the $\mathcal{V}$-functor $\mathbf{y}_X : X \to \mathcal{P}_{\mathcal{V}}X$ is fully faithful, and it is injective if X is separated. There is also a trivial full embedding $\mathcal{P}_{\mathcal{V}}X \to \mathcal{V}^X$ which regards a presheaf on X as an X-indexed family of elements in $\mathcal{V}$. When we precompose this map with $\mathbf{y}_X$, we obtain a full embedding of X into $\mathcal{V}^X$. As a power of $\mathcal{V}$, also $\mathcal{V}^X$ is injective with respect to fully faithful $\mathcal{V}$-functors. □

Example 86 The separated metric space $[0, \infty]$ is a regular cogenerator in $\mathsf{Met}_{\mathrm{sep}}$. With its Euclidean metric (that is, with its coreflective symmetrization), $[0, \infty]$ becomes a regular cogenerator in $\mathsf{Met}_{\mathrm{sym,sep}}$.

Remarks 87 (1) For the sake of completeness, we should mention that the object E is a *generator* in $\mathcal{V}$-Cat, but not *strong* (or, equivalently, not regular); that is, the canonical morphism $X \cdot \mathrm{E} = \coprod_{x\in X} \mathrm{E} \to X$ is surjective for every $X \in \mathcal{V}$-Cat, but in general not an extremal (or regular) epimorphism. However the $\mathcal{V}$-indexed set of objects $G_v = \{0, 1\}$ with $G_v(0, 1) = v \in \mathcal{V}$ forms collectively a strong (and, hence, regular) generator in $\mathcal{V}$-Cat, i.e., the canonical morphism $\coprod_{v\in\mathcal{V}}[G_v, X] \cdot G_v \longrightarrow X$ is an extremal (and, hence, regular) epimorphism, for all $X \in \mathcal{V}$-Cat.

(2) The strong generator $\{G_v \mid v \in \mathcal{V}\}$ in $\mathcal{V}$-Cat plays a key role in establishing another important property that $\mathcal{V}$-Cat inherits from $\mathcal{V}$. First recall that, for an infinite regular cardinal κ, an object A in a category $\mathcal{C}$ is *κ-presentable* if $\mathcal{C}(A, -) : \mathcal{C} \to \mathsf{Set}$ preserves κ-directed colimits. (An ordered set I is κ-directed if every subset of cardinality $< \kappa$ has an upper bound in I.) The category $\mathcal{C}$ is *locally κ-presentable* if $\mathcal{C}$ is cocomplete and has a strong generator formed by a small family of κ-presentable objects Gabriel and Ulmer (1971), Adámek and Rosický (1994).
Here is a path to proving an important result of Kelly and Lack (2001) in the context of our quantale $\mathcal{V}$ (rather than that of an arbitrary symmetric monoidal-closed category $\mathcal{V}$):

- The colimit cocone $(f_i : X_i \to X)_{i\in I}$ of a κ-directed system $(X_i)_{i\in I}$ in $\mathcal{V}$-Cat may be formed by providing the colimit of its shadow in Set with the $\mathcal{V}$-Cat-structure defined by $X(x, y) = \bigvee\{X_i(x', y') \mid i \in I,\ x' \in f_i^{-1}x,\ y' \in f_i^{-1}y\}$ for all $x, y \in X$.
- If $v \in \mathcal{V}$, considered as an object of the thin category $(\mathcal{V}, \leq)$, is κ-presentable, then G_v is a κ-presentable object in $\mathcal{V}$-Cat.
- If the (small) thin category $(\mathcal{V}, \leq)$ is locally κ-presentable, then the category $\mathcal{V}$-Cat is also locally κ-presentable; moreover, if the κ-presentability of elements in $\mathcal{V}$ is stable under tensor products in $\mathcal{V}$, then the κ-presentability of $\mathcal{V}$-categories is also stable under tensor products in $\mathcal{V}$-Cat.

(3) Taking advantage of the countable dense set of rationals in $[0, \infty]_+$, with (2) one concludes that the category Met is locally $\aleph_1$-presentable.

5.6 $\mathcal{P}$-Cocompleteness as an Injectivity Property

In Proposition 83, we showed that the $\mathcal{V}$-category $\mathcal{V}$ is an injective object in $\mathcal{V}$-Cat w.r.t. to fully faithful $\mathcal{V}$-functors. We will show now that this property holds more generally for every separated $\mathcal{P}$-cocomplete $\mathcal{V}$-category X. In fact, separation is not essential, as long as we trade strict commutativity for commutation up to order-equivalence, as follows:

Definition 88 We call a $\mathcal{V}$-category X *pseudo-injective* if for every fully faithful $\mathcal{V}$-functor $f : Y \to Z$ and every $\mathcal{V}$-functor $g : Y \to X$, there is a $\mathcal{V}$-functor $h : Z \to X$ with $hf \simeq g$. If the $\simeq$-sign may be replaced by equality, then X is called *injective*.

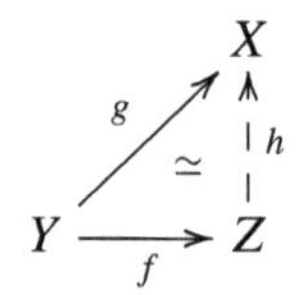

Theorem 89 *A $\mathcal{V}$-category is $\mathcal{P}$-cocomplete if, and only if, it is pseudo-injective.*

Proof If X is pseudo-injective, there is a $\mathcal{V}$-functor $c : \mathcal{P}_{\mathcal{V}}X \to X$ with $c\,\mathbf{y}_X \simeq 1_X$, because $\mathbf{y}_X : X \to \mathcal{P}_{\mathcal{V}}X$ is fully faithful (Theorem 63). This makes X $\mathcal{P}$-cocomplete. Conversely, for X cocomplete and the given $\mathcal{V}$-functors f and g, where f is fully faithful, writing c for $\mathrm{colim}^{(\text{-})}1_X$ and considering the diagram

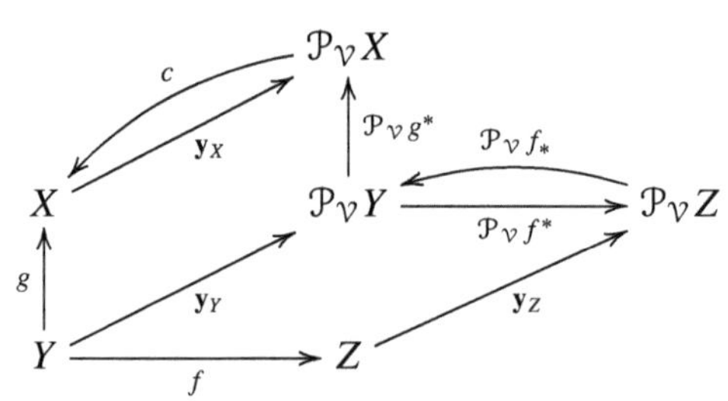

we put $h := c\,(\mathcal{P}_{\mathcal{V}}g^*)(\mathcal{P}_{\mathcal{V}}f_*)\mathbf{y}_Z$. Then, since $\mathbf{y}$ is natural and $\mathcal{P}_{\mathcal{V}}$ functorial, and since f is fully faithful and X $\mathcal{P}$-cocomplete, we successively obtain:

$$\begin{aligned} hf &= c\,(\mathcal{P}_{\mathcal{V}}g^*)(\mathcal{P}_{\mathcal{V}}f_*)(\mathcal{P}_{\mathcal{V}}f^*)\mathbf{y}_Y \\ &= c\,(\mathcal{P}_{\mathcal{V}}g^*)\mathcal{P}_{\mathcal{V}}(f^*\cdot f_*)\mathbf{y}_Y \\ &= c\,(\mathcal{P}_{\mathcal{V}}g^*)\mathbf{y}_Y \\ &= c\,\mathbf{y}_X g \simeq g\,. \end{aligned}$$

□

Since $\mathbf{y}_X$ is fully faithful (Theorem 63) and $\mathcal{P}_{\mathcal{V}}X$ is $\mathcal{P}$-cocomplete (Proposition 81), we conclude:

Corollary 90 *Every separated $\mathcal{V}$-category embeds fully into a separated injective $\mathcal{V}$-category; that is, $\mathcal{V}$-$\mathsf{Cat}_{\mathrm{sep}}$ has enough injectives.*

Remark 91 As shown by Martinelli (2021), $\mathcal{V}$-$\mathsf{Cat}_{\mathrm{sep}}$ has even *injective hulls*. Indeed, using the Isbell Duality, one proves that every separated $\mathcal{V}$-category X admits an *essential* full embedding into a separated injective $\mathcal{V}$-category which, in case $\mathcal{V}=\mathsf{2}$, gives the *MacNeille completion* of X.

The dualization of Theorem 89 reads as follows:

Corollary 92 *A $\mathcal{V}$-category X is $\mathcal{P}^\sharp$-complete if, and only if, X^{op} is pseudo-injective.*

It is easy to show that pseudo-injectivity is closed under arbitrary products and under (pseudo-)retracts in $\mathcal{V}$-Cat. We therefore conclude:

Corollary 93 *$\mathcal{P}$-cocompleteness and $\mathcal{P}^\sharp$-completeness are closed under arbitrary products and under retracts in $\mathcal{V}$-Cat.*

6 Weighted Colimits, Tensors, and the Cocompletion of $\mathcal{V}$-Categories

6.1 Weighted Colimits and Limits

It is time for us to justify the notation $\mathrm{colim}^{\omega}1_X$ used when introducing the notion of $\mathcal{P}$-cocompleteness of a $\mathcal{V}$-category X. We saw in Proposition 79 that the element $c(\omega)=\mathrm{colim}^{\omega}1_X$ in X *represents* the extension $[\omega,1_X^*]:\mathsf{E}\mathrel{\circ\!\!\to}X$ of 1_X^* along the presheaf $\omega:X\mathrel{\circ\!\!\to}\mathsf{E}$, in the sense that $c(\omega)_*=[\omega,1_X^*]$. This begs the question of what happens when 1_X is replaced by any $\mathcal{V}$-functor $h:Z\to X$ (i.e., any *diagram* in X), and when ω is allowed to be any $\mathcal{V}$-distributor $Z\mathrel{\circ\!\!\to}W$ (regarded as the *weight* of the diagram). In particular, when X is $\mathcal{P}$-cocomplete, will we still have a representation of the extension, so that $[\omega,h_*]=q_*$, for some $\mathcal{V}$-functor $q:W\to X$?

Before answering this question, we extend the colimit notation accordingly. Special scenarios motivating the generalized setting will be considered later on.

Definition 94 For $\mathcal{V}$-categories X, Z, W, let $h : Z \to X$ be a $\mathcal{V}$-functor and $\omega : Z \to\!\!\!\!\circ\; W$, $\upsilon : W \to\!\!\!\!\circ\; Z$ be $\mathcal{V}$-distributors.

(1) A *colimit of h weighted by ω* is a $\mathcal{V}$-functor $q : W \to X$ such that $q_* = [\omega, h_*]$; that is, q satisfies the formula $q_* = \bigvee\{\sigma : W \to\!\!\!\!\circ\; X \mid \sigma \cdot \omega \le h_*\}$ or, equivalently,

$$X(qt, x) = \bigwedge_{z \in Z} [\omega(z, t), X(hz, x)]$$

for all $x \in X, t \in W$; one writes $q \simeq \mathrm{colim}^{\omega} h$ in this case.

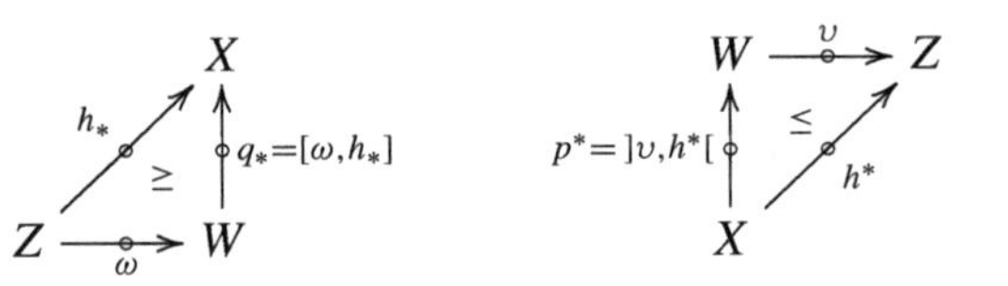

(2) A *limit of h weighted by υ* is a $\mathcal{V}$-functor $p : W \to X$ such that $p^* =]\upsilon, h^*[$; that is, p satisfies the formula $p^* = \bigvee\{\rho : X \to\!\!\!\!\circ\; W \mid \upsilon \cdot \rho \le h^*\}$ or, equivalently,

$$X(x, pt) = \bigwedge_{z \in Z} [\upsilon(t, z), X(x, hz)]$$

for all $x \in X, t \in W$; one writes $p \simeq \lim^{\upsilon} h$ in this case.

(3) The $\mathcal{V}$-category X is *cocomplete* if, for all h, ω as above, there is a chosen colimit of h weighted by ω; likewise, X is *complete* if, for all h, υ as above, there is a chosen limit of h weighted by υ.

Let us note a couple of expected facts, the first of which will be strengthened in Theorem 97.

Corollary 95 *Let X be a $\mathcal{V}$-category.*

(1) *If X is cocomplete, then X is also $\mathcal{P}$-cocomplete, and if X is complete, then X is also $\mathcal{P}^{\sharp}$-complete.*

(2) *X is complete if, and only if, X^{op} is cocomplete; one has the formula:*

$$(\mathrm{colim}^{\omega} h)^{\mathrm{op}} \simeq \lim^{\omega^{\circ}} (h^{\mathrm{op}}).$$

Proof Statement (1) is an immediate consequence of Proposition 79, and (2) may be formally verified with Corollary 43:

$$\begin{aligned}((\mathrm{colim}^{\omega} h)^{\mathrm{op}})^* &= ((\mathrm{colim}^{\omega} h)_*)^{\circ} =]h_*, \omega[^{\circ} =](h_*)^{\circ}, \omega^{\circ}[\\ &=](h^{\mathrm{op}})^* \omega^{\circ}[= (\lim^{\omega^{\circ}} (h^{\mathrm{op}}))^*.\end{aligned}$$

□

Perhaps less expected, but still easily verified, is the fact that it suffices that we just consider (co)limits whose diagram is an identity morphism:

Proposition 96 *For X, h, ω, υ as in* Definition *94, one has the formulae:*

$$\mathrm{colim}^{\omega} h \simeq \mathrm{colim}^{\omega\cdot h^*} 1_X \qquad \textit{and} \qquad \lim^{\upsilon} h \simeq \lim^{h_*\cdot\upsilon} 1_X,$$

with the (co)limit on either side of the equivalence sign existing precisely when the (co)limit on the other side exists.

Proof Concerning the assertion about colimits, it suffices to show $[\omega, h_*] = [\omega \cdot h^*, 1_X^*]$. Indeed, by the adjunction $h_* \dashv h^*$, for all $\rho : W \to\!\!\!\!\!\circ\; X$ one has

$$\rho \le [\omega, h_*] \iff \rho\cdot\omega \le h_* \iff \rho\cdot\omega\cdot h^* \le 1_X^* \iff \rho \le [\omega\cdot h^*, 1_X^*].$$

□

We are now ready to prove that cocompleteness of a $\mathcal{V}$-category is already guaranteed by its $\mathcal{P}$-cocompleteness, i.e., that every weighted colimit in a $\mathcal{V}$-category X may be presented as a colimit of 1_X weighted by a presheaf on X:

Theorem 97 *A $\mathcal{V}$-category X is cocomplete if, and only if, X is $\mathcal{P}$-cocomplete, and X is complete if, and only if, X is $\mathcal{P}^\sharp$-complete.*

Proof After Corollary 95, it suffices to show that every $\mathcal{P}$-cocomplete $\mathcal{V}$-category X is cocomplete. So, we consider $\omega : X \to\!\!\!\!\!\circ\; W$. By hypothesis, for every $t \in W$ we then have the presheaf $\omega_t = t^* \cdot \omega : X \to\!\!\!\!\!\circ\; \mathsf{E}$ and find the colimit of 1_X weighted by ω_t; that is, an element $qt \in X$ with $(qt)_* = [\omega_t, 1_X^*]$. We show that $q : W \to X$ is a $\mathcal{V}$-functor.

Indeed, for all $t, t' \in W$, with the Yoneda Lemma 62 and the $\mathcal{V}$-functor L of the Isbell Duality Theorem 66, since $(qt)_* = \mathbf{y}_X^\sharp(qt)$ and $L\omega_t = [\omega_t, 1_X^*]$ (and likewise for t'), we have

$$X(qt, qt') = \mathcal{P}_\mathcal{V}^\sharp X((qt)_*, (qt')_*) = \mathcal{P}_\mathcal{V}^\sharp(L\omega_t, L\omega_{t'}) \ge \mathcal{P}_\mathcal{V}(\omega_t, \omega_{t'}) \ge W(t, t'),$$

with the last inequality following from the fact that the $\mathcal{V}$-distributor ω may be seen as a $\mathcal{V}$-functor $W \to \mathcal{P}_\mathcal{V} X$, by Theorem 32.

Since for all $t \in W$ and $x \in X$ we can rewrite $(qt)_* x = [\omega_t, 1_X^*]x$ as $q_*(t, x) = [\omega, 1_X^*](t, x)$, we obtain $q \simeq \mathrm{colim}^{\omega} 1_X$. This suffices to make X cocomplete, by Proposition 96. □

With Proposition 81 and Theorem 97 we conclude:

Corollary 98 *For every $\mathcal{V}$-category X, $\mathcal{P}_\mathcal{V} X$ is cocomplete and $\mathcal{P}_\mathcal{V}^\sharp X$ is complete.*

With Corollary 93 we obtain that (co)completeness of $\mathcal{V}$-categories is closed under arbitrary products and retracts. Here is another important stability property:

Proposition 99 *Let $f : X \to Y$ be a dense $\mathcal{V}$-functor. Then Y is cocomplete if X is.*

Proof It suffices to show that, if $q \simeq \mathrm{colim}^{\omega\cdot f_*} 1_X^*$ for $\omega : Y \to\!\!\!\!\circ\; \mathrm{E}$, then $q \simeq \mathrm{colim}^{\omega} 1_Y^*$. Indeed, since $f_* \cdot f^* = 1_Y^*$ by hypothesis (see Definition 47), for all $\sigma : \mathrm{E} \to\!\!\!\!\circ\; Y$ one has

$$\sigma \leq q_* \iff \sigma \cdot \omega \cdot f_* \leq f_* \iff \sigma \cdot \omega \leq 1_Y^*,$$

which implies the assertion. □

Corollary 100 *The separated reflection X_{sep} of a cocomplete $\mathcal{V}$-category X is cocomplete.*

Proof The separated reflection $X \to X_{\mathrm{sep}}$ (see Remark 27(1)) is easily seen to be dense. □

6.2 *Continuous and Cocontinuous $\mathcal{V}$-Functors, Adjoint Functor Theorem*

Next we spell out in detail the obvious notion of (co)limit preservation by a $\mathcal{V}$-functor.

Definition 101 Let $h : Z \to X$ and $f : X \to Y$ be $\mathcal{V}$-functors, and let $\omega : Z \to\!\!\!\!\circ\; W$ and $\upsilon : W \to\!\!\!\!\circ\; Z$ be $\mathcal{V}$-distributors.

(1) If $q \simeq \mathrm{colim}^{\omega} h$ exists in X, one says that $f : X \to Y$ *preserves* $\mathrm{colim}^{\omega} h$ if $\mathrm{colim}^{\omega}(fh)$ exists in Y and is given by fq (up to order-equivalence); that is, if $q_* = [\omega, h_*]$ implies $(fq)_* = [\omega, (fh)_*]$, with the last equation translating to

$$f_* \cdot [\omega, h_*] = [\omega, f_* \cdot h_*]. \qquad (*)$$

(2) Dually, if $p \simeq \lim^{\upsilon} h$ exists in X, one says that $f : X \to Y$ *preserves* $\lim^{\upsilon} h$ if $\lim^{\upsilon}(fh)$ exists in Y and is given by fp (up to order-equivalence); that is, if $p^* =]\upsilon, h^*[$ implies $(fp)^* =]\upsilon, (fh)^*[$, with the last equation translating to

$$]\upsilon, h^*[\, \cdot f^* =]\upsilon, h^* \cdot f^*[\,. \qquad (**)$$

(3) The $\mathcal{V}$-functor f is *(co)continuous* if it preserves all existing (co)limits in X.

Remarks 102 (1) The $\mathcal{V}$-functor $f : X \to Y$ is cocontinuous if, and only if, $f^{\mathrm{op}} : X^{\mathrm{op}} \to Y^{\mathrm{op}}$ is continuous.

(2) For f, h, ω as above, since always $f_* \cdot [\omega, h_*] \cdot \omega \leq f_* \cdot h_*$, the inequality "$\leq$" in $(*)$ comes for free. Likewise for the inequality "$\leq$" in $(**)$.

(3) In the realm of ($\mathcal{P}$-)cocomplete $\mathcal{V}$-categories the notions of $\mathcal{P}$-cocontinuity and cocontinuity of $\mathcal{V}$-functors are equivalent; likewise for ($\mathcal{P}^\sharp$-)complete $\mathcal{V}$-categories and the continuity of their $\mathcal{V}$-functors: see Theorems 78 and 97.

Here are our principal general examples of (co)continuous $\mathcal{V}$-functors, defined on a not necessarily (co)complete $\mathcal{V}$-category:

Theorem 103 *Let X be a $\mathcal{V}$-category.*

(1) *For every $x \in X$, the $\mathcal{V}$-functor $\mathbf{y}_X^\sharp x = X(x, -) : X \to \mathcal{V}$ is continuous, and the $\mathcal{V}$-functor $\mathbf{y}_X x = X(-, x) : X \to \mathcal{V}^{\mathrm{op}}$ is cocontinuous.*

(2) *The $\mathcal{V}$-functor $\mathbf{y}_X : X \to \mathcal{P}_\mathcal{V} X$ is continuous, and $\mathbf{y}_X^\sharp : X \to \mathcal{P}_\mathcal{V}^\sharp X$ is cocontinuous.*

Proof Since $(\mathbf{y}_X^\sharp)^{\mathrm{op}} = \mathbf{y}_{X^{\mathrm{op}}}$, in each, (1) and (2), it suffices to show the first of the two statements, i.e., that the functors $X(x, -)$ and $\mathbf{y}_X$ preserve the limit $p \simeq \lim^\upsilon 1_X$, for every $\upsilon : W \mathrel{-\!\!\circ\!\!\to} X$ where, by Theorem 97, we may assume $W = \mathrm{E}$. Hence, having $p^* =]\upsilon, 1_X[$ with $p \in X$, we must show $X(x, p) =]\upsilon, (\mathbf{y}_X^\sharp x)^*[$ and $(\mathbf{y}_X p)^* =]\upsilon, (\mathbf{y}_X)^*[$.

(1) Since the hypothesis means $X(x', p) = \bigwedge_{z \in X} [\upsilon z, X(x', z)]$ for all $x' \in X$, one can just specialize $x' = x$ to obtain $X(x, p) = \bigwedge_{z \in X} [\upsilon z, (\mathbf{y}_X^\sharp x)^* z]$, which gives the assertion.

(2) For every $\phi \in \mathcal{P}_\mathcal{V} X$, with the Isbell Duality $\mathcal{V}$-functors $L \dashv R$ of Theorem 66, we have

$$(\mathbf{y}_X p)^* \phi = \mathcal{P}_\mathcal{V}(\phi, p^*) = \mathcal{P}_\mathcal{V} X(\phi, R\upsilon) = \mathcal{P}_\mathcal{V}^\sharp X(L\phi, \upsilon) = \bigwedge_{x \in X} [\upsilon x, [\phi, 1_X^*] x].$$

Since $[\phi, 1_X^*] x = \mathcal{P}_\mathcal{V} X(\phi, \mathbf{y}_X x)$ for all $x \in X$, this means

$$(\mathbf{y}_X p)^* \phi = \bigwedge_{x \in X} [\upsilon x, \mathcal{P}_\mathcal{V} X(\phi, \mathrm{y}_X x)] =]\upsilon, (\mathbf{y}_X)^*[\, \phi\,,$$

and $\mathbf{y}_X p \simeq \lim^\upsilon \mathbf{y}_X$ follows.

□

The preservation of (co)limits is closely connected to the notion of adjunction. First we state:

Proposition 104 *Let $f \dashv g : Y \to X$ be adjoint $\mathcal{V}$-functors. Then g is continuous and f is cocontinuous.*

Proof To show the cocontinuity of f, assuming $q_* = [\omega, h_*]$ with ω, h as in Definition 101, we should verify the equivalence

$$\sigma \leq (f \cdot q)_* \iff \sigma \cdot \omega \leq (f \cdot h)_*,$$

for all $\mathcal{V}$-distributors $\sigma : W \nrightarrow Y$. Indeed, $\sigma \leq f_* \cdot q_*$ implies $g_* \cdot \sigma \leq g_* \cdot f_* \cdot q_* \leq q_*$, since $g_* \dashv f_*$ by Proposition 45, and then $q_* = [\omega, h_*]$ gives $g_* \cdot \sigma \cdot \omega \leq h_*$. Using the adjunction $g_* \dashv f_*$ again, we conclude $\sigma \cdot \omega \leq f_* \cdot g_* \cdot \sigma \cdot \omega \leq f_* \cdot h_*$, as desired. For showing the reverse implication, one argues similarly.

Of course, the continuity of g follows dually from the cocontinuity of f. □

Proposition 105 (Adjoint Functor Theorem) *Let the $\mathcal{V}$-category Y be complete. Then a $\mathcal{V}$-functor $g : Y \to X$ admits a left-adjoint $\mathcal{V}$-functor $f \dashv g$ if, and only if, g is continuous.*

Proof After Proposition 104, we need to show only the "if"-part of the claim. Since Y is complete, we may form the limit $f := \lim^{g^*} 1_Y$ and have $g^* \cdot f^* \leq 1_Y^*$ by definition. Since g preserves the limit, one obtains $gf \simeq \lim^{g^*} g$. Now, since trivially $g^* \leq g^*$ implies $1_X^* \leq \lim^{g^*} g$, the inequality $1_X^* \leq f^* \cdot g^*$ follows. This proves $g^* \dashv f^*$; by Proposition 45, this equivalently means $f \dashv g$. □

Corollary 106 *For X cocomplete, a $\mathcal{V}$-functor $f : X \to Y$ admits a right adjoint $\mathcal{V}$-functor if, and only if, f is cocontinuous.*

There is an important addition to Proposition 104 to be made when one of the adjoint functors is fully faithful:

Proposition 107 *Let the fully faithful $\mathcal{V}$-functor $g : Y \to X$ have a left adjoint $f \dashv g$, and consider a $\mathcal{V}$-functor $h : Z \to Y$ and a $\mathcal{V}$-distributor $\omega : Z \nrightarrow W$. Then the existence of $q \simeq \mathrm{colim}^{\omega}(gh)$ in X implies the existence of $\mathrm{colim}^{\omega} h$ in Y, and one has $fq \simeq \mathrm{colim}^{\omega} h$.*

Proof By Proposition 104, the left adjoint f preserves the given colimit q, so that one obtains $fq \simeq \mathrm{colim}^{\omega}(fgh)$. But as g is fully faithful, from $fg \simeq 1_Y$ one easily derives $\mathrm{colim}^{\omega}(fgh) \simeq \mathrm{colim}^{\omega} h$, which confirms the claim. □

The reader will readily dualize Proposition 107, by applying its assertions to $g^{\mathrm{op}} \dashv f^{\mathrm{op}}$, rather than to $f \dashv g$. Then the proposition and its dualization imply:

Corollary 108 *Let $f \dashv g : Y \to X$ be adjoint $\mathcal{V}$-functors.*

(1) *If g is fully faithful and X is cocomplete, then Y is also cocomplete.*
(2) *If f is fully faithful and Y is complete, then X is also complete.*

6.3 Cocompletion and Completion of $\mathcal{V}$-Categories

We denote by

$$\mathcal{V}\text{-}\mathsf{Cat}_{\mathrm{colim}}$$

the category of cocomplete $\mathcal{V}$-categories and their cocontinuous functors. By Theorems 78 and 97, the category $\mathcal{V}\text{-}\mathsf{Cat}_{\mathrm{colim}}$ is equivalent to the category of $\mathcal{P}$-pseudo-algebras and their $\mathcal{P}$-pseudo-homomorphisms, with the structure-forgetting functor to $\mathcal{V}\text{-}\mathsf{Cat}$ becoming the property-forgetting functor:

$$\mathcal{V}\text{-}\mathsf{Cat}_{\mathrm{colim}} \longrightarrow \mathcal{V}\text{-}\mathsf{Cat}.$$

As the former functor has a pseudo-left adjoint, so does the latter. Spelled out in detail, this means:

Theorem 109 (Cocompletion Theorem) *The Yoneda $\mathcal{V}$-functor $\mathbf{y}_X$ of a $\mathcal{V}$-category X is pseudo-universal among $\mathcal{V}$-functors $f : X \to Y$ into a cocomplete $\mathcal{V}$-category Y; that is, up to order-equivalence, every such f factors through $\mathbf{y}_X$, by a cocontinuous $\mathcal{V}$-functor $\tilde{f} : \mathcal{P}_{\mathcal{V}}X \to Y$ that is unique up to order-equivalence. The caveats "up to order-equivalence" are not needed when Y is separated.*

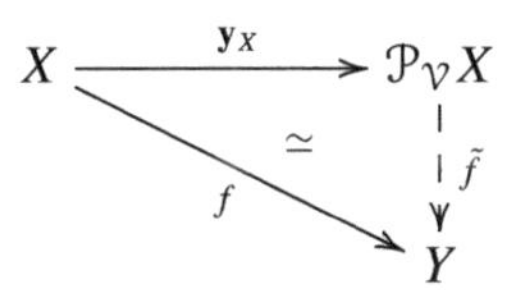

Proof The theorem follows entirely from the general mechanisms of a lax-idempotent monad T on a 2-category $\mathcal{C}$, applied to $\mathcal{C} = \mathcal{V}\text{-}\mathsf{Cat}$ and $T = \mathcal{P}$. In this case, given the $\mathcal{V}$-functor f, these mechanisms produce the factorizing $\mathcal{V}$-functor $\tilde{f} \simeq (\mathrm{colim}^{(\text{-})}1_Y)(\mathcal{P}_{\mathcal{V}}f^*)$, as indicated by the diagram

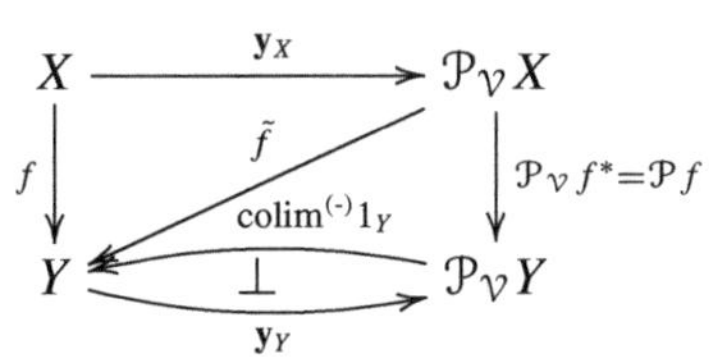

□

Applying the theorem to X^{op} in lieu of X one obtains:

Corollary 110 (Completion Theorem) *The Yoneda $\mathcal{V}$-functor $\mathbf{y}^{\sharp}_X$ of a $\mathcal{V}$-category X is pseudo-universal among $\mathcal{V}$-functors $f : X \to Y$ into a complete $\mathcal{V}$-category Y: up to order-equivalence, every such f factors through $\mathbf{y}^{\sharp}_X$, by a continuous*

$\mathcal{V}$-functor $\overline{f} : \mathcal{P}^\sharp_\mathcal{V} X \to Y$ that is unique up to order-equivalence. The caveats "up to order-equivalence" are not needed when Y is separated.

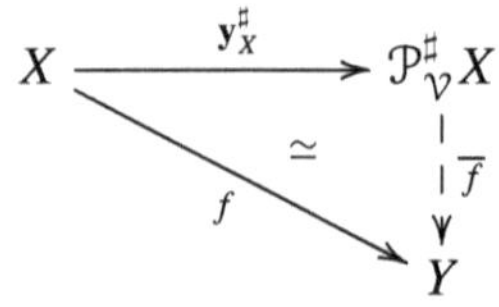

While the uniqueness (up to order-equivalence) of the factorizing $\mathcal{V}$-functor $\tilde{f}$ in the proof of Theorem 109 follows from the general 2-categorical arguments, an immediate alternative reasoning may be based on the following important proposition:

Proposition 111 *Let X be a $\mathcal{V}$-category and ω a $\mathcal{V}$-presheaf on X. Then ω is a colimit of $\mathbf{y}_X$ in $\mathcal{P}_\mathcal{V} X$ weighted by ω; that is, $\omega \simeq \mathrm{colim}^\omega \mathbf{y}_X$.*

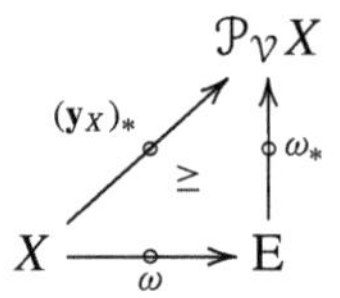

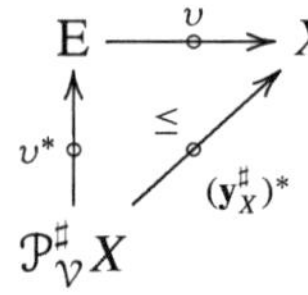

Dually, every $\mathcal{V}$-copresheaf υ on X may be presented as a limit of representable copresheaves on X; that is, $\upsilon \simeq \lim^\upsilon \mathbf{y}^\sharp_X$.

(Note that, again, here ω is treated simultaneously as a $\mathcal{V}$-distributor $X \to\!\!\!\!\!\!\circ\;\; \mathrm{E}$ and as a $\mathcal{V}$-functor $\mathrm{E} \to \mathcal{P}_\mathcal{V} X$; likewise for υ, viewed as a $\mathcal{V}$-distributor $\mathrm{E} \to\!\!\!\!\!\!\circ\;\; X$ and a $\mathcal{V}$-functor $\mathrm{E} \to \mathcal{P}^\sharp_\mathcal{V} X$.)

Proof For the colimit presentation of ω we should show that ω_* satisfies the universal property of the extension $[\omega, (\mathbf{y}_X)_*]$, that is, for all $\Phi : E \to\!\!\!\!\!\!\circ\;\; \mathcal{P}_\mathcal{V} X$, we should confirm the equivalence

$$\Phi \cdot \omega \leq (\mathbf{y}_X)_* \iff \Phi \leq \omega_* .$$

But the LHS, once evaluated at $x \in X$ and $\phi \in \mathcal{P}_\mathcal{V} X$, means (with the Yoneda Lemma)

$$\omega x \otimes \Phi\phi \leq \mathcal{P}_\mathcal{V} X(\mathbf{y}_X x, \phi) = \phi x$$

or, equivalently, $\Phi\phi \leq [\omega x, \phi x]$. Therefore, the LHS translates to $\Phi\phi \leq \mathcal{P}_\mathcal{V} X(\omega, \phi) = \omega_* \phi$ for all $\phi \in \mathcal{P}_\mathcal{V} X$—which is precisely what the RHS says. □

Remark 112 Further to the proof of Theorem 109, as indicated before Proposition 111, any cocontinuous $\mathcal{V}$-functor $g : \mathcal{P}_{\mathcal{V}}X \to Y$ with $g\mathbf{y}_X \simeq f : X \to Y$ must, for all $\omega \in \mathcal{P}_{\mathcal{V}}X$, satisfy

$$g(\omega) \simeq g(\operatorname{colim}^{\omega}\mathbf{y}_X) \simeq \operatorname{colim}^{\omega}(g\mathbf{y}_X) \simeq \operatorname{colim}^{\omega} f$$

and is, therefore, uniquely determined by f, up to order-equivalence.

6.4 *Tensored and Cotensored $\mathcal{V}$-Categories*

For a $\mathcal{V}$-category X and every element $x \in X$, we have the $\mathcal{V}$-functors

$$\mathbf{y}_X^{\sharp}x = X(x, -) : X \longrightarrow \mathcal{V} \quad \text{and} \quad \mathbf{y}_X x = X(-, x) : X^{\mathrm{op}} \longrightarrow \mathcal{V}$$

which are continuous (Theorem 103). It is therefore natural to consider those $\mathcal{V}$-categories X for which either of them have left adjoints.

Definition 113 (1) X is *tensored* (or *copowered*) if, for every $x \in X$, the $\mathcal{V}$-functor $X(x, -)$ has a left adjoint, here denoted by $- \odot x : \mathcal{V} \longrightarrow X$; equivalently, if for all $x \in X$, $u \in \mathcal{V}$ one has a (chosen) element $u \odot x \in X$ (called the *tensor* of u and x) with

$$X(u \odot x, y) = [u, X(x, y)]$$

for all $y \in X$ (see Proposition 30).

(2) X is *cotensored* (or *powered*) if X^{op} is tensored; that is, if $X(-, x)$ has a left adjoint for all $x \in X$, usually denoted by $- \pitchfork x : \mathcal{V} \longrightarrow X^{\mathrm{op}}$, so that its values (called *cotensors*) are (up to order-equivalence) determined by

$$X(y, u \pitchfork x) = [u, X(y, x)]$$

for all $u \in \mathcal{V}$ and $y \in X$.

With the formulae above we see immediately that tensors are very special weighted colimits, and cotensors are special weighted limits:

Lemma 114 *For a $\mathcal{V}$-category X and all $x \in X$, $u \in \mathcal{V}$, one has*

$$u \odot x \simeq \operatorname{colim}^{u} x \quad \textit{and} \quad u \pitchfork x \simeq \lim{}^{u} x,$$

with the term on either side of the $\simeq$-sign existing when the term on the other side exists.

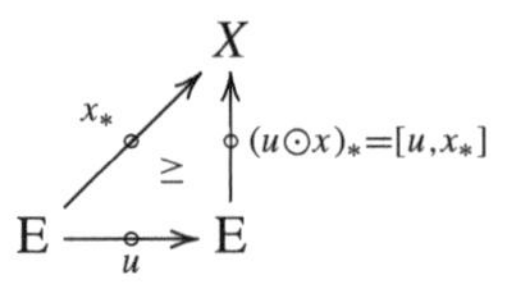

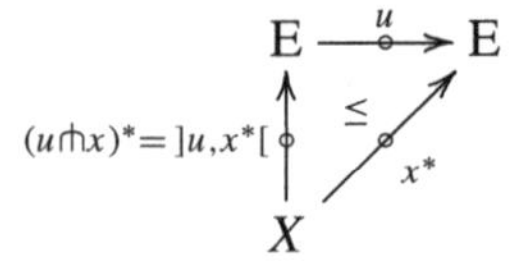

The immediate consequence of the lemma is that a $\mathcal{V}$-category X is tensored when cocomplete, and cotensored when complete. But more importantly, when applying the lemma to E and $\mathcal{P}_{\mathcal{V}}X$ instead of to X, one easily obtains:

Theorem 115 *The $\mathcal{V}$-category $\mathcal{V}$ is tensored and cotensored, with tensors in $\mathcal{V}$ given by $u \odot v = u \otimes v$, and cotensors by $u \pitchfork v = [u, v]$, for all $u, v \in \mathcal{V}$. More generally, for every $\mathcal{V}$-category X, the presheaf category $\mathcal{P}_{\mathcal{V}}X$ is tensored and cotensored, with the tensor $u \odot \phi$ and the cotensor $u \pitchfork \phi$ for $u \in \mathcal{V}$, $\phi \in \mathcal{P}_{\mathcal{V}}X$ given for all $x \in X$ by*

$$(u \odot \phi)x = u \otimes (\phi x) \quad \textit{and} \quad (u \pitchfork \phi)x = [u, \phi x].$$

Remarks 116 (1) Of course, by duality, also $\mathcal{P}^{\sharp}_{\mathcal{V}}X = (\mathcal{P}_{\mathcal{V}}(X^{\mathrm{op}}))^{\mathrm{op}}$ is tensored and cotensored, with tensors of copresheaves over X given by cotensors of presheaves of X^{op}, and cotensors of copresheaves over X given by tensors of presheaves of X^{op}.

(2) By their defining formulae, any existing tensor $u \odot x$ or cotensor $u \pitchfork x$ in a $\mathcal{V}$-category X is, in the underlying order of X, presentable as

$$u \odot x \simeq \bigwedge\{y \in X \mid u \leq X(x, y)\} \quad \text{and} \quad u \pitchfork x \simeq \bigvee\{y \in X \mid u \leq X(y, x)\}.$$

But note that the existence of such infima and suprema may not be sufficient to make X tensored or cotensored: see Example (2).

Examples 117 (1) A non-empty ordered set X, considered as an object of $\mathsf{2\text{-}Cat}$, is tensored if, and only if, X has a least element; dually, X is cotensored precisely when X has a largest element.

(2) For the 3-element quantale $\mathcal{V} = \{\bot, \mathrm{k}, \top\}$ and any non-empty set X considered as a $\mathcal{V}$-category with $X(x, y) = \mathrm{k}$ for all $x, y \in X$, one certainly has all infima and suprema in X (since all elements are order-equivalent), but no element x may admit the tensor $\top \odot x$ or cotensor $\top \pitchfork x$ (since $[\top, X(x, y)] = [\top, \mathrm{k}] = \bot$).

(3) A full $\mathcal{V}$-subcategory X of the $\mathcal{V}$-category $\mathcal{V}$ may fail to be tensored. For instance, consider $\mathcal{V} = [0, \infty]_+$ and $X = [0, \infty)$, as a subobject of $\mathcal{V}$. Then the tensor $\infty \odot x$ cannot exist in X since $[\infty, X(x, y)] = 0$ is constant for all $y \in X$, whereas there is no element $z \in X$ making $X(z, y)$ constant w.r.t. y.

(4) While the interval $[0, \infty]$, as an object of $\mathsf{Met} = [0, \infty]_+\text{-}\mathsf{Cat}$, is tensored and cotensored, this property gets lost after symmetrization, i.e., when we put the Euclidean metric on $[0, \infty]$ (naturally extended to ∞). Indeed, if $1 \odot 1$ existed in $[0, \infty]_{\mathrm{sym}}$, by Remark 116(2) the tensor would have to have value 2, but with that value one cannot satisfy the defining condition of the tensor. Likewise for the non-existence of the cotensor $1 \pitchfork 1$.

6.5 Tensoring as an Action of $\mathcal{V}$

Let us now look at the tensoring operation from an algebraic perspective and show that a tensored $\mathcal{V}$-category may be described as an ordered set X, equipped with a monotone pseudo-action of the monoid $(\mathcal{V}, \otimes, \mathrm{k})$ that preserves suprema taken in $\mathcal{V}$, as specified by the following rules:

Lemma 118 *In a tensored $\mathcal{V}$-category X, when equipped with its induced order one has for all $u, v, u_i \in \mathcal{V}$ $(i \in I)$, $x, y \in X$:*

(1) $\mathrm{k} \odot x \simeq x$;
(2) $u \odot (v \odot x) \simeq (u \otimes v) \odot x$;
(3) $(\bigvee_{i \in I} u_i) \odot x \simeq \bigvee_{i \in I}(u_i \odot x)$ *(in particular, the join on the right exists in X);*
(4$^-$) *if $x \leq y$, then $u \odot x \leq u \odot y$.*

Proof Since, for all $z \in X$, one has $X(\mathrm{k} \odot x, z) = [\mathrm{k}, X(x, z)] = X(x, z)$ and

$$\begin{aligned} X(u \odot (v \odot x), z) &= [u, X(v \odot x, z)] = [u, [v, X(x, z)]] \\ &= [u \otimes v, X(x, z)] = X((u \otimes v) \odot x, z), \end{aligned}$$

rules (1) and (2) follow. Rule (3) is an immediate consequence of the fact that the left-adjoint map $- \odot x : \mathcal{V} \longrightarrow X$ in Ord must preserve suprema. To show (4$^-$), assuming $x \leq y$, it suffices to show $u \odot x \leq z$ whenever $z \in X$ is such that $u \leq X(y, z)$ (see Remarks 116(2)). Indeed, since $X(-, z) : X^{\mathrm{op}} \to \mathcal{V}$ is monotone, we have $u \leq X(y, z) \leq X(x, z)$ and, therefore, $u \odot x \leq z$. □

We denote by $\mathcal{V}$-$\mathsf{Cat}_{\mathrm{ten}}$ the 2-category of tensored $\mathcal{V}$-categories and their tensor-preserving $\mathcal{V}$-functors, i.e., the $\mathcal{V}$-functors $f : X \to Y$ satisfying $f(u \odot x) \simeq u \odot fx$ $(u \in \mathcal{V}, x \in X)$, with 2-cells given by pointwise order, as in $\mathcal{V}$-Cat. Lemma 118 now describes the object assignment of the 2-functor

$$\mathcal{O} : \mathcal{V}\text{-}\mathsf{Cat}_{\mathrm{ten}} \longrightarrow \mathsf{Ord}^{\mathcal{V}}$$

that maps morphisms identically. Here, the 2-category $\mathsf{Ord}^{\mathcal{V}}$ is defined to have as

- objects (pre)ordered sets $(X, \leq)$ which come equipped with an operation $* : \mathcal{V} \times X \longrightarrow X$ satisfying rules (1-4$^-$) of Lemma 118, with $\odot$ traded for $*$ and $\simeq$ denoting order-equivalence, i.e., the coreflective symmetrization of the order $\leq$ of X;
- morphisms are monotone and pseudo-equivariant maps $f : X \to Y$, so that $f(u * x) \simeq u * fx$ for all $u \in \mathcal{V}$, $x \in X$;
- 2-cells are given by pointwise order.

We note that, in establishing the functor $\mathcal{O}$, for every tensored $\mathcal{V}$-category X we had to choose the operation $\odot$ which, unless X is separated, is only determined up to order-equivalence. Now we can state (see also Hofmann and Nora (2018)):

Theorem 119 *The 2-functor $\mathcal{O}$ is an equivalence of 2-categories.*

Proof In order to establish a pseudo-inverse of $\mathcal{O}$, we consider an object $(X, \leq, *)$ in $\mathsf{Ord}^{\mathcal{V}}_{\mathrm{sup}}$. For every $x \in X$, the (by (3)) suprema-preserving map $- * x : \mathcal{V} \to X$ has a right adjoint which, anticipating that it will define a $\mathcal{V}$-category structure on X, we denote by $X(x, -) : X \to \mathcal{V}$. By definition then,

$$X(x, y) = \bigvee\{u \in \mathcal{V} \mid u * x \leq y\},$$

for all $y \in X$, and we have in particular the inequality $X(x, y) * x \leq y$ (representing the counit of the adjunction $(- * x) \dashv X(x, -)$ at y). We show that X(-,-) makes X indeed a $\mathcal{V}$-category. Clearly, $\mathrm{k} \leq X(x, x)$ holds by (1), and with (2) and (4) we see

$$(X(y, z) \otimes X(x, y)) * x = X(y, z) * (X(x, y) * x) \leq X(y, z) * y \leq z,$$

which implies the desired inequality $X(x, y) \otimes X(y, z) \leq X(x, z)$, for all $x, y, z \in X$. Since trivially $(\mathrm{k} \leq X(x, y) \iff x = \mathrm{k} * x \leq y)$, the induced order of the $\mathcal{V}$-category X just defined is precisely the given order of X. Also, X is tensored via $*$, by the definition of its $\mathcal{V}$-category structure.

Next we show that we have now established the object assignment of a 2-functor $\mathcal{F} : \mathsf{Ord}^{\mathcal{V}} \longrightarrow \mathcal{V}\text{-}\mathsf{Cat}_{\mathrm{ten}}$. For that we consider a morphism $f : X \to Y$ in $\mathsf{Ord}^{\mathcal{V}}$ and show that $\mathcal{F}f = f : \mathcal{F}X \longrightarrow \mathcal{F}Y$ is a $\mathcal{V}$-functor. Indeed, using consecutively the unit of the adjunction $(- * fx) \dashv Y(fx, -)$ at $X(x, x')$, the pseudo-equivariance of f, and the counit of the adjunction $(- * x) \dashv X(x, -)$ at x' in conjunction with the monotonicity of $Y(fx, f-)$, we obtain for all $x, x' \in X$ the desired inequality:

$$\begin{aligned} X(x, x') &\leq Y(fx, X(x, x') * fx) \\ &\simeq Y(fx, f(X(x, x') * x)) \\ &\leq Y(fx, fx'). \end{aligned}$$

One clearly has the strict identity $\mathcal{F}\mathcal{O} = \mathrm{Id}_{\mathcal{V}\text{-}\mathsf{Cat}}$, but the composite $\mathcal{O}\mathcal{F}$ is only isomorphic to $\mathrm{Id}_{\mathsf{Ord}^{\mathcal{V}}_{\mathrm{sup}}}$ (since the tensor of a $\mathcal{V}$-category is only determined up to order-equivalence). In any case, as all other needed verifications are trivial, we have completed the proof. □

Remark 120 From Clementino and Fitas (2023) one obtains a description of tensored $\mathcal{V}$-categories as the pseudo-algebras w.r.t. a so-called *ball monad* (see also Goubault-Larrecq (2019)) on $\mathcal{V}$-Cat. Hence, separation granted, one has an explicit presentation of $\mathcal{V}\text{-}\mathsf{Cat}_{\mathrm{ten,sep}}$ as a monadic category over $\mathcal{V}\text{-}\mathsf{Cat}_{\mathrm{sep}}$.

6.6 Conical Suprema and Infima

Let $(x_i)_{i\in I}$ be a family of elements in a $\mathcal{V}$-category X. A supremum x of that family w.r.t. the induced order of X, written as $x \simeq \bigvee_{i\in I} x_i$, is characterized, up to order-equivalence, by satisfying, for all $z \in X$, the condition

$$\mathrm{k} \leq X(x, z) \iff \mathrm{k} \leq \bigwedge_{i\in I} X(x_i, z).$$

This condition is certainly satisfied in case we have $X(x, z) = \bigwedge_{i\in I} X(x_i, z)$ for all $z \in X$, which, however, is a stronger condition than the order-theoretical equivalence.

Example 121 For $\mathcal{V} = [0, \infty]_+$, consider the following 4-point generalized metric space X, with one-way distances as indicated; except for self-distances, all other undeclared distances may be taken to be ∞:

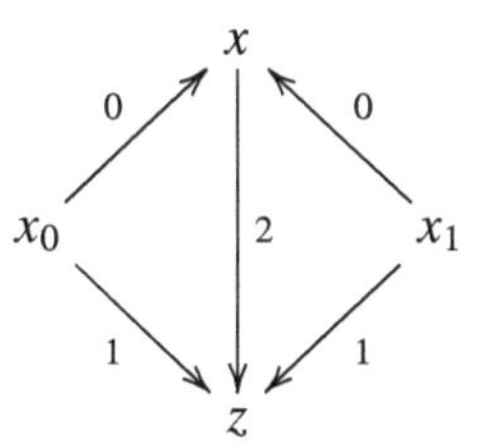

In the underlying order of X, we have $x = x_0 \vee x_1$. However, in $[0, \infty]_+$ we obtain

$$X(x, z) = 2 \neq 1 = \max\{X(x_0, z), X(x_1, z)\} = X(x_0, z) \wedge X(x_1, z).$$

The satisfaction of the stronger condition implying the supremum condition in the induced order deserves a special name, as follows.

Definition 122 Let $x,\ x_i\ (i \in I)$ be elements in a $\mathcal{V}$-category X.

(1) One calls x a *conical supremum* of $(x_i)_{i\in I}$ in X, written as[5] $x \simeq \bigvee^{\triangledown}_{i\in I} x_i$, if

$$X(x, z) = \bigwedge_{i\in I} X(x_i, z) \qquad (*)$$

for all $z \in X$; equivalently, if x is a supremum of $(x_i)_{i\in I}$ in X w.r.t. its induced order and the Yoneda $\mathcal{V}$-functor $\mathbf{y}^{\sharp}_X : X \longrightarrow \mathcal{P}^{\sharp}_{\mathcal{V}} X = [X, \mathcal{V}]^{\mathrm{op}}$ preserves this supremum.

[5] Of course, a conical supremum in X is in particular a supremum in the induced order of X. We use the $\triangledown$-superscript only whenever we want to stress the conical property of the supremum.

(2) Dually, x is a *conical infimum* of $(x_i)_{i\in I}$ in X, written as $x \simeq \bigwedge^{\nabla}_{i\in I} x_i$, if

$$X(z, x) = \bigwedge_{i\in I} X(z, x_i) \qquad (**)$$

for all $z \in X$; equivalently, if x is an infimum of $(x_i)_{i\in I}$ in the induced order of X and the Yoneda $\mathcal{V}$-functor $\mathbf{y}_X : X \to \mathcal{P}_{\mathcal{V}}X = [X^{\mathrm{op}}, \mathcal{V}]$ preserves the infimum.

(3) X is *order-complete* if every family in X has a chosen infimum or, equivalently, a supremum, with respect to its induced order. X is *conically cocomplete* (*conically complete*) if every family in X has a chosen conical supremum (conical infimum, respectively) in X.

As an immediate consequence of the Definition and the rules of Corollary 4 (as extended to $\mathcal{V}$-relations in Remark 13), we can state:

Proposition 123 *All suprema and infima in the $\mathcal{V}$-category $\mathcal{V}$ and, hence, in all (co)presheaf $\mathcal{V}$-categories, are conical.*

Therefore the conical (co)completeness of (co)presheaf $\mathcal{V}$-categories is evident. But this fact follows also from their (co)completeness. Indeed, like in the case of tensors and cotensors, in the Lemma below we specify how we would like to regard conical suprema and infima as special weighted colimits and limits, respectively. Its proof entails only a short routine calculation which we omit.

Lemma 124 *For a $\mathcal{V}$-category X and all $x_i \in X$, let $\omega := \bigvee_i x_i^*$ in $\mathcal{P}_{\mathcal{V}}X$ and $\upsilon := \bigwedge_i (x_i)_*$ in $\mathcal{P}^{\sharp}_{\mathcal{V}}X = [X, \mathcal{V}]^{\mathrm{op}}$. Then*

$$\bigvee^{\nabla}_{i\in I} x_i \simeq \mathrm{colim}^{\omega} 1_X^* \quad \textit{and} \quad \bigwedge^{\nabla}_{i\in I} x_i \simeq \lim{}^{\upsilon} 1_X^*,$$

with the term on either side of the $\simeq$-sign existing when the term on the other side exists.

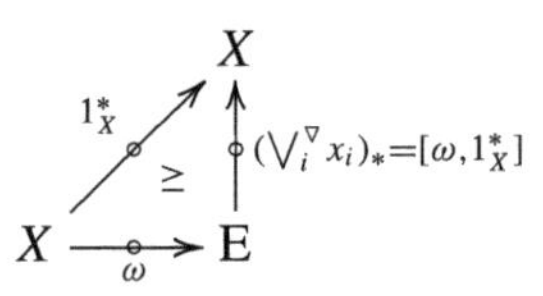

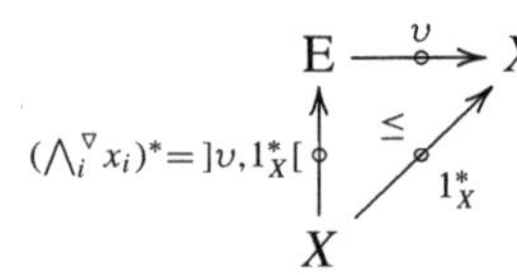

Despite its simplicity, the Lemma is a crucial stepping stone for the presentation of conically cocomplete $\mathcal{V}$-categories as the pseudo-algebras of the $\mathcal{V}$-Hausdorff monad, as given in Theorem 127.

Proposition 125 *Let X be a $\mathcal{V}$-category.*

(1) *X is conically cocomplete if, and only if, X is order-complete and the Yoneda map $\mathbf{y}^\sharp_X : X \to \mathcal{P}^\sharp_\mathcal{V} X$ preserves suprema. The preservation condition is satisfied when X is cotensored.*
(2) *X is conically complete if, and only if, X is order-complete and $\mathbf{y}_X : X \to \mathcal{P}_\mathcal{V} X$ preserves infima. The preservation condition is satisfied when X is tensored.*

Proof The first statement in (1) is an immediate consequence of the definition of conical supremum. For the second statement in (1), observe that the existence of all cotensors in X makes the $\mathcal{V}$-functor $\mathbf{y}_X z = X(-, z) : X^{\mathrm{op}} \to \mathcal{V}$ right adjoint and, in particular, infima preserving, for every $z \in X$. This translates to the preservation of suprema by $\mathbf{y}^\sharp_X$. (2) follows from (1) by dualization. □

Examples 126 (1) In an ordered set $X \in \mathsf{2\text{-}Cat} = \mathsf{Ord}$, every infimum or supremum is conical. Hence, X is already conically (co)complete if X is order-complete.
(2) If $\mathcal{V}$ is non-integral, one finds trivial examples of $\mathcal{V}$-categories that are order-complete, but neither conically complete nor conically cocomplete. For instance, when a non-empty set X carries the $\mathcal{V}$-category structure with $X(x, y) = \mathrm{k}$ for all $x, y \in X$, so that all elements are equivalent in the induced order, then it is trivially order-complete, but it is neither conically complete nor conically cocomplete: while every element in X is both, an infimum and a supremum of the empty family in X, none of these is conical, because of the non-integrality of $\mathcal{V}$.
(3) In Example 130(2), we show that a conically cocomplete (generalized) metric space generally fails to be conically complete.
(4) The induced order of a Fréchet metric space X, considered as an object of $[0, \infty]_+\text{-}\mathsf{Cat}_{\mathrm{sym,sep}}$, is discrete. Hence, X is order-complete only when it is a singleton space, and then it is trivially conically complete and cocomplete.

6.7 The $\mathcal{V}$-Hausdorff Monad and Its Pseudo-Algebras

Following the template provided in Sect. 2.3 by the quantale $[0, \infty]_+$, where we lifted the power-set monad from Set to the (non-symmetric) Hausdorff monad of Met, for our general quantale $\mathcal{V}$, let us now define the $\mathcal{V}$-Hausdorff monad $\mathcal{H}$ of $\mathcal{V}$-Cat, as follows. For every $\mathcal{V}$-category X, simply consider the map

$$\mathbf{j}_X : \mathcal{H}X = \{A \mid A \subseteq X\} \longrightarrow \mathcal{P}X, \quad A \longmapsto X(\text{-}, A) := \bigvee_{x \in A} \mathbf{y}_X x.$$

We provide $\mathcal{H}X$ with the initial structure inherited from the presheaf category $\mathcal{P}_\mathcal{V} X$ along $\mathbf{j}_X$; that is, for all $A, B \subseteq X$ we put $\mathcal{H}X(A, B) := \mathcal{P}X(\mathbf{j}_X A, \mathbf{j}_X B)$ and obtain with the Yoneda Lemma

$$\begin{aligned}\mathcal{H}X(A,B) &= \bigwedge_{z\in X}[X(z,A),X(z,B)]\\ &= \bigwedge_{z\in X}\bigwedge_{x\in A}[X(z,x),X(z,B)]\\ &= \bigwedge_{x\in A}\mathcal{P}X(\mathbf{y}_X x, X(\text{-},B))\\ &= \bigwedge_{x\in A}X(x,B) = \bigwedge_{x\in A}\bigvee_{y\in B}X(x,y)\,.\end{aligned}$$

It is a routine exercise to show that $\mathcal{H}$ can be made a monad of $\mathcal{V}$-Cat whose shadow in Set is the power-set monad. Furthermore, $\mathbf{j}:\mathcal{H}\to\mathcal{P}$ is a morphism of monads, and $\mathcal{H}$ inherits its lax idempotence from the monad $\mathcal{P}$: see Theorem 2 of Akhvlediani et al. (2010) for details, and Stubbe (2010) for a generalization thereof, from quantales to quantaloids.

The monad morphism $\mathbf{j}$ gives us the forgetful functor from the category of $\mathcal{P}$-pseudo-algebras and their $\mathcal{P}$-pseudo-homomorphisms to the category of $\mathcal{H}$-pseudo-algebras and their $\mathcal{H}$-pseudo-homomorphisms. Having seen in Theorems 78 and 97 how to describe (up to equivalence) the former category, it has become obvious with Lemma 124 how to describe the latter category, as follows.

Theorem 127 *The 2-category of $\mathcal{H}$-pseudo-algebras and $\mathcal{H}$-pseudo-homomorphisms is equivalent to the 2-category $\mathcal{V}$-$\mathsf{Cat}_{\nabla\mathrm{sup}}$ of conically cocomplete $\mathcal{V}$-categories and their $\mathcal{V}$-functors that preserve conical suprema. The monad morphism $\mathbf{j}:\mathcal{H}\to\mathcal{P}$ induces the forgetful 2-functor*

$$\mathcal{V}\text{-}\mathsf{Cat}_{\mathrm{colim}} \longrightarrow \mathcal{V}\text{-}\mathsf{Cat}_{\nabla\mathrm{sup}}$$

which, for a cocomplete $\mathcal{V}$-category, remembers only its conical cocompleteness.

For every $\mathcal{V}$-category X we have in particular the "free" conically cocomplete $\mathcal{V}$-category $\mathcal{H}X$. Its structure is given by the monad multiplication, a lifting of the multiplication of the power-set monad in Set. Hence, adopting the language of Theorem 109, we obtain:

Corollary 128 *Let X be a $\mathcal{V}$-category. Then the $\mathcal{V}$-functor $X\to\mathcal{H}X,\ x\mapsto\{x\}$ is pseudo-universal among all $\mathcal{V}$-functors $X\to Y$ with $Y\in\mathcal{V}$-$\mathsf{Cat}_{\nabla\mathrm{sup}}$. The conical supremum of $(A_i)_{i\in I}$ in $\mathcal{H}X$ may be given by $\bigcup_{i\in I}A_i$.*

Remarks 129 (1) Of course, analogously to the definitions of $\mathcal{P}^\sharp,\mathbf{y}^\sharp$ vis-à-vis $\mathcal{P},\mathbf{y}$ one defines the dualizations $\mathcal{H}^\sharp,\mathbf{j}^\sharp$ of $\mathcal{H},\mathbf{j}$ and obtains the corresponding dualizations of the theorem and the corollary above.

(2) The corollary does *not* say that the underlying order of $\mathcal{H}X$ is the inclusion order. In fact, for every $A\subseteq X$, one easily confirms that A is order-equivalent to $\overline{A}:=\{x\in X\mid \mathrm{k}\le X(x,A)\}$. Consequently, for $\mathcal{V}=[0,\infty]_+$ (see Example 130(1)) and a bounded Fréchet metric space X, any non-empty subsets A, B

satisfy $A \leq B$ in the induced order of $\mathcal{H}X$ precisely when A is contained in the topological closure of B in X.

(3) While for $\mathcal{V} = \mathsf{2}$ conical cocompleteness and conical completeness are equivalent properties, as both amount to order-completeness, this is not the case for $\mathcal{V} = [0, \infty]_+$, as Example 130(2) shows.

(4) The Axiom of Choice granted, if the complete lattice $\mathcal{V}$ is (constructively) completely distributive Wood (2004), then the non-symmetric Hausdorff distance for $A, B \subseteq X \in \mathcal{V}\text{-}\mathsf{Cat}$ satisfies the alternative formula:

$$\mathcal{H}X(A, B) = \bigvee_{\phi: A \to B} \bigwedge_{x \in A} X(x, \phi x),$$

where ϕ runs through *all* Set-maps $A \to B$. For $\mathcal{V} = [0, \infty]$, this formula has been used in Lawvere (1973) and, indeed, in the context of *normed categories*, this formula appears very naturally even for general $\mathcal{V}$: see Tholen (2022).

(5) For a treatment of the Gromov distance (as mentioned in Sect. 2.4 in case $\mathcal{V} = [0, \infty]$), for symmetric $\mathcal{V}$-categories in the general context we refer the reader to Akhvlediani et al. (2010).

Examples 130 (1) For $\mathcal{V} = [0, \infty]_+$, $X \in \mathcal{V}\text{-}\mathsf{Cat} = \mathsf{Met}$, and $A, B \subseteq X$, we have the (non-symmetric) Hausdorff distance $\mathcal{H}X(A, B)$ which, after coreflective symmetrization, takes the standard form:

$$(\mathcal{H}X)_{\mathrm{csym}}(A, B) = \max\{\sup_{x \in A} \inf_{y \in B} X(x, y),\ \sup_{y \in B} \inf_{x \in A} X(y, x)\}.$$

Considering $\mathcal{V} = [0, \infty]_{\max}$ instead shows that the same formula gives us an ultrametric on the power-set of X when $X \in \mathsf{UMet}$.

(2) Let $X = \{x, y, z\}$ be the 3-point Fréchet metric space with all non-zero distances being 1. Then, the conically cocomplete (generalized) metric space $\mathcal{H}X \in \mathsf{Met}$ is not conically complete. Indeed, if the sets $B = \{y\}$, $C = \{z\}$ had a conical infimum D, the set D would have to be order-equivalent to $B \cap C = \emptyset$ and, thus, $D = \emptyset$. But for $A = \{x\}$ we have $\mathcal{H}X(A, \emptyset) = \infty$, whereas $\max\{\mathcal{H}X(A, B), \mathcal{H}X(A, C)\} = 1$.

6.8 The Completeness Theorem

In the previous sections, we paid special attention to two particular types of weighted colimits, tensors and conical suprema, for a good reason. Indeed, we finally show that they are the essential building blocks for obtaining all weighted colimits in a $\mathcal{V}$-category:

Proposition 131 *Let $h : Z \to X$ be a $\mathcal{V}$-functor to the tensored $\mathcal{V}$-category X, and let $\omega : Z \nrightarrow W$ be a $\mathcal{V}$-distributor. Then one has the formula*

$$(\mathrm{colim}^{\omega} h)t \simeq \bigvee_{z\in Z}^{\nabla} \omega(z,t) \odot hz$$

for all $t \in W$, with the colimit on the left existing precisely when the conical supremum on the right exists in X for all $t \in W$.

Proof According to $(*)$ of Definition 122(2), the existence of the conical supremum for every $t \in W$ is characterized by

$$X(\bigvee_{z\in Z}^{\nabla} \omega(z,t) \odot h(z), x) = \bigwedge_{z\in Z} X(\omega(z,t) \odot h(z), x) = \bigwedge_{z\in Z} [\omega(z,t), X(h(z), x)]$$

for all $x \in X$. But this means precisely that $q : W \to X$, defined by $qt = \bigvee_{z\in Z} \omega(z,t) \odot hz$, serves as $\mathrm{colim}^{\omega} h$, provided that q is actually a $\mathcal{V}$-functor. Hence, we must just confirm that this provision is indeed satisfied: since $\omega(z,t) \odot hz \le qt$ transcribes to $\omega(z,t) \le X(hz, qt)$, with the $\mathcal{V}$-distributivity of ω one obtains for all $s, t \in W$ and $z \in Z$ the inequality $\omega(z,s) \otimes W(s,t) \le X(hz, qt)$, or $W(s,t) \le [\omega(z,s), X(hz, qt]$. Consequently,

$$W(s,t) \le \bigwedge_{z\in Z} [\omega(z,s), X(hz, qt] = X(qs, qt),$$

as desired. □

Corollary 132 *If the $\mathcal{V}$-category X is cotensored, the limit of $h : Z \to X$ weighted by $\upsilon : W \nrightarrow Z$ exists and is computed by*

$$(\lim^{\upsilon} h)t \simeq \bigwedge_{z\in Z}^{\nabla} \upsilon(t,z) \pitchfork hz,$$

precisely when the conical infimum on the right exists, for every $t \in W$.

Corollary 133 *Let $f : X \to Y$ be a $\mathcal{V}$-functor.*

(1) *For X tensored, f is cocontinuous if, and only if, f preserves all tensors and conical suprema.*
(2) *For X cotensored, f is continuous if, and only if, f preserves all cotensors and conical infima.*

We can now summarize our findings on cocompleteness of a $\mathcal{V}$-category and, moreover, show that it is a self-dual property, as follows:

Theorem 134 (Completeness Theorem) *The following conditions on a $\mathcal{V}$-category X are equivalent:*

(i) *X is $\mathcal{P}$-cocomplete;*
(ii) *X is cocomplete;*
(iii) *X is tensored and conically cocomplete;*
(iv) *X is tensored, cotensored, and order-complete;*
(v) *X is cotensored and conically complete;*
(vi) *X is complete;*
(vii) *X is $\mathcal{P}^{\sharp}$-complete.*

Proof The equivalence of (i) and (ii) has been shown in Theorem 97. With (iv) being a self-dual statement, it suffices to show that the statements (ii)-(iv) are equivalent, since the statements (v), (vi), (vii) are the respective dualizations of (iii), (ii), (i).
(iv)$\Longrightarrow$(iii) follows from Proposition 125.
(iii)$\Longrightarrow$(ii) follows from Proposition 131.
(ii)$\Longrightarrow$(iv): It remains to be shown that X is cotensored when X is cocomplete. But by Theorem 103, for every $x \in X$, the $\mathcal{V}$-functor $X(-,x) : X^{\mathrm{op}} \to \mathcal{V}$ is continuous. With X^{op} being complete, by Theorem 105 this makes $X(-,x)$ have a left adjoint, as required for the existence of all cotensors $u \pitchfork x$ $(u \in \mathcal{V})$. □

Corollary 135 *Every $\mathcal{V}$-presheaf category and $\mathcal{V}$-copresheaf category is both complete and cocomplete.*

6.9 Cocompleteness as an Action of $\mathcal{V}$

When the $\mathcal{V}$-category X is cocomplete, the tensor action of $\mathcal{V}$ on X allows for a strengthening of rule (4^-) of Lemma 118, as follows:

Lemma 136 *In a cocomplete $\mathcal{V}$-category X equipped with its induced order, one has for all $u, v, u_i \in \mathcal{V}$ $(i \in I)$, $x, x_i \in X (i \in I)$:*

(1) $k \odot x \simeq x$;
(2) $u \odot (v \odot x) \simeq (u \otimes v) \odot x$;
(3) $(\bigvee_{i \in I} u_i) \odot x \simeq \bigvee_{i \in I}(u_i \odot x)$;
(4) $u \odot (\bigvee_{i \in I} x_i) \simeq \bigvee_{i \in I}(u \odot x_i)$.

Proof Since suprema in a cocomplete $\mathcal{V}$-category are conical, we have, for all $y \in X$,

$$\begin{aligned} X(\bigvee_{i \in I} u \odot x_i, y) &= \bigwedge_{i \in I} X(u \odot x_i, y) = \bigwedge_{i \in I} [u, X(x_i, y)] \\ &= [u, \bigwedge_{i \in I} X(x_i, y)] = [u, X(\bigvee_{i \in I} x_i, y)], \end{aligned}$$

which, by definition of $\odot$, validates (4). □

We now define the (non-full) sub-2-category $\mathsf{Ord}^{\mathcal{V}}_{\mathrm{sup}}$ of $\mathsf{Ord}^{\mathcal{V}}$ (see Theorem 119):

- objects are complete (pre)ordered sets $(X, \le)$ that come equipped with an operation $* : \mathcal{V} \times X \longrightarrow X$ satisfying rules (1-4) of Lemma 136, with $\odot$ traded for $*$ and $\simeq$ denoting order-equivalence w.r.t. $\le$;
- morphisms are suprema-preserving and pseudo-equivariant maps $f : X \to Y$;
- 2-cells are given by pointwise order.

Comparing this category with $\mathcal{V}\text{-}\mathsf{Cat}_{\mathrm{colim}}$ (as introduced in Sect. 6.3), we can state:

Theorem 137 *The 2-categories $\mathcal{V}\text{-}\mathsf{Cat}_{\mathrm{colim}}$ and $\mathsf{Ord}^{\mathcal{V}}_{\mathrm{sup}}$ are equivalent.*

Proof After Theorem 134 and Corollary 133, it suffices to show that the functor $\mathcal{O}$ of Theorem 119 makes the existing suprema of an object $X \in \mathsf{Ord}^{\mathcal{V}}_{\mathrm{sup}}$ conical in its $\mathcal{V}$-category structure, determined by $(u \le X(x, y) \iff u * x \le y)$ for all $x, y \in X,\ u \in \mathcal{V}$. For $x = \bigvee_{i \in I} x_i$ in the complete ordered set X, this means

$$
\begin{aligned}
u \le X(\bigvee_i x_i,\ y) \iff u * \bigvee_i x_i \le y &\iff \bigvee_i u * x_i \le y \\
&\iff \forall i\ (u * x_i \le y) \iff \forall i\ (u \le X(x_i, y)) \\
&\iff u \le \bigwedge_i X(x_i, y),
\end{aligned}
$$

which implies the desired equality $X(\bigvee_i x_i,\ y) = \bigwedge_i X(x_i, y)$. □

7 The $\mathcal{V}$-Quantification of the $*$-Autonomous Category Sup

The category Sup of complete lattices and suprema-preserving maps distinguishes itself by a number of important properties, such as:

- Sup is *monadic* over both, Set and Pos, i.e., is isomorphic to the Eilenberg–Moore categories of both, the power-set monad on Set and of the down-set monad on Pos.
- Sup is *self-dual*. Indeed, considering every sup-lattice X as an inf-lattice, we obtain the functor $\mathsf{Sup}^{\mathrm{op}} \longrightarrow \mathsf{Sup},\ X \longmapsto X^{\mathrm{op}}$, which assigns to every sup-map $f : X \to Y$ the map $(f^\bullet)^{\mathrm{op}} : Y^{\mathrm{op}} \to X^{\mathrm{op}}$, where $f^\bullet : Y \to X$ is the right adjoint of f in Pos and, thus, an inf-map. Obviously, this functor is an isomorphism.
- Sup is *symmetric monoidal-closed*, with its tensor product $\boxtimes$ classifying its *bi-morphisms*, so that the sup-maps $X \boxtimes Y \to Z$ correspond naturally to the set maps $X \times Y \to Z$ which preserve suprema in each variable separately (see Banaschewski and Nelson (1976), Joyal and Tierney (1984)).

- In fact, as shown in Joyal and Tierney (1984), Sup is $*$-*autonomous* (Barr (1979)), so that its internal hom admits a natural isomorphism $\mathsf{Sup}(X \boxtimes Y,\ Z^{\mathrm{op}}) \cong \mathsf{Sup}(X, (Y \boxtimes Z)^{\mathrm{op}})$ for all sup-lattices X, Y, Z.

In what follows we describe how these properties can be established more generally for the category

$$\mathcal{V}\text{-}\mathsf{Sup} := \mathcal{V}\text{-}\mathsf{Cat}_{\mathrm{colim,sep}},$$

i.e., the category of separated and (co)complete objects in $\mathcal{V}$-Cat with their cocontinuous $\mathcal{V}$-functors. Of course, in case $\mathcal{V} = \mathsf{2}$ we have $\mathcal{V}$-$\mathsf{Sup} = \mathsf{Sup}$. Note that, by the Adjoint Functor Theorem (Proposition 105), *the morphisms in* $\mathcal{V}$-Sup *are precisely the* $\mathcal{V}$-Cat-*morphisms which have a right adjoint.*

7.1 The Monadicity and Self-Duality of $\mathcal{V}$-Sup

Having introduced $\mathcal{P}$-cocomplete $\mathcal{V}$-categories as $\mathcal{P}$-pseudo-algebras, we are not at all surprised that $\mathcal{V}$-Sup is monadic over $\mathcal{V}$-$\mathsf{Cat}_{\mathrm{sep}}$, but monadicity over Set is a different matter:

Theorem 138 (Clementino-Hofmann 2008, Martinelli 2021) *The category* $\mathcal{V}$-Sup *is monadic over both* $\mathcal{V}$-$\mathsf{Cat}_{\mathrm{sep}}$ *and* Set.

Proof The monadicity of the forgetful functor $\mathcal{V}\text{-}\mathsf{Sup} \to \mathcal{V}\text{-}\mathsf{Cat}_{\mathrm{sep}}$ follows immediately from Theorems 78 and 97. To show that the forgetful functor $\mathcal{V}\text{-}\mathsf{Sup} \to \mathsf{Set}$ is monadic, one applies Duskin's (variation of Beck's) monadicity criterion (see Duskin (1969) and Hofmann and Seal (2014)); for details, we must refer to Clementino and Hofmann (2008) and Martinelli (2021). □

The monad induced by the right-adjoint forgetful functor $\mathcal{V}\text{-}\mathsf{Sup} \to \mathsf{Set}$ is the discrete presheaf monad $\mathcal{P}_{\mathrm{d}}$ of Corollary 73. Therefore:

Corollary 139 *The category* $\mathcal{V}$-Sup *is equivalent to the Eilenberg–Moore category of* $\mathcal{P}_{\mathrm{d}}$-*algebras and their homomorphisms.*

Let us illustrate the algebraic nature of $\mathcal{V}$-Sup as a category over $\mathcal{V}$-Cat in the case $\mathcal{V} = [0, \infty]_+$.

Example 140 A separated generalized metric space X is cocomplete if it comes with a left action $((u, x) \longmapsto u \odot x)$ of $[0, \infty]_+$ on X, and with I-ary operations $((x_i)_{i \in I} \longmapsto \bigvee^{\triangledown}_{i \in I} x_i)$ on X (for every set I), satisfying the conditions

$$X(u \odot x, y) = \max\{X(x, y) - u, 0\} \quad \text{and} \quad X(\bigvee_{i \in I}^{\triangledown} x_i, y) = \sup_{i \in I} X(x_i, y),$$

for all $y \in X$. Cocontinuous non-expansive maps $f : X \to Y$ must preserve the action and the operations.

The free cocompletion of a metric space X is given by the set

$$\mathcal{P}_{\mathcal{V}}X = \{\phi : X \to [0,\infty] \mid \forall y \in X\ (\phi x - \phi y \leq X(x,y))\},$$

provided with the sup-metric. According to Theorem 115, in $\mathcal{P}_{\mathcal{V}}X$ we can compute $u \odot \phi$ simply by $(u \odot \phi)x = u + \phi x$, and $\bigvee^{\nabla}_{i\in I}\phi_i$ is given by $(\bigvee^{\nabla}_{i\in I}\phi_i)x = \inf_{i\in I}\phi_i x$, for all $x \in X$.

To show the self-duality of the category $\mathcal{V}$-Sup one can use the same argumentation as in the case of the role model Sup:

Proposition 141 *There is an isomorphism of categories* $\mathcal{V}\text{-Sup}^{\mathrm{op}} \longrightarrow \mathcal{V}\text{-Sup}$ *which assigns to a cocontinuous* $\mathcal{V}$*-functor* $f : X \to Y$ *the* $\mathcal{V}$*-functor* $(f^{\bullet})^{\mathrm{op}} : Y^{\mathrm{op}} \to X^{\mathrm{op}}$, *where* $f \dashv f^{\bullet}$ *in* $\mathcal{V}$-Cat.

Proof As noted above, by the Adjoint Functor Theorem, the cocontinuous $\mathcal{V}$-functor f has a right adjoint $f^{\bullet}$. Hence, as a left adjoint of f^{op}, the $\mathcal{V}$-functor $(f^{\bullet})^{\mathrm{op}}$ is a morphism of $\mathcal{V}$-Sup. □

Example 142 For a cocontinuous map $f : X \to Y$ of (co)complete metric spaces, the map $f^{\bullet} : Y \to X$ assigns to $y \in Y$ the element $\bigvee^{\nabla}\{x \in X \mid Y(fx, y) = 0\}$ in X.

7.2 The Box-Tensor Product Classifying Bi-Morphisms

Following the lead of Joyal and Tierney (1984) and proceeding as in Tholen (2024), like in the case $\mathcal{V} = 2$, for $X, Y \in \mathcal{V}$-Sup we now construct a tensor product $X \boxtimes Y$ which classifies the *bi-morphisms* $f : X \times Y \to Z$, i.e., the set maps with values in $Z \in \mathcal{V}$-Sup such that the maps

$$f(x,-) : Y \to Z \qquad \text{and} \qquad f(-,y) : X \to Z$$

are cocontinuous $\mathcal{V}$-functors, for all $x \in X$ and $y \in Y$. We first confirm that such maps f give $\mathcal{V}$-functors $X \otimes Y \to Z$.

Lemma 143 *Let* X, Y, Z *be* $\mathcal{V}$*-categories. If a map* $f : X \times Y \to Z$ *is such that, for all* $x \in X$ *and* $y \in Y$*, the maps* $f(x,-)$ *and* $f(-,y)$ *are* $\mathcal{V}$*-functors, then* $f : X \otimes Y \to Z$ *is a* $\mathcal{V}$*-functor.*

Proof By hypothesis, for $x, x' \in X$, $y, y' \in Y$, one has $X(x, x') \le Z(f(x, y), f(x', y))$ and $Y(y, y') \le Z(f(x', y), f(x', y'))$. This gives

$$\begin{aligned}(X \otimes Y)((x, y), (x', y')) &= X(x, x') \otimes Y(y, y') \\ &\le Z(f(x, y), f(x', y)) \otimes Z(f(x', y), f(x', y')) \\ &\le Z(f(x, y), f(x', y')).\end{aligned}$$

□

As a consequence of the lemma, for $X, Y, Z \in \mathcal{V}\text{-}\mathsf{Sup}$, we can think of the bi-morphisms $X \times Y \to Z$ as $\mathcal{V}$-functors $X \otimes Y \to Z$ which are cocontinuous $\mathcal{V}$-functors in each variable separately. Hence, taking advantage of the equational characterization of separated and cocomplete $\mathcal{V}$-categories as given by Theorem 137, we should construct a universal bi-morphism $X \otimes Y \longrightarrow X \boxtimes Y$ following standard procedures; that is, we should form the free cocomplete $\mathcal{V}$-category $\mathcal{P}_{\mathcal{V}}(X \otimes Y)$ and find the least congruence relation $\sim$ on $\mathcal{P}_{\mathcal{V}}(X \otimes Y)$ which makes the quotient a cocomplete $\mathcal{V}$-category and the composite map

$$t_{X,Y} : X \otimes Y \xrightarrow{\quad \mathbf{y} \quad} \mathcal{P}_{\mathcal{V}}(X \otimes Y) \xrightarrow{\quad r \quad} X \boxtimes Y \qquad (*)$$

(of the Yoneda map $\mathbf{y} = \mathbf{y}_{X \otimes Y}$ followed by the canonical projection r of $\sim$) a bi-morphism. However, given the cumbersome description of quotient structures in $\mathcal{V}\text{-}\mathsf{Cat}$ (see Corollary 24), establishing the existence of such a relation $\sim$ looks like a daunting task!

Fortunately, in $\mathcal{V}\text{-}\mathsf{Sup}$, there is a very efficient way of describing the congruence relation generated by any relation on any object, as shown in the case $\mathcal{V} = \mathsf{2}$ by Joyal and Tierney in (Joyal and Tierney (1984)). Indeed, their construction extends easily to the general case, as follows.

Proposition 144 *Let $R \subseteq P \times P$ be any relation on a separated (co)complete $\mathcal{V}$-category P, and let the set*

$$Q = \{y \in P \mid \forall (z_1, z_2) \in R \ \ (P(z_1, y) = P(z_2, y))\}$$

be considered as a full $\mathcal{V}$-subcategory of P. Then:

(1) *Q is (co)complete;*
(2) *there is a surjective morphism $r : P \to Q$ in $\mathcal{V}\text{-}\mathsf{Sup}$ whose kernel relation contains R; moreover,*
(3) *any morphism $r' : P \to Q'$ in $\mathcal{V}\text{-}\mathsf{Sup}$ whose kernel relation contains R factors (uniquely) through r, by a morphism $h : Q \to Q'$ in $\mathcal{V}\text{-}\mathsf{Sup}$.*

Proof (1) For all $y \in Q$, $(z_1, z_2) \in Q$ and $u \in \mathcal{V}$ one has

$$P(z_1, u \pitchfork y) = [u, P(z_1, y)] = [u, P(z_2, y)] = P(z_2, u \pitchfork y).$$

This shows that Q is closed under cotensors in P; likewise for conical infima. Consequently, Q is closed under all weighted limits in P and, hence, (co)complete.

(2) Being continuous, the fully faithful inclusion $\mathcal{V}$-functor $i : Q \hookrightarrow P$ has a left adjoint, $r : P \to Q$, exhibiting Q as a full reflective $\mathcal{V}$-subcategory of P, satisfying the strict identity $ri = 1_Q$, by separation in P. By adjunction, we have $Q(rx, y) = P(x, y)$ for all $x \in P$, $y \in Q$, and, in the induced order of P, we can describe r by $rx = \bigwedge^{\triangledown}\{y \in Q \mid x \leq y\}$. Consequently, for all $(z_1, z_2) \in R$ we have

$$rz_1 = \bigwedge\{y \in Q \mid \mathrm{k} \leq P(z_1, y)\} = \bigwedge\{y \in Q \mid \mathrm{k} \leq P(z_2, y)\} = rz_2.$$

(3) For $r' : P \to Q'$ in $\mathcal{V}$-Sup with $r'z_1 = r'z_2$ for all $(z_1, z_2) \in R$ we have the right adjoint $j : Q' \to P$ of r', satisfying $P(x, jy') = Q'(r'x, y')$ for all $x \in P$ and $y' \in Q'$. Consequently, with

$$P(z_1, jy') = Q'(r'z_1, y') = Q'(r'z_2, y') = P(z_2, jy')$$

holding for all $(z_1, z_2) \in R$, we see that the continuous $\mathcal{V}$-functor j takes values in Q, and we obtain the left adjoint $h : Q \to Q'$ of the restriction of j. Moreover, the left adjoints commute as $hr = r'$, since their respective right adjoints commute trivially.

□

Returning to our goal described above, given $X, Y \in \mathcal{V}$-Sup, we consider $P := \mathcal{P}_{\mathcal{V}}(X \otimes Y)$ and, writing $z_1 \, R \, z_2$ for $(z_1, z_2) \in R$, let $R \subseteq P \times P$ be the relation defined by

$$\mathbf{y}(u \odot x, y) \;\; R \;\; u \odot \mathbf{y}(x, y) \;\; R \;\; \mathbf{y}(x, u \odot y),$$

$$\mathbf{y}(\bigvee_{i \in I} x_i, y) \;\; R \;\; \bigvee_{i \in I} \mathbf{y}(x_i, y), \qquad \mathbf{y}(x, \bigvee_{i \in I} y_i) \;\; R \;\; \bigvee_{i \in I} \mathbf{y}(x, y_i),$$

for all $u \in \mathcal{V}$, $x, x_i \in X$, $y, y_i \in Y$, $i \in I$, with any set I. Then Proposition 144 gives us the $\mathcal{V}$-Sup-object $X \boxtimes Y := Q$ and its morphism $r : P \to Q$ satisfying the properties (1-3). These properties now give immediately:

Theorem 145 *Given $X, Y \in \mathcal{V}$-Sup, the map $t_{X,Y} = r\mathbf{y} : X \otimes Y \to X \boxtimes Y$ is a universal bi-morphism into a $\mathcal{V}$-Sup-object: every bi-morphism $f : X \otimes Y \to Z$ with $Z \in \mathcal{V}$-Sup has the form $f = \overline{f} t_{X,Y}$, with a unique $\mathcal{V}$-Sup-morphism $\overline{f} : X \boxtimes Y \to Z$.*

Proof Since Z is (co)complete, by Theorem 109 the bi-morphism f factors as $f = \tilde{f}\mathbf{y}$, with a unique $\mathcal{V}$-Sup-morphism $\tilde{f}$ which, in turn, factors uniquely as $\tilde{f} = \overline{f}r$.

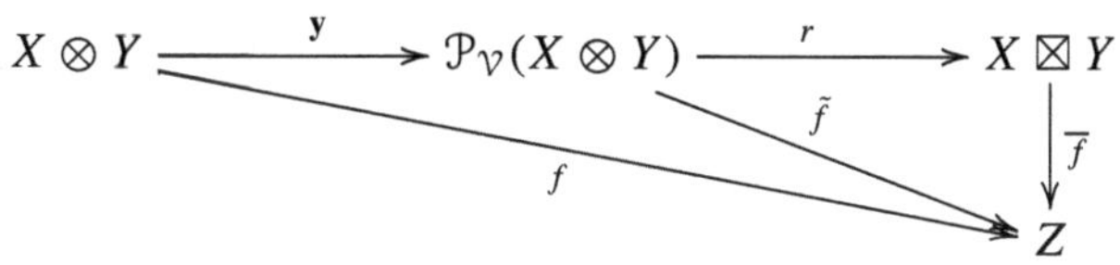

□

7.3 *The Symmetric Monoidal-Closed Category* ($\mathcal{V}$-Sup, ⊠, $\mathcal{V}$)

It is clear that the box-tensor product in $\mathcal{V}$-Sup inherits its natural associativity and commutativity isomorphisms from the corresponding morphisms of the tensor product of $\mathcal{V}$-Cat, *qua* universality of its definition, as indicated by the following diagram:

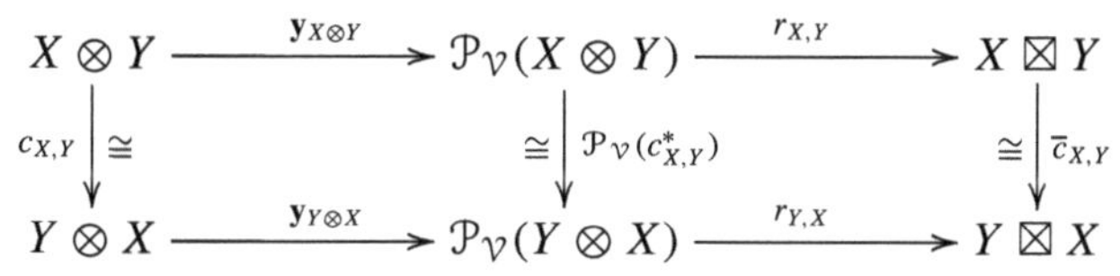

Let us show that the $\mathcal{V}$-category $\mathcal{V} = \mathcal{P}_{\mathcal{V}}\mathrm{E}$ is ⊠-neutral in $\mathcal{V}$-Sup:

Lemma 146 *There is an isomorphism* $\ell_X : \mathcal{V} \boxtimes X \to X$, *natural for all* $X \in \mathcal{V}$-Sup.

Proof It suffices to confirm that the map $\mathcal{V} \otimes X \to X$, $(u, x) \mapsto u \odot x$, is a bi-morphism, universal among all bi-morphisms $\mathcal{V} \otimes X \to Z$ into $\mathcal{V}$-Sup-objects Z. But this property one checks routinely using the rules of Lemma 118. □

Theorem 147 *The category* $\mathcal{V}$-Sup*, provided with the box-tensor product* ⊠ *and the* ⊠*-neutral* $\mathcal{V}$*-category* $\mathcal{V}$*, becomes symmetric monoidal-closed, with the internal hom*

$$[\![X, Y]\!] = \mathcal{V}\text{-}\mathsf{Sup}(X, Y)$$

inheriting its $\mathcal{V}$*-category structure from* $[X, Y] = \mathcal{V}$-$\mathsf{Cat}(X, Y)$*, for all* $X, Y \in \mathcal{V}$-Sup.

Proof The coherence conditions for the monoidal structure are an easy consequence of the fact that the structure has been defined entirely by a universal property. Let us therefore focus on its closedness. For $X, Y \in \mathcal{V}$-Sup, we must show that the full $\mathcal{V}$-subcategory $[\![X, Y]\!]$ of $[X, Y]$ is cocomplete. Indeed, defining tensors and conical suprema pointwise by

$$(u \odot f)x = u \odot fx \qquad \text{and} \qquad (\bigvee_{i\in I}^{\nabla} f_i)x = \bigvee_{i\in I}^{\nabla} f_i x$$

for $u \in \mathcal{V}$, $f, f_i \in [\![X, Y]\!]$ $(i \in I)$ and all $x \in X$, one routinely shows that $u \odot f$ and $\bigvee_{i\in I}^{\nabla} f_i$ are well defined (i.e., preserve tensors and conical suprema) and actually do the job indicated by their names. For example, the computation

$$\begin{aligned}[\![X, Y]\!](u \odot f, g) &= \bigwedge_{x\in X} Y(u \odot fx, gx) = \bigwedge_{x\in X}[u, Y(fx, gx)] \\ &= [u, \bigwedge_{x\in X} Y(fx, gx)] = [u, [\![X, Y]\!](f, g)]\end{aligned}$$

for all $g \in [\![X, Y]\!]$ confirms that $u \odot f$ is indeed the tensor of u and f in $[\![X, Y]\!]$.

In order to show that $[\![X, Y]\!]$ serves as the internal hom in $\mathcal{V}$-Sup, for all $Z \in \mathcal{V}$-Sup, we now sketch how to establish a natural $\mathcal{V}$-Cat-isomorphism

$$[\![Z \boxtimes X, Y]\!] \cong [\![Z, [\![X, Y]\!]]\!]. \tag{1}$$

We let $\mathrm{BiMor}(Z, X; Y)$ be the $\mathcal{V}$-subcategory of $[Z \otimes X, Y]$ given by all bimorphisms $Z \otimes X \to Y$. By Theorem 147 one has the isomorphism $[\![Z \boxtimes X, Y]\!] \to \mathrm{BiMor}(Z, X; Y)$ which actually lives in $\mathcal{V}$-Cat. It then remains to be shown that the natural isomorphism $[Z \otimes X, Y] \to [Z, [X, Y]]$ of Theorem 32 can be restricted as indicated by the diagram (in which the vertical arrows are given by inclusion)

$$\begin{array}{ccccc} & & [Z \otimes X, Y] & \xrightarrow{\quad\cong\quad} & [Z, [X, Y]] \, . \\ & & \uparrow & & \uparrow \\ [\![Z \boxtimes X, Y]\!] & \xrightarrow{\quad\cong\quad} & \mathrm{BiMor}(Z, X; Y) & \xrightarrow{\quad\cong\quad} & [\![Z, [\![X, Y]\!]]\!] \end{array}$$

This, however, is now a routine exercise, as one may use tensors and conical suprema of cocontinuous $\mathcal{V}$-functors, all defined pointwise, to guarantee well-definedness of the restricted isomorphism. $\square$

7.4 ($\mathcal{V}$-Sup, $\boxtimes$, $\mathcal{V}$) Is $*$-Autonomous

We will now combine the isomorphism (1) of the proof of Theorem 147 (exhibiting ($\mathcal{V}$-Sup, $\boxtimes$, $\mathcal{V}$) as symmetric monoidal-closed) with two further fundamental isomorphisms. First, by Theorem 109 one has, for all $X, Z \in \mathcal{V}$-Sup, the natural isomorphism $[Z, X] \cong [\![\mathcal{P}_{\mathcal{V}}Z, X]\!]$, which is easily seen to live in $\mathcal{V}$-Cat. For $Z = \mathrm{E}$, with Lemma 146, we therefore have the natural isomorphism:

$$X \cong [\![\mathcal{V}, X]\!]. \tag{2}$$

It assigns to $x \in X$ the $\mathcal{V}$-functor $- \odot x : \mathcal{V} \to X$; the inverse map evaluates $\phi \in [\![\mathcal{V}, X]\!]$ at k. Furthermore, from Proposition 141 we have the natural isomorphism:

$$[\![X, Y]\!] \cong [\![Y^{\mathrm{op}}, X^{\mathrm{op}}]\!]. \tag{3}$$

Replacing X in (2) by X^{op} and then applying (3), one obtains the natural isomorphism:

$$X^{\mathrm{op}} \cong [\![X, \mathcal{V}^{\mathrm{op}}]\!], \tag{4}$$

which turns out to send $x \in X$ to $X(-, x) : X \to \mathcal{V}^{\mathrm{op}}$.

Since $(X^{\mathrm{op}})^{\mathrm{op}} = X$, and likewise for morphisms, the isomorphism (4) exhibits the $\mathcal{V}$-category $\mathcal{V}^{\mathrm{op}}$ as a *global dualizing object* in $\mathcal{V}$-Sup. This fact, in conjunction with Theorem 147, makes the category $\mathcal{V}$-Sup a $*$-*autonomous category* in the sense of Barr (1979) (see also Pastro and Street (2009)). This result was first established in Chap. 3 of Eklund et al. (2018) from a more order-theoretical than categorical perspective, with a proof that uses the symmetric monoidal-closed structure of Sup, whereas here this structure has been re-established in a generalized context.

Theorem 148 (Duality Theorem) *The functor $[\![-, \mathcal{V}^{\mathrm{op}}]\!] : \mathcal{V}\text{-Sup}^{\mathrm{op}} \to \mathcal{V}\text{-Sup}$ is an equivalence of categories, with pseudo-inverse $[\![-, \mathcal{V}^{\mathrm{op}}]\!]^{\mathrm{op}}$, so that*

$$X \cong [\![[\![X, \mathcal{V}^{\mathrm{op}}]\!], \mathcal{V}^{\mathrm{op}}]\!],$$

naturally in $X \in \mathcal{V}$-Sup.

Proof We must show that, for $f : X \to Y$ in $\mathcal{V}$-Sup, under the natural isomorphism (4) the map $(f^{\bullet})^{\mathrm{op}} : Y^{\mathrm{op}} \to X^{\mathrm{op}}$ corresponds to precomposition by f, i.e., that the following diagram commutes:

$$\begin{array}{ccc} X^{\mathrm{op}} & \xrightarrow{\ \cong\ } & [\![X, \mathcal{V}^{\mathrm{op}}]\!] \\ {\scriptstyle (f^{\bullet})^{\mathrm{op}}}\uparrow & & \uparrow{\scriptstyle -f} \\ Y^{\mathrm{op}} & \xrightarrow{\ \cong\ } & [\![Y, \mathcal{V}^{\mathrm{op}}]\!] \end{array}$$

But the commutativity is just a consequence of the adjunction $X(x, f^{\bullet}y) = Y(fx, y)$. □

Remark 149 The dualizing natural isomorphism $\delta_X : X \to \llbracket \llbracket X, \mathcal{V}^{\mathrm{op}} \rrbracket, \mathcal{V}^{\mathrm{op}} \rrbracket$ is, up to $\boxtimes$-symmetry, simply the transpose of the adjunction counit ε_X, determined by the commutative diagram:

$$\begin{array}{ccc}
\llbracket \llbracket X, \mathcal{V}^{\mathrm{op}} \rrbracket, \mathcal{V}^{\mathrm{op}} \rrbracket \boxtimes \llbracket X, \mathcal{V}^{\mathrm{op}} \rrbracket & \xrightarrow{\ \varepsilon_{\llbracket X, \mathcal{V}^{\mathrm{op}} \rrbracket}\ } & \mathcal{V}^{\mathrm{op}} \\
{\scriptstyle \delta_X \boxtimes 1_{\llbracket X, \mathcal{V}^{\mathrm{op}} \rrbracket}} \uparrow & & \uparrow {\scriptstyle \varepsilon_X} \\
X \boxtimes \llbracket X, \mathcal{V}^{\mathrm{op}} \rrbracket & \xrightarrow{\ \cong\ } & \llbracket X, \mathcal{V}^{\mathrm{op}} \rrbracket \boxtimes X
\end{array}$$

We list two further natural isomorphisms arising easily from the isomorphisms (1-4), exhibiting the smooth interaction of the box-tensor and the internal hom in the self-dual category $\mathcal{V}$-Sup.

Corollary 150 *For all* $X, Y \in \mathcal{V}$-Sup, *one has natural isomorphisms*

$$(X \boxtimes Y)^{\mathrm{op}} \cong \llbracket X, Y^{\mathrm{op}} \rrbracket \quad \text{and} \quad \llbracket X, Y \rrbracket^{\mathrm{op}} \cong X \boxtimes Y^{\mathrm{op}}.$$

Proof In light of $(Y^{\mathrm{op}})^{\mathrm{op}} = Y$, the first isomorphism is a replica of the second isomorphism, which follows from

$$\begin{aligned}
\llbracket X, Y \rrbracket &\cong \llbracket Y^{\mathrm{op}}, X^{\mathrm{op}} \rrbracket && (3) \\
&\cong \llbracket Y^{\mathrm{op}}, \llbracket X, \mathcal{V}^{\mathrm{op}} \rrbracket \rrbracket && (4) \\
&\cong \llbracket Y^{\mathrm{op}} \boxtimes X, \mathcal{V}^{\mathrm{op}} \rrbracket && (1) \\
&\cong (Y^{\mathrm{op}} \boxtimes X)^{\mathrm{op}} && (4) \\
&\cong (X \boxtimes Y^{\mathrm{op}})^{\mathrm{op}}.
\end{aligned}$$

□

In conjunction with (1), the first of the two natural isomorphisms of the corollary gives the natural isomorphism

$$\llbracket X \boxtimes Y, Z^{\mathrm{op}} \rrbracket \cong \llbracket X, (Y \boxtimes Z)^{\mathrm{op}} \rrbracket,$$

which is often used in the axiomatics of a $*$-autonomous category (see the preamble to Sect. 7).

The assignment $(f : X \to Y) \longmapsto (f^{\mathrm{op}} : X^{\mathrm{op}} \to Y^{\mathrm{op}})$ makes the category Sup isomorphic to the category Inf of complete lattices and their infima-preserving maps. This isomorphism extends to an isomorphism

$$\mathsf{Sup}^{\mathcal{V}} \cong \mathsf{Inf}^{\mathcal{V}^{\mathrm{op}}},$$

where the objects of $\mathsf{Inf}^{\mathcal{V}^{\mathrm{op}}}$ are complete lattices with an action of $\mathcal{V}$ such that the conditions (1–4) as in Proposition 136 now hold with $\bigvee$ traded for $\bigwedge$, but note that then "$\bigwedge_{i\in I} u_i$" has to be read in the lattice $\mathcal{V}^{\mathrm{op}}$, i.e., it coincides with $\bigvee_{i\in I} u_i$ taken in $\mathcal{V}$. Still, this viewpoint is useful in the case of $\mathcal{V} = [0, \infty]_+$ since its underlying lattice $([0, \infty], \geq)$ reverts back to the natural order $\leq$ under the isomorphism, as follows.

Example 151 The objects X of the category $\mathsf{Sup}^{[0,\infty]_+}$ may be described isomorphically as complete lattices equipped with an action $+ : [0, \infty] \times X \longrightarrow X$ (denoted suggestively like the monoid operation in $[0, \infty]$), satisfying the following conditions for all $x, x_i \in X,\ u, u_i \in [0, \infty]$ $(i \in I)$:

(1) $0 + x = x$;
(2) $u + (v + x) = (u + v) + x$;
(3) $(\inf_i u_i) + x = \inf_i(u_i + x)$;
(4) $u + \inf_i x_i = \inf_i(u + x_i)$.

Morphisms are infima-preserving and $+$-equivariant maps. The underlying metric of X is then given by $X(x, y) = \inf\{u \in [0, \infty] \mid x \leq u + y\}$, and the dualizing object of X in our $*$-autonomous category takes the form:

$$X^\star := \mathsf{Inf}^{[0,\infty]}(X, [0, \infty]),$$

provided with the infima and the $+$-action defined pointwise in $[0, \infty]$. The induced metric of $X^\star$ is the non-symmetric sup-metric, which is also inherited from $[0, \infty]$. The natural isomorphism $X \to (X^\star)^\star$ is given by evaluation.

8 Cauchy–Lawvere Completion and Its Generalizations

8.1 Cauchy Sequences and Their Convergence

Recall that, in a Fréchet metric space X, a sequence $s = (x_n)_{n\in\mathbb{N}}$ is Cauchy if, and only if,

$$\inf_{N\geq 0}\ \sup_{m,n\geq N} X(x_m, x_n) = 0.$$

For an arbitrary sequence s, we may look at the left-hand side as a value for the extent to which s fails to be a Cauchy sequence. Since s converges to a point $x \in X$ if, and only if,

$$\inf_{N\geq 0}\ \sup_{n\geq N} X(x_n, x) = 0 = \inf_{N\geq 0}\ \sup_{n\geq N} X(x, x_n),$$

we similarly look at these inf-sup terms as values for the extent to which an arbitrary sequence s fails to converge to x. Of course, in case of a generalized (and, hence, generally non-symmetric) metric, we have to distinguish between the two inf-sup terms.

Extrapolating from $[0, \infty]_+$ to our arbitrary quantale $(\mathcal{V}, \otimes, \mathrm{k})$, one defines.

Definition 152 For $X \in \mathcal{V}$-Cat, let $x \in X$ and $s = (x_n)_{n\in\mathbb{N}}$ be a sequence in X. We put

$$\mathrm{Cauchy}_X(s) := \bigvee_{N\geq 0} \bigwedge_{m,n\geq N} X(x_m, x_n),$$

$$\lambda_s(x) := \bigvee_{N\geq 0} \bigwedge_{n\geq N} X(x_n, x), \quad \text{and} \quad \rho_s(x) := \bigvee_{N\geq 0} \bigwedge_{n\geq N} X(x, x_n),$$

and call $\mathrm{Cauchy}_X(s)$ the *Cauchy value of* s and $\lambda_s(x)$ $(\rho_s(x))$ the *left* (*right*, respectively) *convergence value of* s *to* x. We say that

(1) s is a *Cauchy* sequence if $\mathrm{k} \leq \mathrm{Cauchy}_X(s)$;
(2) s *converges to* x (briefly, $s \rightsquigarrow x$) if $\mathrm{k} \leq \bigvee_{N\geq 0}(\bigwedge_{m\geq N} X(x_m, x) \otimes \bigwedge_{n\geq N} X(x, x_n))$;
(3) X is *sequentially Cauchy-complete*, henceforth just *Cauchy-complete*, if every Cauchy sequence in X converges to some point in X.

Remarks 153 (1) These notions are self-dual: since $\mathrm{Cauchy}_X(s) = \mathrm{Cauchy}_{X^{\mathrm{op}}}(s)$, the sequence s is Cauchy (converges to x) in X if, and only if, the respective statement holds in X^{op}.
(2) Unlike in the case $\mathcal{V} = [0, \infty]_+$, the above notions are trivial for the Boolean quantale $\mathcal{V} = 2$: a Cauchy sequence in $X \in$ Ord consists eventually of order-equivalent elements, each of which will serve as the limit of the sequence. Consequently, every $X \in$ Ord is Cauchy-complete.

Before we explore the connections between these terms and notions in the proposition below, let us note the following easy, but useful, fact.

Lemma 154 *Let the ordered set I be (upward) directed, and $(u_i)_i$ and $(v_i)_i$ be I-indexed diagrams in (the thin category) $\mathcal{V}$. Then*

$$\bigvee_{i\in I} u_i \otimes v_i = \bigvee_{i\in I} u_i \otimes \bigvee_{j\in I} v_j\ .$$

Proof One obviously has "$\leq$." For "$\geq$," note that the RHS computes to $\bigvee_{i,j\in I} u_i \otimes v_j$. But for every pair i, j we have some $k \geq i, j$ in I, so that $u_i \otimes v_j \leq u_k \otimes v_k \leq$ LHS follows. □

Proposition 155 *For $X, x, s = (x_i)_{i\in I}$ as above, with $\mathcal{V} = \mathcal{V}$-$\mathsf{Dist}(\mathrm{E}, \mathrm{E})$ one has:*

(1) $E \overset{\lambda_s}{\longrightarrow\!\!\!\!\!\circ\!\!\!\!\!\!\longrightarrow} X$ *and* $X \overset{\rho_s}{\longrightarrow\!\!\!\!\!\circ\!\!\!\!\!\!\longrightarrow} E$, *with* $\lambda_s \cdot \rho_s \leq 1_X^*$.
(2) $\mathrm{Cauchy}_X(s) \otimes \mathrm{Cauchy}_X(s) \leq \rho_s \cdot \lambda_s \leq \mathrm{Cauchy}_X(s)$.
(3) *s is Cauchy* $\iff \mathrm{k} \leq \rho_s \cdot \lambda_s \iff \lambda_s \dashv \rho_s$.
(4) $s \rightsquigarrow x \iff \mathrm{k} \leq \rho_s(x) \otimes \lambda_s(x) \iff x_* = \lambda_s \cdot \rho_s(x) \iff x^* = \lambda_s(x) \cdot \rho_s$.
(5) $s \rightsquigarrow x \Longrightarrow$ *s is Cauchy.*
(6) *(s Cauchy and $\lambda_s \leq x_*$ and $\rho_s \leq x^*$)* $\iff$ *($\lambda_s = x_*$ and $\rho_s = x^*$)* $\Longrightarrow s \rightsquigarrow x$.

Proof (1) The proof that λ_s and ρ_s are $\mathcal{V}$-distributors is straightforward. For $\lambda_s \cdot \rho_s \leq 1_X^*$, with the lemma we have for all $x, y \in X$

$$
\begin{aligned}
(\lambda_s \cdot \rho_s)(x, y) = \rho_s(x) \otimes \lambda_s(y) &= (\bigvee_N \bigwedge_{m\geq N} X(x, x_m)) \otimes (\bigvee_N \bigwedge_{n\geq N} X(x_n, y)) \\
&= \bigvee_N (\bigwedge_{m\geq N} X(x, x_m) \otimes \bigwedge_{n\geq N} X(x_n, y)) \\
&\leq \bigvee_N X(x, x_N) \otimes X(x_N, y) \leq X(x, y).
\end{aligned}
$$

(2) Under the identification $\mathcal{V}$-$\mathsf{Dist}(\mathrm{E}, \mathrm{E}) = \mathcal{V}$, with the lemma we obtain

$$
\begin{aligned}
\mathrm{Cauchy}_X(s) \otimes \mathrm{Cauchy}_X(s) &\leq \bigvee_N \bigwedge_{m\geq N} X(x_m, x_N) \otimes \bigvee_N \bigwedge_{n\geq N} X(x_N, x_n) \\
&\leq \bigvee_N \bigvee_{z\in X} (\bigwedge_{m\geq N} X(x_m, z) \otimes \bigwedge_{n\geq N} X(z, x_n)) \\
&= \bigvee_{z\in X} (\bigvee_N \bigwedge_{m\geq N} X(x_m, z)) \otimes (\bigvee_N \bigwedge_{n\geq N} X(z, x_n)) \\
&= \bigvee_{z\in X} \lambda_s(z) \otimes \rho_s(z) = \rho_s \cdot \lambda_s \\
&= \bigvee_{z\in X} \bigvee_N (\bigwedge_{m\geq N} X(x_m, z)) \otimes (\bigwedge_{n\geq N} X(z, x_n)) \\
&\leq \bigvee_{z\in X} \bigvee_N \bigwedge_{m,n\geq N} X(x_m, z) \otimes X(z, x_n) \\
&\leq \bigvee_N \bigwedge_{m,n\geq N} X(x_m, x_n) = \mathrm{Cauchy}_X(s).
\end{aligned}
$$

(3) Noting that $\mathrm{k} \leq \mathrm{Cauchy}_X(s)$ implies $\mathrm{k} = \mathrm{k} \otimes \mathrm{k} \leq \mathrm{Cauchy}_X(s) \otimes \mathrm{Cauchy}_X(s)$, one derives the assertions from (2) and (1).

(4) The first equivalence follows from the lemma again. Next, assuming $\mathrm{k} \leq \rho_s(x) \otimes \lambda_s(x)$, with (1) one obtains for all $y \in X$

$$x_*(y) \leq \rho_s(x) \otimes \lambda_s(x) \otimes X(x, y) \leq \rho_s(x) \otimes \lambda_s(y) \leq X(x, y) = x_*(y)$$

and, hence, $x_* = \lambda_s \cdot \rho_s(x)$. Evaluating this equality at x takes us back to $s \rightsquigarrow x$. Dually for x^*.

(5) With (4) one has $\mathrm{k} \leq x^*(x) = \lambda_s(x) \otimes \rho_s(x)$, which gives s Cauchy, by (3).

(6) If s is Cauchy and $\lambda_s \leq x_*$ and $\rho_s \leq x^*$, since always $x_* \dashv x^*$, and $\lambda_s \dashv \rho_s$ by (3), we obtain $x_* \leq x_* \cdot \rho_s \cdot \lambda_s \leq x_* \cdot x^* \cdot \lambda_s \leq \lambda_s$ and, dually, $x^* \leq \rho_s$. This gives $\lambda_s = x_*$ and $\rho_s = x^*$ which, in turn, implies $\mathrm{k} \leq X(x, x) \otimes X(x, x) = \lambda_s(x) \otimes \rho_s(x)$ and, hence, $s \rightsquigarrow x$, by (4).

□

In order to be able to reverse the last implication in (6) of the Proposition, we assume $\mathrm{k} = \top$ in $\mathcal{V}$, as follows.

Theorem 156 *Let $\mathcal{V}$ be integral, and let $X, x, s = (x_n)_n$ be as above. Then the following statements are equivalent:*

(i) *$s \rightsquigarrow x$;*
(ii) *$\lambda_s = x_*$ and $\rho_s = x^*$;*
(iii) *s is Cauchy and $\lambda_s = x_*$;*
(iv) *s is Cauchy and $\rho_s = x^*$;*
(v) *s is Cauchy, and $\lim^{\lambda_s} 1_X$, $\mathrm{colim}^{\rho_s} 1_X$ both exist in X and, up to order-equivalence, coincide with x.*

Proof Since $\mathcal{V}$ is integral, one has $\rho_s(x) \otimes \lambda_s(x) \leq \mathrm{k} \otimes \lambda_s(x) = \lambda_s(x)$, and likewise $\rho_s(x) \otimes \lambda_s(x) \leq \rho_s(x)$. Hence, ($\mathrm{k} \leq \rho_s(x) \otimes \lambda_s(x) \iff \mathrm{k} \leq \lambda_s(x)$ and $\mathrm{k} \leq \rho_s(x)$), and (i) $\iff$ (ii) follows with (6) and (4) of the proposition.

(ii) $\iff$ (iii) $\iff$ (iv) follows from (3) and (6) of the proposition.

(ii) $\iff$ (v): By the adjunction $x_* \dashv x^*$, one always has $\mathrm{colim}^{x^*} 1_X \simeq x \simeq \lim^{x_*} 1_X$. This makes (ii)$\Longrightarrow$(v) obvious. Conversely, assuming $\mathrm{colim}^{\rho_s} 1_X \simeq x \simeq \lim^{\lambda_s} 1_X$, since $\lambda_s \cdot \rho_s \leq 1_X^*$ by (1) of the proposition, the (co)limit property implies $\lambda_s \leq x_*$ and $\rho_s \leq x^*$, so that (6) of the proposition gives (ii) when s is Cauchy. □

Corollary 157 *Let $\mathcal{V}$ be integral and X be a $\mathcal{V}$-category with conical suprema and infima. Then, for any sequence $s = (x_n)_n$ in X, $\mathrm{colim}^{\rho_s} 1_X$ and $\lim^{\lambda_s} 1_X$ both exist in X and may be computed as*

$$\mathrm{colim}^{\rho_s} 1_x \simeq \bigvee_{N \geq 0} \bigwedge_{n \geq N} x_n \quad \textit{and} \quad \lim^{\lambda_s} 1_x \simeq \bigwedge_{N \geq 0} \bigvee_{n \geq N} x_n.$$

Furthermore, one has $s \rightsquigarrow x$ if, and only if, s is Cauchy and x is the common value of the above colimit and limit, up to order-equivalence.

Proof Once we have shown the first assertion, the second follows with the theorem. By hypothesis we have the conical infima $x'_N := \bigwedge_{n\geq N} x_n$ for all $N \geq 0$, as well as the conical supremum $x := \bigvee_{N\geq 0} x'_N$. The following computation for all $y \in X$ confirms the existence of $\mathrm{colim}^{\rho_s} 1_X$, given by x:

$$\begin{aligned}\bigwedge_{z\in X}[\rho_s(z), X(z,y)] &= \bigwedge_{z\in X}[\bigvee_{N}\bigwedge_{n\geq N} X(z,x_n), X(z,y)] \\ &= \bigwedge_{N}\bigwedge_{z\in X}[X(z,x'_N), X(z,y)] \\ &= \bigwedge_{N}(\mathcal{P}_{\mathcal{V}}X)(\mathbf{y}_X x'_N, \mathbf{y}_X y) = \bigwedge_{N} X(x'_N, y) = X(x,y).\end{aligned}$$

Likewise one shows the existence of $\lim^{\lambda_s} 1_X$. □

Example 158 For $\mathcal{V} = [0,\infty]_+$ and $X = \mathcal{V}$ we obtain

$$\mathrm{colim}^{\rho_s} 1_X = \inf_{N\geq 0}\sup_{n\geq N} x_n \quad \text{and} \quad \lim^{\lambda_s} 1_X = \sup_{N\geq 0}\inf_{n\geq N} x_n,$$

with the colimit known as the *limit superior* and the limit as the *limit inferior* of the arbitrary sequence $s = (x_n)_n$ in $[0,\infty]$.

The characterization of Cauchy sequences via adjunction as in (3) of the previous proposition, as well as the notion of Cauchy completeness, will be further explored in the following sections. Here we mention only the easily seen fact that, for a sequence to be Cauchy, or to converge, is a well-behaved property vis-à-vis $\mathcal{V}$-functors and change of base (as defined in Proposition 19). We omit the straightforward proofs.

Proposition 159 *Let $s = (x_n)_{n\in\mathbb{N}}$ be a sequence in the $\mathcal{V}$-category X, and let $x \in X$.*

(1) *For every $\mathcal{V}$-functor $f : X \to Y$, if s is Cauchy (converges to x), then $fs = (fx_n)_n$ is Cauchy (converges to fx, respectively).*
(2) *For every lax homomorphism $\varphi : \mathcal{V} \to \mathcal{W}$ which preserves suprema, if s is Cauchy or converges to x, the respective property holds in the $\mathcal{W}$-category $\mathcal{B}_\varphi X$.*

8.2 Lawvere Completeness

The characterization of Cauchy sequences via adjunction as in Proposition 155 is one of the many great insights in Lawvere's paper (Lawvere (1973)) which led him to the following notion of completeness, here named after him.[6]

[6] Many other authors use "Cauchy-complete" instead.

Definition 160 A $\mathcal{V}$-category X is *Lawvere-complete* if every adjunction $\varphi \dashv \psi : X \nrightarrow W$ is represented by a $\mathcal{V}$-functor $f : W \to X$, such that $\varphi = f_*$ and (necessarily then) $\psi = f^*$.

Let us observe immediately that it suffices to consider $W = \mathsf{E}$. Indeed, assuming that these particular $\mathcal{V}$-distributors are represented, and now considering $\varphi \dashv \psi$ with W arbitrary, then for every $t \in W$ the adjunction $\varphi \cdot t_* \dashv t^* \cdot \psi$ is represented by some element in X, $ft \in X$, say. Hence, choice granted, it only remains to be checked that this defines a $\mathcal{V}$-functor $f : W \to X$:

$$W(s,t) = t^* \cdot 1_X^* \cdot s_* \leq t^* \cdot \psi \cdot \varphi \cdot s_* = (ft)^* \cdot (fs)_* = X(fs, ft),$$

for all $s, t \in W$.

Proposition 161 *The following statements are equivalent for a $\mathcal{V}$-category X:*

(i) *X is Lawvere-complete;*
(ii) *for every right-adjoint $\mathcal{V}$-distributor $\psi : X \nrightarrow \mathsf{E}$, the colimit* $\mathrm{colim}^{\psi} 1_X$ *exists in X;*
(iii) *for every left-adjoint $\mathcal{V}$-distributor $\varphi : \mathsf{E} \nrightarrow X$, the limit* $\lim^{\psi} 1_X$ *exists in X.*

Proof For every adjunction $\varphi \dashv \psi : X \nrightarrow \mathsf{E}$, the $\mathcal{V}$-distributor φ is the extension of 1_X^* along ψ, and ψ is the lifting of 1_X^* along φ. Representing these, respectively, as x^* and x_* means precisely $\mathrm{colim}^{\psi} 1_X \simeq x$ and $\lim^{\varphi} 1_X \simeq x$. □

The proposition above trivially gives the first implication below, and the second follows from Proposition 155(4), (6).

Corollary 162 *For every $\mathcal{V}$-category X one has the implications*

$$X\ (co)complete \Longrightarrow X\ Lawvere\text{-}complete \Longrightarrow X\ Cauchy\text{-}complete.$$

Whereas the first implication may *rarely* (unless $\mathcal{V}$ is trivial) be inverted (as it fails already for $\mathcal{V} = \mathsf{2}$, since a non-empty ordered set is always Lawvere-complete, but tensored only if it has a least element, by Example 117(1)), the second implication is *often* a logical equivalence, as we show next. One simply mimics in a general quantale a key property of its role model $[0, \infty]_+$ (which trivially holds for $\mathcal{V} = \mathsf{2}$, but also non-trivially in other relevant quantales).

First we recall that, in our complete lattice $\mathcal{V}$ one calls an element u *totally below* the element v, written as $u \ll v$, if $v \leq \bigvee_{i \in I} z_i$ always implies $u \leq z_j$ for some $j \in I$. Note that this makes $u \ll \bot = \bigvee \emptyset$ impossible. For $\mathcal{V} = \mathsf{2} = \{\bot, \top\}$, we see $(u \ll v \iff v = \top)$, and in $\mathcal{V} = ([0, \infty], \geq)$ one has $(u \ll v \iff u > v)$, with the natural $>$.

Definition 163 We say that $v \in \mathcal{V}$ *is sequentially approximated from totally below* if there is a sequence $(\varepsilon_n)_{n\geq 0}$ in $\mathcal{V}$ such that $\varepsilon_n \leq \varepsilon_{n+1} \ll v$ for all $n \in \mathbb{N}$, and $\bigvee_{n\geq 0} \varepsilon_n = v$.

We are now ready to prove:

Theorem 164 (Hofmann-Reis 2018) *In the integral quantale $\mathcal{V}$, let $\mathrm{k} = \top$ be sequentially approximated from totally below. Then every adjunction $\varphi \dashv \psi : X \multimapdotinv \mathrm{E}$ in $\mathcal{V}$-*Dist *is of the form $\lambda_s \dashv \rho_s$ for some (Cauchy) sequence s in X.*

Proof Since $\varepsilon_n \ll \mathrm{k} \leq \psi \cdot \varphi = \bigvee_{x\in X} \varphi x \otimes \psi x$, for every $n \in \mathbb{N}$ we have some (tacitly chosen) $x_n \in X$ with $\varepsilon_n \leq \varphi x_n \otimes \psi x_n$. We note that, by integrality, this implies $\varepsilon_n \leq \varphi x_n \wedge \psi x_n$, and that, by Lemma 154, $\mathrm{k} = \bigvee_n \varepsilon_n$ gives $\mathrm{k} = \bigvee_n \varepsilon_n \otimes \varepsilon_n$.

We claim that the sequence $s = (x_n)_n$ is Cauchy. Indeed, for all $m, n \geq N \geq 0$ one has with Proposition 155(1)

$$X(x_m, x_n) \geq \psi x_m \otimes \varphi x_n \geq \varepsilon_m \otimes \varepsilon_n \geq \varepsilon_N \otimes \varepsilon_N,$$

which implies $\bigvee_{N\geq 0} \bigwedge_{m,n\geq N} X(x_m, x_n) \geq \bigvee_{N\geq 0} \varepsilon_N \otimes \varepsilon_N = \mathrm{k}$, as desired.

Since $\lambda_s \dashv \rho_s$, it now suffices to show $\lambda_s \leq \varphi$ and (likewise) $\rho_s \leq \psi$ to conclude that (like in the proof of Proposition 155(6)) these two inequalities are actually equalities. Indeed, treating the $\mathcal{V}$-distributor $\varphi : \mathrm{E} \multimapdotinv X$ as a $\mathcal{V}$-functor $X \to \mathcal{V}$, we have for all $x \in X$ and $N \geq 0$

$$\bigwedge_{n\geq N} X(x_n, x) \leq \bigwedge_{n\geq N} [\varphi x_n, \varphi x] \leq [\bigvee_{n\geq N} \varepsilon_n, \varphi x] = [\mathrm{k}, \varphi x] = \varphi x,$$

which implies $\lambda_s \geq \varphi$. Likewise, considering the $\mathcal{V}$-functor $\psi : X^{\mathrm{op}} \to \mathcal{V}$ one obtains $\rho_s \leq \psi$. □

Of course, Theorem 164, in conjunction with Corollary 162, implies:

Corollary 165 *Under the hypothesis of* Theorem *164, a $\mathcal{V}$-category X is Lawvere-complete if, and only if, X is Cauchy-complete.*

We refer the reader to Flagg (1991), Flagg (1997) for other interesting "analytic applications" of the $\ll$-relation in a quantalic setting.

8.3 Restriction of the $\mathcal{V}$-Presheaf Monad with Respect to a Class of Weights

A Cauchy completion of a Fréchet metric space X is a Cauchy-complete Fréchet metric space $\overline{X}$ which contains X as a dense metric subspace. Its underlying set may be taken to be the quotient of the set of all Cauchy sequences in X, modulo the equivalence relation that identifies two Cauchy sequences in X when they have the same limit in the prospective completion; more precisely, this equivalently means that the Cauchy sequences s and t in X are equivalent if, and only if, for every $x \in X$, the right (say) convergence values of s to x and of t to x coincide—briefly: if $\rho_s = \rho_t$. In other words, ρ_s may be interpreted as an equivalence class of Cauchy sequences in X and, hence, as a point of the completion $\overline{X}$.

This is a blueprint for establishing the Cauchy completion of X in the Lawvere terms of the right-adjoint $\mathcal{V}$-distributors ρ_s, which we now show to work in amazing generality, as follows. In fact, noting that the $\mathcal{V}$-presheaf category $\mathcal{P}X$ is given by all weights needed for arbitrary colimits in the $\mathcal{V}$-category X, the idea originating with Clementino and Hofmann (2008) is to consider classes of weights which, when restricted to $\mathcal{P}X$ for every X, lead to the formation of a submonad of $\mathcal{P}$ and allow us to pursue a relativized cocompletion theory, completely analogously to our procedure in the unrestricted situation.

Definition 166 A class Ψ of $\mathcal{V}$-distributors is a *cocompletion class* if it contains all identity distributors, is closed under composition, and if for all $\mathcal{V}$-distributors $\psi : Y \to\!\!\!\!\circ\!\!\!\to X$ it satisfies the composition-cancellation condition:

$$\text{(CCC)}\quad \psi \in \Psi \iff \forall x \in X\ (x^* \cdot \psi \in \Psi)\ .$$

In other words, Ψ is a wide subcategory of $\mathcal{V}$-Dist such that, for every $\mathcal{V}$-category X, the family of distributors $x^* : X \to\!\!\!\!\circ\!\!\!\to \mathrm{E}$ $(x \in X)$ is collectively Cartesian in Ψ with respect to the inclusion functor $\Psi \to \mathcal{V}$-Dist.

The following hands-on criterion avoids having to establish directly the closure under composition for $\mathcal{V}$-distributors in Ψ in full generality:

Lemma 167 *A class Ψ of $\mathcal{V}$-distributors Ψ is a cocompletion class if, and only if, the following conditions hold for all $\mathcal{V}$-distributors $\psi : Y \to\!\!\!\!\circ\!\!\!\to X$ and $\mathcal{V}$-functors $f : Y \to Z$, $g : W \to X$, $h : Z \to Y$:*

(CC1) $f^* \in \Psi$;
(CC2) $\psi \in \Psi \Longrightarrow g^* \cdot \psi \cdot f^* \in \Psi$;
(CC3) $\psi \in \Psi$ *and* $h_* \in \Psi \Longrightarrow \psi \cdot h_* \in \Psi$;
(CC4) $\forall x \in X\ (x^* \cdot \psi \in \Psi) \Longrightarrow \psi \in \Psi$.

Proof For a cocompletion class Ψ, since $1_Z^* \in \Psi$, condition (CCC) gives us $z^* \in \Psi$ for all $z \in Z$ and, in particular, $y^* \cdot f^* = (fy)^* \in \Psi$ for all $y \in Y$, so that $f^* \in \Psi$ follows. This shows (CC1) and (CC2-4) are now trivial consequences of the definition.

Conversely, assuming (CC1-4) for the class Ψ, we first observe for every $\mathcal{V}$-category X that, by (CC1), $\mathbf{y}_X : X \to \mathcal{P}X$ takes values in the full $\mathcal{V}$-subcategory

$$\Psi X := \Psi \cap \mathcal{P}X$$

of $\mathcal{P}X$. Furthermore, denoting the corresponding restricted map by $\mathbf{y}_X^\Psi : X \to \Psi X$, since the Yoneda Lemma gives $\phi^* \cdot (\mathbf{y}_X^\Psi)_* = \phi$ for all $\phi \in \Psi X$, we deduce $(\mathbf{y}_X^\Psi)_* \in \Psi$ with (CC4). Now, in order to show that with $\chi : Z \nrightarrow Y$ and $\psi : Y \nrightarrow X$ in Ψ also $\psi \cdot \chi$ stays in Ψ, using the Yoneda Lemma again we factorize

$$\chi = (Z \xrightarrow{(\mathbf{y}_Z^\Psi)_*} \Psi Z \xrightarrow{\widetilde{\chi}^*} Y),$$

where $\widetilde{\chi} : Y \to \Psi Z,\ y \mapsto y^* \cdot \chi$, corresponds to χ via $[Y, [Z^{\mathrm{op}}, \mathrm{E}]] \cong [Z^{\mathrm{op}} \otimes Y, \mathrm{E}]$ and takes values in ΨZ by (CC1-2). Consequently, with (CC2-3) we obtain that the composite $\mathcal{V}$-distributor $\psi \cdot \chi = \psi \cdot \widetilde{\chi}^* \cdot (\mathbf{y}_Z^\Psi)_*$ lies in Ψ. □

Here are the primary general examples of cocompletion classes:

Examples 168 (1) There is a least and a largest cocompletion class, namely:

$$\Gamma = \{f^* \mid f\ \mathcal{V}\text{-functor}\} \quad \text{and} \quad \mathcal{P} = \{\text{all } \mathcal{V}\text{-distributors}\},$$

respectively, with the apparent double notation for $\mathcal{P}$ (as the class of all $\mathcal{V}$-distributors and the $\mathcal{V}$-presheaf functor) to be explained by the Corollary below.

(2) The class $\Lambda = \{\psi \mid \psi$ is a right-adjoint $\mathcal{V}$-distributor$\}$ is the *Lawvere cocompletion class*. Under the hypotheses of Theorem 164 one has

$$\Lambda = \{\psi : Y \nrightarrow X \mid \forall x \in X\ \exists s \text{ Cauchy sequence in } X\ (x^* \cdot \psi = \rho_s)\}.$$

In the proof of Lemma 167 we took the first step to showing that a cocompletion class Ψ lets us restrict the $\mathcal{V}$-presheaf-functor $\mathcal{P}$ and, in fact, the monad $\mathcal{P}$, to just those presheaves that lie in Ψ, explicitly:

Corollary 169 *Let Ψ be a cocompletion class. Then, for every $\mathcal{V}$-category X, the Yoneda $\mathcal{V}$-functor $\mathbf{y}_X : X \to \mathcal{P}X$ takes values in the full $\mathcal{V}$-subcategory $\Psi X = \Psi \cap \mathcal{P}X$ of $\mathcal{P}X$. Moreover, for every $\mathcal{V}$-functor $f : X \to Y$, the $\mathcal{V}$-functor $\mathcal{P}f$ restricts to a $\mathcal{V}$-functor $\Psi f : \Psi X \to \Psi Y$. This, by abuse of notation, makes Ψ an endofunctor of $\mathcal{V}$-*Cat*, endowed with the restriction $\mathbf{y}^\Psi : \mathrm{Id} \to \Psi$ of the natural Yoneda transformation $\mathbf{y} : \mathrm{Id} \to \mathcal{P}$.*

Proof By (CC2), the $\mathcal{V}$-functor $\mathcal{P}f = (-) \cdot f^*$ maps ΨX into ΨY, making the following diagram commute:

$$\begin{array}{ccccc} X & \xrightarrow{\mathbf{y}_X^{\Psi}} & \Psi X & \xrightarrow{\mathrm{inc}_X^{\Psi}} & \mathcal{P}X \\ {\scriptstyle f}\downarrow & & \downarrow{\scriptstyle \Psi f} & & \downarrow{\scriptstyle \mathcal{P}f} \\ Y & \xrightarrow{\mathbf{y}_Y^{\Psi}} & \Psi Y & \xrightarrow{\mathrm{inc}_Y^{\Psi}} & \mathcal{P}Y \end{array}$$

□

In order to show that Ψ even inherits the monad structure from $\mathcal{P}$, let us first observe that, for every $\mathcal{V}$-functor $f : X \to Y$, the $\mathcal{V}$-functor $\mathcal{P}f = (-) \cdot f^*$ has a right (!) adjoint, $(-) \cdot f_*$. It turns out to be useful to examine the situation when this right adjoint may be restricted to a $\mathcal{V}$-functor $\Psi Y \to \Psi X$; this leads us to define:

Definition 170 A $\mathcal{V}$-functor f is Ψ*-dense* if the $\mathcal{V}$-distributor f_* lies in the class Ψ.

Examples 171 (1) Trivially, every $\mathcal{V}$-functor is $\mathcal{P}$-dense.

(2) By Proposition 45, a $\mathcal{V}$-functor f is Γ-dense if, and only if, f has a right adjoint $\mathcal{V}$-functor.

(3) By definition, for a $\mathcal{V}$-functor $f : X \to Y$ to be Λ-dense means that f_* is right adjoint in $\mathcal{V}$-Dist; equivalently, by (CC4), that $y^* \cdot f_*$ is right adjoint in $\mathcal{V}$-Dist, for all $y \in Y$. Under the hypotheses of Theorem 164, this means that, for every $y \in Y$ there is a sequence $s = (x_n)_n$ in X with $y^* \cdot f_* = \rho_s$, or

$$Y(fx, y) = \bigvee_{N \geq 0} \bigwedge_{n \geq N} X(x, x_n),$$

for all $x \in X$.

Lemma 172 *Let Ψ be a cocompletion class, and let $f : X \to Y$, $g : Y \to Z$ be $\mathcal{V}$-functors. Then:*

(1) *If f has a right adjoint, then f is Ψ-dense.*
(2) *If f and g are Ψ-dense, so is gf.*
(3) *If gf is Ψ-dense, then f is Ψ-dense whenever g is fully faithful, and g is Ψ-dense whenever f is dense (in the sense of* Definition *47).*
(4) $\mathbf{y}_X^{\Psi} : X \to \Psi X$ *is Ψ-dense.*

Proof (1) follows from Proposition 45 and (CC1), and (2) is trivial. (3) From $g_* \cdot f_* \in \Psi$ one obtains with (CC2) $f_* = g^* \cdot g_* \cdot f_* \in \Psi$ when g is fully faithful, and $g_* = g_* \cdot f_* \cdot f^* \in \Psi$ when f is dense. (4) has been shown in the proof of Lemma 167. □

Corollary 173 *The following conditions are equivalent for a $\mathcal{V}$-functor $f : X \to Y$:*

(i) *f is Ψ-dense;*
(ii) *Ψf is left adjoint;*
(iii) *Ψf is Ψ-dense.*

Proof (i)$\Longrightarrow$(ii) Since $f_* \in \Psi$, the right adjoint $(-) \cdot f_*$ of $\mathcal{P}f$ restricts to a $\mathcal{V}$-functor $\Psi Y \to \Psi X$ by (CC3) and becomes a right adjoint of Ψf. (ii)$\Longrightarrow$(iii) follows from Lemma 172(1). (iii)$\Longrightarrow$(i) By Lemma 172(2), (4), the hypothesis (iii) makes $\mathbf{y}_Y^\Psi \cdot f = \Psi f \cdot \mathbf{y}_X^\Psi$ Ψ-dense. Since $\mathbf{y}_Y^\Psi$ is fully faithful, (i) follows with Lemma 172(3). $\square$

Theorem 174 *For every cocompletion class Ψ, the endofunctor Ψ of $\mathcal{V}$-*Cat *inherits the structure of a monad from the $\mathcal{V}$-presheaf monad $\mathcal{P}$, such that the full inclusions* $\mathrm{inc}_X^\Psi : \Psi X \to \mathcal{P}X$ *form a monad morphism $\Psi \to \mathcal{P}$. Like $\mathcal{P}$, also Ψ is lax idempotent.*

Proof After Proposition 169, it suffices to show that, for every $\mathcal{V}$-category X, the monad multiplication $\mathbf{s}_X : \mathcal{P}\mathcal{P}X \to \mathcal{P}X,\ \theta \mapsto \theta \cdot (\mathbf{y}_X)_*$, restricts to a $\mathcal{V}$-functor $\mathbf{s}_X^\Psi :$ $\Psi\Psi X \to \Psi X$, such that the following diagram commutes:

$$\begin{array}{ccc}
\mathcal{P}\mathcal{P}X & \xrightarrow{\ \mathbf{s}_X\ } & \mathcal{P}X \\
{\scriptstyle \mathcal{P}(\mathrm{inc}_X^\Psi)}\uparrow & & \\
\mathcal{P}\Psi X & & \uparrow{\scriptstyle \mathrm{inc}_X^\Psi} \\
{\scriptstyle \mathrm{inc}_{\Psi X}^\Psi}\uparrow & & \\
\Psi\Psi X & \xrightarrow{\ \mathbf{s}_X^\Psi\ } & \Psi X
\end{array}$$

Indeed, given $\theta \in \Psi\Psi X$, so that $\theta : \Psi X \nrightarrow \mathsf{E}$ lies in the class Ψ, using that inc_X^Ψ is (by definition of ΨX) fully faithful, we see that

$$\begin{aligned}
\mathbf{s}_X(\mathcal{P}(\mathrm{inc}_X^\Psi)(\theta)) = \mathbf{s}_X(\theta \cdot (\mathrm{inc}_X^\Psi)^*) &= \theta \cdot (\mathrm{inc}_X^\Psi)^* \cdot (\mathbf{y}_X)_* \\
&= \theta \cdot (\mathrm{inc}_X^\Psi)^* \cdot (\mathrm{inc}_X^\Psi)_* \cdot (\mathbf{y}_X^\Psi)_* \\
&= \theta \cdot (\mathbf{y}_X^\Psi)_*
\end{aligned}$$

lies in Ψ as well, by Lemma 172(4) and (CC3). Hence, $\mathbf{s}_X$ restricts to $\mathbf{s}_X^\Psi$, as desired.

The monad laws for $(\Psi, \mathbf{s}^\Psi, \mathbf{y}^\Psi)$ and its lax idempotency now follow directly from the corresponding laws for $\mathcal{P}$ and its lax idempotency. $\square$

Remark 175 The 2-monads of $\mathcal{V}$-Cat that are presentable as submonads of the presheaf monad $\mathcal{P}$ have recently been characterized as precisely those which are induced by a cocompletion class of $\mathcal{V}$-distributors; see Theorem 2.4 of Clementino and Fitas (2023). Corollary 3.9 of Clementino and Fitas (2023) gives a more general argument for their lax idempotence.

8.4 Ψ-Cocompleteness, Ψ-Injectivity, and Ψ-Cocompletion

With Theorem 174 at our disposal we can now easily relativize some key definitions and results of Sects. 5 and 6 from the monad $\mathcal{P}$ to the monad Ψ, given by a cocompletion class.

Definition 176 (1) A $\mathcal{V}$-category X is *Ψ-cocomplete* if $\mathbf{y}_X^\Psi$ has a pseudo-retraction $c : \Psi X \to X$, so that $c\,\mathbf{y}_X^\Psi \simeq 1_X$. By the lax idempotency of Ψ, such $\mathcal{V}$-functor c is, up to order-equivalence, uniquely determined and makes X a Ψ-pseudo-algebra. A $\mathcal{V}$-functor $f : X \to Y$ with a Ψ-cocomplete $\mathcal{V}$-category Y is called *Ψ-cocontinuous* if f is a Ψ-pseudo-homomorphism.

(2) The $\mathcal{V}$-category X is called *Ψ-pseudo-injective* if X is injective with respect to fully faithful and Ψ-dense $\mathcal{V}$-functors; that is, whenever a $\mathcal{V}$-functor $j : Y \to Z$ satisfies $j^* \cdot j_* = 1_Y^*$ with $j_* \in \Psi$, then for every $\mathcal{V}$-functor $g : Y \to X$ there is a $\mathcal{V}$-functor $h : Z \to X$ with $hj \simeq g$.

Let us note immediately that the proof of Proposition 79 remains intact when relativized to a cocompletion class Ψ. Likewise, Theorem 97 may obviously be relativized, as follows:

Theorem 177 *A $\mathcal{V}$-category X is Ψ-cocomplete if, and only if, a colimit of any diagram in X with weight in Ψ exists in X, and a $\mathcal{V}$-functor $f : X \to Y$ to a Ψ-cocomplete $\mathcal{V}$-category Y is Ψ-cocontinuous if, and only if, it preserves any colimit in X with weight in Ψ. In particular,* $\mathrm{inc}_X^\Psi : \Psi X \to \mathcal{P}X$ *is Ψ-cocontinuous.*

Proof Confirming the last claim, one may again use the commutative diagram of the proof of Theorem 174. □

In its relativized form, Theorem 89 reads as follows:

Theorem 178 *A $\mathcal{V}$-category X is Ψ-cocomplete if, and only if, X is Ψ-pseudo-injective.*

Proof In re-employing the proof of Theorem 89, we just need to note again that $\mathbf{y}_X^\Psi$ is Ψ-dense by Lemma 172(4), and that for f fully faithful and Ψ-dense, Ψf has, like $\mathcal{P}f$, a right-adjoint right-inverse; see the motivation for Definition 170. □

Likewise we have the relativized version of Theorem 109:

Theorem 179 (Clementino and Hofmann 2008) *Let Ψ be a cocompletion class. Then the $\mathcal{V}$-functor $\mathbf{y}_X^\Psi$ of a $\mathcal{V}$-category X is pseudo-universal among $\mathcal{V}$-functors $f : X \to Y$ into a Ψ-cocomplete $\mathcal{V}$-category Y; that is, up to order-equivalence, every such f factors through $\mathbf{y}_X^\Psi$, by a Ψ-cocontinuous $\mathcal{V}$-functor $\tilde{f} : \Psi X \to Y$ that is unique up to order-equivalence. The caveats "up to order-equivalence" are not needed when Y is separated.*

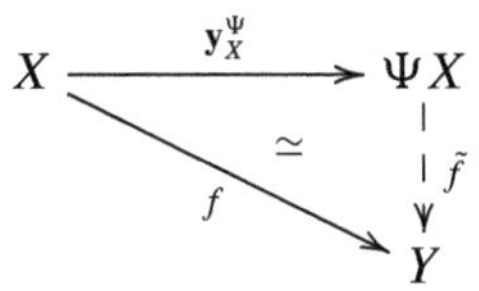

Corollary 180 *Let X be a $\mathcal{V}$-category. Then the $\mathcal{V}$-functor $\mathbf{y}_X^{\Lambda} : X \to \Lambda X$ is pseudo-universal among $\mathcal{V}$-functors $f : X \to Y$ into a Lawvere-complete $\mathcal{V}$-category Y. If $\mathcal{V}$ satisfies the hypotheses of Theorem 164, then*

$$\Lambda X = \{\rho_s \mid s \text{ Cauchy sequence in } X\}$$

is Cauchy-complete and $\mathbf{y}_X^{\Lambda}$ is pseudo-universal among $\mathcal{V}$-functors f into a Cauchy-complete category Y.

Remark 181 We note that Theorem 138 (monadicity of $\mathcal{V}$-Sup over Set) cannot be generalized to the level of an arbitrary cocompletion class Ψ. Indeed, it fails already for $\mathcal{V} = \mathsf{2}$ and $\Psi = \Lambda$: in that case, every poset (i.e., every object of $\mathcal{V}$-$\mathsf{Cat}_{\mathrm{sep}}$) is Lawvere-(and Cauchy-)complete, and every monotone map trivially preserves colimits of Cauchy sequences, but the category Pos (drastically) fails to be monadic over Set (as its forgetful functor does not even reflect isomorphisms).

However, as shown in Clementino and Hofmann (2008) in a context even more general than the setting of this chapter, one may generalize Theorem 138, as follows:

Theorem 182 *If the cocompletion class Ψ contains the $\mathcal{V}$-distributor f_* of every surjective $\mathcal{V}$-functor f, then the Set-valued forgetful functor of $\mathcal{V}$-Sup^{Ψ} is monadic.*

We must refer to Clementino and Hofmann (2008) for the proof. Various applications of the relativized cocompletion approach are also to be found in Clementino and Hofmann (2008).

Acknowledgements I thank the editors for their invitation to contribute to this book. I am very grateful for an important pointer to the literature by the anonymous referee, and for some valuable comments on the axiomatics of a cocompletion class as presented in Sect. 8.4. Cordial thanks go also to Maria Manuel Clementino for further good suggestions, her careful final reading of the chapter, and her and Júlio Neves's involvement in the preparation of its final layout.

References

Adámek J, Herrlich H, Strecker GE (2006) Abstract and concrete categories—the Joy of Cats. Repr Theory Appl Categ (17):1–507. Reprint of the 1990 original [Pure and Applied Mathematics, Wiley, New York]. http://tac.mta.ca/tac/reprints/articles/17/tr17abs.html

Adámek J, Rosický J (1994) Locally presentable and accessible categories. London Mathematical Society Lecture Note Series, vol 189. Cambridge University Press, Cambridge. https://doi.org/10.1017/CBO9780511600579

Akhvlediani A, Clementino MM, Tholen W (2010) On the categorical meaning of Hausdorff and Gromov distances. I. Topol Appl 157(8):1275–1295. https://doi.org/10.1016/j.topol.2009.06.018

Banaschewski B, Nelson E (1976) Tensor products and bimorphisms. Canad Math Bull 19(4):385–402. https://doi.org/10.4153/CMB-1976-060-2

Barr M (1979) $*$-autonomous categories, vol 752. Lecture Notes in Mathematics. Springer, Berlin, Heidelberg, New York

Barr M (1999) $*$-autonomous categories and linear logic. https://www.math.mcgill.ca/barr/papers/scatll.pdf

Bénabou J (1963) Catégories avec multiplication. C R Acad Sci Paris 256:1887–1890

Bénabou J (1965) Catégories relatives. C R Acad Sci Paris 260:3824–3827

Bénabou J (1973) Les distributeurs. Seminaires de Mathématiques Pure, Institut de Mathématiques, Université Catholique de Louvain Rapport no 33

Borceux F (1994) Handbook of categorical algebra 2-categories and structures, encyclopedia of mathematics and its applications, vol 51. Cambridge University Press, Cambridge

Clementino MM, Fitas C (2023) On presheaf submonads of quantale-enriched categories. Quaest Math 46(10):2079–2108. https://doi.org/10.2989/16073606.2022.2144532

Clementino MM, Hofmann D (2008) Relative injectivity as cocompleteness for a class of distributors. Theory Appl Categ 21(12):210–230

Duskin J (1969) Variations on Beck's tripleability criterion. In: Mac Lane S (ed) Reports of the Midwest Category Seminar, III. Lecture notes in mathematics, vol 106. Springer, Berlin, Heidelberg, New York, pp 74–129

Eilenberg S, Kelly GM (1965) Closed categories. In: Eilenberg SMS, Harrison DK, Röhrl H (eds) Proceedings of the conference on categorical algebra (La Jolla, CA, 1965). Springer, New York Inc, New York

Eklund P, Gutiérrez García J, Höhle U, Kortelainen J (2018) Semigroups in complete lattices-quantales, modules and related topics, developments in mathematics, vol 54. Springer, Cham. https://doi.org/10.1007/978-3-319-78948-4

Flagg RC (1992) Completeness in continuity spaces. In: Seely RAG (ed) Category theory 1991 (Montreal, PQ, 1991), CMS conference proceedings, vol 13. American Mathematical Society, Providence, RI, pp 183–199

Flagg RC (1997) Quantales and continuity spaces. Algebra Univers 37(3):257–276. https://doi.org/10.1007/s000120050018

Fréchet M (1960) Sur quelques points du calcul fonctionnel. Rend Circ Matem Palermo 22(1):1–72. https://doi.org/10.1007/BF03018603

Gabriel P, Ulmer F (1971) Lokal präsentierbare Kategorien, vol 221. Lecture Notes in Mathematics. Springer, Berlin, New York

Goubault-Larrecq J (2019) Formal ball monads. Topol Appl 263:372–391. https://doi.org/10.1016/j.topol.2019.06.044

Grandis M (2009) Directed algebraic topology-models of non-reversible worlds, new mathematical monographs, vol 13. Cambridge University Press, Cambridge. https://doi.org/10.1017/CBO9780511657474

Gromov M (1981) Groups of polynomial growth and expanding maps. Publ Math Inst Hautes Études Sci 53:53–73. https://doi.org/10.1007/BF02698687

Hausdorff F (1914) Grundzüge der Mengenlehre. Veit & Comp, Leipzig

Hofmann D, Nora P (2018) Enriched stone-type dualities. Adv Math 330:307–360. https://doi.org/10.1016/j.aim.2018.03.010

Hofmann D, Reis CD (2013) Probabilistic metric spaces as enriched categories. Fuzzy Sets Syst 210:1–21. https://doi.org/10.1016/j.fss.2012.05.005

Hofmann D, Reis CD (2018) Convergence and quantale-enriched categories. Categ Gen Algebr Struct Appl 9(1):77–138)

Hofmann D, Seal GJ, Tholen W (eds) (2014) Monoidal topology—a categorical approach to order, metric, and topology, encyclopedia of mathematics and its applications, vol 153. Cambridge University Press, Cambridge. https://doi.org/10.1017/CBO9781107517288

Joyal A, Tierney M (1984) An extension of the Galois theory of Grothendieck. Mem Am Math Soc 51(309):vii+71. https://doi.org/10.1090/memo/0309

Kelly GM (2005) Basic concepts of enriched category theory, London mathematical society lecture note series, vol 64. Cambridge University Press, Cambridge-New York (1982). Republished in Repr Theory Appl Categ 10

Kelly GM, Lack S (2001) $\mathcal{V}$-CAT is locally presentable or locally bounded if $\mathcal{V}$ is so. Theory Appl Categ 8(23):555–575

Lawvere FW (2002) Metric spaces, generalized logic, and closed categories. Rend Sem Mat Fis Milano 43:135–166 (1973). https://doi.org/10.1007/BF02924844. Republ Repr Theory Appl Categ 1:1–37 (2002)

Mac Lane S (1971) Categories for the working mathematician, graduate texts in mathematics, vol 5. Springer, Berlin, Heidelberg, New York [2nd ed (1994)]

Martinelli E (2021) Actions, injectives and injective hulls in quantale-enriched multicategories. PhD thesis, University of Aveiro, Aveiro

Menger K (1942) Statistical metrics. Proc Nat Acad Sci USA **28**, 535–537. https://doi.org/10.1073/pnas.28.12.535

Mulvey CJ (1986) In: Proceedings of the second topology conference (Taormina, 1984). Rendiconti del Circolo Matematico di Palermo (2) Supply, vol 12, pp 99–104

Pastro C, Street R (2009) Closed categories, star-autonomy, and monoidal comonads. J Algebra 321(11):3494–3520. https://doi.org/10.1016/j.jalgebra.2008.05.004

Pompeiu D (1950) Sur la continuité des fonctions de variables complexes. Ann Fac Sci Toulouse Sci Math Sci Phys (2)7(3):265–315 [(These), Gauthier-Villars, Paris]. http://www.numdam.org/item?id=AFST_1905_2_7_3_265_0

Rosenthal KI (1990) Quantales and their applications, pitman research notes in mathematics series, vol 234. Addison Wesley Longman, Harlow

Rosenthal KI (1996) The theory of Quantaloids, pitman research notes in mathematics series, vol 348. Addison Wesley Longman, Harlow

Stubbe I (2005) Categorical structures enriched in a quantaloid: categories, distributors and functors. Theory Appl Categ 14(1):1–45

Stubbe I (2006) Categorical structures enriched in a quantaloid: tensored and cotensored categories. Theory Appl Categ 16(14):283–306

Stubbe I (2007) $\mathcal{Q}$-modules are $\mathcal{Q}$-suplattices. Theory Appl Categ 19(4):50–60

Stubbe I (2010) "Hausdorff distance" via conical cocompletion. Cah Topol Géom Differ Catég 51(1):51–76

Tholen W (2019) Lax distributive laws for topology. I Cah Topol Géom Différ Catég 60(3):311–364

Tholen W (2022) Another categorical look at monoids, quantales, metrics, etc. Talk given at 5th workshop on categorical algebra—recent developments and future perspectives, Gargnano del Garda (Italy). https://tholen.mathstats.yorku.ca

Tholen W (2024) On the metrical and quantalic versions of the $*$-autonomous category of sup-lattices. Theory Appl Categ 40:Paper No 7, 227–248. https://doi.org/10.1007/s10114-024-2234-y

Walters RFC (1981) Sheaves and cauchy-complete categories. Cahiers Topologie Géom Différentielle 22(3):283–286

Wood RJ (2004) Ordered sets via adjunctions. In: Pedicchio MC, Tholen W (eds) Categorical foundations. Encyclopedia mathematics and its applications, vol 97. Cambridge University Press, Cambridge, pp 5–47

MIX
Papier aus verantwortungsvollen Quellen
Paper from responsible sources
FSC® C105338

If you have any concerns about our products,
you can contact us on
ProductSafety@springernature.com

In case Publisher is established outside the EU,
the EU authorized representative is:
Springer Nature Customer Service Center GmbH
Europaplatz 3, 69115 Heidelberg, Germany

Printed by Libri Plureos GmbH
in Hamburg, Germany